XXth International Conference on Integrable Systems and Quantum Symmetries 2012

(ISQS-20)

Journal of Physics: Conference Series Volume 411

Prague, Czech Republic
17-23 June 2012

ISBN: 978-1-62276-989-6
ISSN: 1742-6588

Printed from e-media with permission by:

Curran Associates, Inc.
57 Morehouse Lane
Red Hook, NY 12571

Some format issues inherent in the e-media version may also appear in this print version.

Printed by Curran Associates, Inc. (2013)

For permission requests, please contact the Institute of Physics at the address below.

Institute of Physics
Dirac House, Temple Back
Bristol BS1 6BE UK

Phone: 44 1 17 929 7481
Fax: 44 1 17 920 0979

techtracking@iop.org

Additional copies of this publication are available from:

Curran Associates, Inc.
57 Morehouse Lane
Red Hook, NY 12571 USA
Phone: 845-758-0400
Fax: 845-758-2634
Email: curran@proceedings.com
Web: www.proceedings.com

TABLE OF CONTENTS

Journal of Physics: Conference Series **411** (2013) 011001 doi:10.1088/1742-6596/411/1/011001

Foreword

The XXth International Conference on Integrable Systems and Quantum Symmetries (ISQS-20), organized by the Department of Mathematics, Faculty of Nuclear Sciences and Physical Engineering, Czech Technical University, Prague, and the Bogoliubov Laboratory of Theoretical Physics of the Joint Institute for Nuclear Research, is part of the successful series of conferences held at the Czech Technical University which began in 1992, and is devoted to problems of mathematical physics related to the theory of integrable systems, quantum groups and quantum symmetries. During the last 20 years each of these conferences has gathered around 60 scientists from all over the world. 31 plenary lectures and contributions presented at ISQS-20 are published in this volume of *Journal of Physics: Conference Series*.

Guest Editors

Čestmír Burdík, Ondřej Navrátil and Severin Pošta

Peer review statement

All papers published in this volume of *Journal of Physics: Conference Series* have been peer reviewed through processes administered by the proceedings Editors. Reviews were conducted by expert referees to the professional and scientific standards expected of a proceedings journal published by IOP Publishing.

Localization of matter fields and non-Abelian gauge fields on domain walls

Masato Arai[1,2], **Filip Blaschke**[2,3], **Minoru Eto**[4], **and Norisuke Sakai**[5]

[1]Fukushima National College of Technology, Iwaki, Fukushima 970-8034, Japan
[2]Institute of Experimental and Applied Physics, Czech Technical University in Prague, Horská 22, 128 00 Prague 2, Czech Republic
[3]Institute of Physics, Silesian University in Opava, Bezručovo nám. 1150/13, 746 01 Opava, Czech Republic
[4]Department of Physics, Yamagata University, Yamagata 990-8560, Japan
[5]Department of Mathematics, Tokyo Woman's Christian University, Tokyo 167-8585, Japan

E-mail: masato.arai@gmail.com, filip.blaschke@fpf.slu.cz, meto@sci.kj.yamagata-u.ac.jp, norisuke.sakai@gmail.com

Abstract. We propose a method for simultaneous localization of non-Abelian gauge fields and matter fields with minimal interaction on domain walls using $U(N)_c$ gauge theory with $SU(N)_L \times SU(N)_R \times U(1)_A$ flavor symmetry. Localization of non-Abelian fields is achieved using field-dependent gauge coupling. We find that effective Lagrangian up to second order of derivatives for low energy fluctuations resembles a chiral model from hadron physics with additional minimal interactions of pions with localized gauge fields together with nonlinear interactions of moduli fields. This result provides a step towards a realistic model building of brane-world scenario using topological solitons.

1. Introduction

In the brane-world scenario [1, 2, 3] it is assumed that all Standard Model (SM) fields are localized on (3+1)-dimensional defect called 3-brane, which is embedded in multidimensional space-time (bulk), where only graviton may propagate freely. To realize this scenario dynamically, we may use a topological soliton. A domain wall is the simplest soliton in (4+1)-dimensional theory.

Generally, bulk fields can have massless and massive modes. It has been shown that massless modes localize naturally on the domain wall [4]. Integrating out all massive modes, one obtains effective field theory describing low energy interactions of these massless modes. However, in the context of field theory, localization of gauge fields, turned out to be difficult [5]. It has been noted that the broken gauge symmetry in the bulk outside of the soliton inevitably makes the localized gauge field massive with the mass of the order of inverse width of the wall [6, 7]. Thus, to localize a massless gauge field, one needs to have the confining phase rather than the Higgs phase in the bulk outside of the soliton.

Recently, a classical picture of the confinement [8, 9] realized by position-dependent gauge coupling has been used to localize the non-Abelian gauge field on domain walls [10]. This position-dependent gauge coupling was naturally introduced on the domain wall background through a scalar-field-dependent gauge coupling function resulting from a cubic prepotential of supersymmetric gauge theories. This mechanism in its simplest form uses two copies of $U(1)_c$ gauge theory, where the difference of respective kink-like profiles of adjoint scalars in both sectors is taken as position-dependent gauge coupling.

However, it is still an interesting problem to make localized matter fields and gauge fields interact. The purpose of this work is to present a (4+1)-dimensional field theory model of localized massless matter fields minimally coupled to the non-Abelian gauge field which is also localized on the domain wall with the (3+1)-dimensional world volume. We derive the low-energy effective field theory of these localized matter and gauge fields and describe full non-linear interaction between moduli fields.

To introduce non-Abelian flavor symmetry (to be gauged eventually) in the domain wall sector, we replace one of the two copies of the $U(1)_c$ gauge theory with the flavor symmetry $U(1)_L \times U(1)_R$ in [11], by $U(N)_c$ gauge theory with the extended flavor (global) symmetry $SU(N)_L \times SU(N)_R \times U(1)_A$. By choosing the coincident domain wall solution for this domain wall sector, we obtain the maximal unbroken non-Abelian flavor symmetry group $SU(N)_{L+R}$ which is preserved in both left and right vacua outside of the domain wall. Therefore we can introduce gauge field for the (subgroup of) the flavor $SU(N)_{L+R}$ symmetry. In order to obtain the field-dependent gauge coupling function, for the gauge field localization mechanism [10], we also introduce a coupling between a scalar field and gauge field strengths inspired by supersymmetric gauge theories, although we do not make the model fully supersymmetric at present. This scalar-field-dependent gauge coupling function gives appropriate profile of position-dependent gauge coupling through the background domain wall solution. With this localization mechanism for gauge field, we find massless non-Abelian gauge fields localized on the domain wall.

In what follows, we only concentrate on presenting our main result. Detailed discussion of involved calculations and other topic outside this main line is presented in [15].

2. Gauge field localization

2.1. The domain wall sector

Let us briefly illustrate the localization mechanism for the gauge fields and the matter fields on the domain walls by using a simplest model in (4+1)-dimensional spacetime: two copies ($i = 1, 2$) of $U(1)$ models, each of which has two flavors (L, R) of charged Higgs scalar fields $H_i = (H_{iL}, H_{iR})$:

$$\mathcal{L}_i = -\frac{1}{4g_i^2}\left(\mathcal{F}^i_{MN}\right)^2 + \frac{1}{2g_i^2}\left(\partial_M \sigma_i\right)^2 + |\mathcal{D}_M H_i|^2 - V_i, \tag{1}$$

$$V_i = \frac{g_i^2}{2}\left(|H_i|^2 - v_i^2\right)^2 + |\sigma_i H_i - H_i M_i|^2 . \tag{2}$$

We use the metric $\eta_{MN} = \mathrm{diag}(+, -, \cdots, -)$, $M, N = 0, 1, \cdots, 4$. The Higgs field H_i is charged with respect to the $U(1)_i$ gauge symmetry and the covariant derivative is given by

$$\mathcal{D}_M H_i = \partial_M H_i + i w^i_M H_i, \tag{3}$$

where w^i_M is the $U(1)_i$ gauge field with the field strength

$$\mathcal{F}^i_{MN} = \partial_M w^i_N - \partial_N w^i_M. \tag{4}$$

Since we want domain walls, we will choose

$$M_i = \mathrm{diag}\,(m_i, -m_i)\,, \quad (m_i > 0), \tag{5}$$

resulting in the $U(1)_{iA}$ flavor symmetry. We have included the neutral scalar fields σ_i in this Abelian-Higgs model. Notice that the same coupling constant g_i appears not only in front of the kinetic terms of the gauge fields and σ_i, but also as the the quartic coupling constant of H_i. We have taken this special relation among the coupling constants only to simplify computations. One may repeat the whole procedure in models with more generic coupling constants without changing essential results.

There are two discrete vacua for each copy i

$$(H_{iL}, H_{iR}, \sigma_i) = (v_i, 0, m_i)\,, \ (0, v_i, -m_i)\,. \tag{6}$$

Thanks to the special choice of the coupling constants in $\mathcal{L}_i$ motivated by the supersymmetry, there are Bogomol'nyi-Prasad-Sommerfield (BPS) domain wall solutions in these models. Let y be the coordinate of the direction orthogonal to the domain wall and we assume all fields depend on only y. The BPS equations are

$$\partial_y \sigma_i + g_i^2 \left(|H_i|^2 - v_i^2\right) = 0, \quad \mathcal{D}_y H_i + \sigma_i H_i - H_i M_i = 0. \tag{7}$$

In order to obtain the domain wall solution interpolating the two vacua in Eq. (6), we impose the boundary conditions :

$$\begin{aligned}(H_{iL}, H_{iR}, \sigma_i) &= (0, v_i, -m_i)\,, \ \ y = -\infty, \\ (H_{iL}, H_{iR}, \sigma_i) &= (v_i, 0, m_i)\,, \ \ \ y = \infty.\end{aligned} \tag{8}$$

Tension T_i of the domain wall is given by a topological charge as

$$T_i = \int_{-\infty}^{\infty} dy \, \left[v_i^2 \partial_y \sigma_i - \partial_y \left((\sigma_i H_i - H_i M_i) H_i^\dagger\right)\right]_{-\infty}^{\infty} = 2 m_i v_i^2. \tag{9}$$

The second equation of the BPS equations (7) can be solved by the moduli matrix formalism [12, 13] with the constant matrix (vector) $H_{i0} = (C_{iL}, C_{iR})$

$$H_i = v_i e^{-\frac{\psi_i}{2}} H_{i0} e^{M_i y}, \quad \sigma_i + i w_i = \frac{1}{2} \partial_y \psi_i. \tag{10}$$

For a given H_{i0}, the scalar function ψ_i is determined by the master equation

$$\partial_y^2 \psi_i = 2 g_i^2 v_i^2 \left(1 - e^{-\psi_i} H_{i0} e^{2M_i y} H_{i0}^\dagger\right). \tag{11}$$

There exists redundancy in the decomposition in Eq. (10), which is called the V-transformation:

$$H_{i0} \to V_i H_{i0}, \quad \psi_i \to \psi_i + 2 \log V_i, \quad V_i \in \mathbb{C}^*. \tag{12}$$

No analytic solutions for the master equation have been found for finite gauge couplings g_i. But we can take a strong gauge coupling limit $g_i \to \infty$ without losing any essential features (see [15] for further details). The generic solutions of the domain wall are generated by the generic moduli matrices (after fixing the V-transformation)

$$H_{i0} = (C_{iL}, C_{iR})\,, \quad C_{iL}, C_{iR} \in \mathbb{C}^*. \tag{13}$$

The complex constants C_{iL}, C_{iR} are free parameters containing the moduli parameters of the BPS solutions. The moduli parameter can be defined by

$$C_i \equiv \sqrt{\frac{C_{iR}}{C_{iL}}} = e^{i\alpha_i} e^{m_i y_i}. \tag{14}$$

The other degrees of freedom in C_{iL}, C_{iR} can be eliminated by the V-transformation in Eq. (12) and has no physical meaning. Then, in the strong gauge coupling limit, the master equation is found to be (where we used coincident walls set-up $H_{i0} = (1,1)$)

$$1 = e^{-\psi_i}\left(e^{2m_i(y-y_i)} + e^{-2m_i(y-y_i)}\right), \tag{15}$$

with the solution

$$\sigma_i = m_i\frac{1-|\phi_i|^2}{1+|\phi_i|^2} = m_i \tanh 2m_i(y-y_i)\,. \tag{16}$$

It is obvious that the real parameter y_i is the translational moduli of the domain wall. The other parameter α_i is an internal moduli which is the Nambu-Goldstone (NG) mode associated with the $U(1)_{iA}$ flavor symmetry spontaneously broken by the domain walls.

Let us next derive the low energy effective theory on the domain wall. We integrate all the massive modes while keeping the massless modes. We use the so-called moduli approximation where the dependence on (3+1)-dimensional spacetime coordinates comes into the effective Lagrangian only through the moduli fields:

$$C_i \to C_i(x^\mu), \quad \phi_i(y) \to \phi_i(y, C_i(x^\mu)) = C_i(x^\mu)^2 e^{-2m_i y}. \tag{17}$$

The effective Lagrangian for the moduli field $C_i(x^\mu) = e^{i\alpha}e^{m_i y_i}$ is given by

$$\mathcal{L}_{i,\text{eff}} = \frac{2m_i v_i^2}{2}(\partial_\mu y_i)^2 + \frac{v_i^2}{m_i}(\partial_\mu \alpha_i)^2, \tag{18}$$

where energy of soliton solution is neglected since it does not contribute to dynamics of moduli. Note that $2m_i v_i^2$ is precisely the domain wall tension. This is the free field Lagrangian.

In order to obtain the massless gauge field to be localized on the domain wall, we need a new gauge symmetry which is unbroken in the bulk. Recently, a new mechanism was proposed to localize gauge fields on domain walls [10].

A key ingredient is the so-called dielectric coupling constant [8, 9]. Let us introduce a new $U(1)$ gauge field a_M which we wish to localize on the domain wall. Since this gauge symmetry should be unbroken in the bulk, we consider the case where all the Higgs fields are neutral under this newly introduced $U(1)$ gauge symmetry. The gauge field a_M is assumed to couple to the neutral scalar fields σ_i only in the following particular combination

$$\mathcal{L} = \mathcal{L}_1 + \mathcal{L}_2 - \frac{\lambda}{2}\left(\frac{\sigma_1}{m_1} - \frac{\sigma_2}{m_2}\right)(\mathcal{G}_{MN})^2, \tag{19}$$

where a real constant λ with the unit mass dimension, in accordance with the (4+1)-dimensional spacetime and the field strength is defined by

$$\mathcal{G}_{MN} = \partial_M a_N - \partial_N a_M. \tag{20}$$

The field-dependent gauge coupling function is given by

$$\frac{1}{4e^2(\sigma)} = \frac{\lambda}{2}\left(\frac{\sigma_1}{m_1} - \frac{\sigma_2}{m_2}\right), \tag{21}$$

which depends on the position y through fields σ_i. Thus the field-dependent gauge coupling function $e(\sigma)$ plays the role of the dielectric coupling constant. The form of Eq. (21) is chosen so that the gauge interaction becomes strongly coupled in the bulk ($\sigma_i \to \pm m_i$ as $y \to \pm\infty$).

	$SU(N)_c$	$U(1)_1$	$U(1)_2$	$SU(N)_L$	$SU(N)_R$	$U(1)_{1A}$	$U(1)_{2A}$	mass
H_{1L}	$\square$	1	0	$\square$	$\mathbf{1}$	1	0	$m_1\mathbf{1}_N$
H_{1R}	$\square$	1	0	$\mathbf{1}$	$\square$	-1	0	$-m_1\mathbf{1}_N$
Σ_1	adj $\oplus$ $\mathbf{1}$	0	0	$\mathbf{1}$	$\mathbf{1}$	0	0	0
H_{2L}	$\mathbf{1}$	0	1	$\mathbf{1}$	$\mathbf{1}$	0	1	m_2
H_{2R}	$\mathbf{1}$	0	1	$\mathbf{1}$	$\mathbf{1}$	0	-1	$-m_2$
Σ_2	$\mathbf{1}$	0	0	$\mathbf{1}$	$\mathbf{1}$	0	0	0

Table 1. Quantum numbers of the domain wall sectors in the chiral model.

Since our Lagrangian has no term linear in a_M, the equations of motion for a_M is trivially solved by $a_M = 0$, and the rest of the equations of motion are explicitly the same as those in the previous subsection. Therefore the domain wall solution in the previous subsection together with $a_M = 0$ is still a solution of the equations of motion. Thus, the (3+1)-dimensional gauge coupling constant is given by

$$\frac{1}{4e_4^2} = \frac{\lambda}{2}\int_{-\infty}^{\infty} dy \left(\frac{\sigma_1}{m_1} - \frac{\sigma_2}{m_2}\right) = \frac{\lambda}{4}\left[\frac{\psi_1}{m_1} - \frac{\psi_2}{m_2}\right]_{-\infty}^{\infty} = \lambda(y_2 - y_1), \tag{22}$$

where we have used the asymptotic behavior $\psi_i \to \log 2\cosh 2m_i(y - y_i)$ as $|y| \to \infty$. The low energy effective Lagrangian is found to be

$$\mathcal{L}_{\text{eff}} = \sum_{i=1,2}\left[\frac{2m_i v^2}{2}(\partial_\mu y_i)^2 + \frac{v_i^2}{m_i}(\partial_\mu \alpha_i)^2\right] - \lambda(y_2 - y_1)(\mathcal{G}_{\mu\nu})^2. \tag{23}$$

Although we succeeded in localizing the massless $U(1)$ gauge field a_μ on the domain walls, the Lagrangian Eq. (23) has no charged matter fields minimally coupled with the localized gauge field a_μ. In the next section, we will give a model with a non-Abelian global symmetry whose unbroken subgroup can be gauged to yield massless localized gauge fields on the domain wall.

3. The chiral model

In this section we study domain walls in the chiral model which is a natural extension of the Abelian-Higgs model in the previous section.

Let us consider the Yang-Mills-Higgs model with $SU(N)_c \times U(1)$ gauge symmetry with $S[U(N)_L \times U(N)_R] = SU(N)_L \times SU(N)_R \times U(1)_A$ flavor symmetry [14, 11]. To localize the gauge field in a simple manner, we again introduce two sectors $\mathcal{L}_1$ and $\mathcal{L}_2$, but only the former is extended to Yang-Mills-Higgs system and the latter is the same form as in (1) with $i = 2$. The matter contents are summarized in Table 1. Since the presence of two factors of $SU(N)$ global symmetry resembles the chiral symmetry of QCD, we call this Yang-Mils-Higgs system as the chiral model.

The Lagrangian is given by

$$\mathcal{L} = \mathcal{L}_1 + \mathcal{L}_2, \tag{24}$$

$$\mathcal{L}_1 = \mathrm{Tr}\left[-\frac{1}{2g_1^2}(F_{1MN})^2 + \frac{1}{g_1^2}(\mathcal{D}_M\Sigma_1)^2 + |\mathcal{D}_M H_1|^2\right] - V_1, \tag{25}$$

$$V_1 = \mathrm{Tr}\left[\frac{g_1^2}{4}\left(H_1 H_1^\dagger - v_1^2\mathbf{1}_N\right)^2 + |\Sigma_1 H_1 - H_1 M_1|^2\right], \tag{26}$$

with $H_1 = (H_{1L}, H_{2L})$. $\mathcal{L}_2$ is the same form as (1) with $i = 2$. Gauge fields of $U(N)_c = (SU(N)_c \times U(1)_1)/Z_{\mathbf{N}}$ are denoted as W_{1M}, and adjoint scalar as Σ_1. The covariant derivative and the field strength are denoted as $\mathcal{D}_M\Sigma_1 = \partial_M\Sigma_1 + i\,[W_{1M}, \Sigma_1]$, $\mathcal{D}_M H_1 = \partial_M H_1 + iW_{1M}H_1$, and $F_{1MN} = \partial_M W_{1N} - \partial_N W_{1M} + i\,[W_{1M}, W_{1N}]$. The mass matrix is given by $M_1 = \mathrm{diag}\,(m_1\mathbf{1}_N, -m_1\mathbf{1}_N)$. Let us note that the chiral model reduces to the Abelian-Higgs model in the limit of $N \to 1$, by deleting all the $SU(N)$ groups.

The role of the second sector is to provide the localization mechanism in the similar way as in the previous section. Therefore, the remaining discussion will be focused solely on the first sector ($i = 1$) and we suppress the index $i = 1$.

There exist $N + 1$ vacua in which the fields develop the following VEV

$$H = (H_L, H_R) = v\left(\begin{array}{cc|cc} \mathbf{1}_{N-r} & & \mathbf{0}_{N-r} & \\ & \mathbf{0}_r & & \mathbf{1}_r \end{array}\right), \tag{27}$$

$$\Sigma = m\begin{pmatrix} \mathbf{1}_{N-r} & \\ & -\mathbf{1}_r \end{pmatrix}, \tag{28}$$

with $r = 0, 1, 2, \cdots, N$. We refer these vacua with the label r. In the r-th vacuum, both the local gauge symmetry $U(N)_c$ and the global symmetry are broken, but a diagonal global symmetries are unbroken (color-flavor-locking)

$$\begin{aligned} &U(N)_c \times SU(N)_L \times SU(N)_R \times U(1)_A \to \\ &SU(N-r)_{L+c} \times SU(r)_L \times SU(r)_{R+c} \times SU(N-r)_R \times U(1)_{A+c}. \end{aligned} \tag{29}$$

The BPS equations for the domain walls can be obtained as:

$$\mathcal{D}_y\Sigma - \frac{g^2}{2}\left(v^2\mathbf{1}_N - HH^\dagger\right) = 0, \tag{30}$$

$$\mathcal{D}_y H + \Sigma H - HM = 0. \tag{31}$$

The tension of the domain wall is given by

$$\begin{aligned} T &= \int_{-\infty}^{\infty} dy\; \partial_y \left\{ \mathrm{Tr}\left[v^2\Sigma - (\Sigma H - HM)\,H^\dagger\right]\right\} \\ &= v^2\;\mathrm{Tr}\left[\Sigma(+\infty) - \Sigma(-\infty)\right]. \end{aligned} \tag{32}$$

Let us concentrate on the domain wall which connects the 0-th vacuum at $y \to \infty$ and the N-th vacuum at $y \to -\infty$. Its tension can be read as

$$T = 2Nv^2m, \tag{33}$$

from Eq. (32). Since there are $N + 1$ possible vacua, the maximal number of walls is N at various positions. The simplest domain wall solution corresponding to the coincident walls is given by making an ansatz that H_L, H_R, Σ and W_y are all proportional to the unit matrix. Then the BPS equations (30) and (31) can be identified with the BPS equations in Eq. (7) in the Abelian-Higgs model. Thus the domain wall solution can be solved as

$$H_L = ve^{-\frac{\psi}{2}}e^{my}\;\mathbf{1}_N, \tag{34}$$

$$H_R = ve^{-\frac{\psi}{2}}e^{-my}\;\mathbf{1}_N, \tag{35}$$

$$\Sigma + iW_y = \frac{1}{2}\partial_y\psi\mathbf{1}_N, \tag{36}$$

where ψ is the solution of the master equation (11) in the Abelian-Higgs model. Eq.(29) shows that the unbroken global symmetry for N-th vacuum ($H_L = 0$, $H_R = v\mathbf{1}_N$ and $\Sigma = -m\mathbf{1}_N$) at the left infinity $y \to -\infty$ is $SU(N)_L \times SU(N)_{R+c} \times U(1)_{A+c}$, whereas that for the 0-th vacuum ($H_L = v\mathbf{1}_N$, $H_R = 0$ and $\Sigma = m\mathbf{1}_N$) at the right infinity $y \to \infty$ is $SU(N)_{L+c} \times SU(N)_R \times U(1)_{A+c}$.

The domain wall solution further breaks these unbroken symmetries because it interpolates the two vacua.

$$U(N)_c \times SU(N)_L \times SU(N)_R \times U(1)_A \to SU(N)_{L+R+c}. \tag{37}$$

This spontaneous breaking of the global symmetry gives NG modes on the domain wall as massless degrees of freedom valued on the coset similarly to the chiral symmetry breaking in QCD :

$$\frac{SU(N)_L \times SU(N)_R}{SU(N)_{L+R+c}} \times U(1)_A. \tag{38}$$

Since our model can be embedded into a supersymmetric field theory [15], these NG modes ($U(N)$ chiral fields) appear as complex scalar fields accompanied with additional N^2 pseudo-NG modes[1].

We will now give the low-energy effective Lagrangian on the domain walls where the massless moduli fields (the matter fields) are localized. The best way to parametrize these massless moduli fields is to use the moduli matrix formalism [12, 13]

$$H_L = ve^{my}S^{-1}, \tag{39}$$
$$H_R = ve^{-my}S^{-1}e^{\phi}, \tag{40}$$
$$\Sigma + iW_y = S^{-1}\partial_y S, \tag{41}$$

where $S \in GL(N, \mathbf{C})$ and $\Omega = SS^\dagger$ is the solution of the following master equation

$$\partial_y\left(\Omega^{-1}\partial_y\Omega\right) = g^2v^2\left(\mathbf{1}_N - \Omega^{-1}\Omega_0\right), \tag{42}$$

where

$$\Omega_0 = e^{2my}\mathbf{1}_N + e^{-2my}e^{\phi}e^{\phi^\dagger}.$$

We have used the V-transformation to identify the moduli e^ϕ, which is a complex N by N matrix. It can be parametrized by an $N \times N$ hermitian matrix $\hat{x}$ and a unitary matrix U as [11]

$$e^\phi = e^{\hat{x}}U^\dagger, \tag{43}$$

where U is nothing but the $U(N)$ chiral fields associated with the spontaneous symmetry breaking Eq. (38) and $\hat{x}$ is the pseudo-NG modes whose existence we promised above.

In the strong gauge coupling limit $g \to \infty$, a solution of master equation is simply $\Omega = \Omega_0$. After fixing the $U(N)_c$ gauge, we obtain

$$S = e^{\hat{x}/2}\sqrt{2\cosh(2my - \hat{x})}. \tag{44}$$

Let us denote, for brevity

$$\hat{y} = 2my - \hat{x}, \tag{45}$$

[1] One of them is actually a genuine NG mode corresponding to the broken translation.

the Higgs fields are then given as

$$H_L = v\frac{e^{\hat{y}/2}}{\sqrt{2\cosh \hat{y}}}, \tag{46}$$

$$H_R = v\frac{e^{-\hat{y}/2}}{\sqrt{2\cosh \hat{y}}}U^\dagger . \tag{47}$$

From this solution, one can easily recognize that eigenvalues of $\hat{x}$ correspond to the positions of the N domain walls in the y direction. Now we promote moduli parameters $\hat{x}$ and U to fields on the domain wall world volume, namely functions of world volume coordinates x^μ. We plug the domain wall solutions $H_{L,R}(y;\hat{x}(x^\mu),U(x^\mu))$ into the original Lagrangian $\mathcal{L}$ in Eq. (25) at $g\to\infty$ and pick up the terms quadratic in the derivatives. Thus the low energy effective Lagrangian is given by

$$\mathcal{L}_{\text{eff}} = \int_{-\infty}^{\infty} dy\, \text{Tr}\left[\partial_\mu H_L \partial^\mu H_L^\dagger + \partial_\mu H_R \partial^\mu H_R^\dagger - v^2 W_\mu W^\mu\right], \tag{48}$$

where

$$W_\mu = \frac{i}{2v^2}\left[\partial_\mu H_L H_L^\dagger - H_L \partial_\mu H_L^\dagger + (L \leftrightarrow R)\right]. \tag{49}$$

Here we have eliminated the massive gauge field W_μ by using the equation of motion. Let us next introduce the gauge fields which are to be massless and localized on the domain walls. Since associated gauge symmetry should not be broken by the domain walls, we can gauge only unbroken symmetry $SU(N)_{L+R+c}$ itself or its subgroup. Let us gauge $SU(N)_{L+R}\equiv SU(N)_V$ and let A_μ^a be the $SU(N)_{L+R}$ gauge field. The Higgs fields are in the bi-fundamental representation of $U(N)_c$ and $SU(N)_{L+R}$. The covariant derivatives of the Higgs fields are modified by

$$\tilde{\mathcal{D}}_M H_{1L} = \partial_M H_{1L} + iW_{1M}H_{1L} - iH_{1L}A_M, \tag{50}$$

$$\tilde{\mathcal{D}}_M H_{1R} = \partial_M H_{1R} + iW_{1M}H_{1R} - iH_{1R}A_M. \tag{51}$$

We now introduce a field-dependent gauge coupling function $g^2(\Sigma)$ for A_M, which is inspired by the supersymmetric model in Ref.[10].

$$\frac{1}{2e^2(\Sigma)} = \frac{\lambda}{2}\left(\frac{\text{Tr}\Sigma_1}{Nm_1} - \frac{\Sigma_2}{m_2}\right). \tag{52}$$

The Lagrangian is given by

$$\mathcal{L} = \tilde{\mathcal{L}}_1 + \mathcal{L}_2 - \frac{1}{2e^2(\Sigma)}\text{Tr}\left[G_{MN}G^{MN}\right]. \tag{53}$$

The $\tilde{\mathcal{L}}_1$ in Eq. (53) is given by Eq. (25) where the covariant derivatives are replaced with those in Eqs. (50) and (51).

We first wish to find the domain wall solutions in this extended model. As before, we make ansatz that all the fields depend on only y and $W_\mu = A_\mu = 0$. Let us first look on the E.O.M. of the new gauge field A_M. It is of the form

$$\mathcal{D}_M G^{MN} = J^N, \tag{54}$$

where J_M stands for the current of A_M. Note that the current J_M is zero, by definition, if we plug the domain wall solutions in the chiral model before gauging the $SU(N)_{L+R}$. This is because the domain wall configurations do not break $SU(N)_{L+R}$. Therefore, $A_M = 0$ is a solution of Eq. (54).

Then, we are left with equation of motion with $A_M = 0$ which are identical to those in the ungauged chiral model in the previous subsections. Therefore the gauged chiral model admits the same domain wall solutions as those (Eqs. (46) and (47)) in the ungauged chiral model.

The next step is to derive the low energy effective theory on the domain wall world-volume in the moduli approximation as before. We just repeat the similar computation presented above. It is convenient to define the covariant derivative only for the flavor $(SU(N)_{L+R})$ gauge interactions as

$$\hat{D}_\mu H = \partial_\mu H - iHA_\mu. \tag{55}$$

Then we obtain the effective Lagrangian of the first sector as

$$\begin{aligned}\mathcal{L}_{1,\text{eff}} &= \int_{-\infty}^{\infty} dy\,\mathrm{Tr}\Big[\hat{D}_\mu H_L(\hat{D}^\mu H_L)^\dagger + \hat{D}_\mu H_R(\hat{D}^\mu H_R)^\dagger - v^2 W_\mu W^\mu \\ &\quad - \frac{1}{2e^2(\Sigma)} G_{MN}G^{MN}\Big],\end{aligned} \tag{56}$$

with

$$W_\mu = \frac{i}{2v^2}\left[\hat{D}_\mu H_L H_L^\dagger - H_L(\hat{D}_\mu H_L)^\dagger + (L \leftrightarrow R)\right]. \tag{57}$$

Eliminating W_μ, we obtain the following expression for the integrand of the effective Lagrangian after some simplification

$$\mathcal{L}_{\text{eff}} = \frac{1}{2v^2}\int_{-\infty}^{\infty} dy\,\mathrm{Tr}\Big[\mathcal{D}_\mu H_{ab}\mathcal{D}^\mu H_{ba}\Big], \tag{58}$$

where we defined fields H_{ab} with the label ab of adjoint representation of the flavor gauge group $SU(N)_{L+R+c}$ and the covariant derivative as

$$\mathcal{D}_\mu H_{ab} = \partial_\mu H_{ab} + i[A_\mu, H_{ab}], \quad H_{ab} \equiv H_a^\dagger H_b, \quad a, b = L, R. \tag{59}$$

The full description of the procedure to obtain the effective Lagrangian in the closed form can be found in [15]. Here we merely state the result:

$$\begin{aligned}\mathcal{L}_{1,\text{eff}} = \frac{v^2}{2m}\mathrm{Tr}\Big[&\mathcal{D}_\mu\hat{x}\,\frac{\cosh(\mathcal{L}_{\hat{x}}) - 1}{\mathcal{L}_{\hat{x}}^2 \sinh(\mathcal{L}_{\hat{x}})}\ln\left(\frac{1+\tanh(\mathcal{L}_{\hat{x}})}{1-\tanh(\mathcal{L}_{\hat{x}})}\right)(\mathcal{D}^\mu\hat{x}) \\ &+ U^\dagger\mathcal{D}_\mu U\,\frac{\cosh(\mathcal{L}_{\hat{x}}) - 1}{\mathcal{L}_{\hat{x}} \sinh(\mathcal{L}_{\hat{x}})}\ln\left(\frac{1+\tanh(\mathcal{L}_{\hat{x}})}{1-\tanh(\mathcal{L}_{\hat{x}})}\right)(\mathcal{D}^\mu\hat{x}) \\ &+ \frac{1}{2}\mathcal{D}_\mu U^\dagger U\,\frac{1}{\tanh(\mathcal{L}_{\hat{x}})}\ln\left(\frac{1+\tanh(\mathcal{L}_{\hat{x}})}{1-\tanh(\mathcal{L}_{\hat{x}})}\right)(U^\dagger\mathcal{D}^\mu U)\Big],\end{aligned} \tag{60}$$

where

$$\mathcal{L}_A(B) = [A, B] \tag{61}$$

is a Lie derivative with respect to A. The covariant derivative $\mathcal{D}_\mu$ is defined by

$$\mathcal{D}_\mu U = \partial_\mu U + i\,[A_\mu, U]\,. \tag{62}$$

Journal of Physics: Conference Series **411** (2013) 012001 doi:10.1088/1742-6596/411/1/012001

By expanding (60), we here illustrate nonlinear interactions of $\hat{x}$ up to fourth orders in the fluctuations $\hat{x}$ and $U - \mathbf{1}$

$$\begin{aligned}\mathcal{L}_{1,\text{eff}} &= \frac{v^2}{2m}\text{Tr}\Big(\mathcal{D}_\mu U^\dagger \mathcal{D}^\mu U + \mathcal{D}_\mu \hat{x}\mathcal{D}^\mu \hat{x} + U^\dagger \mathcal{D}_\mu U\,[\hat{x}, \mathcal{D}^\mu \hat{x}] \\ &\quad -\frac{1}{12}[\mathcal{D}_\mu \hat{x}, \hat{x}]\,[\hat{x}, \mathcal{D}^\mu \hat{x}] + \frac{1}{3}[\mathcal{D}_\mu U^\dagger U, \hat{x}][\hat{x}, U^\dagger \mathcal{D}^\mu U] + \cdots\Big). \end{aligned} \tag{63}$$

In a similar way to Eq.(22), we can define the (3+1)-dimensional non-Abelian gauge coupling e_4 by integrating (52) and find

$$\frac{1}{2e_4^2} = \int dy \frac{1}{2e^2(\Sigma)} = \lambda(y_2 - y_1)\,, \tag{64}$$

where y_i is the wall position for the i-th domain wall sector. Summarizing, we obtain the following effective Lagrangian

$$\mathcal{L}_{\text{eff}} = \mathcal{L}_{1,\text{eff}} + \mathcal{L}_{2,\text{eff}} - \frac{1}{2e_4^2}\text{Tr}\Big[G_{\mu\nu}G^{\mu\nu}\Big], \tag{65}$$

where $\mathcal{L}_{2,\text{eff}}$ is given in (18). This is the main result. We have succeeded in constructing the low energy effective theory in which the matter fields (the chiral fields) and the non-Abelian gauge fields are localized with the non-trivial interaction. We show the profile of "wave functions" of localized massless gauge field and massless matter fields as functions of the coordinate y of the extra dimension in Fig. 1.

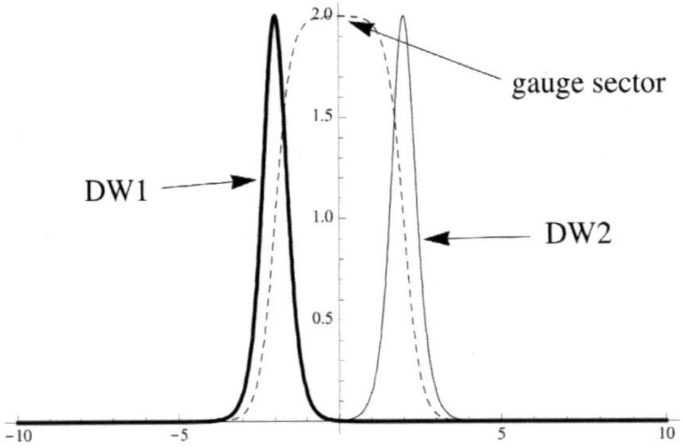

Figure 1. The wave functions of the zero modes. DW1 and DW2 stand for the wave functions of the massless matter fields of the $i = 1$ domain wall and $i = 2$ domain wall, respectively for strong gauge coupling limit $g_i = \infty$ and $m_i = 1$. The gauge fields are localized between the domain walls.

As is seen from Eq.(63), the flavor gauge symmetry $SU(N)_{L+R+c}$ is further (partly) broken and the corresponding gauge field A_μ becomes massive, when the fluctuation $\phi = e^{\hat{x}}U$ develops non-zero vacuum expectation values. Especially, $\hat{x}$ is interesting because its non-vanishing (diagonal) values of the fluctuation has the physical meaning as the separation between walls away from the coincident case. For instance, if all the walls are separated, $SU(N)_{L+R+c}$ is spontaneously broken to the maximal $U(1)$ subgroup $U(1)^{N-1}$. However, if r walls are still coincident and all other walls are separated, we have an unbroken gauge symmetry $SU(r) \times U(1)^{N-r+1}$. Then, a part of the pseudo-NG modes $\hat{x}$ turn to NG modes associated with the further symmetry breaking $SU(N)_{L+R+c} \to SU(r) \times U(1)^{N-r+1}$, so that the total number of zero modes[2] is preserved [11]. These new NG modes, called the non-Abelian cloud, spread between the separated domain walls [11]. The flavor gauge fields eat the non-Abelian cloud and get masses which are proportional to the separation of the domain walls. This is the Higgs mechanism in our model. This geometrical understanding of the Higgs mechanism is quite similar to D-brane systems in superstring theory. So our domain wall system provides a genuine prototype of field theoretical D3-branes.

4. Conclusions and discussion

In this work we have successfully localized both massless non-Abelian gauge fields and massless matter fields in non-trivial representation of the gauge group. Then we studied the low-energy effective Lagrangian and showed that massless localized matter fields interact minimally with localized $SU(N)_{L+R}$ gauge field as adjoint representations. Moreover, full nonlinear interaction between the moduli containing up to the second derivatives, was worked out.

Main result of this paper is the effective Lagrangian (65). The moduli field U appearing in the effective theory, is a chiral $N \times N$ matrix field like a pion, since it is a NG boson of spontaneously broken chiral symmetry. Other moduli in (65), denoted by $N \times N$ Hermitian matrix $\hat{x}$, has the physical meaning of positions of N domain walls as its diagonal elements. We argued that the fluctuations of moduli field $\hat{x}$, can develop VEV corresponding to splitting of walls, and the Higgs mechanism will occur as a result. Namely, the flavor gauge fields get masses by eating the non-Abelian cloud. Therefore, in this model, Higgs mechanism has a geometrical origin like low energy effective theories on D-branes in superstring theory.

Amongst the possible future investigations, we would like to study non-coincident solution to further clarify this geometrical Higgs mechanism. For further discussions see [15].

References

[1] P. Horava and E. Witten, Nucl. Phys. B **460** (1996) 506 [arXiv:hep-th/9510209].
[2] N. Arkani-Hamed, S. Dimopoulos and G. R. Dvali, Phys. Lett. B **429** (1998) 263 [hep-ph/9803315]; I. Antoniadis, N. Arkani-Hamed, S. Dimopoulos and G. R. Dvali, Phys. Lett. B **436** (1998) 257 [hep-ph/9804398].
[3] L. Randall and R. Sundrum, Phys. Rev. Lett. **83** (1999) 3370 [hep-ph/9905221]; Phys. Rev. Lett. **83** (1999) 4690 [hep-th/9906064].
[4] V. A. Rubakov and M. E. Shaposhnikov, Phys. Lett. B **125** (1983) 136.
[5] S. L. Dubovsky and V. A. Rubakov, Int. J. Mod. Phys. A **16** (2001) 4331 [hep-th/0105243].
[6] G. R. Dvali and M. A. Shifman, Phys. Lett. B **396** (1997) 64 [Erratum-ibid. B **407** (1997) 452] [hep-th/9612128].
[7]
[7] N. Maru and N. Sakai, Prog. Theor. Phys. **111** (2004) 907 [arXiv:hep-th/0305222]. Y. Isozumi, K. Ohashi and N. Sakai, JHEP **0311** (2003) 061 [arXiv:hep-th/0310130].
[8] J. B. Kogut and L. Susskind, Phys. Rev. D **9** (1974) 3501.
[9] R. Fukuda, Phys. Lett. B **73** (1978) 305
[10] K. Ohta and N. Sakai, Prog. Theor. Phys. **124** (2010) 71 [arXiv:1004.4078 [hep-th]].
[11] M. Eto, T. Fujimori, M. Nitta, K. Ohashi and N. Sakai, Phys. Rev. D **77** (2008) 125008 [arXiv:0802.3135 [hep-th]].
[12] Y. Isozumi, M. Nitta, K. Ohashi and N. Sakai, "Construction of non-Abelian walls and their complete moduli space," Phys. Rev. Lett. **93** (2004) 161601 [arXiv:hep-th/0404198];
[13] M. Eto, Y. Isozumi, M. Nitta, K. Ohashi and N. Sakai, J. Phys. A A **39** (2006) R315 [hep-th/0602170].
[14] M. Shifman and A. Yung, Phys. Rev. D **70** (2004) 025013 [hep-th/0312257].
[15] M. Arai, F. Blaschke, M. Eto and N. Sakai, "Matter Fields and Non-Abelian Gauge Fields Localized on Walls", ArXiv e-prints, [hep-th/1208.6219].

[2] In Ref.[14] the authors argued that the non-Abelian clouds spreading between walls become massive contrary to the results of [11].

Worldline fuzziness and spacetime noncommutativity

Valerio Astuti

Dipartimento di Fisica, Università "La Sapienza" and Sez. Roma1 INFN, P.le A. Moro 2, 00185 Roma, Italy

E-mail: `valerio.astuti@roma1.infn.it`

Abstract. We here offer a perspective on results obtained in collaboration with Amelino-Camelia and Rosati, which were reported in Refs.[6, 7, 8]. We advocate a formalization of spacetime noncommutativity within the manifestly covariant formulation of quantum mechanics. Using as illustrative example the case of the much-studied κ-Minkowski noncommutative spacetime, we show how such a setup may lead to a crisp characterization of relative locality and of worldline fuzziness in a quantum spacetime.

1. Spacetime noncommutativity and covariant formulation of quantum mechanics

One of the proposal to implement quantum effects in the description of spacetime is that of non-commutative spacetimes, in which coordinates of classical relativistic mechanics are promoted to operators that, in addition, do not commute between them.

One of the most studied such a proposal is the κ-Minkowski noncommutative spacetime[1, 2, 3, 4, 5],

$$[x_j, x_0] = i\ell x_j \ , \qquad [x_j, x_k] = 0 \ , \tag{1}$$

with the only non trivial commutator between the time coordinate and the spatial ones.

Once given such a commutation rule one would like to give also some more "physically characterizing" effects of such spacetimes, but to implement the non trivial commutation rules in the known formalism of classical or even quantum mechanics is quite difficult.

Evidently classical mechanics cannot be our choice, since its formalization provides no room for noncommutativity of coordinates. And even giving a formulation of κ-Minkowski spacetime noncommutativity in a quantum-mechanics setup is not straightforward. This is due to the fact that in κ-Minkowski the time coordinate should be an operator that does not commute with the spatial-coordinate operators, but in the standard setup of quantum mechanics we are not in the situation of time being described by an operator that commutes with the spatial-coordinate operators: in the standard setup of quantum mechanics time is not an observable at all, it just plays the role of evolution parameter.

We believe that it was indeed because of this mismatch between the nature of time in quantum mechanics and the properties of the κ-Minkowski time coordinate that progress did not materialize for formulating some of the observable spacetime consequences of κ-Minkowski that we are here interested in. We here summarize and offer an overall perspective on our results reported in Refs. [6, 7, 8] whose main starting point is recent progress on a manifestly-covariant formulation of quantum mechanics, which matured significantly over the last decade [9, 10, 11, 12].

This formulation gives us the percect environment to implement our non trivial commutators, because space and time coordinates are treated on the same grounds, are both non trivial operators on an hilbert space, so we can seek for a representation of our non-commutative spacetime as a deformation of a standard covariant formulation of quantum mechanics.

In our analysis we will restrict ourselves to a 1 + 1-dimensional spacetime, for ease of explanation. In this case the "kinematical hilbert space" [11] of our theory is simply the space of square-integrable functions of two variables. In the standard, commutative case spatial and time coordinates q_0, q_1 are well-defined multiplicative operators on this hilbert space, and one also has standard momenta π_0,π_1 conjugate to the spacetime coordinates:

$$[\pi_0, q_0] = i\,, \quad [\pi_1, q_1] = -i\,, \quad [\pi_1, q_0] = [\pi_0, q_1] = 0 \tag{2}$$

We want to represent the κ-Minkowski defining commutators,

$$[x_1, x_0] = i\ell x_1\ , \tag{3}$$

on the above defined hilbert space.

We can achieve this goal by posing a relationship between κ-Minkowski coordinates and the phase-space observables of the undeformed, standard covariant formulation of quantum mechanics (the ones of Eq. (2), here viewed merely as formal auxiliary operators [6]):

$$x_1 = e^{\ell\pi_0}\, q_1\ , \quad x_0 = q_0\ . \tag{4}$$

Besides a meaningful description of the noncommutativity of the κ-Minkowski coordinates, this setup leads straightforwardly [6, 8] also to a description of translation transformations in κ-Minkowski, and to the deformation of the whole Poincaré group that suits well as a symmetry group for our non-commutative spacetime - the so-called κ-Poincaré (that we do not show here, [6]):

$$\mathcal{T}_{a^\mu} \triangleright f(x_0, x_1) \longleftrightarrow f(q_0, q_1 e^{\ell\pi_0}) - i a^\mu \left[\pi_\mu, f(q_0, q_1 e^{\ell\pi_0})\right]\ , \tag{5}$$

Studying the whole symmetry algebra of κ-Minkowski, one can obtain two other strongly-characterizing results. The first result [6, 8] concerns the deformation of the mass casimir, that in the covariant formalism is the tool we use to enforce dynamics on our kinematical space. The corresponding operator (Hamiltonian constraint [6, 8, 11]) for the κ-Poincaré group is:

$$\mathcal{H}_\ell = \left(\frac{2}{\ell}\right)^2 \sinh^2\left(\frac{\ell\pi_0}{2}\right) - e^{-\ell\pi_0}\pi_1^2\ . \tag{6}$$

The second result [6, 8] concerns the measure for integration over momenta, needed for evaluating scalar products in the "momentum representation" of our hilbert space; it turns out that the right, invariant measure of integration in the non-commutative case is:

$$d\pi_0 d\pi_1 \longrightarrow d\pi_0 d\pi_1 e^{-\ell\pi_0} \tag{7}$$

2. Fuzzy points, translation transformations and relative locality

States in the kinematical hilbert space are distribution over spacetime, that can be interpreted as the probability amplitude to find a particle in a given point of spacetime. In the standard commutative case we can have - at the kinematical level - states that are arbitrarily close to points: a dirac delta-like distribution in the commutative hilbert space have a sharply definite value for both the time and the spatial coordinates.

This is no more the case in the deformed, noncommutative case: spatial and time coordinates no longer commute, so we cannot expect them to be diagonalizable at the same time.

Journal of Physics: Conference Series **411** (2013) 012002 doi:10.1088/1742-6596/411/1/012002

For the first time in our setting we can give in addition a precise, quantitative statement of what the consequence of this non-commutativity are. A class of states which is well suited for exploring the properties of κ-Minkowski fuzziness is the one of gaussian states.

We denote them by $\Psi_{\overline{q}_0,\overline{q}_1}(\pi_\mu;\overline{\pi}_\mu,\sigma_\mu)$ and they are specified by functions of the variables π_μ parametrized by $\overline{\pi}_\mu$, σ_μ, and $\overline{q}_\mu$:

$$\Psi_{\overline{q}_0,\overline{q}_1}(\pi_\mu;\overline{\pi}_\mu,\sigma_\mu)=Ne^{-\frac{(\pi_0-\overline{\pi}_0)^2}{4\sigma_0^2}-\frac{(\pi_1-\overline{\pi}_1)^2}{4\sigma_1^2}}e^{i\pi_0\overline{q}_0-i\pi_1\overline{q}_1}, \tag{8}$$

where N is a normalization constant. Of course, our main focus of attention will be on establishing how the κ-Minkowski scale ℓ affects the results, since this is going to be our indicator of the difference between classical Minwkoswki spacetime and κ-Minkowski.

Following the analysis we reported in Ref. [6] one finds that the κ-Minkowski scale ℓ turns out to play a particularly significant role in the properties of the coordinate x_1,for which we find

$$\langle x_1\rangle = \langle q_1\rangle\left\langle e^{\ell\pi_0}\right\rangle = \overline{q}_1 e^{\ell\overline{\pi}_0}e^{-\frac{\ell^2\sigma_0^2}{2}}\ , \quad \delta x_1 = e^{\ell\overline{\pi}_0}\left[\frac{1}{4\sigma_1^2}+\overline{q}_1^2\left(1-e^{-\ell^2\sigma_0^2}\right)\right]^{1/2}\ , \tag{9}$$

whereas for the κ-Minkowski time coordinate x_0 one has

$$\langle x_0\rangle = \overline{q}_0 - i\frac{\ell}{2}\ , \qquad \delta x_0 = \frac{1}{2\sigma_0} \tag{10}$$

(comments on the imaginary contribution $-i\ell/2$ to $\langle x_0\rangle$ are given in Ref. [6]).

We argue that in our non-commutative spacetime the notion of point itself is changed: before imposing any dynamic, already at the kinematical level, there are no more sharply definite spacetime points, but they are replaced by non-local distributions on spacetime, as it is intuitively suggested by a spacetime foam picture. In the following, when we talk of points, is to these distributions that we are referring, being them the best replace available to classical points.

Our next task is to act with a translation transformation on our gaussian state and see what changes as a result. For the expectation values here of interest this is equivalently done by acting with the translation transformation on the operators x_1 and x_0. Following again our work in Ref. [6] one finds that

$$\langle \mathcal{T}_{a^\mu}\triangleright x_0\rangle = \overline{q}_0 - a_0 - i\frac{\ell}{2}\ , \quad \delta\left(\mathcal{T}_{a^\mu}\triangleright x_0\right) = \frac{1}{2\sigma_0}\ , \tag{11}$$

and

$$\langle \mathcal{T}_{a^\mu}\triangleright x_1\rangle = \left(\overline{q}_1 - a_1\right)e^{\ell\overline{\pi}_0}e^{-\frac{\ell^2\sigma_0^2}{2}}\ , \quad \delta(\mathcal{T}_{a^\mu}\triangleright x_1) = e^{\ell\overline{\pi}_0}\left[\frac{1}{4\sigma_1^2}+(\overline{q}_1-a_1)^2\left(1-e^{-\ell^2\sigma_0^2}\right)\right]^{1/2}. \tag{12}$$

The interpretation here of course is such that the x_0, x_1 are operators characterizing the distance of a given (fuzzy) point from the frame origin of some observer Alice, and then $\mathcal{T}_{a^\mu}\triangleright x_0$, $\mathcal{T}_{a^\mu}\triangleright x_1$ are the operators that characterize the distance of that point from the frame origin of an observer Bob, purely translated with respect to Alice. Accordingly one can deduce the relation between the mean values and uncertainties in positions among two distant observers in relative rest by comparing (10) to (11) and comparing (9) to (12). This we did in detail in Ref. [6]. We here summarize the main message of that analysis, focusing on the case of two fuzzy points in κ-Minkowski as described by two distant observers: one of the points is near observer Alice, while the other one is near observer Bob, purely translated with respect to Alice. By comparing

(10) to (11) and comparing (9) to (12) one then finds two main features [6]:
(i) the same point appears to be more fuzzy to a distant observer than to a nearby observer,
(ii) the point at Alice is not described as being at Alice in the coordinatization of spacetime of observer Bob, and *vice versa* the point at Bob is not described as being at Bob in the coordinatization of spacetime of observer Alice.
The second feature, (ii), is essentially already known from previous studies of relative locality in the classical limit [14, 15]: one can have consistently relativistic theories where pairs of points found to be coincident by a nearby observer (or, as in the case here considered, a point found to coincide with the origin of that observer) are instead described as noncoincident if one uses the inferences about those points by a distant observer. Feature (i) was established for the first time in our Ref. [6], and is a feature of relative locality for the fuzziness of points in a quantum spacetime. The emerging picture is fully relativistic (though in the sense of deformed relativistic symmetries [16, 17, 18, 19]): all observers in κ-Minkowski perceive their origin as the point of lowest fuzziness and attribute to distant points fuzziness proportional to the distance.

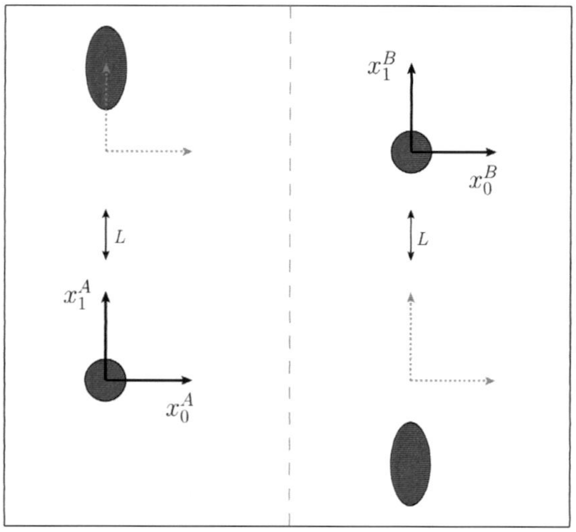

Figure 1. We illustrate the features of relative locality we uncovered for the κ-Minkowski quantum spacetime by considering the case of two distant observers, Alice and Bob, in relative rest (with synchronized clocks). In figure we have only two points in κ-Minkowski, each described by a gaussian state in our Hilbert space. One of the points is at Alice (centered in the spacetime origin of Alice's coordinatization) while the other point is at Bob. The left panel reflects Alice's description of the two points, which in particular attributes to the distant point at Bob larger fuzziness than Bob observes (right panel). And in Alice's coordinatization the distant point is not exactly at Bob. Bob's description (right panel) of the two points is specular, in the appropriately relativistic fashion, to the one of Alice. The magnitude of effects shown would require the distance L to be much bigger than drawable. And for definiteness in figure we assumed $\bar{\pi}_0 \simeq 2\sigma_0$ and $\sigma_1 \simeq \sigma_0$.

3. Fuzziness of κ-Minkowski worldlines

In the previous section we worked on the kinematical Hilbert space. For our next (and here last) task we must progress to the level of the physical Hilbert space [6, 8, 11]. We want now to use κ-Minkowski noncommutativity as a way to model the influence of quantum-gravity effects on the fuzziness of worldlines. This should be viewed in the context of attempts to describe the dynamics of matter particles as effectively occurring in an "environment" of short-distance quantum-gravitational degrees of freedom: for propagating particles with wavelength much larger than the Planck length, when it may be appropriate to integrate out these quantum-gravitational degrees of freedom, the main residual effect of short-distance gravity could indeed be an additional contribution to the fuzziness of worldlines, which one could model with spacetime noncommutativity.

For working on the physical Hilbert space we follow (as we did in Ref. [8]) the prescription adopted in Ref. [11]: we impose the restriction for our states to belong to the kernel of the Hamiltonian constraint $\mathcal{H}_\ell$, Eq.(6), by introducing a new scalar product that projects all the orbit of the gauge transformation generated by $\mathcal{H}_\ell$ on the same state. For massless particles, on which we here focus, this amounts to inserting the Hamiltonian constraint as follows

$$\langle\psi|\phi\rangle_{\mathcal{H}_\ell} = \langle\psi|\delta\left(\mathcal{H}_\ell\right)\Theta(\pi_0)|\phi\rangle \tag{13}$$

where $\langle\rangle$ is the scalar product on the kinematical Hilbert space and $\Theta(\pi_0)$ specifies a restriction [11] to positive-energy solutions of the constraint.

The next problem we must face concerns the identification of an observable suitable for the characterization of the fuzziness of the worldline. The apparently obvious choices, x_1 and x_0, are not suitable for this task, since they are not self-adjoint operators on the *physical* Hilbert space (they do not commute with $\mathcal{H}_\ell$). We propose to remedy this by focusing on the following "intercept operator" $\mathcal{A}$:

$$\mathcal{A} = e^{\ell\pi_0}\left(q_1 - \mathcal{V}q_0 - \frac{1}{2}[q_0,\mathcal{V}]\right) \tag{14}$$

where $\mathcal{V}$ is short-hand for $\mathcal{V} \equiv (\partial\mathcal{H}_\ell/\partial\pi^0)^{-1}\partial\mathcal{H}_\ell/\partial\pi^1$.

One may notice that $\mathcal{A}$ is describable as an ℓ-deformed Newton-Wigner operator [20]. And it is well known that within special-relativistic quantum mechanics there is no better estimator of localization than the Newton-Wigner operator (it can only be questioned for localization comparable to the Compton wavelength of the particle [20], but this merely conceptual limit of ideal localization is not relevant for our purposes here [8]).

Let us focus, for conceptual clarity, on the analysis of the properties of $\mathcal{A}$ for the case of $\Psi_{0,0}$, *i.e.* for $\overline{q}_0 = 0, \overline{q}_1 = 0$. One then easily finds that

$$\langle\Psi_{0,0}|\mathcal{A}|\Psi_{0,0}\rangle_{\mathcal{H}_\ell} = 0$$

so this is a case where the particle intercepts the observer Alice in her origin.

The fuzziness of this intercept, which reflects the fuzziness of the worldline [7, 8] described by $\Psi_{0,0}$, is characterized by [8]

$$\delta\mathcal{A}^2_{[\ell]} = \left(\langle\Psi_{0,0}|\mathcal{A}^2|\Psi_{0,0}\rangle_{\mathcal{H}_\ell}\right)_{[\ell]} \approx \ell\langle\pi_0\rangle\sigma^{-2}/2 \tag{15}$$

where for simplicity we focused on the leading ℓ-dependent contributions (index $[\ell]$) and we assumed that σ_1 is small enough, in comparison to σ_0, $\overline{\pi}_1$, to allow a saddle point approximation in the π_1 integration; then σ (without indices) is the effective gaussian width after the saddle point approximation in π_1: $\sigma^{-2} \equiv \sigma_1^{-2} + <\mathcal{V}>^2\sigma_0^{-2}$.

In the interpretation we proposed in Ref. [8] we describe Eq. (15) as the fuzziness of the worldline "at Alice" (at the point of crossing the origin of Alice's reference frame). It is then interesting to establish whether observers reached by the particle at cosmological distances from Alice will observe bigger fuzziness. We characterize such observers as those who are connected to Alice by a pure translation, so that for them the state of the particle is Ψ_{a_0,a_1}, and are, like Alice, such that $<\mathcal{A}>= 0$, *i.e.* $\langle\Psi_{a_0,a_1}|\mathcal{A}|\Psi_{a_0,a_1}\rangle_{\mathcal{H}_\ell} = 0$. Unsurprisingly this requires $a^1 = \langle\mathcal{V}\rangle a^0$. And for these observers one finds

$$\delta\mathcal{A}^2_{[\ell]} = \left(\langle\Psi_{a^0,\langle\mathcal{V}\rangle a^0}|\mathcal{A}^2|\Psi_{a^0,\langle\mathcal{V}\rangle a^0}\rangle_{\mathcal{H}_\ell}\right)_{[\ell]} \approx \left(\frac{\ell\langle\pi_0\rangle}{2\sigma^2} + \ell^2 a_0^2\sigma^2\right) \tag{16}$$

From this we see that the approach we developed in Refs. [6, 7, 8] leads to the first example of a quantum-spacetime picture (and of an interpretation of relativistic quantum mechanics in such a spacetime) providing the main ingredient of scenarios [21, 22, 23, 24, 25, 26] motivating the idea that quantum-gravity effects could affect worldline fuzziness in ways that inevitably lead to an increase of this fuzziness as the particle propagates. One can in fact interpret our observer Alice, the observer on the worldline for whom the fuzziness of the intercept takes the minimum value, as the observer at the source (where the particle is produced), and then the intercept of the particle worldline with the origin of the reference frames of observers distant from Alice (where the particle could be detected) has bigger uncertainty.

As we stressed in Refs. [7, 8] this formalization of gravity's contribution to worldline fuzziness could be relevant for an ongoing phenomenological effort aimed at finding experimental evidence of this sort of spacetime-fuzziness effects [21, 22, 23, 24, 25, 26].

References

[1] Majid S and Ruegg H 1994 *Phys. Lett.* B **334** 348
[2] Lukierski J, Ruegg H and Zakrzewski W J 1995 *Ann. Phys.* **243** 90
[3] Amelino-Camelia G and Majid S 2000 *Int. J. Mod. Phys.* A **15** 4301 (*Preprint* arXiv:hep-th/9907110)
[4] D'andrea F 2006 *J. Math. Phys.* **47** 062105 (*Preprint* arXiv:hep-th/0503012)
[5] Meljanac S and Kresic-Juric S 2011 *Int. J. Mod. Phys.* A **26** 3385 (*Preprint* arXiv:1004.4647 [math-ph])
[6] Amelino-Camelia G, Astuti V and Rosati G 2012 *Preprint* arXiv:1206.3805 [hep-th]
[7] Amelino-Camelia G, Astuti V and Rosati G 2012 *Preprint* arXiv:1207.2509 [gr-qc]
[8] Amelino-Camelia G, Astuti V and Rosati G 2012 Predictive description of Planck-scale-induced spacetime fuzziness *Preprint* Sapienza University of Rome
[9] Halliwell J J 2001 *Phys. Rev.* D **64** 04408 (*Preprint* arXiv:gr-qc/0008046)
[10] Gambini R and Porto R A 2001 *Phys. Rev.* D **63** 105014
[11] Reisenberger M and Rovelli C 2002 *Phys. Rev.* D **65** 125016 (*Preprint* arXiv:gr-qc/0111016)
[12] Freidel L, Girelli F and Livine E R 2007 *Phys. Rev.* D **75** 105016
[13] Henneaux M and Teitelboim C 1992 *Quantization of Gauge Systems* (Princeton: Princeton University Press)
[14] Amelino-Camelia G, Matassa M, Mercati F and Rosati G 2011 *Phys. Rev. Lett.* **106** 071301 (*Preprint* arXiv:1006.2126 [gr-qc])
[15] Amelino-Camelia G, Freidel L, Kowalski-Glikman J and Smolin L 2011 *Phys. Rev.* D **84** 084010 (*Preprint* arXiv:1101.0931 [hep-th])
[16] Amelino-Camelia G 2002 *Int. J. Mod. Phys.* D **11** 35 (*Preprint* arXiv:gr-qc/0012051)
[17] Kowalski-Glikman J 2001 *Phys. Lett.* A **286** 391 (*Preprint* arXiv:hep-th/0102098)
[18] Magueijo J and Smolin L 2003 *Phys. Rev.* D **67** 044017 (*Preprint* arXiv:gr-qc/0207085)
[19] Amelino-Camelia G 2010 *Symmetry* **2** 230 (*Preprint* arXiv:1003.3942 [gr-qc])
[20] Newton T D and Wigner E P 1949 *Rev. Mod. Phys.* **21** 400
[21] Amelino-Camelia G 1999 *Nature* **398** 216
[22] Ng Y J, van Dam H and Christiansen W A 2003 *Astrophys. J.* **591** L87 (*Preprint* arXiv:astro-ph/0302372)
[23] Lieu R and Hillman L W 2003 *Astrophys. J.* **585** L77
[24] Ragazzoni R, Turatto M and Gaessler W 2003 *Astrophys. J.* **587** L1
[25] Christiansen W A, Ng Y J and van Dam H 2006 *Phys. Rev. Lett.* **96** 051301
[26] Tamburini F, Cuofano C, Della Valle M and Gilmozzi R 2011 *Astron. Astrophys.* **533** A71

XXth International Conference on Integrable Systems and Quantum Symmetries (ISQS-20) IOP Publishing
Journal of Physics: Conference Series **411** (2013) 012003 doi:10.1088/1742-6596/411/1/012003

The sine-Gordon model in the presence of defects

Jean Avan
LPTM, Universite de Cergy-Pontoise (CNRS UMR 8089), F-95302 Cergy-Pontoise, France
E-mail: avan@u-cergy.fr

Anastasia Doikou[1]
Department of Engineering Sciences, University of Patras, GR-26500 Patras, Greece
E-mail: adoikou@upatras.gr

Abstract.
The sine-Gordon model in the presence of dynamical integrable defects is investigated. This is an application of the algebraic formulation introduced for integrable defects in earlier works. The quantities in involution as well as the associated Lax pairs are explicitly extracted. Integrability i also shown using certain sewing constraints, which emerge as suitable continuity conditions.

1. Introduction

The investigation of integrable defects has been a quite challenging problem, and there is a wealth of relevant articles in recent years at both classical and quantum level [1]–[18]. A fully algebraic picture for a description of a Liouville integrable defect was recently proposed [1]. It was exemplified in the case of the continuous non-linear Schrödinger model (NLS) [1], and now the procedure is extend to the sine Gordon model [2].

This procedure will now be applied to the sine-Gordon (SG) model, for which we shall consider a generic dynamical defect (type II) (see also [4, 9]). We must immediately emphasize that this model provides an example where the initial off-shell continuous "conserved" Hamiltonians do not Poisson commute, but will be shown to weakly Poisson-commute once the sewing conditions are implemented, thereby guaranteeing Liouville integrability of the reduced model. By contrast in the NLS case the continuous Hamiltonians strongly (i.e. off-shell) Poisson-commuted. As seen above this does not modify the conclusions on Liouville integrability on-shell.

This paper is mainly based on the analysis of [2], therefore for more details on the explicit algebraic formulation we refer the interested reader to the aforementioned relevant article.

[1] Based on a talk presented by A. Doikou in "Integrable Systems and Quantum Symmetries", Prague, June 2012.

Journal of Physics: Conference Series **411** (2013) 012003 doi:10.1088/1742-6596/411/1/012003

2. General setting

The key point in the description of classical integrable lattice models is the existence of the Lax pair $\mathbb{U}$, $\mathbb{V}$. Define Ψ as being a solution of the following set of equations (see e.g. [20])

$$\frac{\partial \Psi}{\partial x} = \mathbb{U}(x,t,\lambda)\Psi \tag{1}$$

$$\frac{\partial \Psi}{\partial t} = \mathbb{V}(x,t,\lambda)\Psi \tag{2}$$

$\mathbb{U}$, $\mathbb{V}$ being in general $n \times n$ matrices with entries defined as functions of complex valued dynamical fields, their derivatives, and the complex spectral parameter λ. Compatibility of the two aforementioned equations (1), (2) gives rise to the zero curvature condition

$$\dot{\mathbb{U}}(x,t) - \mathbb{V}'(x,t) + \Big[\mathbb{U}(x,t),\ \mathbb{V}(x,t)\Big] = 0, \tag{3}$$

which provides the equations of motion of the system at hand.

As is well known the generating function of the local integrals of motion is given by the expression

$$\mathcal{G} = \ln(trT(L,-L,\lambda)), \tag{4}$$

where the monodromy matrix T is defined as,

$$T(L,-L,\lambda) = P\exp\Big\{\int_{-L}^{L} dx\ \mathbb{U}(x)\Big\}, \tag{5}$$

We now impose that the operator $\mathbb{U}$ satisfy the ultra-local Poisson structure described by the linear algebraic relations

$$\Big\{\mathbb{U}_1(x,\lambda),\ \mathbb{U}_2(y,\mu)\Big\} = \Big[r_{12}(\lambda-\mu),\ \mathbb{U}(x,\lambda) + \mathbb{U}_2(y,\mu)\Big]\delta(x-y) \tag{6}$$

It is then straightforward to show that T satisfies the fundamental quadratic algebra:

$$\Big\{T_1(\lambda),\ T_2(\mu)\Big\} = \Big[r_{12}(\lambda-\mu),\ T_1(\lambda)\ T_2(\mu)\Big]. \tag{7}$$

$r_{12}(\lambda-\mu)$ is the so-called classical r-matrix assumed here to be a non-dynamical skew-symmetric solution of the classical Yang-Baxter equation.

Let us now particularize our analysis to the sine-Gordon model. In this case the $\mathbb{U}$ operator of the Lax pair is a 2×2 matrix and is given by [20]:

$$\mathbb{U}(x,t,u) = \frac{\beta}{4i}\pi(x,t)\sigma^z + \frac{mu}{4i}e^{\frac{i\beta}{4}\phi\sigma^z}\sigma^y e^{-\frac{i\beta}{4}\phi\sigma^z} - \frac{mu^{-1}}{4i}e^{-\frac{i\beta}{4}\phi\sigma^z}\sigma^y e^{\frac{i\beta}{4}\phi\sigma^z} \tag{8}$$

$u \equiv e^{\lambda}$, $\sigma^{x,y,z}$ are the 2×2 Pauli matrices, and the associated classical r-matrix in this case is given by the familiar form [20]:

$$r(\lambda) = \frac{\beta^2}{8\sinh\lambda}\begin{pmatrix} \frac{\sigma^z+1}{2}\cosh\lambda & \sigma^- \\ \sigma^+ & \frac{-\sigma^z+1}{2}\cosh\lambda \end{pmatrix}. \tag{9}$$

Stating that the Lax operator $\mathbb{U}$ satisfies the linear Poisson algebra (6) is equivalent to setting that ϕ, π are canonical conjugates, i.e.

$$\Big\{\phi(x),\ \pi(y)\Big\} = \delta(x-y). \tag{10}$$

Let us now apply the generic defect construction to the sine-Gordon model.

Journal of Physics: Conference Series **411** (2013) 012003 doi:10.1088/1742-6596/411/1/012003

3. The sine-Gordon model with defect

The main ingredient in the description of integrable defects is the modified monodromy matrix. More precisely, in the presence of an integrable defect the monodromy matrix of the field theory is modified (see also [6, 11, 1]), and takes the generic form

$$\begin{aligned} T(L,-L,\lambda) &= T^{+}(L,x_0^{+},\lambda)\,\tilde{L}(\lambda)\,T^{-}(x_0^{-},-L,\lambda) \\ &= P\exp\Big\{\int_{x_0^{+}}^{L} dx\;\mathbb{U}^{+}(x)\Big\}\,\tilde{L}(\lambda)\,P\exp\Big\{\int_{-L}^{x_0^{-}} dx\;\mathbb{U}^{-}(x)\Big\} \end{aligned} \tag{11}$$

Assuming that the defect Lax matrix $\tilde{L}$ also satisfies the quadratic Poisson algebra (7) T given in (11) satisfies (7).

A consistent generic parametrization of an integrable defect of the so-called Type-II or dynamical will be considered. We assume the following form for the classical $\tilde{L}$ matrix

$$\tilde{L}(\lambda) = \begin{pmatrix} e^{\lambda}V - e^{-\lambda}V^{-1} & \bar{a} \\ a & e^{\lambda}V^{-1} - e^{-\lambda}V \end{pmatrix}. \tag{12}$$

Requiring that $\tilde{L}$ satisfies the algebraic relation (7), one then extracts the following Poisson relations between the defect fields:

$$\begin{aligned} \Big\{V,\;\bar{a}\Big\} &= \frac{\beta^2}{8}V\,\bar{a}, \\ \Big\{V,\;a\Big\} &= -\frac{\beta^2}{8}Va, \\ \Big\{\bar{a},\;a\Big\} &= \frac{\beta^2}{4}(V^2 - V^{-2}) \end{aligned} \tag{13}$$

From these Poisson brackets one naturally extracts a cyclic variable $C_0 = V^2 + V^{-2} + \bar{a}a$ identified as the Casimir of a deformed $\mathfrak{sl}_2$. This variable Poisson-commute with all other dynamical quantities and can therefore be fixed to some particular value c_0. We shall nevertheless keep the redundant three-parameter expression for $\tilde{L}$ for reasons of form simplicity in the explicit expressions.

We shall express the term of order u in $\mathbb{U}$ independently of the fields, after applying a suitable gauge transformation [20]

$$T^{\pm}(x,y,\lambda) = \Omega^{\pm}(x)\,\tilde{T}^{\pm}(x,y)\,(\Omega^{\pm}(y))^{-1}, \qquad \Omega^{\pm} = e^{\frac{i\beta}{4}\phi^{\pm}\sigma^z}, \tag{14}$$

The gauge transformed operator $\tilde{\mathbb{U}}$ is expressed as:

$$\tilde{\mathbb{U}}^{\pm}(x,t,u) = \frac{\beta}{4i}\mathfrak{f}^{\pm}\sigma^z + \frac{mu}{4i}\sigma^y - \frac{mu^{-1}}{4i}e^{-\frac{i\beta}{2}\phi^{\pm}\sigma^z}\sigma^y e^{\frac{i\beta}{2}\phi^{\pm}\sigma^z} \tag{15}$$

where we define

$$\mathfrak{f}^{\pm}(x,t) = \pi^{\pm}(x,t) + \phi^{\pm'}(x,t). \tag{16}$$

We consider the following convenient decomposition for $\tilde{T}$, as $|u| \to \infty$ [20],

$$\tilde{T}^{\pm}(x,y,\lambda) = (1 + W^{\pm}(x,\lambda))\,e^{Z^{\pm}(x,y,\lambda)}\,(1 + W^{\pm}(y,\lambda))^{-1} \tag{17}$$

$W^{\pm}$ is an off-diagonal matrix and $Z^{\pm}$ is purely diagonal. They are expanded as:

$$W^{\pm} = \sum_{k=0}^{\infty} \frac{W^{\pm(k)}}{u^k}, \qquad Z^{\pm} = \sum_{k=-1}^{\infty} \frac{Z^{\pm(k)}}{u^k} \tag{18}$$

Note that $\tilde{T}$ naturally satisfies the gauged Lax equation:

$$\frac{\partial \tilde{T}^{\pm}}{\partial x} = \tilde{\mathbb{U}}^{\pm}(x,\lambda)\tilde{T}^{\pm}(x,y,\lambda) \tag{19}$$

Inserting expressions (17), (18) in (19) one identifies the matrices $W^{\pm(k)}$ and $Z^{\pm(k)}$. More precisely, we end up with an equation for the off-diagonal matrix:

$$\frac{\partial W^{\pm}}{\partial x} + W^{\pm}\tilde{\mathbb{U}}_D^{\pm} - \tilde{\mathbb{U}}_D^{\pm} W^{\pm} + W^{\pm}\tilde{\mathbb{U}}_A^{\pm} W^{\pm} - \mathbb{U}_A^{\pm} = 0 \tag{20}$$

where the indices D, A denote the diagonal and anti-diagonal part of the Lax operator $\tilde{\mathbb{U}}^{\pm}$, leading to typical Riccati-type equations for the entries of $W^{\pm}$ [2].

Similarities with corresponding equations emerging in [12, 18] from the inverse scattering point of view are apparent as expected, given that one essentially solves the same fundamental equations. Liouville integrability is guaranteed within the present approach by construction, at least formally, whereas in the methodology of [12, 18] only the conservation of the charges for a singled-out time-evolution is shown through the zero curvature condition i.e. explicit use of the equations of motion.

It is sufficient for our purposes here to identify only the first few terms of the expansions. Indeed based on equation (19) we conclude (see also [20]):

$$\begin{aligned} &W^{\pm(0)} = i\sigma_1, \quad W^{\pm(1)} = -\frac{i\beta}{m}\mathfrak{f}^{\pm}(x)\sigma_1, \\ &W^{\pm(2)} = \frac{2i\beta\mathfrak{f}^{\pm'}}{m^2}\,\sigma_2 - i\sin(\beta\phi^{\pm})\,\sigma_2 - \frac{\beta^2(\mathfrak{f}^{\pm})^2}{2im^2}\,\sigma_1. \end{aligned} \tag{21}$$

We also need to identify the diagonal elements $Z^{\pm(n)}$. In particular from equation (1) we extract the following expressions:

$$\begin{aligned} Z^{+(-1)} &= -\frac{im(L-x_0)}{4}\sigma_3, \qquad Z^{-(-1)} = -\frac{im(L+x_0)}{4}\sigma_3 \\ Z^{+(1)} &= \frac{m}{4}\begin{pmatrix} -\int_{x_0^+}^{L} dx\; W_{21}^{+(2)}(x) & \\ & \int_{x_0^+}^{L} dx\; W_{12}^{+(2)}(x) \end{pmatrix} - \frac{m}{4}\begin{pmatrix} -i\int_{x_0}^{L} dx\; e^{-i\beta\phi^+} & \\ & i\int_{x_0^+}^{L} dx\; e^{i\beta\phi^+} \end{pmatrix}, \\ Z^{-(1)} &= \frac{m}{4}\begin{pmatrix} -\int_{-L}^{x_0^-} dx\; W_{21}^{-(2)}(x) & \\ & \int_{-L}^{x_0^-} dx\; W_{12}^{-(2)}(x) \end{pmatrix} - \frac{m}{4}\begin{pmatrix} -i\int_{-L}^{x_0^-} dx\; e^{-i\beta\phi^-} & \\ & i\int_{-L}^{x_0^-} dx\; e^{i\beta\phi^-} \end{pmatrix}. \end{aligned} \tag{22}$$

Notice that for $-iu \to \infty$ the leading contribution comes from the $Z_{11}^{\pm}$ elements. This observation will be subsequently quite useful.

We shall first derive the associated local integrals of motion. In particular, the energy and momentum in the presence of defect will be obtained. Let us first recall the generating function of the local integrals of motion

$$\mathcal{G}(\lambda) = \ln\left[trT^{+}(L,x_0,\lambda)\;\tilde{L}(x_0,\lambda)\;T^{-}(x_0,L,\lambda)\right] \tag{23}$$

Schwartz boundary conditions are imposed at the end point of the system $\pm L$. Recalling also the ansatz for the monodromy matrices we conclude that the generating function:

$$\mathcal{G}(\lambda) = \ln tr\Big[e^{Z^+(L,x_0)}(1+W^+(x_0))^{-1}(\Omega^+(x_0))^{-1}\tilde{L}(x_0)\Omega^-(x_0)(1+W^-(x_0))e^{Z^-(x_0,-L)}\Big] \quad (24)$$

Choosing to consider the $-iu \to \infty$ behavior we take into account the leading contribution for the $Z^{\pm}_{11}$ terms, then the generating function takes the form reads as:

$$\mathcal{G}(\lambda) = Z^+_{11} + Z^-_{11} + \ln\Big[(1+W^+(x_0))^{-1}(\Omega^+(x_0))^{-1}\tilde{L}(x_0)\Omega^-(x_0)(1+W^-(x_0))\Big]_{11} \quad (25)$$

Expanding the latter expression in powers of u^{-1} we obtain the following:

$$\mathcal{G}(\lambda) = \sum_{m=0}^{\infty}\frac{I^{(m)}}{u^m}. \quad (26)$$

Recalling now the expression for the generating function of integrals of motion we conclude that

$$\begin{aligned} I^{(1)} &= -\frac{m}{4i}\int_{-L}^{x_0^-}dx\,\Big(-\frac{\beta^2}{2m^2}\mathfrak{f}^{-2}(x)+\cos(\beta\phi^-(x))\Big) - \frac{m}{4i}\int_{x_0^+}^{L}dx\,\Big(-\frac{\beta^2}{2m^2}\mathfrak{f}^{+2}(x)+\cos(\beta\phi^+(x))\Big) \\ &+ \frac{i}{\mathcal{D}}\Big(e^{-\frac{i\beta}{4}(\phi^+(x_0)+\phi^-(x_0))}\bar{a} - e^{\frac{i\beta}{4}(\phi^+(x_0)+\phi^-(x_0))}a\Big) + \frac{\beta}{2m\mathcal{D}}\Big(\mathfrak{f}^+(x_0)+\mathfrak{f}^-(x_0)\Big)\mathcal{A} \end{aligned} \quad (27)$$

where we define:

$$\begin{aligned} \mathcal{D} &= e^{-\frac{i\beta}{4}(\phi^+(x_0)-\phi^-(x_0))}V + e^{\frac{i\beta}{4}(\phi^+(x_0)-\phi^-(x_0))}V^{-1}, \\ \mathcal{A} &= e^{-\frac{i\beta}{4}(\phi^+(x_0)-\phi^-(x_0))}V - e^{\frac{i\beta}{4}(\phi^+(x_0)-\phi^-(x_0))}V^{-1}. \end{aligned} \quad (28)$$

If we now perform the same expansion for $\lambda \to -\infty$, we basically end up with a similar expression, by simply exploiting the fundamental symmetry of the monodromy matrix:

$$T(u^{-1},\phi,\ \pi,V,a,\bar{a}) = T(-u,\ -\phi,\ \pi,V^{-1},a,\bar{a}). \quad (29)$$

More precisely, one concludes that:

$$\begin{aligned} I^{(-1)} &= \frac{m}{4i}\int_{-L}^{x_0^-}dx\,\Big(-\frac{\beta^2}{2m^2}\hat{\mathfrak{f}}^{-2}(x)+\cos(\beta\phi^-(x))\Big) - \frac{m}{4i}\int_{x_0^+}^{L}dx\,\Big(-\frac{\beta^2}{2m^2}\hat{\mathfrak{f}}^{+2}(x)+\cos(\beta\phi^+(x))\Big) \\ &- \frac{i}{\mathcal{D}}\Big(e^{\frac{i\beta}{4}(\phi^+(x_0)+\phi^-(x_0))}\bar{a} - e^{-\frac{i\beta}{4}(\phi^+(x_0)+\phi^-(x_0))}a\Big) + \frac{\beta}{2m\mathcal{D}}\Big(\hat{\mathfrak{f}}^+(x_0)+\hat{\mathfrak{f}}^-(x_0)\Big)\mathcal{A} \end{aligned} \quad (30)$$

where we define

$$\hat{\mathfrak{f}}^{\pm}(\phi,\ \pi) = \mathfrak{f}^{\pm}(-\phi,\ \pi). \quad (31)$$

Any combination of the quantities $I^{(1)}$, $I^{(-1)}$ can be picked as one of the charges in involution. In particular the defect-extended form of the sine-Gordon Hamiltonian is defined as:

$$\begin{aligned} \mathcal{H} &= \frac{2im}{\beta^2}(I^{(1)} - I^{(-1)}) \\ &= \int_{-L}^{x_0^-}dx\,\Big(\frac{1}{2}(\pi^{-2}(x)+\phi^{-'2}(x)) - \frac{m^2}{\beta^2}\cos(\beta\phi^-(x))\Big) \\ &+ \int_{x_0^+}^{L}dx\,\Big(\frac{1}{2}(\pi^{+2}(x)+\phi^{+'2}(x)) - \frac{m^2}{\beta^2}\cos(\beta\phi^+(x))\Big) \\ &- \frac{4m}{\beta^2\mathcal{D}}\cos\frac{\beta}{4}(\phi^+(x_0)+\phi^-(x_0))\,\Big(\bar{a}-a\Big) + \frac{2i}{\beta\mathcal{D}}\Big(\phi^{+'}(x_0)+\phi^{-'}(x_0)\Big)\mathcal{A} \end{aligned} \quad (32)$$

and we also identify the sine-Gordon momentum as:

$$\begin{aligned}\mathcal{P} &= \frac{2im}{\beta^2}\Big(I^{(1)} + I^{(-1)}\Big) \\ &= \int_{-L}^{x_0^-} dx\ \phi^{-'}(x)\pi^-(x) + \int_{x_0^+}^{L} dx\ \phi^{+'}(x)\pi^+(x) \\ &+ \frac{4mi}{\beta^2\mathcal{D}}\sin\frac{\beta}{4}(\phi^+(x_0)+\phi^-(x_0))\ \Big(\bar{a}+a\Big) + \frac{2i}{\beta\mathcal{D}}\Big(\pi^+(x_0)+\pi^-(x_0)\Big)\mathcal{A}. \end{aligned} \quad (33)$$

Explicit computation of the Poisson bracket $\{\mathcal{H}, \mathcal{P}\}$ now yields a number of non-zero terms; we shall come back to this issue after deriving the sewing conditions.

The next step is the derivation of the time components of the associated Lax pairs. Expressions of the time component $\mathbb{V}$ of the Lax pair are known (see e.g. [21, 20]). The generic expressions for the bulk left and right theories as well as the defect points are given as [1, 22]:

$$\begin{aligned} \mathbb{V}^+(x,\lambda,\mu) &= t^{-1}(\lambda)tr_a\Big(T_a^+(L,x,\lambda)r_{ab}(\lambda-\mu)T_a^+(x,x_0,\lambda)\tilde{L}_a(x_0,\lambda)T_a^-(x_0,-L,\lambda)\Big) \\ \mathbb{V}^-(x,\lambda,\mu) &= t^{-1}(\lambda)tr_a\Big(T_a^+(L,x_0,\lambda)\tilde{L}_a(x_0)T_a^-(x_0,x,\lambda)r_{ab}(\lambda-\mu)T_a^-(x,-L,\lambda)\Big) \\ \tilde{\mathbb{V}}^+(x_0,\lambda,\mu) &= t^{-1}(\lambda)tr_a\Big(T_a^+(L,x_0,\lambda)r_{ab}(\lambda-\mu)\tilde{L}_a(x_0,\lambda)T_a^-(x_0,-L,\lambda)\Big) \\ \tilde{\mathbb{V}}^-(x_0,\lambda,\mu) &= t^{-1}(\lambda)tr_a\Big(T_a^+(L,x_0,\lambda)\tilde{L}_a(x_0,\lambda)r_{ab}(\lambda-\mu)T_a^-(x_0,-L,\lambda)\Big). \end{aligned} \quad (34)$$

In order to identify the Lax pair associated to the Hamiltonian and momentum it is necessary to formulate the expansion of $\mathbb{V}$ in both negative and positive powers of u.

The first order contribution in the u^{-1} expansion of the bulk $\mathbb{V}^\pm$ operator (we have set the second spectral parameter $v \equiv e^\mu$) reads:

$$\mathbb{V}^{\pm(1)} = \frac{\beta^2}{8}\left(\frac{\beta}{2m}\sigma^z(\pi^\pm + \phi^{\pm'}) + iv\Big(\sigma^- e^{-\frac{i\beta}{2}\phi^\pm} - \sigma^+ e^{\frac{i\beta}{2}\phi^\pm}\Big)\right) \quad (35)$$

The first order contribution in the u expansion reads:

$$\hat{\mathbb{V}}^{\pm(1)} = \frac{\beta^2}{8}\left(\frac{\beta}{2m}\sigma^z(\pi^\pm - \phi^{\pm'}) - iv^{-1}\Big(\sigma^- e^{\frac{i\beta}{2}\phi^\pm} - \sigma^+ e^{-\frac{i\beta}{2}\phi^\pm}\Big)\right) \quad (36)$$

Subtracting these two expressions and multiplying by $-\frac{2im}{\beta^2}$ we obtain the time component of the Lax pair associated to the Hamiltonian: ($\Omega^\pm = e^{\frac{i\beta}{4}\phi^\pm\sigma^z}$)

$$\mathbb{V}_{\mathcal{H}}^\pm = \frac{\beta}{4i}\phi^{\pm'}\sigma^z + \frac{vm}{4i}\Omega^\pm\sigma^y(\Omega^\pm)^{-1} + \frac{v^{-1}m}{4i}(\Omega^\pm)^{-1}\sigma^y\Omega^\pm \quad (37)$$

Adding now (35), (36), after multiplying with $-\frac{2im}{\beta^2}$, provides the time component of the Lax pair associated to the momentum:

$$\mathbb{V}_{\mathcal{P}}^\pm = \frac{\beta}{4i}\pi^\pm\sigma^z + \frac{vm}{4i}\Omega^\pm\sigma^y(\Omega^\pm)^{-1} - \frac{v^{-1}m}{4i}(\Omega^\pm)^{-1}\sigma^y\Omega^\pm \quad (38)$$

The next step is the derivation of the relevant Lax pairs for the defect point from the left and the right, based on the expression (34). Indeed, after some cumbersome but quite straightforward computations, and after defining:

$$w^\pm = -\frac{i\beta}{m}\mathfrak{f}^\pm, \qquad \hat{w}^\pm = \frac{i\beta}{m}\hat{\mathfrak{f}}^\pm, \quad (39)$$

we conclude from the expansion in powers of u^{-1}:

$$\begin{aligned}\tilde{\mathbb{V}}^{+(1)} &= \frac{i\beta^2}{8}\mathcal{D}^{-2}\sigma^z\Big[w^+ + w^- + e^{\frac{i\beta}{2}\phi^-}Va + e^{-\frac{i\beta}{2}\phi^-}V^{-1}\bar{a}\Big] \\ &+ \frac{i\beta^2}{4}\mathcal{D}^{-1}v\Big[\sigma^- e^{-\frac{i\beta}{4}(\phi^+ + \phi^-)}V^{-1} - \sigma^+ e^{\frac{i\beta}{4}(\phi^+ + \phi^-)}V\Big], \end{aligned} \tag{40}$$

whereas the expansion in powers of u leads to:

$$\begin{aligned}\hat{\tilde{\mathbb{V}}}^{+(1)} &= -\frac{i\beta^2}{8}\mathcal{D}^{-2}\sigma^z\Big[\hat{w}^+ + \hat{w}^- - e^{\frac{i\beta}{2}\phi^-}V\bar{a} - e^{-\frac{i\beta}{2}\phi^-}V^{-1}a\Big] \\ &- \frac{i\beta^2}{4}\mathcal{D}^{-1}v^{-1}\Big[\sigma^- e^{\frac{i\beta}{4}(\phi^+ + \phi^-)}V - \sigma^+ e^{-\frac{i\beta}{4}(\phi^+ + \phi^-)}V^{-1}\Big] \end{aligned} \tag{41}$$

Similarly, the corresponding expressions for $\tilde{\mathbb{V}}^{-(1)}$, $\hat{\tilde{\mathbb{V}}}^{-(1)}$ are given below:

$$\begin{aligned}\tilde{\mathbb{V}}^{-(1)} &= \frac{i\beta^2}{8}\mathcal{D}^{-2}\sigma^z\Big[w^+ + w^- - e^{\frac{i\beta}{2}\phi^+}V^{-1}a - e^{-\frac{i\beta}{2}\phi^+}V\bar{a}\Big] \\ &+ \frac{i\beta^2}{4}\mathcal{D}^{-1}v\Big[\sigma^- e^{-\frac{i\beta}{4}(\phi^+ + \phi^-)}V - \sigma^+ e^{\frac{i\beta}{4}(\phi^+ + \phi^-)}V^{-1}\Big] \end{aligned} \tag{42}$$

$$\begin{aligned}\hat{\tilde{\mathbb{V}}}^{-(1)} &= -\frac{i\beta^2}{8}\mathcal{D}^{-2}\sigma^z\Big[\hat{w}^+ + \hat{w}^- + e^{\frac{i\beta}{2}\phi^+}V^{-1}\bar{a} + e^{-\frac{i\beta}{2}\phi^+}Va\Big] \\ &- \frac{i\beta^2}{4}\mathcal{D}^{-1}v^{-1}\Big[\sigma^- e^{\frac{i\beta}{4}(\phi^+ + \phi^-)}V^{-1} - \sigma^+ e^{-\frac{i\beta}{4}(\phi^+ + \phi^-)}V\Big]. \end{aligned} \tag{43}$$

We are now in a position to apply the scheme elaborated in [1]. The first manifest observation from the continuity conditions

$$\begin{aligned}&\tilde{\mathbb{V}}^{+(1)}(x_0) \to \mathbb{V}^{+(1)}(x_0^+), \qquad x_0^+ \to x_0 \\ &\tilde{\mathbb{V}}^{-(1)}(x_0) \to \mathbb{V}^{-(1)}(x_0^-), \qquad x_0^- \to x_0 \end{aligned} \tag{44}$$

(similar continuity conditions apply for the "hatted" quantities, but are omitted for brevity), is that:

$$V = e^{\frac{i\beta}{4}(\phi^+ - \phi^-)}. \tag{45}$$

and will be hereafter denoted as "first sewing condition S_1 ". Remember that from the very beginning one has already fixed the Casimir C_0 to some value c_0 independently of any sewing requirement. This can be seen as an "order zero condition S_0 " without any dependance in the bulk variables and yields a first-class constraint Poisson-commuting with all dynamical variables.

After imposing (45) the time components of the Lax pairs on the defect point take the following simple expressions:

$$\tilde{\mathbb{V}}^{\pm(1)} = \frac{\beta^2}{8}\left(\frac{\beta}{4m}\sigma^z\Big(\pi^+ + \phi^{+'} + \pi^- + \phi^{-'}\Big) \pm \frac{i\sigma^z}{4}\mathbb{M} + iv\Big(\sigma^- e^{-\frac{i\beta}{2}\phi^\pm} - \sigma^+ e^{\frac{i\beta}{2}\phi^\pm}\Big)\right) \tag{46}$$

and the first term in the u expansion provides:

$$\hat{\tilde{\mathbb{V}}}^{\pm} = \frac{\beta^2}{8}\left(\frac{\beta}{4m}\sigma^z\Big(\pi^+ - \phi^{+'} + \pi^- - \phi^{-'}\Big) \pm \frac{i\sigma^z}{4}\hat{\mathbb{M}} - iv^{-1}\Big(\sigma^- e^{\frac{i\beta}{2}\phi^\pm} - \sigma^+ e^{-\frac{i\beta}{2}\phi^\pm}\Big)\right) \tag{47}$$

where we define:

$$\mathbb{M} = e^{-\frac{i\beta}{4}(\phi^+ + \phi^-)}\bar{a} + e^{\frac{i\beta}{4}(\phi^+ + \phi^-)}a$$
$$\hat{\mathbb{M}} = e^{\frac{i\beta}{4}(\phi^+ + \phi^-)}\bar{a} + e^{-\frac{i\beta}{4}(\phi^+ + \phi^-)}a \qquad (48)$$

Continuity conditions on the Lax pair as also described in (44) give rise to the following sewing conditions on the defect point x_0 associated to the momentum and the Hamiltonian respectively:

$$S_2: \qquad \pi^+(x_0) - \pi^-(x_0) = \frac{im}{\beta}\cos\frac{\beta}{4}(\phi^+(x_0) + \phi^-(x_0))\left(a + \bar{a}\right)$$
$$S_2': \qquad \phi^{+'}(x_0) - \phi^{-'}(x_0) = \frac{m}{\beta}\sin\frac{\beta}{4}(\phi^+(x_0) + \phi^-(x_0))\left(\bar{a} - a\right) \qquad (49)$$

the prime denotes the derivative with respect to x.

It is instructive to point out that comparison of the extracted charges (32), (33), and the latter equations (49) with similar results obtained for instance in [11, 12] reveal manifest discrepancies. We shall further comment on this matter in the discussion section.

Consistency of the sewing conditions S_1, S_2, S_2' can now be checked by computing their Poisson brackets with the first two Hamiltonians $\mathcal{H}, \mathcal{P}$. Indeed one gets:

$$\left\{\mathcal{H}, S_1\right\} = -\frac{i\beta}{4}(\pi^+(x_0) - \pi^-(x_0))S_1 + \frac{i\beta}{4}S_2 V + o(\mathcal{D} - 2)$$
$$\left\{\mathcal{P}, S_1\right\} = -\frac{i\beta}{4}(\phi^{+'}(x_0) - \phi^{-'}(x_0))S_1 + \frac{i\beta}{4}S_2' V + o(\mathcal{D} - 2) \qquad (50)$$

We recall that on-shell $\mathcal{D} \approx 2$; $\mathcal{A} \approx 0$.

Consider now the Poisson brackets of $\mathcal{H}$, $\mathcal{P}$ with S_2, S_2'. One easily obtains that they are given by expressions of the following form:

$$\left\{\mathcal{P}, S_2\right\} = (\pi^{+'}(x_0) - \pi^{-'}(x_0))\ F(\pi^+(x_0) + \pi^-(x_0), \phi^{+'}(x_0) + \phi^{-'}(x_0), \phi^+(x_0), \phi^-(x_0), V, a, \bar{a})$$
$$\left\{\mathcal{H}, S_2\right\} = (\phi^{+''}(x_0) - \phi^{-''}(x_0))\ G(\pi^+(x_0) + \pi^-(x_0), \phi^{+'}(x_0) + \phi^{-'}(x_0), \phi^+(x_0), \phi^-(x_0), V, a, \bar{a}) \qquad (51)$$

where F and G are given functions to be computed specifically. Poisson brackets with S_2' are given by similar expressions exchanging $\mathcal{H}$ and $\mathcal{P}$. Note that (contrary to the non-linear Schroedinger case) no term proportional to the singular contribution $\delta(0)$ arise, they fully cancel in the Poisson brackets. It is therefore to be expected that the finite terms on the r.h.s. of both PB's will yield the third sewing conditions S_3, S_3' which will respectively take the form (expected from general arguments) $(\pi^{+'}(x_0) - \pi^{-'}(x_0)) = -F$ and $(\phi^{+''}(x_0) - \phi^{-''}(x_0) = -G$. Explicit derivation of these sewing conditions from higher terms in the expansion of the $\mathbb{V}$ operators is technically quite cumbersome but we conjecture that they will coincide with the rhs of (51).

Let us now reconsider the Poisson bracket $\{\mathcal{H}, \mathcal{P}\}$. It turns out from explicit computations that one has in fact:

$$\left\{\mathcal{H},\ \mathcal{P}\right\} \approx 0, \qquad (52)$$

i.e. the Poisson bracket vanishes provided that the constraints S_1, S_2, S_2' be satisfied.

Assuming that the Hamiltonians $\mathcal{H}$ and $\mathcal{P}$ weakly preserve all constraints (as already established for S_1 and conjectured for S_2, S_2') we deduce that the momentum and Hamiltonian Dirac commute. Our construction of Type II defect is thus compatible with a statement of Liouville-integrability on-shell. A general formal argument based on the underlying Poisson structure has developed in [1] establishing the property of weak preservation of constraints.

Equations of motion

To extract the associated equations of motion for the left and right bulk theories as well as the defect point one needs to employ the zero curvature condition expressed as:

$$\dot{\mathbb{U}}^{\pm}(x,t) - \mathbb{V}^{\pm\prime}(x,t) + \Big[\mathbb{U}^{\pm}(x,t), \mathbb{V}^{\pm}(x,t)\Big] = 0 \qquad x \neq x_0. \tag{53}$$

As usual the dot denotes derivative with respect to t.

On the defect point in particular the zero curvature condition is formulated as (this is also transparent when discussing the continuum limit of discrete theories (see e.g. [1])

$$\dot{\tilde{L}}(x_0) = \tilde{\mathbb{V}}^{+}(x_0)\tilde{L}(x_0) - \tilde{L}(x_0)\tilde{\mathbb{V}}^{-}(x_0), \tag{54}$$

and describes explicitly the jump occurring across the defect point.

The equations of motion are obtained via the zero curvature conditions as described above or (equivalently thanks to the sewing conditions) via the Hamiltonian equations i.e.

$$\begin{aligned} &\dot{\phi}^{\pm} = \Big\{\mathcal{H},\ \phi^{\pm}\Big\}, \qquad \dot{\pi}^{\pm} = \Big\{\mathcal{H},\ \pi^{\pm}\Big\}, \\ &\dot{\mathrm{e}} = \Big\{\mathcal{H},\ \mathrm{e}\Big\}, \qquad \mathrm{e} \in \Big\{a,\ \bar{a},\ V\Big\} \end{aligned} \tag{55}$$

bear also in mind that

$$\Big\{\pi^{\pm},\ \mathrm{e}\Big\} = \Big\{\phi^{\pm},\ \mathrm{e}\Big\} = 0. \tag{56}$$

For the left and right bulk theories the familiar equations of motion for the sine-Gordon model arise

$$\ddot{\phi}^{\pm}(x,t) - \phi^{\pm\prime\prime}(x,t) + \frac{m^2}{\beta}\sin(\beta\phi^{\pm}(x,t)) = 0 \tag{57}$$

On the defect point the time evolution of the defect degrees of freedom are obtained as:

$$\begin{aligned} \dot{a} \ &= \ -\frac{m}{2\mathcal{D}^2}\ \mathcal{A}\ a\ \cos\frac{\beta}{4}(\phi^{+}+\phi^{-})\ \Big(\bar{a}-a\Big) - \frac{m}{\mathcal{D}}\ \cos\frac{\beta}{4}(\phi^{+}+\phi^{-})\ \Big(V^2 - V^{-2}\Big) \\ &- \ \frac{\beta i}{\mathcal{D}^2}\ a\ \Big(\phi^{+\prime} + \phi^{-\prime}\Big) \end{aligned} \tag{58}$$

$$\begin{aligned} \dot{\bar{a}} \ &= \ \frac{m}{2\mathcal{D}^2}\ \mathcal{A}\ \bar{a}\ \cos\frac{\beta}{4}(\phi^{+}+\phi^{-})\ \Big(\bar{a}-a\Big) - \frac{m}{\mathcal{D}}\ \cos\frac{\beta}{4}(\phi^{+}+\phi^{-})\ \Big(V^2 - V^{-2}\Big) \\ &+ \ \frac{i\beta}{\mathcal{D}^2}\ \bar{a}\ \Big(\phi^{+\prime} + \phi^{-\prime}\Big) \end{aligned} \tag{59}$$

$$\dot{V} = \frac{m}{2\mathcal{D}}\ V\ \cos(\frac{\beta}{4}(\phi^{+}+\phi^{-}))\ \Big(a+\bar{a}\Big). \tag{60}$$

With this we conclude our presentation on the sine-Gordon model in the presence of integrable dynamical defects.

Acknowledgments
Material of this article was also presented by A. Doikou in "Integrable systems and Quantum Symmetries", Prague, June 2012.

References

[1] J. Avan and A. Doikou, JHEP **01** (2012) 040, arXiv:1110.4728 [hep-th].
[2] J. Avan and A. Doikou, arXiv:1205.1661 [hep-th].
[3] G. Delfino, G. Mussardo and P. Simonetti, Phys. Lett. **B328** (1994) 123, hep-th/9403049; G. Delfino, G. Mussardo and P. Simonetti, Nucl. Phys. **B432** (1994) 518, hep-th/9409076.
[4] E. Corrigan and C. Zambon, J. Phys. A: Math. Theor. **43** (2010) 345201, arXiv:1006.0939 [hep-th].
[5] R. Konik and A. LeClair, Nucl. Phys **B538** (1999) 587; hep-th/9793985.
[6] P. Bowcock, E. Corrigan and C. Zambon, JHEP **08** (2005) 023, hep-th/0506169.
[7] F. Nemes, Semiclassical analysis of defect sine-Gordon theory, Int. J. Mod. Phys. **A 25** (2010) 4493; arXiv:0909.3268 [hep-th].
[8] E. Corrigan and C. Zambon, J. Phys. **A 42** (2009) 304008, arXiv:0902.1307 [hep-th]; P. Bowcock, E. Corrigan and C. Zambon, JHEP **01** (2004) 056, hep-th/0401020
[9] E. Corrigan and C. Zambon, JHEP **07** (2007) 001, arXiv:0705.1066 [hep-th]; E. Corrigan and C. Zambon, J. Phys. **A 42** (2009) 475203; arXiv:0908.3126 [hep-th].
[10] E. Corrigan and C. Zambon, Nonlinearity **19** (2006) 1447, nlin/0512038.
[11] I. Habibullin and A. Kundu, Nucl. Phys. **B795** (2008) 549, arXiv:0709.4611 [hep-th].
[12] V. Caudrelier, IJGMMP **5** No. 7 (2008) 1085, arXiv:0704.2326 [math-ph].
[13] Z. Bajnok, A. George, Int. J. Mod. Phys. A21 (2006) 1063, hep-th/0404199; Z. Bajnok, J. Stat. Mech. 0606 (2006) P06010, hep-th/0601107; Z. Bajnok, Zs. Simon, Nucl. Phys. B802 (2007) 307, arXiv:0712.4292 [hep-th].
[14] R. Weston, *An Algebraic Setting for Defects in the XXZ and Sine-Gordon Models*, arXiv:1006.1555 [math-ph].
[15] M. Mintchev, E. Ragoucy and P. Sorba, Phys. Lett. **B547** (2002) 313, hep-th/0209052; M. Mintchev, E. Ragoucy and P. Sorba, J. Phys. **A36** (2003) 10407, hep-th/0303187.
[16] V. Caudrelier, M. Mintchev and E. Ragoucy, J. Phys. **A37** (2004) L367, hep-th/0404144.
[17] A. Doikou, Nucl. Phys. **B854** (2012) 153, arXiv:1106.1602, [hep-th].
[18] A.R. Aguirre, T.R. Araujo, J.F. Gomes and A.H. Zimerman , *Type-II Bäcklund Transformations via Gauge Transformations*, arXiv:1110.1589 [hep-th]; A.R. Aguirre, *Inverse scattering approach for massive Thirring models with integrable type-II defects*, arXiv:1111.5249 [math-ph].
[19] J.M. Maillet; New integrable canonical structures Nucl. Phys. **B269** (1986) 54
[20] L.D. Faddeev and L.A. Takhtakajan, *Hamiltonian Methods in the Theory of Solitons*, (1987) Springer-Verlag.
[21] M.A. Semenov-Tjan-Shanskii, Funct. Anal. Appl. **17** (1983), 259.
[22] J. Avan and A. Doikou, Nucl. Phys. **B812** (2009) 481, arXiv:0809.2734 [hep-th]

Journal of Physics: Conference Series **411** (2013) 012004 doi:10.1088/1742-6596/411/1/012004

The Boundary Conformal Field Theories of the 2D Ising critical points

Smain Balaska and Nahed Sihem Bounoua

Laboratoire de Physique Theorique d'Oran, Departement de Physique. Universite d'Oran, BP 1524 El'Manouer 31000 Oran. Algéria.

E-mail: sbalaska@yahoo.com

Abstract. We present a new method to identify the Boundary Conformal Field Theories (BCFTs) describing the critical points of the Ising model on the strip. It consists in measuring the low-lying excitation energies spectra of its quantum spin chain for different boundary conditions and then to compare them with those of the different boundary conformal field theories of the (A_2, A_3) minimal model
PACS numbers :05.50.+q, 05.70.JK, 11.25.HF

1. Introduction

The critical properties of the two-dimensional (2D) Ising model on the plane are known to be described by the (A_2, A_3) conformal model, which is a unitary minimal conformal field theory of the ADE classification. On two dimensional geometries with boundaries they should be described by conformal field theories (CFTs) defined on the half plane or equivalently on the strip (for review see [1] [3], [4],[5],[6] and [7]). For each boundary conditions, there exists a boundary conformal field theory (BCFT) having a spectrum formed by one or more of the Verma modules of the corresponding CFT on the plane. The main object of this work is a numerical study of the Ising linear quantum chain at criticality to identify the BCFTs of the (A_2, A_3) conformal model.

The paper is organized as follows. In the first section we introduce briefly the subject of boundary conformal field theories and give their predictions (boundary states, partition functions, operators contents and energy spectrum) concerning the (A_2, A_3) conformal model. In section two we introduce the methods used to make the identification of the different BCFTs for the 2D Ising model with the different allowed boundary conditions. The last section is consacred for presenting our numerical data for the quantum linear chain realizations of the Ising singularity of the 2D Ising model with different boundary conditions and to identify them with BCFTs candidates of (A_2, A_3) conformal model.

2. BCFT predictions

A conformal field theory constructed on the complex upper half plane will be strongly constrained by the presence of the real axe. In this case only symmetries preserving this boundary are concerned. So that, the independence of the holomorphic and the anti-holomorphic components of the Virasoro algebra is broken and only the holomorphic symmetries for example are considered. As a consequence, the partition function of a BCFT on the half plane is no more a sesquilinear combination of the Virasoro characters, as it was the case in the plane geometry. The partition functions $\mathcal{Z}_{\alpha|\beta}$ describing the evolution of the system between the boundary states $|\alpha\rangle$ and $|\beta\rangle$ become linear on the characters, as follows ([1], [3]):

$$\mathcal{Z}_{\alpha|\beta}(q) = \sum_j n_{j\alpha}^{\beta} \chi_j(q) \tag{1}$$

in this equation, the indices j runs over all the characters $\chi_j(q)$ of the Virasoro algebra and the multiplicities $n_{j\alpha}{}^{\beta}$ are positive integers giving the operator content of the BCFT. They are obtained from the Verlinde formula [9] :

$$n_{j\alpha}{}^{\beta} = \sum_i \frac{S_{ji} S^*_{\alpha i} S_{\beta i}}{S_{1i}} \tag{2}$$

The allowed physical boundary states are some combinations of Ishibashi states [2] of the following form :

$$|\alpha\rangle = \sum_j \frac{S_{\alpha j}}{\sqrt{S_{1j}}} |j\rangle\rangle \tag{3}$$

For the (A_2, A_3) conformal model we have three representations generated by the three primary fields

$$\phi_{(1,1)} = I \quad , \quad \phi_{(1,2)} = \sigma \quad \text{and} \quad \phi_{(2,1)} = \epsilon$$

Their conformal weights are respectively $h_{1,1} = 0$, $h_{1,2} = \frac{1}{16}$ and $h_{2,1} = \frac{1}{2}$. In this case the modular matrix is

$$S = \frac{1}{2} \begin{pmatrix} 1 & 1 & \sqrt{2} \\ 1 & 1 & -\sqrt{2} \\ \sqrt{2} & -\sqrt{2} & 0 \end{pmatrix}$$

Applying the relation (3) we obtain the three allowed boundary states

$$\begin{aligned} |I\rangle &= \frac{1}{\sqrt{2}}|I\rangle\rangle + \frac{1}{\sqrt{2}}|\epsilon\rangle\rangle + \frac{1}{(2)^{\frac{1}{4}}}|\sigma\rangle\rangle \\ |\epsilon\rangle &= \frac{1}{\sqrt{2}}|I\rangle\rangle + \frac{1}{\sqrt{2}}|\epsilon\rangle\rangle - \frac{1}{(2)^{\frac{1}{4}}}|\sigma\rangle\rangle \\ |\sigma\rangle &= |I\rangle\rangle - |\epsilon\rangle\rangle \end{aligned}$$

From (1) and (2) we obtain the allowed partition functions

$$\begin{aligned} \mathcal{Z}_{\sigma|\sigma} &= \chi_{(1,1)} + \chi_{(2,1)} = \chi_I + \chi_\epsilon \\ \mathcal{Z}_{I|I} &= \mathcal{Z}_{\epsilon|\epsilon} = \chi_{(1,1)} = \chi_I \\ \mathcal{Z}_{\sigma|I} &= \mathcal{Z}_{\sigma|\epsilon} = \chi_{(1,2)} = \chi_\sigma \\ \mathcal{Z}_{\epsilon|I} &= \mathcal{Z}_{I|\epsilon} = \chi_{(2,1)} = \chi_\epsilon \end{aligned}$$

The normalized Low-Lying excitation spectra of these four CFT candidates are given in table(1)

Table 1. excitation spectra of the BCFTs candidates in the(A_2, A_3)minimal model

CFT of $h_{1,1}$ verma module								
excitation energy	2	3	4	5	6	7	8	9
degeneracy	1	1	2	2	3	3	5	5
CFT of $h_{1,2}$ verma module								
excitation energy	1	2	3	4	5	6	7	8
degeneracy	1	1	2	2	3	4	5	6
CFT of $h_{2,1}$ verma module								
excitation energy	1	2	3	4	5	6	7	8
degeneracy	1	1	1	2	2	3	4	5
CFT of $(h_{1,1} \oplus h_{2,1})$ verma module								
excitation energy	1	3	4	5	6	7	8	9
degeneracy	1	1	1	1	1	1	2	2

3. The Methods

The Hamiltonian limit of the statistical 2D Ising model provides a quantum spin chain model having the same critical behavior. We will focus on the Ising singularity of such quantum spin chain model including special boundary magnetic fields at the first and the last sites. That is, different boundary conditions are imposed on the quantum spin chain. We will determine its low lying excitation spectrum for different chain lengths and at special values of the spin-spin coupling constant which we will call "pseudo-critical" values. The phenomenological renormalization group (PRG) will fixe these special values that define the Ising singularity (see [8] and the references therein).
For a given boundary condition, the numerical measurements of the spectrum for different chain lengths lead to series of values. When fitted to leading scaling behavior, these values will correspond to the energy levels at criticality.
In the last step, this will be compared with the BCFT predictions given in table(1).
The general form of the hamiltonian describing the quantum Ising chain with magnetic fields at the first and the last sites is given by

$$H = -\sum_{n=1}^{N-1} t\sigma_z(n)\sigma_z(n+1) - \sum_{n=1}^{N} \sigma_x(n) - h_1\sigma_z(1) - h_2\sigma_z(N) \tag{4}$$

In equation (4), N represents the number of sites in the chain, $\sigma_x(n)$ and $\sigma_z(n)$ are the 2x2 Pauli spin matrices at site "n", h_1 and h_2 are respectively the external magnetic fields applied at the first and the last sites and "t" is the ferromagnetic spin-spin coupling. We distinguish four cases:

(i) The free-free boundary conditions. In this case, one sets $h_1 = h_2 = 0$.
(ii) The fixed-parallel boundary conditions. In this case one sets:$h_1 = h_2 = h_0$.
(iii) The fixed anti-parallel boundary conditions. In this cases one sets, $h_1 = -h_2 = h_0$
(iv) The free-fixed boundary conditions. In this cases one sets:$h_1 = 0$ and $h_2 = h_0$

At the Ising singularity, the special coupling constants solve the PRG equation [13]

$$[N-1]m(t(N), N-1) - [N]m(t(N), N) = 0 \tag{5}$$

where $m(t, N)$ is the energy gap $[E_1(t, N) - E_0(t, N)]$. The energies $E_0(t, N)$ and $E_1(t, N)$ are the energies of the respective ground state "0" and first excited state "1" for the Ising quantum spin chain of length N. As $N \to \infty$, the special values of t solving the PRG equation, converge to the Ising singularity. The scaling behavior of physical quantities at solutions of the PRG equation provide the scaling behavior of the same physical quantities near the corresponding critical point.

4. Numerical Check of the BCFT Predictions

For all the boundary conditions, we have measured the energy spectrum of the Ising quantum chains of lengths $N = 6$ to $N = 12$. The study of these spectra for different values of N permits to us to obtain the series describing the energy levels. Then we have extrapolated them for $N \to \infty$ to obtain the energy spectra at criticality.
For both the free-free and fixed-parallel boundary conditions the solutions $t_c(N)$ of the PRG equation are presented respectively in tables (2) and (3). They converge toward the critical value $t_c(\infty) = 1$ when $(N \longrightarrow \infty)$.

Table 2. *The critical values of t solving the PRG equation for free-free B.Cs*

$N =$	4	5	6	7	8	9	10	11	12
t_c	1.09269	1.05685	1.03846	1.02775	1.02097	1.01640	1.01318	1.01083	1.00905

Table 3. *The critical values of t solving the PRG equation for Fixed Parallel B.Cs*

$N =$	5	6	7	8	9	10	11	12
t_c	1.03284	1.04577	1.04302	1.03743	1.03199	1.02739	1.02353	1.02036

The measured low lying excitation energies spectra corresponding to the first case are normalized, organized in series and presented with the extrapolated values (obtained using the BST algorithm [14])in table (4). For the second case see table (7) for the spectrum at criticality. For the fixed anti-parallel boundary conditions we observed a decreasing and an increasing values of the pseudo-critical coupling constants (see table (5)) which prove that in this case we didn't reach the asymptotic regime with lengths below $N = 12$. The extrapolated values of the low lying excitation energies corresponding to this case are given in table.
In the special case of free-fixed boundary condition, the PRG equation doesn't admit any exact solution for finite N. But we remark that for the special value of $t = 1$, the PRG equation admits a minimum which converges to 0 when N goes to infinity. This means that the value $t = 1$ is the critical value for infinite chain. We give in table(6) the corresponding values of the PRG equation for finite size chain at $t = 1$ and in table (7)we present the corresponding extrapolated low lying excitation energies spectra.

In conclusion the analysis of the numerical measurement of the energy excitation spectra for the Ising linear chains with different boundary conditions confirm that these models at criticality are in the same universality class as those of the boundary conformal field theories of the (A_2, A_3) minimal model.
Note also that in this case all the singularities have been considered.

Table 4. *Measured low lying excitation energies of the Ising quantum spin chain with free-free B.Cs*

	N=6	N=7	N=8	N=9	N=10	N=11	N=12	$N = \infty$
1^{st}Excitation	1.00000	1.00000	1.00000	1.00000	1.00000	1.00000	1.00000	1.00000
2 //	3.22593	3.19176	3.16683	3.14777	3.13270	3.12047	3.11033	3.00185
3 //	4.22593	4.19176	4.16683	4.14777	4.13270	4.12047	4.11033	4.00194
4 //	5.20053	5.19231	5.18240	5.17229	5.16261	5.15357	5.14526	5.00334
5 //	6.20053	6.19231	6.18240	6.17229	6.16261	6.15357	6.14526	6.00314
6 //	6.86405	6.95879	7.01551	7.05058	7.07267	7.08664	7.09540	7.00484
7 //	7.86405	7.95879	8.01551	8.05058	8.07267	8.08664	8.09540	8.00480
8 //	8.42647	8.38407	8.34923	8.32007	8.29531	8.27405	8.25560	8.00528
9 //	8.12516	8.41845	8.60753	8.73470	8.82314	8.88630	8.93238	9.00406
10//	9.42647	9.38407	9.34923	9.32007	9.29531	9.27405	9.25560	9.00520

Table 5. *The critical values of t solving the PRG equation for Fixed Anti-parallelB.Cs*

$N =$	5	6	7	8	9	10	11	12
t_c	1.17704	1.03605	0.99940	0.98770	0.98421	0.98380	0.98442	0.98552

Table 6. *The value of the PRG equation at t=1 for free-fixed B.Cs*

N	5	6	7	8	9	10	11	12
PRG	0.05930	0.03284	0.02012	0.01325	0.00921	0.00667	0.00500	0.00384

Table 7. The low lying Energy Excitation at Criticality for the different B.Cs

Fixed Parallel B.Cs									
2.0000	3.0043	3.9983	4.0362	5.0098	5.0618	6.0325	6.0272	6.0068	
Free-Fixed B.Cs									
1.0000	2.0006	3.0087	3.0007	4.0428	4.0088	5.0847	5.0409	5.0052	
Fixed AntiParallel B.Cs									
1.0000	2.0254	3.0155	3.9984	4.0857	5.0265	5.0843			
Free-Free B.Cs									
1.0000	3.0018	4.0019	5.0033	6.0031	7.0048	8.0048	8.0052	9.0040	9.0052

References

[1] J. L. Cardy, *Nucl. Phys. B 324 (1989) 581*

[2] N. Ishibashi, *The Boundary and Crosscap States in Conformal Field Theories,* Mod. Phys. Lett. A 4 (1987) 251-264.

[3] J.-B. Zuber , " *CFT, BCFT, ADE and all that*", *Proc.* of the bariloche Summer School, "Quantum Symmetries in Theoretical Physics and Mathematics", Jan 2000, eds R. Coquereaux, A. Garcia and R. Trinchero, (preprint hep-th/0006151).

[4] V. B. Petkova and J.-B. Zuber , " *Conformal Boundary Conditions and what they teach us*", (preprint hep-th/0103007). *Proc.* of Budapest School and Conferences on Nonperturbative Quantum Field Theoretic Methods and their Aplications, August 2000, World Scientific Publishing Company.

[5] R. E. Behrend, P. A. Pearce, V. B. Petkova, and J.-B. Zuber, *On the classification of bulk and boundary*

conformal field theories, Phys. Lett. **B 444** (1998)163-166.
[6] R. E. Behrend, P. A. Pearce, V. B. Petkova, and J.-B. Zuber, *Boundary conditions in rational conformal field theories*, Nucl. Phys. **B570** (2000) 525-589, [hep-th/9908036].
[7] R. E. Behrend, P. A. Pearce, and J.-B. Zuber, *Integrable boundaries, conformal boundary conditions and A -D -E fusion rules,* J. Phys. A 31 (1998) L 763-770, [hep-th/9807142].
[8] M. Henkel, *Conformal Invariance and Critical Phenomena*, Springer-Verlag, Germany, 1999.
[9] E. Verlinde, Nucl. Phys. B300 [FS22] (1988) 360.
[10] S. Kawai, *Coulomb-gas approach for boundary conformal field theory*, Nucl. Phys. **B630** (2002) 203221, [hep-th/0201146].
[11] S. Kawai, *Free-field realisation of boundary states and boundary correlation functions of minimal models*, J. Phys. **A36** (2003) 6547.
[12] S.balaska, J. MCcabe and T. Wydro, *Excitation Spectrum at the Yang-Lee Edge Singularity of 2D Ising Model on the Strip,* Int. J. Mod. Phys. B. Vol. 22, N 28 (2008) 4967-4973. ArXiv:0710.5068 v1
[13] C. Itzykson, H. Saleur, and J.-B. Zuber, Europhys. Lett. 2 (1986) 91.
[14] M. Henkel and G. Schuetz, J. of Phys A : Math and General 21 (1988) 2617-2633.

Asymptotic form of level density distributions for a class of inhomogeneous 1D vertex models

P Banerjee[1] and B Basu-Mallick[2]

Theory Division, Saha Institute of Nuclear Physics,
1/AF Bidhan Nagar, Kolkata 700 064, India

E-mail: [1]pratyay.banerjee@saha.ac.in
[2]bireswar.basumallick@saha.ac.in

Abstract. Level density distributions for some Haldane-Shastry like spin chains associated with A_{N-1} root system have been computed recently by Enciso *et al.*, using the connection of these spin chains with inhomogeneous one-dimensional vertex models. The energy functions of such vertex models are given by specific polynomials of first or second degree. Here we consider a much broader class of one-dimensional vertex models whose energy functions are given by arbitrary polynomials of any possible degree and show that the level densities for this class of vertex models asymptotically follow the Gaussian distribution for large number of vertices.

One-dimensional (1D) quantum integrable spin chains are usually classified into following two types depending on their range of interaction. One such class consists of spin chains having only local interactions, like nearest or next-to-nearest neighbour interaction. Isotropic and anisotropic versions of Heisenberg spin-$\frac{1}{2}$ chain, Hubbard model etc. are prominent examples of this type of spin models with short range interaction [1–3]. However, it is also possible to construct 1D quantum integrable spin systems like Haldane-Shastry (HS) spin chain [4,5], Polychronakos spin chain (also known as Polychronakos-Frahm (PF) spin chain) [6–8] and Frahm-Inozemstsev (FI) spin chain chain [9,10], for which all constituent spins mutually interact with each other through long range forces. This class of spin chains with long range interactions have attracted a lot of attention in recent years due to their exact solvability and applicability in a wide range of subjects like generalized exclusion statistics, fractional quantum Hall effect, SUSY Yang-Mills theory and Yangian quantum groups [11–24]. For the simplest case of A_{N-1} root system, the Hamiltonians of $su(m)$ invariant, ferromagnetic type HS, PF and FI spin chains can be written in a general form like [4–10]

$$H = \sum_{1\leq j<k\leq N} \frac{1-P_{jk}}{f(\xi_j-\xi_k)} \,. \tag{1}$$

Here ξ_j denotes the position of the j-th lattice site, spins can take m possible values on each lattice site, P_{jk} represents the exchange operator which interchanges the spins on the j-th and k-th lattice sites and the strength of this exchange interaction is determined by the function $f(\xi_j-\xi_k)$. Supersymmetric generalizations of the spin chains (1) have also been studied in the literature [12,25–27].

As is well known, the Hamiltonians of 1D quantum integrable spin chains with short range interaction are related to the transfer matrices of some exactly solvable 2D vertex models in statistical mechanics [28,29]. However, there exists no simple relation between the energy levels of 1D spin chains of short range interaction and the energy functions of corresponding 2D vertex models. On the other hand, a direct connection between the energy levels of 1D quantum integrable spin chains with long range interaction and energy functions of 1D classical vertex models can be inferred from some works in the literature [20, 22, 30]. In this context it should be noted that, being motivated by the Yangian quantum group symmetry of HS like spin chains with long range interactions, a large class of 1D inhomogeneous vertex models have been proposed recently [31]. Moreover, as an application of a rather general type of equivalence relation, it has been shown that the energy levels of HS, PF and FI spin chains associated with the A_{N-1} root system exactly coincide with those of three vertex models belonging to the above mentioned class, whose energy functions depend on the vertex index through specific polynomials of first or second degree. Such a relation between spin chains with long range interactions and 1D vertex models might be applied to explore various physical properties of these spin chains. For example, by using the transfer matrix approach, it has been proved analytically that the level densities of the three vertex models related to HS, PF and FI spin chains asymptotically tend to the Gaussian distribution as the number of vertices become very large [30]. Hence, it automatically follows that the level densities of the HS, PF and FI spin chains also tend to the Gaussian distribution with the increase in number of lattice sites. However, it is worth noting that, the energy functions for the class of 1D inhomogeneous vertex models proposed in Ref. [31] could be taken as arbitrary functions of the vertex index. Therefore, it is interesting to ask whether the Gaussian nature of level density distribution is a generic feature for all of these vertex models. In a recent paper the present authors have considered a large subclass of these vertex models, where the energy functions are taken as arbitrary polynomials of any possible degree, and shown that the corresponding level densities always follow the Gaussian pattern [32]. The aim of this short review article is to recapitulate some main results of Refs. [30] and [31], and subsequently discuss how these results have been applied in Ref. [32] to find out the asymptotic form of level densities for 1D vertex models with arbitrary polynomial type energy functions.

At first let us discuss how, being motivated by the Yangian quantum group symmetry of HS like spin chains with long range interactions, a class of 1D inhomogeneous vertex models have been proposed in Ref. [31]. For the sake of simplicity, we shall restrict ourselves to the special case of nonsupersymmetric $su(m)$ ferromagnetic spin chains. A class of irreducible representations of the $Y(sl_m)$ Yangian algebra, known as 'border strips' or 'motifs', completely span the Hilbert space of HS like spin systems with long range interaction [12, 17–24, 33]. For the case of a spin system with N number of lattice sites, border strips are denoted by $\langle k_1, k_2, \ldots, k_r \rangle$, where each k_i is a positive integer satisfying the relation $\sum_{i=1}^{r} k_i = N$. Thus the Hilbert space of this spin chain may be decomposed as

$$\mathcal{V} = \sum_{k_1+k_2+\ldots+k_r=N} \oplus \; V_{\langle k_1,k_2,\ldots,k_r \rangle} \,, \tag{2}$$

where $V_{\langle k_1,k_2,\ldots,k_r \rangle}$ denotes the irreducible vector space associated with $\langle k_1, k_2, \ldots, k_r \rangle$. Let us assume that there exists a class of Yangian invariant spin systems, for which the complete set of energy levels can be written in the following way:

$$E_{\langle k_1,k_2,\ldots,k_r \rangle} = \sum_{l=1}^{r-1} \mathcal{E}_N(j)|_{j=K_l}, \tag{3}$$

where $K_l = \sum_{i=1}^{l} k_i$ and $\mathcal{E}_N(j)$ is an arbitrary function of the discrete variables j and N. Due to Yangian symmetry of this class spin systems, the multiplicity of the eigenvalue $E_{\langle k_1,k_2,\ldots,k_r \rangle}$ in Eq. (3) coincides with the dimensionality of the vector space $V_{\langle k_1,k_2,\ldots,k_r \rangle}$. It should be noted that, all energy levels in the spectra of $su(m)$ ferromagnetic HS, PF and FI spin chains can be generated through Eq. (3) as special cases, where $\mathcal{E}_N(j)$'s are given by 'dispersion relations' of the form [17–24, 30]

$$\mathcal{E}_N(j) = \begin{cases} j(N-j), & \text{for the HS chain} \\ j, & \text{for the PF chain} \\ j(c+j-1), \text{ with } c>0, & \text{for the FI chain.} \end{cases} \tag{4}$$

Next, we consider some 1D classical vertex models with $(N+1)$ number of vertices, which are connected through N number of intermediate bonds. The 'colour' of each of these bonds can be chosen in m number of different ways. Therefore, any state for such a vertex model can be represented by a path configuration of finite length:

$$\vec{s} \equiv \{s_1, s_2, \ldots, s_N\}, \tag{5}$$

where $s_i \in \{1, 2, \cdots, m\}$ denotes the colour of the i-th bond. We define an 'energy function' corresponding to the state $\vec{s}$ as

$$E(\vec{s}) = \sum_{j=1}^{N-1} \mathcal{E}_N(j)\delta(s_j, s_{j+1}), \tag{6}$$

where $\delta(x, y)$ is taken as

$$\delta(x,y) = \begin{cases} 0, & \text{if } x \geq y \\ 1, & \text{if } x < y\,. \end{cases} \tag{7}$$

A crucial point to note here is that, the same function $\mathcal{E}_N(j)$ has appeared in Eqs. (6) and (3). It can be shown that, for any possible functional form of $\mathcal{E}_N(j)$, the spectrum of the $su(m)$ ferromagnetic spin chain generated by eigenvalues (3) (along with the degeneracy factors of these energy levels) completely matches with that of the corresponding vertex model generated by energy functions (6) [31]. In particular, the full spectra of $su(m)$ ferromagnetic HS, PF and FI spin chains can be reproduced from the energy function (6), by substituting to it the corresponding forms of dispersion relations (4).

Applying the transfer matrix approach it can be shown analytically that the level densities of the three vertex models, associated with the dispersion relations (4) of HS, PF and FI spin chains, follow the Gaussian distribution for large number of lattice sites [30]. However, one can easily construct a much larger class of vertex models, by taking $\mathcal{E}_N(j)$ in Eq. (6) as an arbitrary polynomial function of the discrete variables j and N:

$$\mathcal{E}_N(j) = \sum_{\alpha=0}^{r}\sum_{\beta=0}^{s} f_{\alpha,\beta}\, N^{\alpha} j^{\beta}, \tag{8}$$

where r and s are some non-negative integers and $f_{\alpha,\beta}$'s are arbitrary real numbers [32]. For the sake of uniquely expressing a polynomial type dispersion relation in the form (8) with minimum possible values of r and s, we assume that (without any loss of generality) there exist at least one value of $\beta \in \{0, \ldots, s\}$ such that $f_{r,\beta} \neq 0$ and at least one value of $\alpha \in \{0, \ldots, r\}$ such that $f_{\alpha,s} \neq 0$. The degree of the two-variable polynomial $\mathcal{E}_N(j)$ in Eq. (8) may be denoted by χ, which is an integer taking value within the range $0 \leqslant \chi \leqslant r+s$. Evidently, $f_{\alpha,\beta}$'s appearing in Eq. (8) satisfy the conditions

$$\begin{aligned} f_{\alpha,\beta} &= 0, \text{ for } \alpha+\beta > \chi, \\ &\neq 0, \text{ for at least one pair of } \alpha,\, \beta \text{ with } \alpha+\beta=\chi\,. \end{aligned} \tag{9}$$

By using these conditions, it is easy to show that

$$\chi \geqslant max\{r, s\}. \tag{10}$$

It may be noted that the general form of dispersion relation (8) yields all dispersion relations in (4) as special cases. Therefore, it is interesting to ask whether the Gaussian nature of level density distribution is a generic feature for all 1D vertex models defined through the energy function (6) and the polynomial type dispersion relation (8).

For the purpose of finding out an answer to the above mentioned question, in the following we shall briefly discuss the transfer matrix approach used in Ref. [30] for calculating the level density distributions of the three vertex models related to ferromagnetic $su(m)$ HS, PF and FI spin chains. A key role in this approach is played by the normalised characteristic function, which may be defined as

$$\hat{\phi}_N(t) = \langle e^{\frac{it(E(\vec{s})-\mu)}{\sigma}} \rangle, \tag{11}$$

where $E(\vec{s})$ is the energy function (6) of the vertex model, μ and σ denote the mean and the variance of this energy function, respectively, and the notation $\langle \mathcal{O} \rangle$ represents the average of $\mathcal{O}$ over all possible states of the vertex model. If this characteristic function satisfies the asymptotic relation

$$\lim_{N\to\infty} \hat{\phi}_N(t) = e^{-\frac{t^2}{2}}, \tag{12}$$

then it can be shown that [34] at $N \to \infty$ limit the normalised level density approaches Gaussian distribution of the form

$$G(E) = \frac{1}{\sqrt{2\pi\sigma^2}} e^{-\frac{(E-\mu)^2}{2\sigma^2}}. \tag{13}$$

Therefore, instead of directly working with the level density of a vertex model, one can study the $N \to \infty$ limit of the related characteristic function. Inserting the form of $E(\vec{s})$ given in Eq. (6) to Eq. (11), it is easy to express $\hat{\phi}_N(t)$ as

$$\hat{\phi}_N(t) = m^{-N} e^{-\frac{i\mu t}{\sigma}} Z_N(e^{\frac{it}{\sigma}}), \tag{14}$$

where the 'partition function' $Z_N(e^{\frac{it}{\sigma}})$ is defined through a product of transfer matrices as

$$Z_N(e^{\frac{it}{\sigma}}) = m^{N-1} \sum_{n,n'=1}^{m} \left[T(\omega_1)T(\omega_2)\ldots T(\omega_{N-1})\right]_{nn'}. \tag{15}$$

Here the transfer matrix $T(\omega_j)$ is given by $T(\omega_j) = T(\omega)|_{\omega=\omega_j}$, with

$$T(\omega) = \frac{1}{m} \begin{pmatrix} 1 & \omega^m & \cdots & \omega^m \\ \vdots & \ddots & \ddots & \vdots \\ \vdots & & \ddots & \omega^m \\ 1 & \cdots & \cdots & 1 \end{pmatrix}, \tag{16}$$

$\omega_j = e^{i\gamma_j t}$ and

$$\gamma_j = \frac{\mathcal{E}_N(j)}{m\sigma}. \tag{17}$$

From the above discussions it is clear that, the $N \to \infty$ limit of γ_j would play a significant role in determining the asymptotic form of $\hat{\phi}_N(t)$. In this context it should be noted that, the asymptotic form of a function can be estimated in several possible ways. For example, one may use the 'big O notation' like $f(N) = O(N^a)$, which implies that

$$\lim_{N\to\infty} |N^{-a} f(N)| = \tau\,, \tag{18}$$

where τ is a non-negative real number. More generally, the notation $f(N) = O(N^a)$ is used when $f(N)$ is asymptotically bounded above by N^a (up to a constant factor). On the other hand, in 'big Θ notation', $f(N) = \Theta(N^a)$ implies that

$$\lim_{N\to\infty} |N^{-a} f(N)| = \tau\,, \tag{19}$$

where τ is a positive real number. More generally, the notation $f(N) = \Theta(N^a)$ is used when $f(N)$ is asymptotically bounded above and below by N^a (up to constant factors) [35–37]. Hence, $f(N) = \Theta(N^a)$ automatically implies $f(N) = O(N^a)$, but the reverse is not true. A key result of Ref. [30] is to show that, if the asymptotic forms of γ_j and $\gamma_j - \gamma_{j+1}$ are given by

$$\gamma_j = O(N^{-1/2}) \tag{20}$$

and

$$\gamma_j - \gamma_{j+1} = O(N^{-3/2}) \tag{21}$$

for $j \in \{1, 2, \ldots, N-1\}$, then the characteristic function $\hat{\phi}_N(t)$ in Eq. (14) satisfies the relation (12) and, consequently, the normalized level density approaches the Gaussian distribution (13) at $N \to \infty$ limit. The mean and the variance of the energy function $E(\vec{s})$ in Eq. (6) can be expressed as

$$\mu \equiv \langle E(\vec{s})\rangle = \frac{1}{2}\left(1 - \frac{1}{m}\right) \sum_{j=1}^{N-1} \mathcal{E}_N(j), \tag{22}$$

and

$$\sigma^2 \equiv \langle (E(\vec{s}) - \mu)^2 \rangle = \left(1 - \frac{1}{m^2}\right)\left[\frac{1}{4}\sum_{j=1}^{N-1} \mathcal{E}_N(j)^2 - \frac{1}{6}\sum_{j=1}^{N-2} \mathcal{E}_N(j)\mathcal{E}_N(j+1)\right]. \tag{23}$$

Computing σ exactly by using (23) for the three vertex models related to dispersion relations (4), and substituting those expressions of σ in Eq. (17), it is found that the asymptotic forms (20) and (21) are satisfied for all of these cases. Thus it is proved that, the level densities of the three vertex models related to ferromagnetic $su(m)$ HS, PF and FI spin chains asymptotically follow the Gaussian distribution.

Next, our goal is to show that the asymptotic forms of γ_j and $\gamma_j - \gamma_{j+1}$ are in fact universal, i.e., Eqs. (20) and (21) are satisfied even if $\mathcal{E}_N(j)$ is chosen as an arbitrary polynomial function of the form (8). Instead of finding out the full expression of σ^2, at present our strategy is to directly compute the order of σ^2 by using Eq. (23). To this end, let us examine the order of $\mathcal{E}_N(j)$ in two possible cases: i) $j = \Theta(1)$, where $N \to \infty$ limit is taken by keeping j fixed, and ii) $j = \Theta(N)$, where $N \to \infty$ limit is taken in such a way that j/N tends to a fixed value. Using Eq. (8), we easily obtain the order of $\mathcal{E}_N(j)$ for case i) as

$$\mathcal{E}_N(j) = \Theta(N^r). \tag{24}$$

Next we consider the case ii), which is defined through the limiting procedure

$$\lim_{N\to\infty} \frac{j}{N} = \rho\,, \tag{25}$$

where ρ is a real parameter taking value within the range: $0 < \rho \leqslant 1$. Due to Eq. (25), one may write $\mathcal{E}_N(j) \approx \mathcal{E}_N(\rho N)$ for large values of N. Substituting ρN in the place of j in Eq. (8), we obtain

$$\mathcal{E}_N(\rho N) = \sum_{\alpha=0}^{r}\sum_{\beta=0}^{s} f_{\alpha,\beta}\, \rho^{\beta} N^{\alpha+\beta}. \tag{26}$$

Let us define a new variable as $\xi \equiv \alpha + \beta$. Making a change of the summation variables in Eq. (26), it can be expressed as

$$\mathcal{E}_N(\rho N) = \sum_{\xi=0}^{r+s} g_\xi(\rho) N^{\xi}, \tag{27}$$

where $g_\xi(\rho)$ is a polynomial of ρ given by

$$g_\xi(\rho) = \sum_{\beta=max\{0,\xi-r\}}^{min\{\xi,s\}} f_{\xi-\beta,\beta}\, \rho^{\beta}. \tag{28}$$

Since from Eq. (9) it follows that $g_\xi(\rho) = 0$ for $\xi > \chi$, one can rewrite Eq. (27) as

$$\mathcal{E}_N(\rho N) = \sum_{\xi=0}^{\chi} g_\xi(\rho) N^{\xi}. \tag{29}$$

With the help of Eqs. (25) and (29), we obtain

$$\lim_{N\to\infty} |N^{-\chi}\mathcal{E}_N(j)| = \lim_{N\to\infty} |N^{-\chi}\mathcal{E}_N(\rho N)| = |g_\chi(\rho)|\,. \tag{30}$$

Since $g_\chi(\rho)$ is a polynomial of ρ, it is bounded within the range $0 < \rho \leqslant 1$. If $g_\chi(\rho) \neq 0$ for some value of ρ, then by using Eqs. (19) and (30) we obtain the order of $\mathcal{E}_N(j)$ as

$$\mathcal{E}_N(j) = \Theta(N^{\chi}). \tag{31}$$

On the other hand, if $g_\chi(\rho) = 0$ for some value of ρ, then Eq. (29) reduces to $\mathcal{E}_N(\rho N) = \sum_{\xi=0}^{\tilde{\chi}} g_\xi(\rho) N^{\xi}$, where $\tilde{\chi} < \chi$ and $g_{\tilde{\chi}}(\rho) \neq 0$. Hence, for the case $g_\chi(\rho) = 0$, we obtain the order of $\mathcal{E}_N(j)$ as

$$\mathcal{E}_N(j) = \Theta(N^{\tilde{\chi}})\,, \tag{32}$$

where $\tilde{\chi} < \chi$. Replacing ξ by χ in Eq. (28) and using Eq. (10), one can get an explicit expression for $g_\chi(\rho)$ as

$$g_\chi(\rho) = \sum_{\beta=\chi-r}^{s} f_{\chi-\beta,\beta}\, \rho^{\beta}. \tag{33}$$

Since $g_\chi(\rho)$ is a polynomial in ρ of degree less than or equal to s, it can have at most s number of distinct zeros within the interval $0 < \rho \leqslant 1$. For such zero points of $g_\chi(\rho)$ the order of $\mathcal{E}_N(j)$ is evidently given by Eq. (32). However, the zero points of $g_\chi(\rho)$ clearly form a set of zero measure within the range $0 < \rho \leqslant 1$, and, therefore, the order of $\mathcal{E}_N(j)$ is given by Eq. (31) for generic values of ρ. Let us now compare the orders of $\mathcal{E}_N(j)$ given in Eqs. (24), (31) and (32) for various possible cases. Since from Eq. (10) it follows that $\chi \geq r$, the order of $\mathcal{E}_N(j)$ given by (31) is clearly the highest one among all these cases. Furthermore, it can be shown that, there exists at least one value of ρ within the range $0 < \rho \leqslant 1$ for which Eq. (31) is satisfied and this highest order of $\mathcal{E}_N(j)$ is preserved even if the index j is varied within a range of order $\Theta(N)$ [32].

By following the above mentioned procedure, one can also find out the asymptotic form for the difference of energy functions like $\mathcal{E}'_N(j) \equiv \mathcal{E}_N(j+1) - \mathcal{E}_N(j)$. With the help of Eq. (8), we express $\mathcal{E}'_N(j)$ as

$$\mathcal{E}'_N(j) = \sum_{\alpha=0}^{r'} \sum_{\beta=0}^{s'} f'_{\alpha,\beta}\, N^\alpha j^\beta, \tag{34}$$

where $r' \leq r$, $s' = s - 1$, and $f'_{\alpha,\beta}$ is given by

$$f'_{\alpha,\beta} = \sum_{\nu=\beta+1}^{s} {}^{\nu}C_\beta\, f_{\alpha,\nu}\,, \tag{35}$$

with ${}^{\nu}C_\beta \equiv \frac{\nu!}{\beta!(\nu-\beta)!}$. Since $\mathcal{E}'_N(j)$ in Eq. (34) is expressed in the same form as $\mathcal{E}_N(j)$ in Eq. (8), our analysis for finding out the order of $\mathcal{E}_N(j)$ would also be applicable to the case of $\mathcal{E}'_N(j)$. Let us denote the degree of $\mathcal{E}'_N(j)$ in Eq. (34) by χ'. With the help of Eq. (35) and the analog of Eq. (9) for the present case, it is easy to find that

$$\chi' \leqslant \chi - 1. \tag{36}$$

In analogy with the polynomial $g_\chi(\rho)$ in Eq. (33), one can define the polynomial $\tilde{g}_{\chi'}(\rho)$ for the present case as

$$\tilde{g}_{\chi'}(\rho) = \sum_{\beta=\chi'-r'}^{s'} f'_{\chi'-\beta,\beta}\, \rho^\beta. \tag{37}$$

Proceeding in the same way as we have done to find out the maximum possible order of $\mathcal{E}_N(j)$ in Eq. (31), it is easy to show that the order of $\mathcal{E}'_N(j)$ becomes maximum when j satisfies the relations $j = \Theta(N)$ and $\tilde{g}_{\chi'}(\rho) \neq 0$, and this order of $\mathcal{E}'_N(j)$ is given by

$$\mathcal{E}'_N(j) = \Theta(N^{\chi'})\,. \tag{38}$$

Finally, for the purpose of computing the orders of σ^2, γ_j and $\gamma_j - \gamma_{j+1}$ for any polynomial type dispersion relation of the form (8), we rewrite Eq. (23) as

$$\sigma^2 = \frac{1}{12}\left(1 - \frac{1}{m^2}\right)\left[\sum_{j=1}^{N-1} \mathcal{E}_N^2(j) - 2\sum_{j=1}^{N-2} \mathcal{E}_N(j)\, \mathcal{E}'_N(j) + 2\,\mathcal{E}_N(N-1)^2\right]. \tag{39}$$

Applying Eqs. (31), (38) and (36), and also noting that Eq. (31) continues to be satisfied if the index j is varied in some range of order $\Theta(N)$ [32], it is easy to see that the dominant contribution in the r.h.s. of Eq. (39) comes from the positive definite terms within the first summation. Thus, we obtain the order of σ^2 as

$$\sigma^2 = \Theta(N^{2\chi+1}). \tag{40}$$

Since the parameter τ in Eq. (19) takes strictly nonzero value, it is possible to 'invert' the relation (40) and calculate the order of $1/\sigma$ as

$$\frac{1}{\sigma} = \Theta(N^{-\chi-\frac{1}{2}}). \tag{41}$$

Next, for the case $j = \Theta(N)$, $g_\chi(\rho) \neq 0$, we use Eqs. (17), (31) and (41) to find out the order of γ_j as

$$\gamma_j = \Theta(N^{-\frac{1}{2}}) . \tag{42}$$

Since the order of $\mathcal{E}_N(j)$ becomes maximum for the above mentioned case, it is evident that the order of γ_j can not exceed the value given in Eq. (42) for any possible choice of j. Translating this result in terms of the 'big O' notation defined in (18), we find that the asymptotic form (20) of γ_j is satisfied for all values of j. Subsequently, for calculating the order of $\gamma_j - \gamma_{j+1}$, we proceed in exactly the same way and consider the case $j = \Theta(N)$, $\tilde{g}_{\chi'}(\rho) \neq 0$. Using Eqs. (17), (38) and (41), we obtain the order of $\gamma_j - \gamma_{j+1}$ for this case as

$$\gamma_j - \gamma_{j+1} = \Theta(N^{\chi'-\chi-\frac{1}{2}}) . \tag{43}$$

As the order of $\gamma_j - \gamma_{j+1}$ can not exceed the value given in the above equation for any possible choice of j, and from Eq. (36) it follows that $\chi' - \chi - \frac{1}{2} \leq -\frac{3}{2}$, we find that the asymptotic form (21) of $\gamma_j - \gamma_{j+1}$ is satisfied for all values of j. Since the asymptotic forms (20) and (21) do not depend at all on the parameters or coefficients present in the dispersion relation (8), they are indeed universal in nature. Consequently, we are able to establish that the level density for all 1D vertex models associated with the polynomial type dispersion relation (8) follow the Gaussian distribution (13) at $N \to \infty$ limit.

Next, we wish to compare the above mentioned result with numerical calculations, by choosing some specific form of the dispersion relation (8). The numerical data for the level density corresponding to any given dispersion relation can be obtained by starting from the energy function $E(\vec{s})$ defined in Eq. (6). In general, one can choose the state $\vec{s}$ of a vertex model in many different ways leading to the same value of the energy function, say, E_i. Let us denote the number of such different states for a vertex model (where the colour of each bond can be chosen in m possible ways) by $D^{(m)}(E_i)$. Choosing values of N and m within some range, and using a symbolic software package like Mathematica, it is possible to obtain values of all E_i and $D^{(m)}(E_i)$ for a given dispersion relation. Consequently one can draw a histogram for the corresponding level density and compare it with the Gaussian distribution. Furthermore, for the purpose of eliminating the effect of local fluctuations, one may also use these data to calculate the cumulative level density given by

$$F(E) = \frac{1}{m^N} \sum_{E_i \leq E} \mathcal{D}^{(m)}(E_i) , \tag{44}$$

and check whether this $F(E)$ agrees with the error function:

$$C(E) = \frac{1}{\sqrt{2\pi\sigma^2}} \int_{-\infty}^{E} e^{-\frac{(x-\mu)^2}{2\sigma^2}} dx = \frac{1}{2}\left[1 + \mathrm{erf}\left(\frac{\mathrm{E}-\mu}{\sqrt{2}\sigma}\right)\right] . \tag{45}$$

To get some specific form of dispersion relation, let us choose $r = 1$ and $s = 2$ in Eq. (8), and set the values of three coefficients viz. $f_{0,0}$, $f_{1,0}$ and $f_{1,2}$ equal to zero. Thus, we obtain

$$\mathcal{E}_N(j) = f_{0,1}j + f_{0,2}j^2 + f_{1,1}Nj , \tag{46}$$

where it is assumed that $f_{0,1}$, $f_{0,2}$ and $f_{1,1}$ are nonzero real parameters. Furthermore, we take a particular case of the dispersion relation (46) by setting $f_{0,1} = 1$, $f_{0,2} = -1$, $f_{1,1} = 1$, and also choose the values of m and N for the corresponding vertex model as $m = 2$, $N = 40$. Then we plot a histogram for the level density of the above mentioned vertex model in Fig. 1. It is evident that this histogram matches very well with the Gaussian distribution $G(E)$ in Eq. (13),

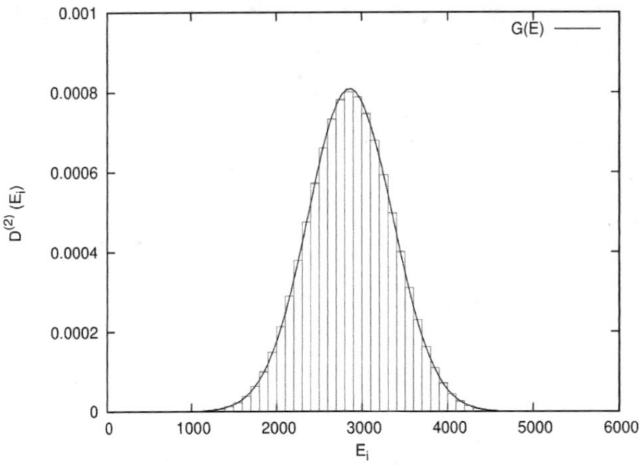

Figure 1. Continuous curve represents the Gaussian distribution $G(E)$ and the histogram represents the normalized level density for the dispersion relation (46) with coefficients $f_{0,1} = 1$, $f_{0,2} = -1$, $f_{1,1} = 1$, drawn for the case $N = 40$ and $m = 2$.

where the values of μ and σ^2 are obtained by using Eqs. (22) and (23) respectively. Inserting the numerical values of $D^{(m)}(E_i)$ in Eq. (44), subsequently we calculate the cumulative level density $F(E)$ for the above mentioned case. Again, it turns out that this $F(E)$ is in excellent agreement with the error function $C(E)$ given in Eq. (45). We also compute the mean square error (MSE) for the deviation of $F(E)$ with respect to $C(E)$ by choosing some equidistant points of E, and find this MSE to be as low as 4.10×10^{-7}. In Table 1 we present the values of such MSE, which are calculated by taking different values of N (while keeping m unchanged) and also by taking different sets of values of the three parameters in the dispersion relation (46). From this table

Table 1. MSE ($\times 10^{-7}$) for cumulative level density for the dispersion relation (46) with $m = 2$

Sets of Parameters $f_{0,1}$	$f_{0,2}$	$f_{1,1}$	$N = 20$	$N = 40$	$N = 50$	$N = 60$	$N = 70$
1	1	1	81.76	11.82	6.56	4.08	2.74
1	-1	1	37.04	4.10	2.29	1.44	0.98
1	2	1	101.52	14.38	7.95	4.93	4.93
1	-1	2	24.63	4.09	2.33	1.48	1.00
1	-1	3	30.75	5.10	2.90	1.83	1.24

we find that the MSE decreases steadily with the increase of N, for any particular choice of these three parameters. This numerical result clearly indicates that the level densities of the corresponding vertex models asymptotically follow the Gaussian distribution.

Let us now summarize the results discussed in this article and also mention about some future directions of investigation. We have analytically shown that the level densities for a large class of 1D vertex models, whose energy functions depend on the vertices through arbitrary polynomials of the form (8), asymptotically follow the Gaussian distribution. We have also obtained a general expression (40) for the order of variance of the energy function associated with this class of vertex models. These results would be very useful if some new $Y(sl_m)$ Yangian

Journal of Physics: Conference Series **411** (2013) 012005 doi:10.1088/1742-6596/411/1/012005

symmetric quantum spin chain is discovered in future, whose spectrum can be written in the form (3) with $\mathcal{E}_N(j)$ being a polynomial function of j and N. In future, we also hope to explore the density of spacings between consecutive energy levels of unfolded spectra for the above mentioned class of vertex models. It should be noted that, for calculating the density of spacings, the energy levels are usually taken from the unfolded spectrum which is generated from the raw spectrum through a transformation involving its level density distribution [38]. Since the level densities of all 1D vertex models associated with polynomial type dispersion relations satisfy the Gaussian distribution, one can easily construct the corresponding unfolded spectra and analyze their density of spacings. Finally, it might be interesting to study various thermodynamical properties for this type of vertex models by using the transfer matrix approach.

References

[1] Sutherland B 2004 *Beautiful Models: 70 Years of Exactly Solved Quantum Many-Body Problems* (Singapore: World Scientific)
[2] Korepin V, Bogoliubov N and Izergin A 1993 *Quantum Inverse Scattering Method and Correlation Functions, Cambridge Monographs on Mathematical Physics* (Cambridge: Cambridge University Press)
[3] Essler F H L, Frahm H, Göhmann F, Klümper A and Korepin V E 2005 *The One-Dimensional Hubbard Model* (Cambridge: Cambridge University Press)
[4] Haldane F 1988 *Phys. Rev. Lett.* **60** 635
[5] Shastry B 1988 *Phys. Rev. Lett.* **60** 639
[6] Polychronakos A 1993 *Phys. Rev. Lett.* **70** 2329
[7] Frahm H 1993 *J. Phys. A* **26**
[8] Polychronakos A 1994 *Nucl. Phys. B* **419** 553
[9] Frahm H and Inozemstsev V 1994 *J. Phys. A.* **27**
[10] Barba J, Finkel F, González-López A and Rodríguez M 2010 *Nucl. Phys. B* **839** 499
[11] Haldane F 1991 *Phys. Rev. Lett.* **67** 937
[12] Haldane F 1994 *Proc. 16th Taniguchi Symp., Kashikojima, Japan, 1993* (Springer)
[13] Bernevig B A and Haldane F 2009 *Phys. Rev. Lett.* **102** 066802
[14] Beisert N and Staudacher M 2005 *Nucl. Phys. B* **727** 1
[15] Bargheer T, Beisert N and Loebbert F 2009 *J. Phys. A* **42** 285205
[16] Serban D 2011 *J. Phys. A* **44** 124001
[17] Haldane F D M, Ha Z, Talstra J, Benard D and Pasquier V 1992 *Phys. Rev. Lett.* **69** 2021
[18] Benard D, Gaudin M, Haldane F D M and Pasquier V 1993 *J. Phys. A* **26** 5219
[19] Hikami K 1995 *Nucl. Phys. B* **441** 530
[20] Kirilov A N, Kuniba A and Nakanishi T 1997 *Commun. Math. Phys.* **185** 441
[21] Hikami K and Basu-Mallick B 2000 *Nucl. Phys. B* **566** 511
[22] Hikami K 2001 *Physics and Combinatorics 2000* (Singapore: World Scientific) pp 22–48
[23] Basu-Mallick B, Bondyopadhaya N, Hikami K and Sen D 2007 *Nucl. Phys. B* **782** 276
[24] Basu-Mallick B, Bondyopadhaya N and Sen D 2008 *Nucl. Phys. B* **795** 596
[25] Basu-Mallick B 1999 *Nucl. Phys. B* **540** 679
[26] Basu-Mallick B, Ujino H and Wadati M 1999 *Jour. Phys. Soc. Jpn.* **68** 3219
[27] Basu-Mallick B and Bondyopadhaya N 2006 *Nucl. Phys. B* **757** 280
[28] Sutherland B 1970 *J. Math. Phys.* **11** 3183
[29] Baxter R J 1971 *Phys. Rev. Lett.* **26** 834
[30] Enciso A, Finkel F and González-López A 2010 *Phys. Rev. E* **82** 051117
[31] Basu-Mallick B, Bondyopadhaya N and Hikami K 2010 *SIGMA* **6** 091
[32] Banerjee P and Basu-Mallick B 2012 *J. Math. Phys* **53** 083301
[33] Ha Z 1996 *Quantum Many-body Systems in one Dimension* (*Advances in Statistical Mechanics* vol 12) (Singapore: World Scientific)
[34] Feller W 1971 *An Introduction to Probability Theory and its Applications* 3rd ed vol 2 (New York: John Wiley and Sons)
[35] de Bruijn N G 1958 *Asymptotic Methods in Analysis* (Amsterdam: North-Holland)
[36] Jeffreys H and Jeffreys B 2000 *Methods of Mathematical Physics* 3rd ed (England: Cambridge University Press)
[37] Knuth D 2005 *ACM SIGACT News* **8**(2) 18
[38] Haake F 2001 *Quantum Signatures of Chaos* 2nd ed (Berlin: Springer-Verlag)

Journal of Physics: Conference Series **411** (2013) 012006 doi:10.1088/1742-6596/411/1/012006

Semiclassical energy of the $AdS_4 \times \mathbb{CP}^3$ folded string

Matteo Beccaria[1,a], **Guido Macorini**[2,b], **CarloAlberto Ratti**[1,c] **and Saulius Valatka**[3,d]

[1] Dipartimento di Matematica e Fisica "Ennio De Giorgi", Universita' del Salento & INFN, Via Arnesano, 73100 Lecce, Italy

[2] Niels Bohr International Academy and Discovery Center, Niels Bohr institute, Blegdamsvej 17 DK-2100 Copenhagen, Denmark

[3] Mathematics Department, Kings College London, The Strand, London WC2R 2LS, UK

E-mail: [a] `matteo.beccaria@le.infn.it`, [b] `macorini@nbi.ku.dk`, [c] `carloalberto.ratti@le.infn.it`, [d] `saulius.valatka@kcl.ac.uk`

Abstract. We consider the classical solution describing a folded type IIA string in the background $AdS_4 \times \mathbb{CP}^3$. The string is spinning in AdS and has angular momentum in $\mathbb{CP}^3$. In the 't Hooft limit, this is the gravity dual of twist operators in the ABJM superconformal theory. We quantize the classical solution by algebraic curve methods and determine the first semiclassical correction to the energy. An integral representation is given, valid for all values of the charges. We analyze its properties in the special regimes associated with a short or long string. Finally, we investigate various properties of the leading term of the energy for short strings (the so-called slope).

1. Introduction

The AdS/CFT correspondence is an extremely fruitful theoretical idea that played a central role in the past. In its simplest instance, it links the integrability of the world-sheet σ-model for type IIB superstring to the strong coupling behaviour of four dimensional gauge theories. In the specific case of planar $\mathcal{N} = 4$ Super Yang-Mills theory, integrability methods determine the spectrum of anomalous dimensions as a function of the 't Hooft coupling λ, from weak to strong coupling. More precisely, the spectrum was shown to be captured by suitable Y-system equations [1, 2, 3, 4, 5]. These equations are expected to be general and valid for any operator, including short ones. The first attempt of explorations of the flow of the anomalous dimension concerned the simplest $\mathfrak{sl}(2)$ sector of the theory and the simplest operator present in this sector, the Konishi operator. An important step has been the determination of the analytical strong coupling expansion of the Konishi anomalous dimension, *i.e.* the energy of its gravity dual state. The sub-leading coefficient in the string theory expansion was found in [6, 7, 8]. A highly nontrivial observation which allows one to reproduce the one-loop result almost without efforts was made recently in [9]. The second nontrivial strong coupling expansion coefficient was derived analytically in [10] by using the one-loop expression for a general (S, J) folded string [11].

The folded string is indeed a very useful theoretical laboratory. At the classical level, it can be described in terms of the scaled charges $\mathcal{S} = S/\sqrt{\lambda}$, $\mathcal{J} = J/\sqrt{\lambda}$. The spin $\mathcal{S}$ is associated with rotation around the center of AdS_5 while $\mathcal{J}$ is an angular momentum on the sphere. At small $\mathcal{S}$, the string is short and admits a near-flat space description. At large $\mathcal{S}$, the string stretches and reaches the boundary of AdS with the characteristic scaling $E \sim \log \mathcal{S}$ of its energy. Quantum corrections in the semiclassical approximation are fully under control and can be studied by algebraic curve tools.

Remarkably, very similar integrability structures as well as folded string classical solutions are known to be available in the ABJM theory [12]. A comprehensive review of integrability methods in ABJM can be found in [13]. For our purposes it is important to remind that also ABJM presents a $\mathfrak{sl}(2)$-like sector. At strong coupling and large spin these operators behave quite similarly to the corresponding ones in $AdS_5 \times S^5$. In particular their dual string state is a folded string rotating in AdS_3 and with angular momentum J in $\mathbb{CP}^3$ [14, 15, 16]. At weak coupling, they are composite operators in totally different theories. The quantized folded string on $AdS_4 \times \mathbb{CP}^3$ have been studied in various limits. The large spin limit has been studied at leading order in two regimes: At J/S fixed [14] and at $J \sim \log S$ [17]. The short string limit has never been addressed. On the numerical side, the analysis of [18] explored the Thermodynamic Bethe Ansatz (TBA) equations of $AdS_4 \times \mathbb{CP}^3$, although in a small window of the coupling reaching at most $\lambda \sim \mathcal{O}(1)$. In paper [29], a complete analysis of the one-loop semiclassical energy of folded string in $AdS_4 \times \mathbb{CP}^3$ has been presented. Here we review the main results found there.

The plan of the paper is the following: in Section 2 we summarize the general features of the quantization through algebraic curve in $AdS_4 \times \mathbb{CP}^3$. In Section 3 we describe the specific features of the algebraic curve associated with the folded string solution. In Section 4 we explain how to compute the semi-classical energy of such solutions. The short string limits and the slope functions are analyzed respectively in Sections 5 and 6 while the long string limit is computed in Section 7. Our final prediction for short states in ABJM can be found in Section 8.

2. Algebraic curve quantization in $AdS_4 \times \mathbb{CP}^3$

In this section we give a compact self-contained summary of the results of [19] using the language of off-shell fluctuation energies [20]. We shall work in the algebraic curve regularization [17] and write all equations in terms of the σ-model coupling g. For large g, it is related to the 't Hooft coupling by $\lambda = N/k = 8\,g^2$, but, contrary to the $AdS_5 \times S^5$ case, this relation will get corrections at finite g. The classical algebraic curve for $AdS_4 \times \mathbb{CP}^3$ is a 10-sheeted Riemann surface. The spectral parameter moves on it and we shall consider 10 symmetric quasi momenta $q_i(x)$

$$(q_1, q_2, q_3, q_4, q_5) = (-q_{10}, -q_9, -q_8, -q_7, -q_6). \tag{1}$$

They can have branch cuts connecting the sheets with $q_i^+ - q_j^- = 2\,\pi\,n_{ij}$. In the terminology of [19], the physical polarizations (ij) can be split into *heavy* and *light* ones and are summarized in the following table:

	AdS_4	Fermions	$\mathbb{CP}^3$
heavy	$(1,10)(2,9)(1,9)$	$(1,7)(1,8)(2,7)(2,8)$	$(3,7)$
light		$(1,5)(1,6)(2,5)(2,6)$	$(3,5)(3,6)(4,5)(4,6)$

Inversion symmetry imposes these other constraints on the quasi momenta

$$q_1(x) = -q_2(1/x), \qquad q_3(x) = 2\,\pi\,m - q_4(1/x), \qquad q_5(x) = q_5(1/x), \tag{2}$$

where $m \in \mathbb{Z}$ is a winding number. Thus, only 3 out of 10 quasi momenta are independent.

2.1. Semiclassical quantization

Semiclassical quantization is achieved by perturbing quasi-momenta introducing extra poles that shift the quasi-momenta $q_i \to q_i + \delta q_i$. The asymptotic expression of δq_i are related to the number N_{ij} of extra fluctuations and to the energy correction δE. The off-shell frequencies $\Omega^{ij}(x)$ are defined in order to have

$$\delta E = \sum_{n,ij} N_n^{ij}\, \Omega^{ij}(x_n^{ij}), \tag{3}$$

where the sum is over all pairs $(ij) \equiv (ji)$ of physical polarizations and integer values of n with $q_i(x_n^{ij}) - q_j(x_n^{ij}) = 2\,\pi\, n$. By linear combination of frequencies and inversion (as in the AdS_5/CFT_4 case), we can derive all the off-shell frequencies in terms of two fundamental ones $\Omega_A(x) = \Omega^{15}(x)$, $\Omega_B(x) = \Omega^{45}(x)$. Their explicit expressions turn out to be

$$\begin{aligned}
&\Omega^{29} = 2\left[-\Omega_A(1/x) + \Omega_A(0)\right], && \Omega^{1,10} = 2\,\Omega_A(x),\\
&\Omega^{19} = \Omega_A(x) - \Omega_A(1/x) + \Omega_A(0), && \Omega^{37} = \Omega_B(x) - \Omega_B(1/x) + \Omega_B(0),\\
&\Omega^{35} = \Omega^{36} = -\Omega_B(1/x) + \Omega_B(0), && \Omega^{45} = \Omega^{46} = \Omega_B(x),\\
&\Omega^{17} = \Omega_A(x) + \Omega_B(x), && \Omega^{18} = \Omega_A(x) - \Omega_B(1/x) + \Omega_B(0),\\
&\Omega^{27} = \Omega_B(x) - \Omega_A(1/x) + \Omega_A(0), && \Omega^{28} = -\Omega_A(1/x) + \Omega_A(0) - \Omega_B(1/x) + \Omega_B(0),\\
&\Omega^{15} = \Omega^{16} = \Omega_A(x), && \Omega^{25} = \Omega^{26} = -\Omega_A(1/x) + \Omega_A(0).
\end{aligned}$$

3. The folded string in $AdS_4 \times \mathbb{CP}^3$

We present the algebraic curve for the folded string in $AdS_4 \times \mathbb{CP}^3$ closely following the notation of [17]. In terms of the semiclassical variables $\mathcal{S} = \frac{S}{4\pi g}$ and $\mathcal{J} = \frac{J}{4\pi g}$, the energy of the folded string can be expanded according to

$$E = 4\,\pi\, g\, \mathcal{E}_0(\mathcal{J},\mathcal{S}) + E_1(\mathcal{J},\mathcal{S}) + \mathcal{O}\left(\frac{1}{g}\right), \tag{4}$$

where the small $\mathcal{S}$ expansion of the classical contribution $\mathcal{E}_0$ reads

$$\mathcal{E}_0 = \mathcal{J} + \frac{\sqrt{\mathcal{J}^2+1}}{\mathcal{J}}\,\mathcal{S} - \frac{\mathcal{J}^2+2}{4\,\mathcal{J}^3(\mathcal{J}^2+1)}\,\mathcal{S}^2 + \frac{3\,\mathcal{J}^6 + 13\,\mathcal{J}^4 + 20\,\mathcal{J}^2 + 8}{16\,\mathcal{J}^5\,(\mathcal{J}^2+1)^{5/2}}\,\mathcal{S}^3 + \cdots. \tag{5}$$

The quasi momenta are closely related to those of the $AdS_5 \times S^5$ folded string since motion is still in $AdS_3 \times S^1$ and the $\mathbb{CP}^3$ part of the background plays almost no role. The only non trivial case is

$$\begin{aligned}
q_1(x) \;=\; \pi\, f(x) \Bigg\{ &-\frac{J}{4\,\pi\, g}\left(\frac{1}{f(1)\,(1-x)} - \frac{1}{f(-1)(1+x)}\right) - \frac{4}{\pi\,(a+b)(a-x)(a+x)} \times \\
&\times \left[(x-a)\,\mathbb{K}\left(\frac{(b-a)^2}{(b+a)^2}\right) + 2\,a\,\Pi\left(\frac{(b-a)(a+x)}{(a+b)(x-a)}\middle|\frac{(b-a)^2}{(b+a)^2}\right)\right]\Bigg\} - \pi,
\end{aligned} \tag{6}$$

where $1 < a < b$, $f(x) = \sqrt{x-a}\,\sqrt{x+a}\,\sqrt{x-b}\,\sqrt{x+b}$ and

$$S = 2\,g\,\frac{ab+1}{ab}\left[b\,\mathbb{E}\left(1-\frac{a^2}{b^2}\right) - a\,\mathbb{K}\left(1-\frac{a^2}{b^2}\right)\right], \quad J = \frac{4\,g}{b}\sqrt{(a^2-1)(b^2-1)}\,\mathbb{K}\left(1-\frac{a^2}{b^2}\right),$$

$$E = 2\,g\,\frac{ab-1}{ab}\left[b\,\mathbb{E}\left(1-\frac{a^2}{b^2}\right) + a\,\mathbb{K}\left(1-\frac{a^2}{b^2}\right)\right]. \tag{7}$$

The other independent quasi-momenta are $q_3(x) = \frac{J}{2\,g}\,\frac{x}{x^2-1}$, $q_5(x) = 0$. The above expressions are valid for a folded string with minimal winding. Adding winding is trivial at the classical level, but requires non trivial changes at the one-loop level [10].

The independent off-shell frequencies can be determined by the methods of [20] and read

$$\Omega_A(x) = \frac{1}{ab-1}\left(1 - \frac{f(x)}{x^2-1}\right), \qquad \Omega_B(x) = \frac{\sqrt{a^2-1}\,\sqrt{b^2-1}}{ab-1}\,\frac{1}{x^2-1}. \tag{8}$$

4. Integral representation for the one-loop correction to the energy

The one-loop shift of the energy is given in full generality by the following sum of zero point energies

$$E_1 = \frac{1}{2}\sum_{n=-\infty}^{\infty}\sum_{ij}(-1)^{F_{ij}}\,\omega_n^{ij}, \qquad \omega_n^{ij} = \Omega^{ij}(x_n^{ij}), \tag{9}$$

where the sum over ij is over the $8_B + 8_F$ physical polarizations and x_n^{ij} is the unique solution to the equation $q_i(x_n^{ij}) - q_j(x_n^{ij}) = 2\,\pi\,n$ under the condition $|x_n^{ij}| > 1$ [1]. The explicit sum over the infinite number of on-shell frequencies requires some care and a definite prescription since the sums are not absolutely convergent due to cancellations between bosonic and fermionic contributions (see [29] for an exhaustive discussion about this point).

In an alternative approach [6, 10], the infinite sum (9) can be evaluated by contour integration in the complex plane. The result is quite similar to the $AdS_5 \times S^5$ one and reads

$$E_1 = E_1^{\text{anomaly},1} + 2E_1^{\text{anomaly},2} + E_1^{\text{dressing}} + E_1^{\text{wrapping}}, \tag{10}$$

where, by defining $x(z) = z + \sqrt{z^2-1}$, we have

$$E_1^{\text{anomaly},1} = 2\int_a^b \frac{dx}{2\,\pi\,i}\left[\Omega^{1,10}(x) - \Omega^{1,10}(a)\right]\partial_x \log\sin q_1(x), \tag{11}$$

$$E_1^{\text{anomaly},2} = 2\times 2\int_a^b \frac{dx}{2\,\pi\,i}\left[\Omega^{1,5}(x) - \Omega^{1,5}(a)\right]\partial_x \log\sin\frac{q_1(x)}{2}, \tag{12}$$

$$E_1^{\text{dressing}} = \sum_{ij}(-1)^{F_{ij}}\int_{-1}^{1}\frac{dz}{2\,\pi\,i}\,\Omega^{ij}(z)\,\partial_z\frac{i\,[q_i(z) - q_j(z)]}{2}, \tag{13}$$

$$E_1^{\text{wrapping}} = \sum_{ij}(-1)^{F_{ij}}\int_{-1}^{1}\frac{dz}{2\,\pi\,i}\,\Omega^{ij}(z)\,\partial_z\log(1 - e^{-i\,(q_i(z)-q_j(z))}), \tag{14}$$

As in $AdS_5 \times S^5$, the labeling of the various contributions reminds their physical origin. In particular, dressing and wrapping contributions are separated so that the asymptotic contribution is divided from finite size effects. As in $AdS_5 \times S^5$, the anomaly terms are special contributions arising from the deformation of contours and ultimately due to the presence of the algebraic curve cuts. The representation (9) is a compact formula for E_1 and can be evaluated numerically with minor effort. We shall now analyze the short and long string limit.

5. Short string limit

The short string limit is generically $\mathcal{S} \to 0$. Regarding $\mathcal{J}$, we shall consider two cases. The first amounts to keeping $\mathcal{J}$ fixed, expanding in the end each coefficient of powers of $\mathcal{S}$ at small $\mathcal{J}$. This is the procedure worked out in [10] in $AdS_5 \times S^5$. In the second case, we shall keep the ratio $\rho = \mathcal{J}/\sqrt{\mathcal{S}}$ fixed as in [26]. The two expansions are related, but not equivalent and provide useful different information.

Fixed $\mathcal{J}$ expansion

[1] If it happens that for some ij and n the above equation has no solution, then we shall say that the polarization (ij) has the missing mode n. Missing modes can be treated according to the procedure discussed in [20].

Journal of Physics: Conference Series **411** (2013) 012006 doi:10.1088/1742-6596/411/1/012006

After a straightforward computation, our main result is

$$\begin{aligned} E_1 &= \left(-\frac{1}{2\mathcal{J}^2}+\frac{\log(2)-\frac{1}{2}}{\mathcal{J}}+\frac{1}{4}+\mathcal{J}\left(-\frac{3\zeta(3)}{8}+\frac{1}{2}-\frac{\log(2)}{2}\right)-\frac{3\mathcal{J}^2}{16}+\right. \\ &\left.+\mathcal{J}^3\left(\frac{3\zeta(3)}{16}+\frac{45\zeta(5)}{128}-\frac{1}{2}+\frac{3\log(2)}{8}\right)+\cdots\right)\mathcal{S}+ \\ &+\left(\frac{3}{4\mathcal{J}^4}+\frac{\frac{1}{2}-\log(2)}{\mathcal{J}^3}-\frac{1}{8\mathcal{J}^2}+\frac{\frac{1}{16}-\frac{3\zeta(3)}{4}}{\mathcal{J}}-\frac{1}{8}+\mathcal{J}\left(\frac{69\zeta(3)}{64}+\frac{165\zeta(5)}{128}-\frac{27}{32}+\frac{\log(2)}{2}\right)+\right. \\ &\left.+\frac{3\mathcal{J}^2}{8}+\mathcal{J}^3\left(-\frac{163\zeta(3)}{128}-\frac{405\zeta(5)}{256}-\frac{875\zeta(7)}{512}+\frac{235}{128}-\log(2)\right)+\cdots\right)\mathcal{S}^2+ \\ &+\left(-\frac{5}{4\mathcal{J}^6}+\frac{\frac{3\log(2)}{2}-\frac{3}{4}}{\mathcal{J}^5}+\frac{\frac{9\zeta(3)}{16}+\frac{1}{16}}{\mathcal{J}^3}+\frac{1}{16\mathcal{J}^2}+\frac{\frac{45\zeta(3)}{64}+\frac{75\zeta(5)}{256}-\frac{7}{32}+\frac{\log(2)}{8}}{\mathcal{J}}+\frac{11}{64}+\right. \\ &+\mathcal{J}\left(-\frac{89\zeta(3)}{32}-\frac{745\zeta(5)}{256}-\frac{3815\zeta(7)}{2048}+2-\frac{33\log(2)}{32}\right)-\frac{465\mathcal{J}^2}{512}+ \\ &\left.+\mathcal{J}^3\left(\frac{5833\zeta(3)}{1024}+\frac{1585\zeta(5)}{256}+\frac{98035\zeta(7)}{16384}+\frac{259455\zeta(9)}{65536}-\frac{405}{64}+\frac{775\log(2)}{256}\right)+\cdots\right)\mathcal{S}^3+\cdots \end{aligned} \tag{15}$$

This expansion is rather similar to the one derived in [10] for $AdS_5 \times S^5$, but there are two remarkable differences: 1) The leading terms at small $\mathcal{J}$ are $\mathcal{O}(\mathcal{S}^n/\mathcal{J}^{2n})$. Instead, they were $\mathcal{O}(\mathcal{S}^n/\mathcal{J}^{2n-1})$ in $AdS_5\times S^5$. Also, there are terms with all parities in $\mathcal{J}$ while in $AdS_5\times S^5$, there appear only terms odd under $\mathcal{J}\to-\mathcal{J}$. The additional terms imply that if one scales $\mathcal{J}\sim\sqrt{\mathcal{S}}$, there is a constant contribution in the short string limit. This is different from $AdS_5 \times S^5$ where the energy correction vanishes like $\sqrt{\mathcal{S}}$ in this regime; 2) There are terms proportional to $\log(2)$. These terms can be removed by redefining $g \longrightarrow g-\frac{\log 2}{4\pi}$. This redefinition connects the coupling in the algebraic curve regularization with the coupling in the world-sheet regularization and agrees with the calculation in [21] for the circular string solution as well as in [22] for the giant magnon.

Fixed $\rho=\mathcal{J}/\sqrt{\mathcal{S}}$ expansion

The result in this limit is

$$E_1=-\frac{1}{2}\,\mathcal{C}(\rho,\mathcal{S})+a_{01}(\rho)\sqrt{\mathcal{S}}+a_{1,1}(\rho)\,\mathcal{S}^{3/2}+\mathcal{O}(\mathcal{S}^{5/2}), \tag{16}$$

where

$$a_{1,0}(\rho) = \frac{2\log(2)-1}{2\sqrt{\rho^2+2}}, \tag{17}$$

$$a_{1,1}(\rho) = -\frac{\log(2)\left(2\rho^4+6\rho^2+3\right)}{4\left(\rho^2+2\right)^{3/2}}+\frac{8\rho^4+25\rho^2+16}{16\left(\rho^2+2\right)^{3/2}}-\frac{3\left(\rho^2+3\right)\zeta(3)}{8\sqrt{\rho^2+2}}. \tag{18}$$

$\mathcal{C}$ is related to the branch cut endpoints by the formula

$$\mathcal{C}=\frac{\sqrt{(a^2-1)\,(b^2-1)}}{1-a\,b}+1. \tag{19}$$

Its expansion at small $\mathcal{S}$ with fixed $\rho=\mathcal{J}/\sqrt{\mathcal{S}}$ is

$$\mathcal{C}=1-\frac{\rho}{\sqrt{\rho^2+2}}-\frac{2\,\rho^3+5\,\rho}{4\,(\rho^2+2)^{3/2}}\,\mathcal{S}+\frac{\rho\,(12\,\rho^6+68\,\rho^4+126\,\rho^2+73)}{32\,(\rho^2+2)^{5/2}}\,\mathcal{S}^2+\cdots \tag{20}$$

Notice also that upon expanding at small $\mathcal{J}$ the expansion of $\mathcal{C}$ at fixed $\mathcal{J}$ we get precisely the terms with even/odd $\mathcal{J}$ exponents in the coefficients of the odd/even powers of $\mathcal{S}$ in (15).

Expanding E_1 at large ρ we partially resum the calculation at fixed $\mathcal{J}$. Just to give an example, from the expansion

$$-\frac{1}{2}\left(1-\frac{\rho}{\sqrt{\rho^2+2}}\right)=-\frac{1}{2\rho^2}+\frac{3}{4\rho^4}-\frac{5}{4\rho^6}+\frac{35}{16\rho^8}-\frac{63}{16\rho^{10}}+\cdots, \tag{21}$$

we read the coefficients of all terms $\sim \mathcal{S}^n/\mathcal{J}^{2n}$. The first ones are of course in agreement with (15). From the large ρ expansion of $a_{11}(\rho)$ we can read the coefficients of all terms $\sim \mathcal{S}^n/\mathcal{J}^{2n-3}$ [29].

6. The slope function

The one-loop correction E_1 tends to zero linearly with $\mathcal{S}$ when $\mathcal{S}\to 0$ at fixed $\mathcal{J}$. The slope ratio

$$\sigma(\mathcal{J})=\lim_{\mathcal{S}\to 0}\frac{E_1(\mathcal{S},\mathcal{J})}{\mathcal{S}}, \tag{22}$$

is an important quantity related to the conjectures in [9] [2] (see also [23, 24]). It is known that it does not receive dressing corrections both in $AdS_5\times S^5$ and in $AdS_4\times\mathbb{CP}^3$ since such contributions start at order $\mathcal{S}^2$ [9]. It also does not receive wrapping corrections in AdS_5. Instead, in the case of AdS_4 the slope has a non vanishing wrapping contribution. For instance, a rough evaluation at $\mathcal{J}=1$ gives a definitely non zero value around -0.042.

Indeed, an analytical calculation shows that the wrapping contribution to the slope in $AdS_4\times\mathbb{CP}^3$ is exactly

$$\sigma^{\rm wrap}(\mathcal{J})=\sum_{n=-\infty}^{\infty}\sigma_n=-\frac{1}{2\mathcal{J}}\sum_{n=-\infty}^{\infty}\frac{(-1)^n}{\sqrt{\mathcal{J}^4+(n^2+1)\mathcal{J}^2+n^2}}. \tag{23}$$

To analyze the small $\mathcal{J}$ limit it is convenient to split this contribution into the $n=0$ term plus the rest. The result is very intriguing. For the $n=0$ term, we find

$$\sigma^{\rm wrap}_{n=0}=-\frac{1}{2\mathcal{J}^2\sqrt{\mathcal{J}^2+1}}=-\frac{1}{2\mathcal{J}^2}+\frac{1}{4}-\frac{3\mathcal{J}^2}{16}+\frac{5\mathcal{J}^4}{32}-\frac{35\mathcal{J}^6}{256}+\frac{63\mathcal{J}^8}{512}+\cdots. \tag{24}$$

This is precisely the set of terms even under $\mathcal{J}\to-\mathcal{J}$ in the full slope which is the first term of (15). Similarly, we can consider the rest of $\sigma^{\rm wrap}$ and expand at small $\mathcal{J}$. We find

$$\begin{aligned}\sum_{n\neq 0}\sigma^{\rm wrap}_n &= \frac{\log(2)}{\mathcal{J}}+\mathcal{J}\left(-\frac{3\zeta(3)}{8}-\frac{\log(2)}{2}\right)+\mathcal{J}^3\left(\frac{3\zeta(3)}{16}+\frac{45\zeta(5)}{128}+\frac{3\log(2)}{8}\right)+\\ &+\mathcal{J}^5\left(-\frac{9\zeta(3)}{64}-\frac{45\zeta(5)}{256}-\frac{315\zeta(7)}{1024}-\frac{5\log(2)}{16}\right)+O\left(\mathcal{J}^6\right).\end{aligned} \tag{25}$$

Comparing again with (15), we see that we are reproducing all the irrational terms of the slope, involving zeta functions or log(2). The remaining terms are the same as in $AdS_5\times S^5$ [3],

$$\sigma(\mathcal{J})-\sigma^{\rm wrap}(\mathcal{J})=-\frac{1}{2\mathcal{J}}+\frac{\mathcal{J}}{2}-\frac{\mathcal{J}^3}{2}+\cdots. \tag{26}$$

[2] The exact slope mentioned in [9] is the coefficient of S in the expansion of E^2. This line of analysis is suggested by the simplicity of the marginality condition in $AdS_5\times S^5$(see [25] for a general discussion). Here, it is simpler to discuss the quantity $\sigma(\mathcal{J})$.

[3] This is due to the fact that the BAE are essentially the same as for $\mathfrak{sl}(2)$ sector in $AdS_5\times S^5$.

Thus, we are led to the following expression for the one-loop full slope

$$\sigma(\mathcal{J}) = -\frac{1}{2\,\mathcal{J}}\left[\frac{1}{\mathcal{J}^2+1} + \sum_{n=-\infty}^{\infty}\frac{(-1)^n}{\sqrt{\mathcal{J}^4+(n^2+1)\,\mathcal{J}^2+n^2}}\right]. \tag{27}$$

The above analysis of the slope is a confirmation that the various terms in (15) are organized in the expected way. The asymptotic contribution is precisely the same as in $AdS_5\times S^5$, while wrapping is different and is exponentially suppressed for large operators [29].

7. Long string limit

The large S behaviour of the one-loop energy E_1 can be computed starting from the integral representation (15). Let us first summarize the result valid for $AdS_5\times S^5$ from [6]. We scale $\mathcal{J}$ with $\mathcal{S}$ for $\mathcal{S}\gg 1$ according to

$$\mathcal{J} = \frac{\ell}{\pi}\,\log\left(\frac{8\,\pi\,\mathcal{S}}{\sqrt{\ell^2+1}}\right), \tag{28}$$

where we assume $\ell>0$. Then, the one-loop energy correction can be written

$$E_1^{AdS_5} = f_{10}^{AdS_5}(\ell)\,\log\left(\frac{8\,\pi\,\mathcal{S}}{\sqrt{\ell^2+1}}\right) + f_{11}^{AdS_5}(\ell) + \frac{c^{AdS_5}}{\log\left(\frac{8\,\pi\,\mathcal{S}}{\sqrt{\ell^2+1}}\right)} + \cdots, \tag{29}$$

with

$$\begin{aligned} f_{10}^{AdS_5}(\ell) &= \frac{\sqrt{\ell^2+1}+2\,(\ell^2+1)\,\log\left(\frac{1}{\ell^2}+1\right)-(\ell^2+2)\,\log\left(\frac{\sqrt{\ell^2+2}}{\sqrt{\ell^2+1}-1}\right)-1}{\pi\sqrt{\ell^2+1}}, \qquad (30)\\ f_{11}^{AdS_5}(\ell) &= \frac{2\left(\log\left(1-\frac{1}{(\ell^2+1)^2}\right)+2\sqrt{\ell^2+1}\cot^{-1}\left(\sqrt{\ell^2+1}\right)+2\coth^{-1}\left(\sqrt{\ell^2+1}\right)-2\ell\cot^{-1}(\ell)\right)}{\pi\sqrt{\ell^2+1}},\\ c^{AdS_5}(\ell) &= -\frac{\pi}{12\,(\ell^2+1)}. \end{aligned}$$

The expansion in $AdS_4\times\mathbb{CP}^3$ can be derived in the same way as in [6] and the result is

$$E_1^{AdS_4} = f_{10}^{AdS_4}(\ell)\,\log\left(\frac{8\,\pi\,\mathcal{S}}{\sqrt{\ell^2+1}}\right) + f_{11}^{AdS_4}(\ell) + \frac{c^{\mathrm{AdS_4}}}{\log\left(\frac{8\,\pi\,\mathcal{S}}{\sqrt{\ell^2+1}}\right)} + \cdots, \tag{31}$$

with

$$f_{10}^{AdS_4}(\ell) = \frac{1}{2}\,f_{10}^{AdS_5}(\ell),\quad f_{11}^{AdS_4}(\ell) = \frac{1}{2}\,f_{11}^{AdS_5}(\ell),\quad c^{AdS_4}(\ell) = 2\,c^{AdS_5}(\ell) = -\frac{\pi}{6\,(\ell^2+1)}. \tag{32}$$

Notice that the simple $\frac{1}{2}$ rule for the leading two terms is in agreement with the result of [27]. The correction $\sim 1/\log\mathcal{S}$ comes from the anomaly terms. It is twice bigger than in SYM. For a clear explanation of this fact see [29].

8. Relation with marginality condition

Let us define $\Lambda\equiv\lambda$ in $AdS_5\times S^5$, and $\Lambda = 16\,\pi^2\,g^2$ in $AdS_4\times\mathbb{CP}^3$. The role of Λ is to emphasize the close analogy between the expressions in the two cases. For the folded string in $AdS_5\times S^5$, the energy admits the following expansion

$$E^2 = J^2 + \left(A_1\sqrt{\Lambda}+A_2+\frac{A_3}{\sqrt{\Lambda}}+\cdots\right)S + \left(B_1+\frac{B_2}{\sqrt{\Lambda}}+\frac{B_3}{\Lambda}+\cdots\right)S^2+\cdots,$$

where the following exact formula for the constants A_i has been conjectured in [9]:

$$A_1\sqrt{\Lambda}+A_2+\frac{A_3}{\sqrt{\Lambda}}+\cdots=2\sqrt{\Lambda}\,Y_J(\sqrt{\Lambda}),\qquad Y_J(x)=\frac{d}{dx}\log I_J(x). \tag{33}$$

It is also know that $B_1=\frac{3}{2}$ and $B_2=\frac{3}{8}-3\,\zeta(3)$ [10] .

The expansion (33) is very convenient since all powers of S have a coefficient with an expansion at large Λ starting with a more and more suppressed term. The simplicity of (33) is a special feature of the folded string with two cusps. If winding is allowed, it is known that such structure is lost in [10] (see also [26, 28]).

For the folded string in $AdS_4\times\mathbb{CP}^3$, the expansion with fixed $\mathcal{J}$ has the general form [10] $E=\sqrt{\Lambda}\ \mathcal{E}_0+\sum_{\ell=0}^{\infty}\frac{1}{(\sqrt{\Lambda})^\ell}\sum_{p=1}^{\infty}\sum_{q=-2p}^{\infty}v_{pq}^{(\ell)}\,\mathcal{J}^q\,\mathcal{S}^p$, where the classical energy $\mathcal{E}_0$ is given in (5) and the semiclassical computation provides $v_{pq}^{(0)}$ according to the results in (15).

Expanding E^2, we find that (33) takes the following form

$$\begin{aligned}E^2-J^2=&\\&+\left[\left(2-\frac{1}{J}\right)\sqrt{\Lambda}+\left(\frac{2v_{1,-2}^{(1)}}{J}-1+2\log(2)\right)+\sqrt{\frac{1}{\Lambda}}\left(\frac{2v_{1,-2}^{(2)}}{J}+2v_{1,-1}^{(1)}+J^2+\frac{J}{2}\right)+\cdots\right]S+\\&+\left[\left(\frac{1}{4J^4}+\frac{1}{2J^3}\right)\Lambda+\sqrt{\Lambda}\left(-\frac{v_{1,-2}^{(1)}}{J^4}+\frac{2v_{1,-2}^{(1)}}{J^3}+\frac{2v_{2,-4}^{(1)}}{J^3}+\frac{1}{2J^3}-\frac{\log(2)}{J^3}\right)+\cdots\right]S^2+\cdots\end{aligned} \tag{34}$$

This structure is different from (33) since higher powers of S are not associated with terms that are more and more suppressed at large Λ. This is possible since the new terms not present in (33) are associated with suitable inverse powers of J. The same phenomenon happens for the winded folded string in $AdS_5\times S^5$ [10]. Here wrapping corrections are responsible for these terms.

8.1. Prediction for short states

We can provide a prediction for the strong coupling expansion of the energy of short states that in principle could be tested by TBA calculations. To this aim, we can start from our results at fixed $\rho=\mathcal{J}/\sqrt{\mathcal{S}}$, and re-expand at large Λ the sum of the (scaled) classical energy $\mathcal{E}_0$ and the one-loop contribution (16). The result is

$$E=(4\,\pi\,g)^{1/2}\sqrt{2\,S}-\frac{1}{2}+\frac{\sqrt{2\,S}}{(4\,\pi\,g)^{1/2}}\left(\frac{J\,(J+1)}{4\,S}+\frac{3\,S}{8}-\frac{1}{4}+\frac{1}{2}\log(2)\right)+\cdots. \tag{35}$$

The same expansion can be written in terms of the redefined coupling $g\longrightarrow g-\frac{\log 2}{4\pi}$ [21, 22] The result is again eq.(35) without the log(2) term.

References

[1] N. Gromov, V. Kazakov, and P. Vieira, *Phys.Rev.Lett.* **103** (2009) 131601, [arXiv:0901.3753].
[2] G. Arutyunov and S. Frolov, *JHEP* **0903** (2009) 152, [arXiv:0901.1417].
[3] D. Bombardelli, D. Fioravanti, and R. Tateo, *J.Phys.A* **A42** (2009) 375401, [arXiv:0902.3930].
[4] N. Gromov, V. Kazakov, A. Kozak, and P. Vieira, *Lett.Math.Phys.* **91** (2010) 265–287, [arXiv:0902.4458].
[5] G. Arutyunov and S. Frolov, *JHEP* **0905** (2009) 068, [arXiv:0903.0141].
[6] N. Gromov, D. Serban, I. Shenderovich, and D. Volin, *JHEP* **1108** (2011) 046, [arXiv:1102.1040].
[7] R. Roiban and A. Tseytlin, *Nucl.Phys.* **B848** (2011) 251–267, [arXiv:1102.1209].
[8] B. C. Vallilo and L. Mazzucato, *JHEP* **1112** (2011) 029, [arXiv:1102.1219].
[9] B. Basso, [arXiv:1109.3154].

[10] N. Gromov and S. Valatka, JHEP **1203**, 058 (2012) [arXiv:1109.6305].
[11] S. Frolov and A. A. Tseytlin, *JHEP* **0206** (2002) 007, [hep-th/0204226].
[12] O. Aharony, O. Bergman, D. L. Jafferis, and J. Maldacena, *JHEP* **0810** (2008) 091, [arXiv:0806.1218].
[13] T. Klose, Lett. Math. Phys. **99**, 401 (2012) [arXiv:1012.3999].
[14] T. McLoughlin and R. Roiban, *JHEP* **0812** (2008) 101, [arXiv:0807.3965].
[15] L. F. Alday, G. Arutyunov, and D. Bykov, *JHEP* **0811** (2008) 089, [arXiv:0807.4400].
[16] C. Krishnan, *JHEP* **0809** (2008) 092, [arXiv:0807.4561].
[17] N. Gromov and V. Mikhaylov, *JHEP* **0904** (2009) 083, [arXiv:0807.4897].
[18] F. Levkovich-Maslyuk, JHEP **1205**, 142 (2012) [arXiv:1110.5869].
[19] N. Gromov and P. Vieira, *JHEP* **0902** (2009) 040, [arXiv:0807.0437].
[20] N. Gromov, S. Schafer-Nameki, and P. Vieira, *JHEP* **0812** (2008) 013, [arXiv:0807.4752].
[21] T. McLoughlin, R. Roiban, and A. A. Tseytlin, *JHEP* **0811** (2008) 069, [arXiv:0809.4038].
[22] M. C. Abbott, I. Aniceto, and D. Bombardelli, *JHEP* **1012** (2010) 040, [arXiv:1006.2174].
[23] B. Basso, [arXiv:1205.0054].
[24] N. Gromov, [arXiv:1205.0018].
[25] A. A. Tseytlin, *Nucl.Phys.* **B664** (2003) 247–275, [hep-th/0304139].
[26] M. Beccaria and G. Macorini, JHEP **1202**, 092 (2012) [arXiv:1201.0608].
[27] M. Beccaria and G. Macorini, *JHEP* **0909** (2009) 017, [arXiv:0905.1030].
[28] M. Beccaria, C. A. Ratti and A. A. Tseytlin, J. Phys. A A **45**, 155401 (2012) [hep-th/1201.5033].
[29] M. Beccaria, G. Macorini, C. Ratti and S. Valatka, JHEP **1205**, 030 (2012) [Erratum-ibid. **1205**, 137 (2012)] [arXiv:1203.3852].

XXth International Conference on Integrable Systems and Quantum Symmetries (ISQS-20) IOP Publishing
Journal of Physics: Conference Series **411** (2013) 012007 doi:10.1088/1742-6596/411/1/012007

Two dimensional symmetric and antisymmetric exponential functions and orthogonal polynomials

Agata Bezubik[1], **Jiří Hrivnák**[2], **Severin Pošta**[3]

[1] Insitute of Mathematics, University of Białystok, Akademicka 2, 15-267 Białystok, Poland
[2] Department of physics, Faculty of nuclear sciences and physical engineering, Czech Technical University in Prague, Břehová 7, 115 19 Prague, Czech Republic
[3] Department of mathematics Faculty of nuclear sciences and physical engineering, Czech Technical University in Prague, Trojanova 13, 120 00 Prague, Czech Republic

E-mail: agatab@math.uwb.edu.pl, jiri.hrivnak@fjfi.cvut.cz, severin.posta@fjfi.cvut.cz

Abstract. The symmetric and antisymmetric exponential functions of two variables, based on the permutation group S_2, are considered. Explicit formulas for the corresponding families of the orthogonal polynomials are derived and the properties of these orthogonal polynomials, such as their generating functions, continuous and discrete orthogonality, are presented. A connection of these polynomials with characteristic polynomials of some particular matrices is shown.

1. Introduction

The complex valued functions on the Euclidean space $\mathbb{R}^n$, symmetric or antisymmetric with respect to the permutation group S_n, are dealt with in quantum theory or theory of integrable systems. The functions considered in this paper form a family of special functions called the symmetric/antisymmetric exponential functions. They are introduced in general for any integer n in [9]. By bringing forward the construction of the associated orthogonal polynomials, we extend the study of their two dimensional case from [5]. After recalling the symmetric and antisymmetric exponential functions of two variables, we focus on families of the associated symmetric and antisymmetric orthogonal polynomials. A detailed study of the properties of the three variable exponential functions and the corresponding orthogonal polynomials is contained in the forthcoming paper [2].

Note that families of orthogonal polynomials of a similar construction are also explored elsewhere — see for example [1, 3, 7, 8, 14, 15, 19, 20, 21]. The symmetric and antisymmetric exponential functions are closely related to the symmetric [11], antisymmetric [12] and $E-$orbit functions [13] and to the corresponding orthogonal polynomials [4, 12, 16, 17, 18]. Through the inclusion of the alternaning subgroup $Alt_n \subset S_n$, they are also related to alternating exponential functions [6, 10].

Journal of Physics: Conference Series **411** (2013) 012007 doi:10.1088/1742-6596/411/1/012007

2. Two dimensional symmetric exponential functions and corresponding orthogonal polynomials

The two dimensional symmetric exponential functions $E^+_{(k,l)}: \mathbb{R}^2 \to \mathbb{C}$, with $k,l \in \mathbb{Z}$ and $k \geq l$, are defined by the formula

$$E^+_{(k,l)}(x,y) = \begin{vmatrix} e^{2\pi ikx} & e^{2\pi iky} \\ e^{2\pi ilx} & e^{2\pi ily} \end{vmatrix}^+ = e^{2\pi i(kx+ly)} + e^{2\pi i(ky+lx)}. \tag{1}$$

Due to their periodicity and symmetry, the functions $E^+_{(k,l)}$ can be considered on the closure of the fundamental domain $F(S_2^{\text{aff}})$ only ([9]), where

$$F(S_2^{\text{aff}}) = \{(x,y) \in (0,1) \times (0,1) \mid x > y\}.$$

For detailed properties of these functions, we refer to the paper [5].

An explicit calculation of the corresponding orthogonal polynomials is based on the observation that multiplying two symmetric exponential functions gives a decomposition rule

$$E^+_{(k,l)}(x,y)E^+_{(m,n)}(x,y) = E^+_{(k+m,l+n)}(x,y) + E^+_{(k+n,l+m)}(x,y). \tag{2}$$

Introducing the following substitution for the two symmetric exponential functions,

$$X = E^+_{(1,0)}, \quad Y = E^+_{(1,1)}, \tag{3}$$

two–variable polynomials $P^+_{(k,l)}(X,Y)$ labeled by the pairs (k,l), with $k \geq l$ and $k,l \geq 0$, are determined by the equation

$$P^+_{(k,l)}(X(x,y),Y(x,y)) = E^+_{(k,l)}(x,y).$$

Using the decomposition rule (2), we obtain the following table for the several lowest polynomials $P^+_{(k,l)}(X,Y)$:

$$P^+_{(0,0)} = 2, \qquad P^+_{(3,2)} = \frac{1}{4}XY^2,$$

$$P^+_{(1,0)} = X, \qquad P^+_{(3,3)} = \frac{1}{4}Y^3,$$

$$P^+_{(1,1)} = Y, \qquad P^+_{(4,0)} = X^4 - 2X^2Y + \frac{1}{2}Y^2,$$

$$P^+_{(2,0)} = X^2 - Y, \qquad P^+_{(4,1)} = \frac{1}{4}XY(2X^2 - 3Y),$$

$$P^+_{(2,1)} = \frac{XY}{2}, \qquad P^+_{(4,2)} = \frac{1}{4}Y^2(X^2 - Y),$$

$$P^+_{(2,2)} = \frac{Y^2}{2}, \qquad P^+_{(4,3)} = \frac{1}{8}XY^3,$$

$$P^+_{(3,0)} = X^3 - \frac{3}{2}XY, \qquad P^+_{(4,4)} = \frac{1}{8}Y^4.$$

$$P^+_{(3,1)} = \frac{1}{2}Y(X^2 - Y),$$

Below we determine directly the explicit formula and the generating function for the polynomials $P^+_{(k,l)}$. Note that in view of (2) we have, for any admissible n and k,

$$P^+_{(n,k)}P^+_{(0,0)} = P^+_{(n,k)} + P^+_{(n,k)} = 2P^+_{(n,k)},$$
$$P^+_{(1,1)}P^+_{(n-1,k-1)} = P^+_{(n,k)} + P^+_{(n,k)} = 2P^+_{(n,k)},$$

therefore $P^+_{(n,k)} = \frac{Y}{2} P^+_{(n-1,k-1)}$. Iterating this relation, we see that

$$P^+_{(n,k)} = \left(\frac{Y}{2}\right)^k P^+_{(n-k,0)}. \tag{4}$$

The recurrence relation (4) gives us

$$P^+_{(n+1,0)} P^+_{(1,0)} = P^+_{(n+1,1)} + P^+_{(n+2,0)} = \frac{Y}{2} P^+_{(n,0)} + P^+_{(n+2,0)},$$

and thus the polynomials $P^+_{(k,l)}$ fulfill the following two-step recurrence relations

$$P^+_{(n+1,0)} X = \frac{Y}{2} P^+_{(n,0)} + P^+_{(n+2,0)}. \tag{5}$$

The equation (5) is a linear difference equation for $P^+_{(n,0)}$ with constant coefficients and satisfying initial conditions $P^+_{(0,0)} = 2$, $P^+_{(1,0)} = X$. The polynomial

$$P^+_{(n,0)}(X,Y) = \frac{1}{2^n}\left(\left(X - \sqrt{X^2 - 2Y}\right)^n + \left(X + \sqrt{X^2 - 2Y}\right)^n\right)$$

is an explicit solution of the difference equation (5). Now we can derive the following explicit formula for the polynomials $P^+_{(n,k)}$,

$$P^+_{(n,k)}(X,Y) = \frac{Y^k}{2^n}\left(\left(X - \sqrt{X^2 - 2Y}\right)^{n-k} + \left(X + \sqrt{X^2 - 2Y}\right)^{n-k}\right).$$

Applying the formulae for the symmetric polynomials given in (4) and (5), we obtain the generating function G_+ for $P^+_{(n,k)}$

$$G_+(X,Y,r,s) = \sum_{n=0}^{+\infty}\sum_{k=0}^{+\infty} P^+_{(n,k)}(X,Y) r^n s^k = \frac{4(Xr + Xs - rsY - 2)}{(r^2Y - 2rX + 2)(-s^2Y + 2sX - 2)}. \tag{6}$$

Note that the Jacobian J of the mapping (3) is equal to

$$J(x,y) = -8\pi^2 e^{2i\pi(x+y)}(e^{2i\pi x} - e^{2i\pi y}).$$

By changing the variables, the Jacobian J may be expressed as

$$J(X,Y) = -4\pi^2 Y\sqrt{X^2 - 2Y}.$$

Using the continuous orthogonality of the $E^+_{(k,l)}$ functions [5, 9], the continuous orthogonality of the polynomials $P^+_{(n,k)}$ follows,

$$\int_{F_2} P^+_{(n,k)}(X,Y)\overline{P^+_{(n',k')}(X,Y)} \frac{1}{-4\pi^2 Y\sqrt{X^2 - 2Y}} \mathrm{d}X\mathrm{d}Y = G^+_{nk}\delta_{nn'}\delta_{kk'}, \tag{7}$$

where the symbol G^+_{kl} is defined by

$$G^+_{nk} = \begin{cases} 2, & n = k \\ 1, & \text{otherwise,} \end{cases}$$

and F_2 is the transformed fundamental region $F(S_2^{\mathrm{aff}})$ via injective mapping (3):

$$F_2 = \{(X,Y) \in \mathbb{C}^2 \,||Y| = 2, 2X = Y\overline{X}, |X| < 2\}.$$

Similarly, the discrete orthogonality relation from [5] yields the following formula for the discrete orthogonality of the polynomials $P^+_{(n,k)}$, $P^+_{(n',k')}$ with $n,k,n',k' \in \{0,1,\dots,N-1\}$,

$$\sum_{\left\{ \substack{r,s=0 \\ r\geq s} \right\}}^{N-1} (G^+_{rs})^{-1} P^+_{(n,k)}\left(e^{2\pi i\frac{r}{N}} + e^{2\pi i\frac{s}{N}}, 2e^{2\pi i(\frac{r}{N}+\frac{s}{N})}\right) \overline{P^+_{(n',k')}\left(e^{2\pi i\frac{r}{N}} + e^{2\pi i\frac{s}{N}}, 2e^{2\pi i(\frac{r}{N}+\frac{s}{N})}\right)} =$$
$$= G^+_{nk} N^2 \delta_{nn'}\delta_{kk'}.$$

Note an interesting connection of the polynomials $P^+_{(n,k)}$ with the characteristic polynomials of the following matrix. Let us consider $n \times n$ matrix $A^{(n)}$ with entries

$$A^{(n)}_{ij} = \min\{2i-1, 2j-1\}, \text{ i.e. } A^{(n)} = \begin{pmatrix} 1 & 1 & 1 & 1 & \dots \\ 1 & 3 & 3 & 3 & \dots \\ 1 & 3 & 5 & 5 & \dots \\ 1 & 3 & 5 & 7 & \dots \\ \dots & \dots & \dots & \dots & \dots \end{pmatrix}.$$

By taking into account a suitable multiple of characteristic polynomials of $A^{(n)}$ multiplied by scalar X^2 in variable Y, a modified characteristic polynomial $P_n(X,Y)$ may be expressed as

$$P_n(X,Y) = \frac{1}{2^{n-1}} \det(X^2 A^{(n)} - YI),$$

where I is the $n \times n$ identity matrix. Computing several first P_n's, we obtain

$$\begin{aligned} P_1(X,Y) &= X^2 - Y = P^+_{(2,0)}(X,Y), \\ P_2(X,Y) &= \tfrac{1}{2}(2X^4 - 4X^2Y + Y^2) = P^+_{(4,0)}(X,Y), \\ P_3(X,Y) &= \tfrac{1}{4}(4X^6 - 12X^4Y + 9X^2Y^2 - Y^3) = P^+_{(6,0)}(X,Y), \quad \text{etc.} \end{aligned}$$

From (2), by induction, we can derive the relation

$$P_n(X,Y) = P^+_{(2n,0)}(X,Y), \quad n \geq 0.$$

3. Two dimensional antisymmetric exponential functions and the corresponding orthogonal polynomials

The two dimensional antisymmetric exponential functions $E^-_{(k,l)} : \mathbb{R}^2 \to \mathbb{C}$, with $k,l \in \mathbb{Z}$ and $k > l$, are defined similarly to the symmetric case by the formula

$$E^-_{(k,l)}(x,y) = \begin{vmatrix} e^{2\pi ikx} & e^{2\pi iky} \\ e^{2\pi ilx} & e^{2\pi ily} \end{vmatrix} = e^{2\pi i(kx+ly)} - e^{2\pi i(ky+lx)}.$$

Due to their periodicity and antisymmetry, the functions $E^-_{(k,l)}$ can be considered on the fundamental domain $F(S_2^{\mathrm{aff}})$ only. For detailed properties of these functions, we refer to the paper [5].

The corresponding orthogonal polynomials are constructed in a similar manner as in the symmetric case. Note that the antisymmetric exponential functions fulfill the similar decomposition rule as the symmetric functions

$$E^-_{(k,l)}(x,y)E^-_{(m,n)}(x,y) = E^+_{(k+m,l+n)}(x,y) - E^+_{(k+n,l+m)}(x,y). \tag{8}$$

Two–variable polynomials $P^-_{(k,l)}$ with the elementary variables (3) are labeled by the pairs (k,l), with $k>l$ and $k,l \geq 0$, and determined by the equation

$$P^-_{(k,l)}(X(x,y),Y(x,y)) = \frac{E^-_{(k,l)}(x,y)}{E^-_{(1,0)}(x,y)}.$$

Below we present a list of several lowest antisymmetric polynomials $P^-_{(k,l)}$:

$$P^-_{(1,0)} = 1,$$
$$P^-_{(2,0)} = X,$$
$$P^-_{(2,1)} = \frac{Y}{2},$$
$$P^-_{(3,0)} = X^2 - \frac{Y}{2},$$
$$P^-_{(3,1)} = \frac{XY}{2},$$
$$P^-_{(3,2)} = \frac{Y^2}{4},$$
$$P^-_{(4,0)} = X^3 - XY,$$
$$P^-_{(4,1)} = \frac{1}{4}Y\left(2X^2 - Y\right),$$
$$P^-_{(4,2)} = \frac{XY^2}{4},$$
$$P^-_{(4,3)} = \frac{Y^3}{8},$$
$$P^-_{(5,0)} = \frac{1}{4}\left(4X^4 - 6X^2Y + Y^2\right),$$
$$P^-_{(5,1)} = \frac{1}{2}XY\left(X^2 - Y\right),$$
$$P^-_{(5,2)} = \frac{1}{8}Y^2\left(2X^2 - Y\right),$$
$$P^-_{(5,3)} = \frac{XY^3}{8},$$
$$P^-_{(5,4)} = \frac{Y^4}{16}.$$

Applying similar arguments as in Section 2, one can derive the explicit formula for the polynomials $P^-_{(k,l)}$,

$$P^-_{(k,l)}(X,Y) = \frac{Y^l}{2^k\sqrt{X^2-2Y}}\left(\left(\sqrt{X^2-2Y}+X\right)^{k-l} - \left(X - \sqrt{X^2-2Y}\right)^{k-l}\right).$$

Their generating function, an analogy of (6), has the form

$$G_-(X,Y,r,s) = \sum_{k=0}^{+\infty}\sum_{l=0}^{+\infty} P^-_{(k,l)}(X,Y)r^k s^l = \frac{4(s-r)}{(r^2Y - 2rX + 2)\left(-s^2Y + 2sX - 2\right)}.$$

The antisymmetric polynomials $P^-_{(k,l)}$ are continuously orthogonal with respect to the scalar product

$$\int_{F_2} P^-_{(k,l)}(X,Y)\overline{P^-_{(k',l')}(X,Y)}\frac{\sqrt{X^2-2Y}}{2\pi^2Y^2}\mathrm{d}X\mathrm{d}Y = \delta_{kk'}\delta_{ll'}$$

and their discrete orthogonality relation, for $k,l,k',l' \in \{0,1,\ldots,N-1\}$, is expressed by the following formula

$$\sum_{\substack{r,s=0 \\ r>s}}^{N-1} P^-_{(k,l)}\left(e^{2\pi i\frac{r}{N}}+e^{2\pi i\frac{s}{N}},2e^{2\pi i(\frac{r}{N}+\frac{s}{N})}\right)\overline{P^-_{(k',l')}\left(e^{2\pi i\frac{r}{N}}+e^{2\pi i\frac{s}{N}},2e^{2\pi i(\frac{r}{N}+\frac{s}{N})}\right)}\times$$
$$\left(-e^{-2i\pi(\frac{r}{N}+\frac{s}{N})}\left(e^{\frac{2i\pi r}{N}}-e^{\frac{2i\pi s}{N}}\right)^2\right)=N^2\delta_{kk'}\delta_{ll'}.$$

Acknowledgments
AB thanks the organizers of the ISQS-20 in Prague for the opportunity to speak at the conference and for their exceptional hospitality. SP acknowledges support of P201/10/1509, project of the Grant Agency of the Czech Republic, and CTU Project SGS12/198/OHK4/3T/14. JH gratefully acknowledges support by the Ministry of Education of Czech Republic (project MSM6840770039).

References
[1] I. Area, E. Godoy, J. Rodal, *On a class of bivariate second-order linear partial difference equations and their monic orthogonal polynomial solutions,* J. Math. Anal. Appl. **389** (2012), 165–178
[2] A. Bezubik, J. Hrivnák, J. Patera, S. Pošta, *Three-variable symmetric and antisymmetric exponential functions and orthogonal polynomials* (in preparation)
[3] H. Berens, H. J. Schmid, Y. Xu, *Multivariate Gaussian cubature formulae,* Arch. Math. (Basel) **64** (1995), 26–32
[4] G. J. Heckman, E. M. Opdam, *Root systems and hypergeometric functions. I.* Compositio Math. **64** no. 3, (1987), 329–352
[5] J. Hrivnák, J. Patera, *Two-dimensional symmetric and antisymmetric generalizations of exponential and cosine functions,* J. Math. Phys **51** (2010) 023515
[6] J. Hrivnák, J. Patera, S. Pošta, *Three-variable exponential functions of the alternating group,* J. Phys. A: Math. Theor. **45** (2012) 045201–045210
[7] P. Iliev, Y. Xu, *Discrete orthogonal polynomials and difference equations of several variables,* Adv. Math. **212** (2007), 1–36
[8] S. Karlin, J. L. McGregor, *Determinants of orthogonal polynomials,* Bull. Amer. Math. Soc. **68** (1962) 204–209
[9] A. Klimyk, J. Patera: *(Anti)symmetric multivariate exponential functions and corresponding Fourier transforms,* J. Phys. A: Math. Theor. **40** (2007), 10473–10489
[10] A. Klimyk, J. Patera, *Alternating group and multivariate exponential functions,* in *Groups and Symmetries; From Neolithic Scots to John McKay,* AMS-CRM Proceedings and Lectures Notes Series, **47**, 233-246, eds. J. Harnad and P. Winternitz; arXiv:0907.0601
[11] A. Klimyk, J. Patera, *Orbit functions,* Symmetry Integrability Geom. Methods Appl. **2** (2006), Paper 006
[12] A. Klimyk, J. Patera, *Antisymmetric orbit functions,* Symmetry Integrability Geom. Methods Appl. **3** (2007), Paper 023
[13] A. Klimyk, J. Patera, *E-orbit functions,* Symmetry Integrability Geom. Methods Appl. **4** (2008), Paper 002
[14] T. H. Koornwinder, *Two-variable analogues of the classical orthogonal polynomials,* in: Theory and Application of Special Functions (Proc. Advanced Sem., Math. Res. Center, Univ. Wisconsin, Madison, WI, 1975), in: Math. Res. Center Univ. Wisconsin Publ., vol. 35, Academic Press, New York (1975) 435–495
[15] M. A. Kowalski, *The recursion formulas for orthogonal polynomials in n variables,* SIAM J. Math. Anal. **13** (1982) 309–315
[16] R. V. Moody, J. Patera, *Cubature formulae for orthogonal polynomials in terms of elements of finite order of compact simple Lie groups,* Adv. in Appl. Math. **47** (2011) 509-535
[17] M. Nesterenko, J. Patera, A. Tereszkiewicz, *Orthogonal polynomials of compact simple Lie groups,* Int. J. Math. Math. Sci. (2011), Art. ID 969424
[18] M. Nesterenko, J. Patera, M. Szajewska, A. Tereszkiewicz, *Orthogonal polynomials of compact simple Lie groups: branching rules for polynomials,* J. Phys. A **43** (2010) 495207–495233
[19] J. Rodal, I. Area, E. Godoy, *Linear partial difference equations of hypergeometric type: orthogonal polynomial solutions in two discrete variables,* J. Comput. Appl. Math. **200** (2007), 722–748

[20] Y. Xu, *Second-order difference equations and discrete orthogonal polynomials of two variables,* Int. Math. Res. Not. (2005), 449–475
[21] Y. Xu, *Recurrence formulas for multivariate orthogonal polynomials,* Math. Comp. **62** (1994), 687–702

Journal of Physics: Conference Series **411** (2013) 012008 doi:10.1088/1742-6596/411/1/012008

Semiclassical 3-point function in WZW AdS_3 model[1]

P. Bozhilov[a], **P. Furlan**[b,c], **V.B. Petkova**[a] **and M. Stanishkov**[a]

[a]) *Institute for Nuclear Research and Nuclear Energy, Bulgarian Academy of Sciences, Sofia, Bulgaria*

[b]) *Dipartimento di Fisica dell'Università di Trieste, Italy,*

[c]) *Istituto Nazionale di Fisica Nucleare (INFN), Sezione di Trieste, Italy*

1. Introduction

The AdS/CFT conjecture implies that the correlation functions in the boundary quantum field theory can be computed alternatively in string theory, i.e., by the methods of a two-dimensional theory. One may start with a semiclassical computation - for large 't Hooft coupling $\lambda >> 1$ (i.e., large worldsheet coupling $\sqrt{\lambda}$) the string path integral for the correlation functions is evaluated at the saddle-point of the action in the presence of sources, created by proper "heavy" vertex operators. In the dual conformal theory they correspond to operators with dimensions of the order of $\sqrt{\lambda}$. Some recent results for correlators of three heavy operators were obtained by Janik and Wereszczynski [1]. They consider strings in $AdS_2 \times S^5$ and compute the AdS_2 contribution to the 3-point correlation function, exploiting the integrability of the model; see also [2].

Classically the relevant 2d actions are conformally invariant and presumably this symmetry is preserved on quantum level. This motivates the elaboration of semiclassical techniques in conformal field models. Here we review the derivation in [3] of the semiclassical 3-point correlator in the (Euclidean) AdS_3 model described by a WZW action. The alternatively derived formula reproduces the "heavy charge" classical limit of the known quantum 3-point constant, exploited in the applications to the superstring on $AdS_3 \times S^3 \times M^4$ in the NS-NS background [4], [5].

2. Preliminaries on Euclidean AdS$_3$

The target space of the theory is the Euclidean AdS$_3$:

$$-X_{-1}^2 + \sum_{i=1}^{3} X_i^2 = -1 \,, \tag{1}$$

a coset $\simeq SL(2,\mathbb{C})/SU(2)$, parametrised in $SL(2,\mathbb{C})$ as

$$g(X) = X_{-1}\mathbf{1}_2 - X_i\sigma_i = \begin{pmatrix} 1 & \gamma \\ 0 & 1 \end{pmatrix} \begin{pmatrix} e^{-\phi} & 0 \\ 0 & e^{\phi} \end{pmatrix} \begin{pmatrix} 1 & 0 \\ \bar{\gamma} & 1 \end{pmatrix} . \tag{2}$$

The WZW classical action given in terms of the coordinates $\phi(z,\bar{z})\,,\gamma(z,\bar{z})\,,\bar{\gamma}(z,\bar{z})$ reads

$$S_{AdS} = \frac{k}{\pi} \int d^2z (\partial_z\phi\, \partial_{\bar{z}}\phi + \partial_z\bar{\gamma}\, \partial_{\bar{z}}\gamma e^{2\phi}) \,. \tag{3}$$

[1] Talk given by V.B. Petkova.

The equations of motion imply the conservation of the affine currents

$$\partial_{\bar z} J(z) = 0 = \partial_z \bar J(\bar z)\,,$$

equivalent to two chiral first order equations for $g(z, \bar z)$

$$k\,\partial_z g + J(z) g = 0\,, \quad k\,\partial_{\bar z} g + g \bar J^a(\bar z) t^a = 0\,. \tag{4}$$

The group $SL(2,\mathbb{C})$ acts as: $g \to \omega g \omega^+$; e.g. the translations shift $\gamma \to \gamma - x$. The fundamental vertex operator is defined as projection on a shifted group element

$$V = V_{j=\frac{1}{2}}(z,\bar z; x, \bar x) = e^{-\phi} + |\gamma - x|^2 e^{\phi} = (1, -x) g(z,\bar z) \begin{pmatrix} 1 \\ -\bar x \end{pmatrix} \tag{5}$$

and the generators t^a of the algebra $sl(2)$ are realised by differential operators with respect to the isospin variable x. The general vertex operators are given by

$$V_j(z,\bar z, x, \bar x) = (e^{-\phi(z,\bar z)} + |\gamma(z,\bar z) - x|^2 e^{\phi(z,\bar z)})^{2j}\,. \tag{6}$$

In the AdS/CFT correspondence the isospin variables $x, \bar x$ play the role of the space-time coordinates of the dual theory and the expression (6), looked as a function of $(x, \bar x)$ and of the AdS_3 variables $(\phi, \gamma, \bar\gamma)$, represents the kernel of the integral boundary-to-bulk intertwining operator with boundary conformal dimension $\triangle = -2j$

$$\Phi(\phi,\gamma,\bar\gamma) = \int d^2x\, (e^{-\phi} + |\gamma - x|^2 e^{\phi})^{-\triangle} \phi_0(x)\,. \tag{7}$$

In the quantum theory $k \to k-2 = 1/b^2$ and furthermore a curvature term is added creating a "charge at infinity". The world sheet (Sugawara) scaling dimension of the vertex operator is

$$\delta^{Su}(j) = -b^2\, j(j+1) \quad = -\tfrac{\alpha}{b} + \alpha(Q-\alpha)\,, \quad \alpha = -jb\,,\; Q = \tfrac{1}{b} + b\,. \tag{8}$$

Here $\triangle_L(\alpha) = \alpha(Q-\alpha)$ is the scaling dimension in the related, via Drinfeld- Sokolov reduction, $c > 25$ Virasoro (Liouville) theory. Accordingly the WZW 3-point constant is closely related to DOZZ formula in Liouville theory and is given up to an overall constant by [6]

$$C(j_1,j_2,j_3) \quad = \tfrac{\Upsilon_b(b)}{\Upsilon_b(\alpha_{123}-b)} \prod_{i=1}^3 \tfrac{\Upsilon_b(2\alpha_i)}{\Upsilon_b(\alpha_{123}-2\alpha_i)}\,, \quad \alpha_{123} = \alpha_1+\alpha_2+\alpha_3 \tag{9}$$

where $\Upsilon_b(x)$, expressed in terms of Barnes double Gamma function $\Gamma_b(x)$, admits the integral representation

$$\log \Upsilon_b(x) = \int_0^\infty \frac{dt}{t}\Big(\big(\frac{Q}{2} - x\big)^2 e^{-t} - \frac{\sinh^2(\frac{Q}{2} - x)\frac{t}{2}}{\sinh b\frac{t}{2}\,\sinh\frac{t}{2b}}\Big)\,. \tag{10}$$

• The **semiclassical** limit corresponds to $b^2 = 1/(k-2) \to 0$ and "heavy" charges

$$\triangle = -2j = \frac{2\eta}{b^2}\,, \quad \eta - \text{finite}\,, \tag{11}$$

so that $b^2 \delta^{Su}(j) \to -\eta^2$. The limit of the 3-point function is reduced to that computed for Liouville theory in [7], using that the function Υ_b goes to

$$b^2 \log \Upsilon_b(\frac{\eta}{b}) \to F(\eta): \quad = \int_{1/2}^{\eta} dx\, \log\gamma(x) \quad, \gamma(x) = \tfrac{\Gamma(x)}{\Gamma(1-x)}\,. \tag{12}$$

We shall reproduce this limit by an alternative derivation - it follows and generalises the approach of [7] in the Liouville case which we will first sketch.

3. Liouville case main steps

The solutions of the classical equations of motion for the Liouville field

$$\partial\,\bar\partial\varphi(z,\bar z) = \pi\mu b^2 e^{2\varphi(z,\bar z)} \tag{13}$$

can be recovered from the solutions of the second order chiral equations with the conserved energy - momentum tensor $\partial_{\bar z} T = 0 = \partial_z \bar T$,

$$\partial_z^2 e^{-\varphi(z,\bar z)} = ((\partial_z\varphi)^2 - \partial_z^2\varphi) e^{-\varphi(z,\bar z)} = -\hat T(z)\, e^{-\varphi(z,\bar z)}\,. \tag{14}$$

It is given by the $b\to 0$ limit of the quantum tensor

$$b^2 T(z) = b^2(-(\partial_z\phi)^2 + Q\partial_z^2\phi) \to \hat T(z)\,, \;\; (\varphi = b\phi\,, Q = b + \frac{1}{b})\,. \tag{15}$$

The equation (13) ensures the conservation of the classical energy-momentum tensor in (14) and vice versa. One looks for a solution of (13) in the presence of (three) sources created by vertex operators $V_{\alpha_i} = e^{2\alpha_i\phi} = e^{\frac{2\eta_i\varphi}{b^2}}$ with "heavy" charges $\alpha_i = \eta_i/b$, i.e., finite η_i , for $b\to 0$.

The semiclassical correlator is determined by the stationary point of the full action with the contribution of the sources accounted for

$$< V_{\alpha_1}(z_1,\bar z_1) V_{\alpha_2}(z_2,\bar z_2) V_{\alpha_3}(z_3,\bar z_3) > \sim e^{-\frac{1}{b^2} S^{\rm cl}[\varphi;\eta_i]}\,. \tag{16}$$

The solutions of (14) in the presence of sources are reviewed, e.g., in [8], in relation to the uniformization of the three punctured sphere problem. The classical tensor $\hat T_L$ is given through the limit $b^2\to 0$ of the normalised correlator of T with the three vertex operators,

$$\hat T_L(z;z_a) = \lim_{b\to 0} b^2 T(z;z_a) =$$

$$\sum_i \left(\frac{h_i}{(z-z_i)^2} + \frac{\partial_i}{z-z_i}\log\frac{1}{(z_{12})^{h_{12}^3}(z_{23})^{h_{23}^1}(z_{13})^{h_{13}^2}}\right) =: (\partial_z w)^2\, \hat T_L(w) \tag{17}$$

where

$$w = \frac{(z-z_1)z_{23}}{(z-z_2)z_{13}} \tag{18}$$

and $h_{ij}^k = h_i + h_j - h_k$, for the Liouville classical dimensions $h_i = \lim_{b\to 0} b^2\alpha_i(Q-\alpha_i) = \eta_i(1-\eta_i)$. The solution for $e^{-\varphi(z,\bar z)}$ - now as a function also of the coordinates $z_i, \bar z_i$ of the sources is given, up to a prefactor, by a monodromy invariant diagonal combination

$$e^{-\varphi(z,\bar z;z_i,\bar z_i)} = \frac{\sqrt{\pi\mu b^2}}{(1-2\eta_1)}\frac{|(z-z_2)^2||z_{13}|}{|z_{23}||z_{12}||N_L|}(|G_1(w)|^2 - N_L^2|G_2(w)|^2) = |y_1|^2 - |y_2|^2 \tag{19}$$

of holomorphic and anti-holomorphic solutions (hypergeometric functions). The permutation symmetry determines the relative constant N_L while the overall constant is fixed by the Liouville equation (13) for $\varphi(z,\bar z;z_i,\bar z_i)$; the check of (13) given (19) requires the knowledge of the Wronskian of the two solutions, $A = y_1\partial_z y_2 - y_2\partial_z y_1$, a constant due to equation (14).

The chiral equation solved by (19) is the classical version of the Belavin-Polyakov-Zamolodchikov (BPZ) equation, resulting from the decoupling of a level 2 singular vector in the Vir module of (fundamental) h.w. $\triangle_L(\alpha) = \triangle_L(-b/2)$, whence the solution (19) is identified with the semi-classical limit $b \to 0$ of the normalised 4-point function of the quantum vertex operator (of "light" charge) $V_{-\frac{b}{2}}$

$$e^{-\varphi(z,\bar{z};x_i;\eta_i)} \sim$$

$$\lim_{b\to 0} < V_{-\frac{b}{2}}(z,\bar{z})V_{\alpha_1}(z_1)V_{\alpha_2}(z_2)V_{\alpha_3}(z_3) > / < V_{\alpha_1}(z_1)V_{\alpha_2}(z_2)V_{\alpha_3}(z_3) >$$

• In the last step, performed in [7], the classical solution for $\varphi(z;z_i)$ is used to compute the saddle point action with added three sources; close to their positions the solution behaves as

$$\varphi \sim -2\eta_1 \log|z-z_i|^2 + X_i + \ldots.$$

where $X_i = X_i(|z_{ij}|;\eta_i)$ is the finite part and

$$X_i = -\partial_{\eta_i} S_{cl}\,, i = 1,2,3\,. \tag{20}$$

The integration of this system of equations recovers the semiclassical 3-point function in (16).

4. The WZW AdS_3 semiclassical case

In this case the BPZ equation is replaced by the (classical) Knizhnik-Zamolodchikov (KZ) equation. In the presence of three sources at (z_i, x_i) the chiral equation for the field $V_{j=1/2}(z,\bar{z};x,\bar{x})$ as a function of the coordinates $\{z_i,\bar{z}_i,x_i,\bar{x}_i\}$ of the three sources becomes

$$(\partial_z - \hat{J}^a(z;z_i;x_i)t^a)V(z,x;z_i,x_i) = 0\,, \tag{21}$$

where the current is defined through the classical limit of its 4-point function normalised with the 3-point function $< V_{j_1}V_{j_2}V_{j_3} >$

$$\hat{J}^a(z;z_i,x_i): \;= \lim_{b\to 0} \tfrac{b^2}{<V_{j_1}V_{j_2}V_{j_3}>} \textstyle\sum_i \tfrac{t^a_i}{z-z_i} < V_{j_1}V_{j_2}V_{j_3} > \,. \tag{22}$$

The $sl(2)$ generators t^a are represented by the standard differential operators in x_i. The projective invariance allows to write

$$V(z,x;z_i,x_i) = \frac{|(x-x_2)^2||x_{13}|}{|x_{12}||x_{23}|}\hat{V}(w,y)\,, \qquad y = \frac{(x-x_1)x_{23}}{(x-x_2)x_{13}} \tag{23}$$

and one obtains the equation (written for the chiral constituents of $\hat{V}$)

$$\begin{aligned}\partial_w G(w,y) \;&= -(\eta_{13}^2(\tfrac{(y-w)^2}{w(w-1)} - 1) + 2\eta_1(\tfrac{y-w}{w}+1) + 2\eta_3(\tfrac{y-w}{w-1}+1))\partial_y G(w,y) \\ &\quad + (\tfrac{\eta_{13}^2(y-w)}{w(w-1)} + \tfrac{\eta_1}{w} + \tfrac{\eta_3}{w-1})G(w,y)\,.\end{aligned} \tag{24}$$

The KZ equation (24) is equivalent to a pair of differential equations for the components G_0, G_2 of

$$G(w,y) = G_0(w) + (y-w)G_2(w)\,, \tag{25}$$

Journal of Physics: Conference Series **411** (2013) 012008 doi:10.1088/1742-6596/411/1/012008

or, in matrix form for the vector $G = (G_0, G_2)^t$

$$\frac{d}{dw}G(w) - \begin{pmatrix} \frac{\eta_1}{w} + \frac{\eta_3}{w-1} & 1-\eta_{123} \\ \frac{\eta_{13}^2}{w(w-1)} & -(\frac{\eta_1}{w} + \frac{\eta_3}{w-1}) \end{pmatrix} G(w) = 0\,. \tag{26}$$

The component $G_0(w) = G(w, y = w)$ satisfies the second order equation (14) with the Liouville energy-momentum tensor $\hat{T}_L(w)$ in (17) - a manifestation of the Drinfeld-Sokolov reduction of the WZW theory.

Combining the left and right solutions of (24) one obtains for V

$$V(z,x;z_i,x_i) = \frac{|x-x_2|^2|x_{13}|}{|x_{23}||x_{12}|N_{WZW}}(|G^{(+)}(w,y)|^2 + N^2_{WZW}|G^{(-)}(w,y)|^2)\,, \tag{27}$$

where the components of

$$G^{(\pm)}(w,y) = G_0^{(\pm)}(w) + (y-w)G_2^{(\pm)}(w) \tag{28}$$

are expressed by hypergeometric functions. The relative constant is related to that in the Liouville case

$$N^2_{\rm WZW} = \frac{\gamma(\eta^1_{23})\gamma(2\eta_1)^2}{\gamma(\eta_{123})\gamma(\eta^3_{12})\gamma(\eta^2_{13})} = -\frac{(1-2\eta_1)^2}{(\eta_{123}-1)^2}N^2_L \tag{29}$$

and is determined from the requirement of permutation invariance. The leading singularities of the solutions in (27) for $z \to z_1$ ($w \to 0$) are given by

$$G^{(+)}(w,y) \sim w^{\eta_1}(1 - \frac{\eta^2_{13}}{2\eta_1}y)\,, \quad G^{(-)}(w,y) \sim w^{-\eta_1}(y - \frac{\eta^1_{23}}{1-2\eta_1}w)\,. \tag{30}$$

• Given the solution for $V(z,x;z_i,x_i)$ depending on the isospin variables $x, \bar{x}$ one can write it as a power series in, say, $(x-x_1)$ and then extract the expressions for the matrix elements of $g(X)$ in (2). The result is

$$X_{-1} + X_3 = e^{\phi} = \textstyle\sum_\pm |u_2^{(\pm)}|^2\,,$$

$$X = -X_1 + iX_2 = \gamma e^{\phi} = -\textstyle\sum_\pm u_1^{(\pm)}\bar{u}_2^{(\pm)}\,, \quad \bar{X} = \bar{\gamma}e^{\phi} = -\textstyle\sum_\pm \bar{u}_1^{(\pm)}u_2^{(\pm)}$$

$$X_{-1} - X_3 = e^{-\phi} + |\gamma|^2 e^{\phi} = \textstyle\sum_\pm |u_1^{(\pm)}|^2 \tag{31}$$

where

$$u_1^{(\pm)} = (\tfrac{x_{13}x_{12}}{x_{23}})^{1/2}\frac{1}{(N_{WZW})^{\pm\frac{1}{2}}}(G_0^{(\pm)}(w) - wG_2^{(\pm)}(w))$$

$$u_2^{(\pm)} = (\tfrac{x_{13}x_{12}}{x_{23}})^{-1/2}\frac{1}{(N_{WZW})^{\pm\frac{1}{2}}}G_2^{(\pm)}(w) \tag{32}$$

subject of the relation

$$|u_1^{(+)}u_2^{(-)} - u_1^{(-)}u_2^{(+)}|^2 = |\det h|^2 = 1\,. \tag{33}$$

Rewritten in a matrix form (30) reads

$$g(X) = hh^+ , \quad h = \begin{pmatrix} u_1^{(+)} & u_1^{(-)} \\ u_2^{(+)} & u_2^{(-)} \end{pmatrix} . \tag{34}$$

The solution gives for $\eta_1 > 0$ the leading contribution to ϕ near the source at $z = z_1$

$$2\phi \sim -2\eta_1 \log|z - z_1|^2 + X_1 , \tag{35}$$

where X_1 is the finite part

$$X_1 = 2\eta_1 \log|\frac{z_{13}z_{12}}{z_{23}}|^2 - \log|\frac{x_{13}x_{12}}{x_{23}}|^2 - \log\frac{\gamma(\eta_{123})\gamma(\eta_{12}^3)\gamma(\eta_{13}^2)}{\gamma(\eta_{23}^1)\gamma(2\eta_1)^2} . \tag{36}$$

Due to the symmetry of the classical solution analogous formulae hold in the vicinity of all three singular points with finite parts X_i.

• Adapting the derivation in [7] in the Liouville case we add to the classical action $S^{(cl)} = b^2 S'_{AdS}$ terms which account for the three vertex insertions

$$\hat{S}^{(cl)} = S^{(cl)} + S^{(src)},$$

$$S^{(src)} = -\int d^2z \sum_i \delta^2(z - z_i, \bar{z} - \bar{z}_i)(2\eta_i\phi(z,\bar{z}) + 2\eta_i^2 \log|z - z_i|^2)$$

$$= -\sum_i \eta_i X_i . \tag{37}$$

Here it is assumed that the integration in the first term $S^{(cl)}$ is on the Riemann sphere with the points of insertion of the three sources excluded. The action is regularised, the logarithmic terms added compensate the singularities of the solution (35).

At the saddle point of $\hat{S}^{(cl)}[\phi, \gamma, \bar{\gamma}]$, i.e., on a solution of the classical equations, only the source term in the action contributes to the derivative of the full action with respect to any of the charges $\eta_i \Rightarrow$

$$\frac{\partial}{\partial \eta_i}\hat{S}^{(cl)} = -X_i \, , i = 1, 2, 3. \tag{38}$$

This set of equations for $\hat{S}^{(cl)}$ integrates and one reproduces the 3-point function

$$e^{-\frac{\hat{S}^{(cl)}}{b^2}} = C^{(cl)}(\eta_1, \eta_2, \eta_3)\frac{|x_{13}|^{2j_{13}^2}|x_{12}|^{2j_{12}^3}|x_{23}|^{2j_{23}^1}}{|z_{13}|^{2\delta_{13}^2}|z_{12}|^{2\delta_{12}^3}|z_{23}|^{2\delta_{23}^1}} , \tag{39}$$

$$\text{where } \delta_{ik}^l = \delta_i + \delta_k - \delta_l \, , \quad \delta_i = -\frac{\eta_i^2}{b^2} \, , \; j_{ik}^l = j_i + j_k - j_l \, , \quad j_i = -\frac{\eta_i}{b^2} . \tag{40}$$

The expression obtained for $C^{(cl)}(\eta_1, \eta_2, \eta_3)$ reproduces up to an overall constant the classical limit of the quantum 3-point function (9)

$$-b^2 \log C(-\tfrac{\eta_1}{b^2}, -\tfrac{\eta_2}{b^2}, -\tfrac{\eta_3}{b^2}) \to -b^2 \log C^{(cl)}(\eta_1, \eta_2, \eta_3) =$$

$$F(\eta_{123}) - F(0) + \sum_i (F(\eta_{123} - 2\eta_i) - F(2\eta_i)) \tag{41}$$

with the function $F(\eta)$ given in (12).

• The semiclassical formula in [1] has the same structure but in terms of a different function, instead of $F(x)$. We note that the monodromy coefficients computed in [1], taken at the particular point of trivial spectral parameter, give an expression, compatible with a particular, differently normalised, basis of conformal blocks. Although the model considered in [1] is different, this coincidence remains to be explained.

Summary and discussion

We have presented an alternative derivation of the semiclassical 3-point function in the AdS_3 WZW model. It recovers its coordinate and isospin dependence and reproduces the $b \to 0$ limit of the quantum 3-point constant when all three charges (i.e., the conformal dimensions of the dual theory) are large $\triangle_i \sim 2\eta_i/b^2$.

Problem : extend the semiclassical computation to higher rank theories as the WZW related AdS_5, respectively S^5 model. For the 3-point constant of three scalar operators the fundamental $j = 1/2$ field is replaced by the sl(4) fundamental weight $\lambda = (0, 1, 0)$.

Our consideration is not directly applicable to the string on $AdS_5 \times S^5$ since the bosonic part of the supersymmetric action [9] has no B field (WZW term), whence no conserved chiral currents; rather the bosonic part is given by the standard sigma model - the realm of recent investigations based on integrability. The conformal invariance is ensured by the WZ term involving fermions. To make 2d CFT methods effective in the string part of the AdS/CFT correspondence one needs to develop relevant supersymmetry technique. One may start with the simpler $AdS_3 \times S^3$ model with R-R background, see, e.g., [10] for a recent discussion.

Acknowledgments

This work was supported in part by the Bulgarian National Science Fund, grant DO 02-257.

References

[1] R. Janik and A. Wereszczynski, Correlation functions of three heavy operators - the AdS_5 contribution, *JHEP* **1112** (2011) 095, arXiv:1109.6262.

[2] Y. Kazama, S. Komatsu, On holographic three point functions for GKP strings from integrability, *JHEP* **1201** (2012), 110, arXiv:hep-th/1110.3949.

[3] P. Bozhilov, P. Furlan, V.B. Petkova and M. Stanishkov, On the semiclassical 3-point function in AdS_3, *Phys. Rev.* **D 86** (2012) 066005, arXiv:1204.1322.

[4] M.R. Gaberdiel and I. Kirsch, Worldsheet correlators in AdS(3)/CFT(2), *JHEP* **0704** (2007), 050, hep-th/0703001.

[5] A. Dabholkar and A. Pakman, Exact chiral ring of $\mathrm{AdS}_3/\mathrm{CFT}_2$, *Adv. Theor. Math. Phys.* **13** (2009) 409, arXiv:hep-th/0703022.

[6] J. Teschner, On structure constants and fusion rules in the $SL(2,\mathbb{C})/SU(2)$ WZNW model *Nucl. Phys.* **B 546** (1999) 390, hep-th/9712256.

[7] A. Zamolodchikov and Al. Zamolodchikov, Structure constants and conformal bootstrap in Liouville field theory, *Nucl. Phys.* **B 477** (1996) 577, hep-th/9506136.

[8] P. Ginsparg and G. Moore, Lectures on 2D gravity and 2D string theory, hep-th/9304011.

[9] R.R. Metsaev and A.A. Tseytlin, Type IIB superstring action in $AdS_5 \times S^5$ background, *Nucl. Phys.* **B 533** (1998) 109, hep-th/9805028.

[10] A. Cagnazzo, K. Zarembo, B-field in AdS_3/CFT_2 correspondence and integrability, arXiv:1209.4049.

XXth International Conference on Integrable Systems and Quantum Symmetries (ISQS-20) IOP Publishing
Journal of Physics: Conference Series **411** (2013) 012009 doi:10.1088/1742-6596/411/1/012009

Grassmann Numbers and Clifford-Jordan-Wigner Representation of Supersymmetry

Sultan Catto[1,2,a], Yoon S. Choun[1], Yasemin Gürcan[3], Amish Khalfan[4] and Levent Kurt[1]

[1] Physics Departments, The Graduate School and University Center, The City University of New York, New York, NY 10010
[2] The Rockefeller University, 1230 York Avenue, New York, NY 10021-6399
[3] Manhattanville College, New York, NY 10577
[4] Yeshiva University, New York, NY 10033

Abstract. The elementary particles of Physics are classified according to the behavior of the multi-particle states under exchange of identical particles: bosonic states are symmetric while fermionic states are antisymmetric. This manifests itself also in the commutation properties of the respective creation operators: bosonic creation operators commute while fermionic ones anticommute. It is natural therefore to study bosons using commuting entities (e.g. complex variables), whereas to describe fermions, anticommuting variables are more naturally suited. In this paper we introduce these anticommuting- and at first sight unfamiliar- variables (Grassmann numbers) and investigate their properties. In particular, we briefly discuss differential and integral calculus on Grassmann numbers. Work supported in part by DOE contracts No. DE-AC-0276-ER 03074 and 03075; NSF Grant No. DMS-8917754.

1. Fermionic Analysis

1.1. Grassmann Numbers

Grassmann numbers (θ_m) are defined to satisfy the following anticommutation rule:

$$\{\theta_m, \theta_n\} \equiv \theta_m\theta_n + \theta_n\theta_m \equiv 0 \tag{1}$$

Note that this implies (by setting $m = n$) that the square of a Grassmann number is zero. In addition Grassmann numbers are associative:

$$[\theta_l, \theta_m, \theta_n] \equiv (\theta_l\theta_m)\theta_n - \theta_l(\theta_m\theta_n) \equiv 0 \tag{2}$$

This second property is usually implied and not written explicitly, but it is important since it is possible to construct a system that satisfies eq.(1) but doesn't satisfy eq.(2). It also allows the θ's to have a matrix representation: for instance, in field theory, a fermion operator such as an electron operator can be represented by an infinite matrix. In general, the matrix representation can be either finite or infinite.

This abstract number system was introduced by Grassmann[1] who didn't give a representation for it. He introduced the system in terms of differential operators in the process of inventing differential forms, which provide an example of Grassmann variables. For instance, we see from Maxwell's equations, expressed in terms of differential forms,

$$df = 0 \quad \text{and} \quad f = dA \quad \Rightarrow d^2 = 0 \tag{3}$$

where A is the four-potential, f is the electromagnetic field tensor and d is the differential operator of the form calculus. We therefore see that d is also a Grassmann operator, although it can't be represented by a finite matrix.

There are simpler systems with finite number of θ's which can be represented by finite matrices. This was first shown by Clifford[2][3] in 1878. Clifford also discovered Clifford algebra; in fact, he invented the algebra in order to find a matrix representation of Grassmann numbers. We will see this explicitly below.

1.2. Grassmann Numbers and the Clifford Algebra

Clifford was a young English mathematician who became a professor at King's college at a very early age. He was years ahead of his time in many respects- he thought that gravitation should be represented as a curvature of Riemannian space, long before Einstein. He also thought that these anticommutation relations must have a deep physical meaning, and was looking to develop them when he died at the age of thirty, postponing the advancement of physics by thirty years or more.

Clifford found that for every Grassmann system, there was a conjugate Grassmann system that obeyed the equations given above, but did not anticommute with the original set of Grassmann numbers. That is, Clifford's conjugate system consists of π's such that:

$$\theta_m \leftrightarrow \pi_m \tag{4}$$

$$\{\pi_m, \pi_n\} = 0 \tag{5}$$

$$[\pi_l, \pi_m, \pi_n] = 0 \tag{6}$$

but where:

$$\{\theta_m, \pi_n\} = \delta_{mn} \tag{7}$$

We recognize this last relation as the algebra of fermion operators (i.e. the Heisenberg uncertainty principle for fermions) where the π's are the momentum conjugates to the fermion operator θ's. Thus Clifford invented the algebra of fermionic operators long before the development of quantum theory. This principle also contains the Pauli exclusion principle. If we identify fermionic creation operators with Grassmann numbers, $\theta^2 = 0$ implies that a single state cannot contain two identical fermions. We will see that in general the total fermionic wave function will naturally be antisymmetrized.

As discussed above, associativity allows us to represent Grassmann numbers in terms of matrices[4] [5][6][7]. Clifford developed a representation of θ's and π's in terms of the elements of the quaternion algebra. Pauli discovered his matrices not knowing that they were essentially quaternions (and therefore the building blocks of the Grassmann matrix representation). He has a footnote to the effect that his friend, Pascual Jordan told him that the matrices he used are quaternion units. Then, around 1928, Jordan and Wigner[8] found a matrix representation of the fermion operators. They introduced fermion creation and annihilation operators which satisfied the Grassmann algebra exactly. They were looking for a fermionic matrix representation, and rediscovered the Clifford algebra, not knowing that Clifford had done it 50 years previously.

We will employ the Jordan-Wigner matrix representation to explicitly demonstrate the connection between the Clifford and Grassmann algebras. This name is unfair to Clifford,

Journal of Physics: Conference Series **411** (2013) 012009 doi:10.1088/1742-6596/411/1/012009

so we will refer to it here as the Clifford-Jordan-Wigner (CJW) representation. We will also see how this leads to the Berezin calculus.

We start with the Clifford algebra. A special case of this was constructed by Dirac, known as the Dirac algebra, which corresponds to a Clifford algebra with two sets of quaternion units. (The CJW representation can be built out of any number of quaternion units).

The Clifford algebra consists of non-anti-commuting associative elements:

$$\{\gamma_\mu, \gamma_\nu\} = 2\delta_{\mu\nu} \tag{8}$$

$$[\gamma_\mu, \gamma_\nu, \gamma_\lambda] = 0 \tag{9}$$

where $\mu, \nu = 1, \cdots, M$. Again, the associative condition is usually not written, but is necessary for the existence of a matrix representation.

For the purposes of the CJW representation we divide Clifford algebras into two classes: $M = 2N$ (even) and $M = 2N + 1$ (odd). We will show that this is an unnecessary distinction, since every even Clifford algebra ($2N$) corrsponds to an odd algebra with ($2N + 1$) defining elements.

For the case of an even algebra, we divide the set in half:

$$\gamma_1, \cdots, \gamma_N \qquad \text{and} \qquad \gamma_{N+1}, \cdots, \gamma_{2N} \tag{10}$$

We can now explicitly construct two Grassmann sets, by letting:

$$\theta_m = \frac{1}{2}(\gamma_m + i\gamma_{N+m}) \tag{11}$$

and

$$\pi_m = \frac{1}{2}(\gamma_m - i\gamma_{N+m}) \tag{12}$$

It is clear from the anticommutation relations satisfied by the γ_m's (note the normalization factor) that the θ's and π's will independently form Grassmann algebras, but will not mutually anticommute- they will satisfy the conjugate relations above eq.(7). This is the connection between the Grassmann and Clifford Algebras - the Clifford Algebra has embedded in it two sets of conjugate Grassmann Algebras, which satisfy the conjugate relations above.

We can always form one or more γ that anticommutes with the first $2N$ by forming the product of all $2N$ γ's:

$$\gamma_{2N+1} = \varepsilon(N)\gamma_1\gamma_2 \cdots \gamma_{2N} \tag{13}$$

where $\varepsilon(N) = \pm 1, \pm i$ is a normalization factor discussed below. It is clear that $\{\gamma_{2N+1}, \gamma_\mu\} = 0$ from the anticommutation relations eq.(8) and the fact that there are even number of γ's. We will also use:

$$(\gamma_{2N+1})^2 = 1 \tag{14}$$

For the case of an odd number of members in the Clifford algebra, we perform the same construction on the first $2N$ of them. Then we identify γ_{2N+1} as the final anticommuting member. Note that we cannot pull the same trick here and construct another matrix which anticommutes with all $2N+1$ γ's from a product of these - if we use all $2N+1$ in the product, it will contain an odd number of γ's and won't anticommute with them. If we use an even subset to form a product, any member of the algebra not in the subset will commute with the new γ, rather than anticommute. In fact, the product of all $2N+1$ γ's will be proportional to the unit matrix.

1.3. Clifford - Jordan - Wigner Construction

A systematic method of constructing representations of Clifford Algebras in N dimensions was devised by Clifford in 1878 and rediscovered by Jordan and Wigner 50 years later. We will now consider the first few cases in detail, and then give the general scheme.

For the $N = 1$ case, we define

$$\gamma_1 = \sigma_1 = \begin{pmatrix} 0 & 1 \\ 1 & 0 \end{pmatrix} \tag{15}$$

$$\gamma_2 = \sigma_2 = \begin{pmatrix} 0 & -i \\ i & 0 \end{pmatrix} \tag{16}$$

$$\gamma_3 = \sigma_3 = -i\sigma_1\sigma_2 = \begin{pmatrix} 1 & 0 \\ 0 & -1 \end{pmatrix} \tag{17}$$

These are the familiar Pauli matrices which satisfy:

$$[\sigma_i, \sigma_j] = 2\epsilon_{ijk}\sigma_k \tag{18}$$

and

$$(\sigma_i)^2 = 1 \tag{19}$$

The two conjugate Grassmann algebras contain one element each:

$$\theta = \theta_1 = \frac{1}{2}(\sigma_1 + i\sigma_2) = \begin{pmatrix} 0 & 1 \\ 0 & 0 \end{pmatrix} \tag{20}$$

and

$$\pi = \pi_1 = \frac{1}{2}(\sigma_1 - i\sigma_2) = \begin{pmatrix} 0 & 0 \\ 1 & 0 \end{pmatrix} \tag{21}$$

where, clearly,

$$\theta^2 = \pi^2 = 0 \tag{22}$$

and

$$\{\theta, \pi\} = \begin{pmatrix} 1 & 0 \\ 0 & 1 \end{pmatrix} = I \tag{23}$$

A mathematical digression is in order. We see that for $N = 1$, the γ's are the generators of $SO(2)$. In general, for $2N + 1$, the γ's will be generators of $SO(2N)$. For more physically interesting cases, such as systems invariant under $SO(3, 1)$, we must also deal with antihermitian matrices. If we modify the Clifford anticommutation relation eq.(8) to

$$\{\gamma_\mu, \gamma_\nu\} = 2\eta_{\mu\nu} \tag{24}$$

where $\eta_{\mu\nu}$ is a metric with signature (p, q), we will have p hermitian and q antihermitian γ's, which are invariant under $SO(p, q)$. This corresponds, for the Dirac algebra, to taking $i\gamma_4 \rightarrow \gamma_0$.

If γ has a matrix representation, then a similarity transformation

$$\gamma \rightarrow S\gamma S^{-1} \tag{25}$$

Journal of Physics: Conference Series **411** (2013) 012009 doi:10.1088/1742-6596/411/1/012009

is another matrix representation of the Clifford Algebra. Similarity transformations do not in general preserve positivity conditions. If γ's are hermitian ($\gamma^\dagger = \gamma$), then in order to preserve hermiticity, the matrices must be transformed by a unitary matrix:

$$\gamma \to U\gamma U^{-1} \tag{26}$$

where

$$UU^\dagger = 1 \tag{27}$$

If hermiticity is unimportant, S can be any matrix. From the construction above, if γ_{2N+1} is to be hermitian, then it is necessary to put a factor of i in front of the product of $2N$ γ matrices. It's always possible to find a matrix representation where the θ's and π's are chosen to be real (this corresponds to Majorana representation) although in this representation they will not be hermitian, and thus cannot represent observables.

Incidentally, the $N = 1$ case shows clearly the connection between Pauli matrices, the Clifford Algebra and quaternions. We construct the three antihermitian matrices from the hermitian Pauli matrices:

$$e_i = -i\sigma_i \tag{28}$$

These satisfy

$$e_i e_j = -\delta_{ij} + \epsilon_{ijk} e_k \tag{29}$$

This is precisely the quaternion algebra. It also follows from eq.(8) that

$$\{e_i, e_j\} = -\delta_{ij} \tag{30}$$

so that two antihermitian matrices give us a metric $(0, 2)$ as expected, and we see that Pauli matrices are just the hermitian version of quaternion units.

The special case discussed by Dirac[10] in 1928 corresponds to $N = 2$, or four elements in the algebra: γ_1, γ_2, γ_3, γ_4. These are the Euclidean analog of the more familiar Dirac matrices, where $\gamma_0 = i\gamma_4$. Dirac also constructed the fifth anticommuting matrix,

$$\gamma_5 = \gamma_1\gamma_2\gamma_3\gamma_4 = i\gamma_0\gamma_1\gamma_2\gamma_3 \tag{31}$$

Dirac needed 4×4 matrices to describe the case $N = 2$. For the general case we will need $2^N \times 2^N$ dimensional matrices.

For the $N = 2$ case, Dirac took two sets of $N = 1$ Clifford algebras (i.e. Pauli matrices, σ_i, ρ_i) which mutually commute:

$$[\sigma_i, \rho_j] = 0 \tag{32}$$

The resulting Clifford matrices will have dimension $2^2 = 4$. we construct them first by taking $\sigma_i = \sigma_i \times I$ and $\rho_i = I \times \rho_i$:

$$\sigma_i = \begin{pmatrix} \sigma_i & o \\ 0 & \sigma_i \end{pmatrix} = \sigma_i I \ , \tag{33}$$

$$\rho_1 = \begin{pmatrix} 0 & I \\ I & 0 \end{pmatrix}, \rho_2 = \begin{pmatrix} 0 & -iI \\ iI & 0 \end{pmatrix}, \rho_3 = \begin{pmatrix} I & 0 \\ 0 & -I \end{pmatrix} \tag{34}$$

where

$$I = \begin{pmatrix} 1 & 0 \\ 0 & 1 \end{pmatrix} \tag{35}$$

The σ's will obviously commute with the ρ's. for the $N = 2$ case, if we identify

$$\sigma_i^{(1)} = \sigma_i \tag{36}$$

$$\sigma_i^{(2)} = \rho_i \tag{37}$$

we can set up the table

$$\begin{array}{cccc} \gamma_1 & \sigma_1^{(1)} & \sigma_2^{(1)} & \gamma_3 \\ \gamma_2 & \sigma_3^{(1)}\sigma_1^{(2)} & \sigma_3^{(1)}\sigma_2^{(2)} & \gamma_4 \\ & \sigma_3^{(1)}\sigma_3^{(2)} & \gamma_5 & \end{array} \tag{38}$$

where the γ's are identified with the product of γ's in the column they are closest to.

We have used these matrix representations for N=2 case in supersymmetric chiral field theories developed earlier (see Catto references in the biblography). The matrix forms allow separation of equations and make calculations clear and more easier. Particular generalization for quark-diquark symmetries, multiquark systems and their mass predictions as well as applications to multi instanton processes were shown in Catto and Gürsey's many papers.

There are actually many different representations possible. Dirac used a unitary transformation of these, such that his γ's were

$$\gamma_1 \cdots \gamma_5 \rightarrow \rho_1, \rho_3, \rho_2\sigma_i \tag{39}$$

Other possibilities are

$$\gamma_1 \cdots \gamma_5 \rightarrow \sigma_1, \sigma_3, \sigma_3\rho_i \tag{40}$$

$$\gamma_1 \cdots \gamma_5 \rightarrow \rho_1, \rho_2, \rho_3\sigma_i \tag{41}$$

Dirac identified γ_4 with ρ_3; Weyl used $\gamma_5 = \rho_3$. All of these are specific representations of the algebra, but the theory is independent of the one used. It is only when an explicit representation of a wavefunction is needed that a particular representation will be used.

We can continue this construction technique for the $N = 3$ case. We use three sets of commuting Pauli matrices (i.e. $\sigma \times I \times I$, $I \times \sigma \times I$, $I \times I \times \sigma$):

$$\begin{array}{cccc} \gamma_1 & \sigma_1^{(1)} & \sigma_2^{(1)} & \gamma_4 \\ \gamma_2 & \sigma_3^{(1)}\sigma_1^{(2)} & \sigma_3^{(1)}\sigma_2^{(2)} & \gamma_5 \\ \gamma_3 & \sigma_3^{(1)}\sigma_3^{(2)}\sigma_1^{(3)} & \sigma_3^{(1)}\sigma_3^{(2)}\sigma_2^{(3)} & \gamma_6 \\ & \sigma_3^{(1)}\sigma_3^{(2)}\sigma_3^{(3)} & \gamma_7 & \end{array} \tag{42}$$

Recall that σ_1, σ_3 are real, σ_2 is imaginary and all three are hermitian. We see that all the entries in the first column are real and hermitian, while all the entries in the second column are imaginary and hermitian. Thus $\gamma_{1,2,3,7}$ are real and symmetric, while $\gamma_{4,5,6}$ are imaginary and antisymmetric.

When these are plugged into the relations defining the θ's and π's in terms of the γ's (eqs.(11,12) it is clear that, by construction, the θ's and π's will all be real. This procedure can be extended indefinitely, yielding the Clifford - Jordan - Wigner construction:

$$\begin{array}{cccc} \gamma_1 & \sigma_1^{(1)} & \sigma_2^{(1)} & \gamma_{N+1} \\ \gamma_2 & \sigma_3^{(1)}\sigma_1^{(2)} & \sigma_3^{(1)}\sigma_2^{(2)} & \gamma_{N+2} \\ \gamma_3 & \sigma_3^{(1)}\sigma_3^{(2)}\sigma_1^{(3)} & \sigma_3^{(1)}\sigma_3^{(2)}\sigma_2^{(3)} & \gamma_{N+3} \\ \vdots & \vdots & \vdots & \vdots \\ \gamma_N & \sigma_3^{(1)}\cdots\sigma_3^{(N-1)}\sigma_1^{(N)} & \sigma_3^{(N-1)}\cdots\sigma_3^{(N-1)}\sigma_2^{(N)} & \gamma_{2N} \\ & \sigma_3^{(1)}\sigma_3^{(2)}\cdots\sigma_3^{(N)} & \gamma_{2N+1} & \end{array} \tag{43}$$

where, in general, we will have $N+1$ real and N imaginary matrices. We can also allow $N \to \infty$ to have infinite matrices which obey Grassmann Algebra conditions.

There is an alternative formula for γ_{2N+1} in terms of the θ's and π's, rather than the γ's:

$$\gamma_{2N+1} = \prod_{m=1}^{N} [\theta_m, \pi_m] = (-i)^{N^2} \gamma_1\gamma_2 \cdots \gamma_{2N} \tag{44}$$

This identity including the prefactor can be derived explicitly, although we will not do it here. As an example, for $N = 1$,

$$\gamma_3 = [\frac{\sigma_1 + i\sigma_2}{2}, \frac{\sigma_1 - i\sigma_2}{2}] = \frac{-i}{2}[\sigma_1, \sigma_2] = -i\sigma_1\sigma_2 = \sigma_3 \tag{45}$$

1.4. Calculus of Grassmann Numbers

We will now begin the study of the calculus of Grassmann numbers, also known as the the Berezin calculus. Specifically, we will construct a differential operator representation of conjugate Grassmann (i.e. Heisenberg) algebras. Since we will construct this by analogy to the bosonic case, let's consider the bosonic Heisenberg Algebra. We have elements x_i and their conjugates p_i such that they obey the uncertainty relation:

$$[x_i, x_j] = [p_i, p_j] = 0 \tag{46}$$

$$[x_i, p_j] = i\delta_{ij} \tag{47}$$

With this definition, we see that the p's can be represented as derivatives:

$$p_j \leftrightarrow -i\frac{\partial}{\partial x_j} \tag{48}$$

where the operators now act on the space of all functions of x. If we want to work without "i"s, we can use the associated creation and annihilation operators:

$$a_i = \frac{1}{\sqrt{2}}(x_i + ip_i) \tag{49}$$

$$a_i^\dagger = \frac{1}{\sqrt{2}}(x_i + ip_i) \tag{50}$$

So that

$$[a_i, a_j] = [a_i^\dagger, a_j^\dagger] = 0 \tag{51}$$

$$[a_i, a_j^\dagger] = \delta_{ij} \tag{52}$$

The $[a_j^\dagger]$ can also be represented as derivative operators:

$$a_j^\dagger \leftrightarrow \frac{\partial}{\partial a_j} \tag{53}$$

where the operators act on the space of all functions of a.

For Grassmann Algebras we have anticommutation relations, rather than commutation relation, but Berezin showed how it was possible to represent the π's as derivatives of θ_i's.

We begin by considering a simple algebra with two θ's. Then the most general function of these will be

$$F(\theta_1, \theta_{2)} = A + B^1\theta_1 + B^2\theta_2 + C\theta_1\theta_2 \tag{54}$$

All higher powers of θ's in the power series expansion vanish as a result of $\theta_i^2 = 0$. We see then that if $N < \infty$, F is a polynomial, terminating after a finite number of terms. Eq.(1) also allows us to write

$$F(\theta_1, \theta_{2)} = A + B^1\theta_1 + B^2\theta_2 - C\theta_1\theta_2 \tag{55}$$

For the moment we will assume that A, B, C are ordinary numbers which commute with θ_i. We define left and right derivatives:

$$\frac{\partial F}{\partial \theta_1} = B^1 + C\theta_2 \tag{56}$$

$$F\frac{\overleftarrow{\partial}}{\partial \theta_1} = B^1 - C\theta_2 \tag{57}$$

where there is a minus sign difference. In everything that follows, we will discuss only the left derivative, but we could have equivalently selected the right derivative and performed and analogous derivation. We can first show that the derivatives satisfy a Grassmann Algebra. First, operating on the eq. (55) from the left we have

$$\frac{\partial F}{\partial \theta_2} = B^2 - C\theta_1 \tag{58}$$

Differentiating eq.(56) with respect θ_2 and eq.(58) with respect to θ_1 we get

$$\frac{\partial}{\partial \theta_1}\frac{\partial F}{\partial \theta_2} = -C \quad \text{and} \quad \frac{\partial}{\partial \theta_2}\frac{\partial F}{\partial \theta_1} = C \tag{59}$$

Thus we have:

$$\{\frac{\partial}{\partial \theta_1}\frac{\partial}{\partial \theta_2} + \frac{\partial}{\partial \theta_2}\frac{\partial}{\partial \theta_1}\}F = 0 \tag{60}$$

It is also obvious that:

$$\frac{\partial}{\partial \theta_1}\frac{\partial}{\partial \theta_1}F = \frac{\partial}{\partial \theta_2}\frac{\partial}{\partial \theta_2}F = 0 \tag{61}$$

All these lead to the fact that the derivatives form a Grassmann Algebra, when acting on the space of all functions:

$$\{\frac{\partial}{\partial \theta_i}, \frac{\partial}{\partial \theta_j}\}F(\theta_1, \theta_2) = 0 \tag{62}$$

Now we check that the derivatives anticommute with the $\theta's$:

$$\theta_1 \frac{\partial}{\partial \theta_1} F = B^1 \theta_1 + C\theta_1\theta_2 \tag{63}$$

and

$$\frac{\partial}{\partial \theta_1} \theta_1 F = A + B^2 \theta_2 \tag{64}$$

so that

$$\{\theta_1, \frac{\partial}{\partial \theta_1}\} F = A + B^1\theta_1 + B^2\theta_2 + C\theta_1\theta_2 = F \tag{65}$$

By going through all the remaining relations in the same way we find that the derivatives are the conjugates of the θ's.

$$\{\theta_i, \frac{\partial}{\partial \theta_j}\} F = \delta_{ij} F \tag{66}$$

We have shown that the derivatives form a conjugate set for the case of two θ's, and it generalizes to any number by induction.

We can work backwards from the Grassmann sets to reconstruct the Clifford Algebra, in terms of the θ's and $\frac{\partial}{\partial \theta}$'s. We can invert eq.(11) and (12) so that:

$$\gamma_m = \theta_m + \frac{\partial}{\partial \theta_m} \tag{67}$$

and

$$\gamma_{N+m} = -i(\theta_m - \frac{\partial}{\partial \theta_m}) \tag{68}$$

To construct the final anticommuting matrix γ_{2N+1}, we use eq.(44):

$$\gamma_{2N+1} = \prod_{m=1}^{N} [\theta_m, \frac{\partial}{\partial \theta_m}] \tag{69}$$

This completes the brief discussion of the Berezin differential calculus.

Now that we have discussed the differential calculus we can consider integration. This is actually rather difficult. We will follow in Berezin's footsteps, and give a set of rules which, when applied, yield the right result. Along the way we will try to motivate some of the more seemingly mysterious aspects of this procedure.

We start by generating the calculus for a set of two Grassmann numbers, which can be easily extrapolated to an arbitrary set. We have fundamentally,

$$\int d\theta_i = 0 \tag{70}$$

$$\int \theta_i d\theta_i = 1 \tag{71}$$

where this is for each i (no sum).

At this point we don't say whether this is a definite or an indefinite integral. We just write an integral without worrying about it - we cannot take θ from $-\infty$ to $+\infty$; it doesn't make sense. We will discuss below what we mean by integrals over Grassmann variables.

Journal of Physics: Conference Series **411** (2013) 012009 doi:10.1088/1742-6596/411/1/012009

We define multiple integrals by:

$$\int d\theta_1 d\theta_2 = \int d\theta_1 \int d\theta_2 \tag{72}$$

We also use the fact that the differentials anticommute with one another and with the θ's:

$$d\theta_m d\theta_n = -d\theta_n \theta_m \tag{73}$$

and

$$d\theta_m \; \theta_n = -\theta_n \; d\theta_m \tag{74}$$

We consider again the most general function of two Grassmann numbers, eq.(54):

$$F(\theta_1, \theta_2) = A + B^1\theta_1 + B^2\theta_2 + C\theta_1\theta_2 \tag{75}$$

When we integrate this over both variables, by eq.(70) above it is clear that only the term with both $\theta_1$and θ_2 will be non-zero:

$$\int F d\theta_1 d\theta_2 = C \int \theta_1\theta_2 d\theta_1 d\theta_2 = -C \int \theta_1 d\theta_1 \int \theta_2 d\theta_2 = -C \tag{76}$$

where the minus sign is due to eq.(74).

We note here the interesting fact that

$$\int F d\theta_1 d\theta_2 = \frac{\partial}{\partial\theta_1}\frac{\partial}{\partial\theta_2} F = -C \tag{77}$$

so that differentiation and integration are essentially the same thing. This result is true for all N.

We will now discuss the connection between Berezin integration and complex integration. Recall that the γ's are defined to be hermitian. Hermitian operators correspond to observables and have real eigenvalues - the γ's are the generalizations of real numbers.

In the CJW representation, the γ_j's are real, while the γ_{j+N}'s are imaginary, but all have real eigenvalues. The θ's are real in this representation, but are not hermitian, so they do not correspond to observables. They have complex eigenvalues and are the generalization of complex numbers. This hints at the possibility of analogies between Berezin rules and the rules of integration of complex variables. For instance, consider Cauchy's theorem:

$$\frac{1}{2\pi i}\oint_C F(z)dz = 0 \tag{78}$$

where this is true if $F(z)$ is analytic inside the contour C. For example with $F(z) = 1$,

$$\frac{1}{2\pi i}\oint_C dz = 0 \tag{79}$$

This can be compared to the Berezin rule (eq.(70)) above:

$$\int d\theta = 0 \tag{80}$$

The general correspondence will be:

$$\frac{1}{2\pi i}\oint \quad \text{for } z \quad \leftrightarrow \quad \int \quad \text{for } \theta \tag{81}$$

Is there a complex analog of eq.(71) $\int \theta d\theta = 1$? We could naively try:

Journal of Physics: Conference Series **411** (2013) 012009 doi:10.1088/1742-6596/411/1/012009

$$\oint_C zdz \tag{82}$$

but this is identically zero. On the other hand,

$$-\frac{1}{2\pi i}\oint_C \frac{dz}{1-z} = 1 \tag{83}$$

if the contour C encloses the point $z = 1$. Can we find an analog of this in the Berezin Calculus? Consider the integral:

$$-\int d\theta \frac{1}{1-\theta} \tag{84}$$

where we choose $d\theta$ to the left of $\frac{1}{1-\theta}$(remember that the order matters for Grassmann variables). θ has no inverse ($\theta^2 = 0$) but $\frac{1}{1-\theta}$ does:

$$\frac{1}{1-\theta} = 1 + \theta \tag{85}$$

This can be seen either by multiplying directly, or by expanding $\frac{1}{1-\theta}$in a power series where all terms of θ^2can higher are zero. We can write:

$$-\int d\theta \frac{1}{1-\theta} = -\int d\theta(1+\theta) = -\int d\theta\theta = \int \theta d\theta = 1 \tag{86}$$

We see that the second Berezin rule is the analog of contour integration. This is why we didn't write limits on integrals of Grassmann variables, or determine whether they were definite or indefinite.

We can consider this issue a little further What do we man by $d\theta$? In the case of Riemannian integration we divide an interval $L = x_n - x_o$ into N parts and put

$$x_j = x_0 + j\epsilon\ , \qquad \epsilon = L/N \tag{87}$$

Then the Riemann integral is defined as the limit of a sum, namely

$$\begin{aligned}\int_{x_0}^{x_N} F(x)dx &= \lim_{N\to\infty}\sum_{j=1}^{N} F(x_j)(x_j - x_{j-1}) \\ &= \lim_{N\to\infty}\sum_{j=1}^{N} F(x_j)\Delta x \end{aligned} \tag{88}$$

where

$$\Delta x = \epsilon = L/N \tag{89}$$

In the case of Grassmann variables we can start by defining the difference of two Grassmann numbers θ_j and θ_{j-1} by

$$\Delta\theta = \theta_j - \theta_{j-1} \tag{90}$$

Since θ_j and θ_{j-1} anticommute, $\Delta\theta$ also anticommutes with both θ_j and θ_{j-1}. Indeed,

$$\{\theta_j - \theta_{j-1}, \theta_j\} = \{\theta_j - \theta_{j-1}, \theta_{j-1}\} = 0 \tag{91}$$

We generalize this by inserting $N-1$ anticommuting Grassmann numbers between θ_0 and θ_N:

$$\theta_j = \theta_0 + j\Delta\theta, \quad \Delta\theta = (\theta_N - \theta_0)/N \tag{92}$$

Let us also define the magnitude of a Grassmann number θ represented by a $k \times k$ matrix by the equation

$$\mid \theta \mid^2 = \frac{Tr(\theta^\dagger \theta)}{k} \tag{93}$$

We see that if θ is a Grassmann number with conjugate π rather then $\lambda\theta$ is also a Grassmann number with conjugate $\lambda^{-1}\pi$. We also have

$$\mid \lambda\theta \mid = \mid \lambda \mid\mid \theta \mid \tag{94}$$

so that

$$\mid \Delta\theta \mid = \frac{1}{N} \mid \theta_N - \theta_0 \mid \tag{95}$$

and as $N \to \infty$ the magnitude of $\Delta\theta$ tend to zero just like Δx in the Riemannian integral.

After these preliminaries it is clear that there is a possibility of defining the Berezin integral as a sum of terms

$$\phi = \sum_j F(\theta_j)\Delta\theta \tag{96}$$

defined on a contour in the θ plane with anticommuting points and taken in the limit when the number N of points on the contour increases indefinitely so that the magnitude of the Grassmann number $\Delta\theta$ tends to zero.

Note that when $N \to \infty$, the Grassmann numbers θ_j becomes infinite matrices. In this case θ can vary quasicontinuosly between θ_0 and θ_N and behave like a continuous bosonic variable x. To each θ_j corresponds the function $F(\theta_j)$ which in the $N \to \infty$ limit can be regarded as a continuous function of the fermionic variable θ in the given interval.

1.5. Digression: Grassmann Numbers and Fermion Annihilation and Creation Operators

We now make a point concerning the identification of Grassmann numbers as fermion creation and annihilation operators. The relation between conjugate Grassmann sets is the same, you will recall, as the relation between fermionic annihilation and creation operators:

$$\{b_i, b_j\} = \{b_i^\dagger, b_j^\dagger\} = 0 \tag{97}$$

$$\{b_i, b_j^\dagger\} = \delta_{ij} \tag{98}$$

There is a complete symmetry between b and $b^\dagger$; how we decide which is a creation and which an annihilation operator? We are only able to decide after we define the vacuum state by:

$$b_i \mid 0 >= 0 \tag{99}$$

$$b_j^\dagger \mid 0 >= \mid j > \tag{100}$$

where $\mid j >$ is a one particle state with label j. Eqs.(97) and (98) tell us that b is an annihilation and $b^\dagger$ is a creation operator. We can extend eq.(97) to any number of annihilation operators

$$b_i b_j ... b_n \mid 0 >= 0 \tag{101}$$

But if we act with creation operators on the vacuum we don't get 0. Instead, extending eq.(98) and using the anticommutation relations:

$$b_i^\dagger b_j^\dagger \mid 0 >= -b_j^\dagger b_i^\dagger \mid 0 >= \mid ij >= - \mid ji > \tag{102}$$

Thus the two-particle state is totally antisymmetric. Similarly, we can construct n-particle states:

$$b_{i^{(1)}}^\dagger, b_{i^{(2)}}^\dagger ... b_{i^{(n)}}^\dagger \mid 0 >= \mid i^{(1)} i^{(2)} ... i^{(n)} > \tag{103}$$

which will also be totally antisymmetric. Anticommunitation of the Grassmann numbers leads directly to the Pauli exclusion principle for fermionic states:

$$b_i^\dagger b_i^\dagger \mid 0 >= 0 \tag{104}$$

such that we cannot have to identical particles in the same state. Dirac discovered this relation between the Pauli exclusion principle, the antisymmetry of the state, and the anticommutation relations.

We see that in order to interpret Grassmann variables as creation and annihilation operators, we need to find a vacuum. Is it possible for us to construct the vacuum? To do this we will consider an alternate formulation of quantum mechanics.

In quantum mechanics, physical systems are described in the language of states and operators which act on states. Quantum operators can be represented by matrices or by differential operators. States are then represented by vectors (column matrices). In the Dirac notation, $\mid \alpha >$ is a state and $M_{\alpha\beta}$ are the matrix elements of an operator M. If we have an orthonormal basis of states, we can describe these matrix elements by:

$$M_{\alpha\beta} =< \alpha \mid M \mid \beta > \tag{105}$$

If M is a hermitian operator, it can always be written as the sum of projection operators:

$$M = \sum_i \lambda_i \mid i >< i \tag{106}$$

where the λ_i are real and

$$P_i = \mid i >< i \mid \tag{107}$$

is a projection operator which projects out state $\mid i >$. All eigenstates are orthonormal:

$$< i \mid j >= \delta_{ij} \tag{108}$$

If we expand a general state:

$$\mid \alpha >= \sum_i \Psi_i \mid i > \tag{109}$$

then P_i projects out the component along the ith eigenstate:

Journal of Physics: Conference Series **411** (2013) 012009 doi:10.1088/1742-6596/411/1/012009

$$\begin{aligned} P_i \mid \alpha &= \mid i >< i \mid \alpha > \\ &= \mid i >< i \mid \sum_j \Psi_j \mid j > \\ &= \mid i > \sum_j \Psi_j < i \mid j > \\ &= \mid i > \Psi_i \end{aligned} \tag{110}$$

By construction, P_i is Hermitian:

$$P_i = \mid i >< i \mid = P_i^\dagger \tag{111}$$

Projection operators have the unique trait that they are idempotent -they equal themselves when squared:

$$P_i^2 = \mid i >< i \mid i >< i \mid = P_i \tag{112}$$

We can thus associate a projection operator with each state: rather then representing states by vectors, we can represent them by their projection operators. Then instead of having operators and observables on a different footing, we have hermitian operators that represent observables, and hermitian projection operators that represent states. This formulation is useful because both operators and states can be represented by hermitian matrices.

As an example, consider a two state (i.e. spin) system. The most general state will be a linear combination of spin up and spin down:

$$\mid \Psi >= a \mid\uparrow> + b \mid\downarrow> \tag{113}$$

$$<\uparrow\mid\downarrow>= 0 \qquad\qquad <\uparrow\mid\uparrow>=<\downarrow\mid\downarrow>= 1 \tag{114}$$

$$\mid a \mid^2 + \mid b \mid^2 = 1 \tag{115}$$

We usually represent these as column vectors:

$$\mid\uparrow>= \begin{pmatrix} 1 \\ 0 \end{pmatrix} \qquad \mid\downarrow>= \begin{pmatrix} 0 \\ 1 \end{pmatrix} \tag{116}$$

$$\mid \Psi >= a \begin{pmatrix} 1 \\ 0 \end{pmatrix} + b \begin{pmatrix} 0 \\ 1 \end{pmatrix} = \begin{pmatrix} a \\ b \end{pmatrix} \tag{117}$$

Alternatively, we could express these states by their projection operators. The projection operators for the up state is:

$$P_\uparrow = \mid\uparrow><\uparrow\mid = \begin{pmatrix} 1 \\ 0 \end{pmatrix} \begin{pmatrix} 1 & 0 \end{pmatrix} = \begin{pmatrix} 1 & 0 \\ 0 & 0 \end{pmatrix} \tag{118}$$

Equivalently, the down state projection operator is:

$$P_\downarrow = \mid\downarrow><\downarrow\mid = \begin{pmatrix} 0 \\ 1 \end{pmatrix} \begin{pmatrix} 0 & 1 \end{pmatrix} = \begin{pmatrix} 0 & 0 \\ 0 & 1 \end{pmatrix} \tag{119}$$

These can be expressed in terms of the Pauli matrices:

$$P_\uparrow = \frac{1+\sigma_3}{2}, \qquad P_\downarrow = \frac{1-\sigma_3}{2} \tag{120}$$

Now, getting away from quantum mechanics, consider the case of a single Grassmann number θ with a conjugate π, and $N = 1$. In the CJW construction, θ and π are represented by linear combinations of the Pauli matrices:

$$\theta = \frac{\sigma_1 + \sigma_2}{2} = \begin{pmatrix} 0 & 1 \\ 0 & 0 \end{pmatrix} \leftrightarrow b \tag{121}$$

$$\pi = \frac{\sigma_1 - i\sigma_2}{2} = \begin{pmatrix} 0 & 0 \\ 1 & 0 \end{pmatrix} \leftrightarrow b^\dagger \tag{122}$$

where we show that we wish to interpret θ as an annihilation and π has a creation operator. We need the vacuum state, which will satisfy

$$\theta \mid 0 >= 0 \tag{123}$$

with normalization

$$< 0 \mid 0 >= 1 \tag{124}$$

In terms of matrices:

$$\begin{pmatrix} 0 & 1 \\ 0 & 0 \end{pmatrix} \begin{pmatrix} a_1 \\ a_2 \end{pmatrix} = 0 \tag{125}$$

which gives (keeping in mind eq.(124))

$$\begin{pmatrix} a_2 \\ 0 \end{pmatrix} = 0 \quad \Rightarrow \quad a_2 = 0 \quad \Rightarrow \quad a_1 = 1 \tag{126}$$

so that the vacuum state is

$$\mid 0 >= \begin{pmatrix} 1 \\ 0 \end{pmatrix} \tag{127}$$

Can we descibe the vacuum state by a projection operator? We have:

$$P^{(0)} = \mid 0 >< 0 \mid = \begin{pmatrix} 1 \\ 0 \end{pmatrix} \ (0 \ \ 1) \ = \begin{pmatrix} 1 & 0 \\ 0 & 0 \end{pmatrix} = \frac{1+\sigma_3}{2} \tag{128}$$

Using the representation of b from eq.(122) we have:

$$bP^{(0)} = 0 \tag{129}$$

and its hermitian conjugate:

$$P^{(0)} b^\dagger = 0 \tag{130}$$

Eq.(129) correspond to eq.(97), which is one of the defining relations of the vacuum. Eq.(130) is the mathematical representation of the statement that the 1-particle state is orthogonal to the vacuum:

$$b^\dagger \mid 0 >= \begin{pmatrix} 0 & 0 \\ 1 & 0 \end{pmatrix} \begin{pmatrix} 1 \\ 0 \end{pmatrix} = \begin{pmatrix} 0 \\ 1 \end{pmatrix} \tag{131}$$

Journal of Physics: Conference Series **411** (2013) 012009 doi:10.1088/1742-6596/411/1/012009

which is clearly orthogonal to the vacuum (eq.(127)).

We can now identify this with our spin system. The vacuum corresponds to the spin up state, and one particle corresponds to the spin down state. This state is obtained from the vacuum by acting with a fermionic creation operator defined in eq.(122).

This generalizes very nicely when we use the CJW construction of N dimensions (eq.(43)). We can generalize the vacuum projection operator to:

$$|\,0><0\,| = P^{(0)} = \Omega = \frac{1+\sigma_3^{(1)}}{2}\frac{1+\sigma_3^{(2)}}{2}\dots\frac{1+\sigma_3^{(N)}}{2} \tag{132}$$

The general form for this will be an $(N+1)\times(N+1)$ matrix:

$$\begin{pmatrix} 1 & 0 & 0 & . & . & . & . & \\ 0 & 0 & & & & & & \\ 0 & & . & & & & & \\ . & & & . & & & & \\ . & & & & . & & & \\ . & & & & & . & & \\ . & & & & & & . & \\ & & & & & & & 0 \end{pmatrix} \tag{133}$$

This vacuum state projection operator satisfies relation that are generalizations of eqs.(129) and (130):

$$\theta_i\Omega = 0\,, \quad \Omega\pi_i = 0 \tag{134}$$

These can be checked using the explicit form of θ and π in terms of the Pauli matrices from the CJW construction (eq.(43)) inserted into eqs.(11) and (12) and by using the form of Ω from eq.(132).

Consider a state with r particles in it. We can represent this by:

$$|\,r> = \pi_{i(1)}\pi_{i(2)}\dots\pi_{i(n)}\,|\,0> \tag{135}$$

To build a projection operator, we need to take $|\,r><r\,|$. Using the fact that, by construction and by identification as creation and annihilation operators, $\pi^\dagger = \theta$, we have:

$$P_r = |\,r><r\,| = \pi_{i(1)}\dots\pi_{i(n)}\Omega\theta_{i(n)}\dots\theta_{i(1)} \tag{136}$$

Here we have a general projection operator that clearly demonstrates the symmetry between θ and π.

We have shown by construction that the vacuum state always exists, allowing us to interpret a system of Grassmann numbers as the annihilation operator of some particle. Identifying the vacuum completes the connection between fermions and Grassmann numbers.

1.6. Spinors with Grassmann Components

We have seen that N Grassmann numbers can be constructed out of the Clifford algebra generated by $2N$ matrices γ_μ which satisfy the anticommutation relation eq.(24). This relation is invariant under $SO(p,q)$ which becomes $SO(2N)$ in the hermitian case. Note that the θ_j ($j = 1, ..., N$) and π_j belong respectively to the (N) and $(\overline{N})$ representations of the $SU(N)$ subgroup of $SO(2N)$. Passage from the Clifford numbers to the Grassmann numbers corresponds to the decomposition of the rotation group $SO(2N)$ with respect to its $SU(N)$ subgroup. The generators of the Lie algebra of this rotation group are the spin matrices

$$S_{\mu\nu} = \frac{1}{4i}[\gamma_\mu, \gamma_\nu] \tag{137}$$

where $\gamma, \nu = 1, ..., 2N$. As we have seen from the CJW construction, these are $M \times M$ matrices, where

$$M = 2^N \tag{138}$$

since N commuting sets of Pauli matrices must be accommodated in the γ matrices. The corresponding $SO(2N)$ group element associated with an M-dimensional representation is

$$R(\omega) = e^{\frac{i}{2}\omega^{\mu\nu}S_{\mu\nu}} \tag{139}$$

which acts on the M-dimensional column ψ, the spinor representation of $SO(2N)$. When a rotation of 2π is performed around any axis, the spinor changes sign, so that strictly speaking it is a representation of the double covering of $SO(2N)$ which is called Spin($2N$).

In relativistic local quantum field theory there is a theorem (the Spin-Statistics Theorem) which states that if a local field is associated with an integer spin representation of the Poincaré group it must be quantized as a boson field, while a local field corresponding to a half-integer spin representation of the same space-time group must be quantized as a fermion. It follows that symmetrical or antisymmetrical tensor fields like scalar or vector fields are bosonic, while spinor fields ($s = 1/2$ or $s = 3/2$) are fermionic.

In relativistic theories, spinor fields like the neutrino field or the electron-positron Dirac field $\psi(x)$ must have Grassmann components that vary "continuously" with x.

In the $2N$ dimensions the 2^N-dimensional spinor representation Ψ is reducible since γ_{2N+1} commutes with the $SO(2N)$ generators $S_{\mu\nu}$. It decomposes into left handed and right handed irreducible spinors (Weyl spinors) given by

$$\Psi_L = \frac{1}{2}(1 + \gamma_{2N+1})\Psi \tag{140}$$

$$\Psi_R = \frac{1}{2}(1 - \gamma_{2N+1})\Psi \tag{141}$$

each having dimensions $M/2$.

In the $2N+1$ dimensions the $SO(2N+1)$ generators consist of the $(2N)^2$generators $S_{\mu\nu}$ from eq.(137) and the $2N$ additional generators

$$S_{2N+1,\nu} = \frac{1}{4i}[\gamma_{2N+1}, \gamma_\nu] \tag{142}$$

where $\nu = 1, ..., 2N$. Spin($2N$+1) also has a 2^N-dimensional representation Ψ but it is irreducible.

When the spinor representation of the $O(p, q)$ is real, it is called Majorana representation. In field theory it can have as components real Grassmann numbers. The Majorana condition depends on the dimension as well as the signature of the metric and cannot always be imposed.

As an example let us mention that in Minkowski space $O(3, 1)$ or its covering group $SL(2, C)$ can have either a 2-component complex Weyl spinor or a 4-component real Majorana spinor as representations. In the (9+1)-dimensional Minkowski space a spinor can be Weyl and Majorana spinor simultaneously so that an irreducible spinor of Spin(9,1) has only 16 real components instead of 32 complex components. In (1+1) dimensions we can choose

$$\gamma_0 = i\sigma_2, \quad \gamma_1 = \sigma_1, \quad \gamma_3 = \sigma_3 \tag{143}$$

Here there is no rotation; there is, however, a boost generator

$$S_{01} = \frac{1}{4i}[\gamma_0, \gamma_1] = \frac{i}{2}\sigma_3 = -\frac{i}{2}\gamma_3 \tag{144}$$

The corresponding group element

$$B = \exp(i\alpha S_{01}) = \exp(\frac{1}{2}\sigma_3\alpha) \tag{145}$$

is real thus

$$\Psi = \begin{pmatrix} \psi_1 \\ \psi_2 \end{pmatrix} \tag{146}$$

can be taken as a real Majorana spinor. It is reducible, though, since we are in an even dimension. We have

$$\Psi_L = \frac{1}{2}(1+\gamma_3)\Psi = \frac{1}{2}(1+\sigma_3)\Psi = \begin{pmatrix} \psi_1 \\ 0 \end{pmatrix} \tag{147}$$

$$\Psi_R = \frac{1}{2}(1-\gamma_3)\Psi = \frac{1}{2}(1-\sigma_3)\Psi = \begin{pmatrix} 0 \\ \psi_2 \end{pmatrix} \tag{148}$$

as irreducible componets of Ψ. Hence in (1+1) dimensions an irreducbile spinor has one component $\psi_1 = \theta$, θ being a real Grassman number. In that sense $O(1,1)$ is like $O(9,1)$ since the Weyl and Majorana conditions can be imposed simultaneously. This is an illustration of the 8-fold Bott periodicity[12]. For the next Weyl-Majorana spinor we have to got to $O(17,1)$. The properties of the spinor in (1+1) and (9+1) Minkowski spaces are crucial in superstring theories.

Consider a function of Grassmann numbers such as eq.(54)

$$F(\theta_1, \theta_2) = A + B^\alpha\theta_\alpha + C\theta_1\theta_2 \quad (\alpha = 1, 2) \tag{149}$$

We are now in a position to define the invariance properties of F under Lorentz transformations when the θ_α are considered to be the Grassmann components of a left-handed Weyl Spinor in (3+1) dimensions. Let

$$\theta = \begin{pmatrix} \theta_1 \\ \theta_2 \end{pmatrix} \tag{150}$$

Under the $SL(2, C)$, the covering group of $SO(3, 1)$, θ transforms as

$$\theta' = L\theta \tag{151}$$

with L a unimodular complex 2×2 matrix

$$Det\, L = 1, \quad L = e^{i\frac{\vec{\sigma}}{2}\cdot(\vec{\omega} - i\vec{\nu})} \tag{152}$$

where $\vec{\omega}$ and $\vec{\nu}$ are respectively the rotation and boost parameters of the Lorentz group. Define

$$\widehat{\theta} = -i\sigma_2\theta^* = \begin{pmatrix} -\theta_2^* \\ \theta_1^* \end{pmatrix} \tag{153}$$

It is easy to show that the $\widehat{\theta}$ transforms as a right-handed spinor, so that

$$\widehat{\theta}' = L^{\dagger-1}\widehat{\theta} \tag{154}$$

Hence

$$\widehat{\theta}^\dagger\theta' = \widehat{\theta}^\dagger L^{-1}L\theta = \widehat{\theta}^\dagger\theta \tag{155}$$

and

$$\widehat{\theta}^\dagger\theta = \begin{pmatrix}-\theta_2 & \theta_1\end{pmatrix}\begin{pmatrix}\theta_1\\ \theta_2\end{pmatrix} = [\theta_1,\theta_2] = 2\theta_1\theta_2 \tag{156}$$

is invariant under Lorentz transformations.

By introducing a metric in spinor space, $\epsilon^{\alpha\beta}$

$$\epsilon^{\alpha\beta} = (i\sigma_2)^{\alpha\beta} \tag{157}$$

we can also write

$$\widehat{\theta}^\dagger\theta = \theta_\alpha\theta^\alpha = \theta_\alpha\epsilon^{\alpha\beta}\theta_\beta = \theta^T(i\sigma_2)\theta \tag{158}$$

displaying the invariance properties of this quantity. Hence if B_α transforms like a spinor we can write

$$F(\theta_1,\theta_2) = A + B_\alpha\epsilon^{\alpha\beta}\theta_\beta + \frac{1}{2}C\theta_\alpha\epsilon^{\alpha\beta}\theta_\beta \tag{159}$$

so that if A is a scalar, $F(\theta_1,\theta_2)$ is a relativistic scalar. Because we have used a left-handed spinor, F is called chiral superfield when A, B_α and C are functions of space-time coordinate x. In order for F to be a bosonic scalar $B_\alpha(x)$ must also be Grassmann valued.

If the θ_α in relativistic field theories transform like scalars (or vectors) they contradict the spin-statistics theorem and behave like unphysical degrees of freedom that must disappear in the expressions of physical observables. In that case they are called ghost fields and are used as auxiliary quantities to cancel the effects of unphysical bosonic degrees of freedom in order to give positive energies and positive probabilities.

Acknowledgments

The development of the ideas presented here is due in a large part to stimulating discussions with our colleagues and friends including Vladimir Akulov, Francesco Iachello, Ramzi Khuri, Carlos Moreno and Pierre Ramond. One of us (SC) would like to thank Professor Cestmir Burdik for invitation to Prague, Czech Republic and for his kind hospitality.

References

[1] H. Grassmann, *Ausdehnunglehre* (1844).
[2] W.K. Clifford, *London Math. Soc. Proc. (1876).*
[3] W.K. Clifford, *American J. Math.* (1878).
[4] S. Catto and F. Gürsey, *Nuovo Cimento A86, 201 (1985).*
[5] S. Catto and F. Gürsey, *Nuovo Cimento A99, 685 (1988).*
[6] S. Catto and Y. Choun, *Acta Polytechnica 51, 77(2011).*
[7] S. Catto, H.Y. Cheung and F. Gürsey, *Mod. Phys. Lett. A6, 3485 (1992).*
[8] P. Jordan and E.P. Wigner, *Z. f. Phys.*47, (631) 1928.
[9] P. Jordan, *Z. f. Phys.*80, 285(1933).
[10] P.A.M. Dirac, *Proc. Roy. Soc. London* A117, 610 (1928).
[11] P.A.M. Dirac, *Pro. Roy. Irish Acad.*50, 261 (1944).
[12] R. Bott, *Ann. of Math. 70, 313 (1959).*
[13] C. Chevalley and R.D. Schafer, *Proc. Natl. Acad. Sci. U.S., 36, 137 (1950).*
[14] R.D. Schafer, *"An Introduction to Non-associative Algebras." (Academic Press, New York, 1966).*
[15] J.R. Faulkner, *Memoirs of the Am. Math. Soc. No.104 (Providence, RI 1970).*

XXth International Conference on Integrable Systems and Quantum Symmetries (ISQS-20) IOP Publishing
Journal of Physics: Conference Series **411** (2013) 012010 doi:10.1088/1742-6596/411/1/012010

Matrix generalizations of integrable systems with Lax integro-differential representations

O. Chvartatskyi[1], Yu. Sydorenko[1]
[1]Department of Mechanics and Mathematics, Ivan Franko National University of Lviv, 1, Universytetska str., Lviv, 79000, Ukraine
E-mail: alex.chvartatskyy@gmail.com, y_sydorenko@franko.lviv.ua

Abstract.
We present (2+1)-dimensional generalizations of the k-constrained Kadomtsev-Petviashvili (k-cKP) hierarchy and corresponding matrix Lax representations that consist of two integro-differential operators. Additional reductions imposed on the Lax pairs lead to matrix generalizations of Davey-Stewartson systems (DS-I,DS-II,DS-III) and (2+1)-dimensional extensions of the modified Korteweg-de Vries and the Nizhnik equation. We also present an integro-differential Lax pair for a matrix version of a (2+1)-dimensional extension of the Chen-Lee-Liu equation.

1. Introduction

In the modern theory of nonlinear integrable systems, algebraic methods play an important role (see the survey in [1]). Among them are the Zakharov-Shabat dressing method [2–4], Marchenko's method [5], and the approach based on Darboux-Crum-Matveev transformations [6, 7]. Algebraic methods allow to omit analytical difficulties that arise in the investigation of corresponding direct and inverse scattering problems for nonlinear equations. A significant contribution to such methods has also been made by the Kioto group [8–12]. In particular, they investigated scalar and matrix hierarchies for nonlinear integrable systems of Kadomtsev-Petviashvili type (KP hierarchy).

The KP hierarchy is of fundamental importance in the theory of integrable systems and shows up in various ways in mathematical physics. Several extensions and generalizations of it have been obtained. For example, the multi-component KP hierarchy contains several physically relevant nonlinear integrable systems, including the Davey-Stewartson equation, the two-dimensional Toda lattice and the three-wave resonant interaction system. There are several equivalent formulations of this hierarchy: matrix pseudo-differential operator (Sato) formulation, τ-function approach via matrix Hirota bilinear identities, multi-component free fermion formulation. Another kind of generalization is the so-called "KP equation with self-consistent sources" (KPSCS), discovered by Melnikov [13–17]. In [18–22], k-symmetry constraints of the KP hierarchy were investigated, which have connections with KPSCS. The resulting k-constrained KP (k-cKP) hierarchy contains physically relevant systems like the nonlinear Schrödinger equation, the Yajima-Oikawa system, a generalization of the Boussinesq equation, and the Melnikov system. Multi-component generalizations of the k-cKP hierarchy were considered in [23]. In papers [24–26] the differential type of the gauge transformation operator was applied to the constrained KP hierarchy at first. A modified k-constrained KP (k-cmKP) hierarchy was proposed in [21,27,28]. It contains, for example, the vector Chen-Lee-Liu and the modified KdV (mKdV) equation. Multi-component versions of the Kundu-Eckhaus and Gerdjikov-Ivanov equations were also obtained in [27], via gauge transformations of the k-cKP, respectively the k-cmKP hierarchy.

Moreover, in [29,30], (2+1)-dimensional extensions of the k-cKP hierarchy were introduced. In [31,32], exact solutions for some representatives of the (2+1)-dimensional k-cKP hierarchy were obtained by dressing binary transformations. (2+1)-dimensional extensions of k-cKP and k-cmKP and their dressings with the help of differential transformations were investigated in [33,34]. The (2+1)-dimensional k-cKP hierarchy in particular contains the DS-III system and a (2+1)-dimensional extension of the mKdV equation. A corresponding Lax representation of the (2+1)-dimensional k-cKP hierarchy consists of one differential and one integro-differential operator. Our aim was to generalize Lax representations of the (2+1)-dimensional k-cKP hierarchy to the case of two integro-differential operators, in order to obtain Lax representations for matrix generalizations of Davey-Stewartson systems DS-I, DS-II, DS-III, and their higher order counterparts. We also present a Lax representation, with two integro-differential operators, for a (2+1)-dimensional generalization of the Chen-Lee-Liu equation, which has been obtained in [28].

This work is organized as follows. In Section 2 we present a short survey of results on constraints for KP hierarchies and their (2+1)-dimensional generalizations. In Sections 3 and 4 we consider integro-differential Lax representations that generalize corresponding representations for the (2+1)-dimensional k-cKP hierarchy. As a result of additional reductions, we obtain matrix generalizations of Davey-Stewartson and (2+1)-dimensional mKdV equations that generalize corresponding systems in the (2+1)-dimensional k-cKP hierarchy. In Section 5, by application of a gauge transformation, we obtain a Lax representation for a (2+1)-dimensional matrix extension of the Chen-Lee-Liu equation. In the final section, we discuss the obtained results and mention problems for further investigations.

2. k-constrained KP hierarchy and its extensions

To make this paper somewhat self-contained, we briefly introduce the KP hierarchy [1], its k-symmetry constraints (k-cKP hierarchy), and the extension of the k-cKP hierarchy to the (2+1)-dimensional case [29,30]. A Lax representation of the KP hierarchy is given by

$$L_{t_n} = [B_n, L], \qquad n \geq 1, \tag{1}$$

where $L = D + U_1 D^{-1} + U_2 D^{-2} + \ldots$ is a scalar pseudodifferential operator, $t_1 := x$, $D := \frac{\partial}{\partial x}$, and $B_n := (L^n)_+ := (L^n)_{\geq 0} = D^n + \sum_{i=0}^{n-2} u_i D^i$ is the differential operator part of L^n. The consistency condition (zero-curvature equations), arising from the commutativity of the flows, are

$$B_{n,t_k} - B_{k,t_n} + [B_n, B_k] = 0. \tag{2}$$

Let B_n^τ denote the formal transpose of B_n, i.e. $B_n^\tau := (-1)^n D^n + \sum_{i=0}^{n-2} (-1)^i D^i u_i^\top$, where $^\top$ denotes the matrix transpose. We use curly brackets to denote the action of an operator on a function whereas, for example, $B_n\, q$ means composition of the operator B_n and the operator of multiplication by the function q. The k-cKP hierarchy [18–22] is given by

$$L_{t_n} = [B_n, L], \quad q_{t_n} = B_n\{q\}, \quad -r_{t_n} = B_n^\tau\{r\}, \quad n = 2, 3, \ldots, \tag{3}$$

with the k-symmetry reduction

$$L_k := L^k = B_k + q D^{-1} r. \tag{4}$$

It admits the Lax representation (here $k \in \mathbb{N}$ is fixed)

$$[B_k + qD^{-1}r, \partial_{t_n} - B_n] = 0. \tag{5}$$

Below we will also use the formal adjoint $B_n^* := \bar{B}_n^{\tau} = (-1)^n D^n + \sum_{i=0}^{n-2}(-1)^i D^i u_i^*$ of B_n, where * denotes Hermitian conjugation (complex conjugation and transpose).

In [23], multi-component (vector) generalizations of the k-cKP hierarchy were introduced,

$$\begin{gathered} L^k = B_k + \sum_{i=1}^{l}\sum_{j=1}^{l} q_i m_{ij} D^{-1} r_j = B_k + \mathbf{q}\mathcal{M}_0 D^{-1}\mathbf{r}^{\top}, \\ \mathbf{q}_{t_k} = B_k\{\mathbf{q}\},\ \mathbf{r}_{t_k} = -B_k^{\tau}\{\mathbf{r}\}, \end{gathered} \tag{6}$$

where $\mathbf{q} = (q_1, \ldots, q_l)$ and $\mathbf{r} = (r_1, \ldots, r_l)$ are vector functions, $\mathcal{M}_0 = (m_{ij})_{i,j=1}^{l}$ is a constant $l \times l$ matrix. A corresponding Lax representation is given by

$$[L_k, M_n] = 0, \quad L_k = B_k + \mathbf{q}\mathcal{M}_0 D^{-1}\mathbf{r}^{\top}, \quad M_n = \partial_{t_n} - B_n. \tag{7}$$

For $k = 1$, this is a multi-component generalization of the AKNS hierarchy. For $k = 2$ and $k = 3$, one obtains vector generalizations of the Yajima-Oikawa and Melnikov [14,15] hierarchies, respectively.

In [21, 27, 28], a k-constrained modified KP (k-cmKP) hierarchy was introduced and investigated. Its Lax representation has the form

$$[\tilde{B}_k + \mathbf{q}\mathcal{M}_0 D^{-1}\mathbf{r}^{\top} D, \partial_{t_n} - \tilde{B}_n] = 0, \tag{8}$$

where $\tilde{B}_k = D^k + \sum_{j=1}^{k-1} w_{kj} D^j$. For $k = 1, 2, 3$, this leads to vector generalizations of the Chen-Lee-Liu, the modified multi-component Yajima-Oikawa and Melnikov hierarchies.

An essential extension of the k-cKP hierarchy is its (2+1)-dimensional generalization [29,30], given by

$$\begin{aligned} &[L_k, M_n] = 0, \quad L_k = \beta_k \partial_{\tau_k} - B_k - \mathbf{q}\mathcal{M}_0 D^{-1}\mathbf{r}^{\top}, \quad M_n = \alpha_n \partial_{t_n} - A_n, \\ &B_k = D^k + \sum_{j=0}^{k-2} u_j D^j, \quad A_n = D^n + \sum_{i=0}^{n-2} v_i D^i, \\ &u_j = u_j(x, \tau_k, t_n),\ v_i = v_i(x, \tau_k, t_n), \quad \alpha_n, \beta_k \in \mathbb{C}, \end{aligned} \tag{9}$$

where u_j and v_i are scalar functions, $\mathbf{q}$ and $\mathbf{r}$ are l-component vector-functions. An equivalent system is

$$\begin{aligned} &\alpha_n B_{k,t_n} = \beta_k A_{n,\tau_k} + [A_n, B_k] + \left([A_n, \mathbf{q}\mathcal{M}_0 D^{-1}\mathbf{r}^{\top}]\right)_{\geq 0}, \\ &\alpha_n \mathbf{q}_{t_n} = A_n\{\mathbf{q}\}, \quad \alpha_n \mathbf{r}_{t_n} = -A_n^{\tau}\{\mathbf{r}\}. \end{aligned}$$

We list some members of this (2+1)-dimensional generalization of the k-cKP hierarchy:

(1) $k = 1$, $n = 2$. Then (9) has the form

$$[L_1, M_2] = 0, \quad L_1 = \beta_1 \partial_{\tau_1} - D - \mathbf{q}\mathcal{M}_0 D^{-1}\mathbf{r}^{\top}, \quad M_2 = \alpha_2 \partial_{t_2} - D^2 - v_0, \tag{10}$$

and it is equivalent to the following system,

$$\begin{aligned} &\alpha_2 \mathbf{q}_{t_2} = \mathbf{q}_{xx} + v_0 \mathbf{q}, \quad \alpha_2 \mathbf{r}_{t_2} = -\mathbf{r}_{xx} - v_0 \mathbf{r}, \\ &\beta_1 v_{0,\tau_1} = v_{0,x} - 2(\mathbf{q}\mathcal{M}_0 \mathbf{r}^{\top})_x. \end{aligned} \tag{11}$$

After the reduction $\beta_1 \in \mathbb{R}$, $\alpha_2 \in i\mathbb{R}$, $\mathbf{r} = \bar{\mathbf{q}}$, $\mathcal{M}_0 = \mathcal{M}_0^*$, $v_0 = \bar{v}_0$, the operators L_1 and M_2 in (10) are skew-Hermitian and Hermitian, respectively, and (11) becomes the DS-III system

$$\alpha_2 \mathbf{q}_{t_2} = \mathbf{q}_{xx} + v_0 \mathbf{q}, \quad \beta_1 v_{0,\tau_1} = v_{0,x} - 2(\mathbf{q}\mathcal{M}_0 \mathbf{q}^*)_x. \tag{12}$$

(2) $k=1$, $n=3$. Now (9) becomes

$$[L_1, M_3]=0,\quad L_1=\beta_1\partial_{\tau_1}-D-\mathbf{q}\mathcal{M}_0D^{-1}\mathbf{r}^\top,\quad M_3=\alpha_3\partial_{t_3}-D^3-v_1D-v_0. \tag{13}$$

After the additional reduction $\alpha_3, \beta_1\in\mathbb{R}$, $\mathcal{M}_0=\mathcal{M}_0^*$, $v_1=\bar{v}_1$, $\bar{v}_0+v_0=v_{1x}$, the operators L_1, M_3 in (13) are skew-Hermitian, and the Lax equation (13) is equivalent to the following (2+1)-dimensional generalization of the mKdV system,

$$\begin{aligned}\alpha_3\mathbf{q}_{t_3}&=\mathbf{q}_{xxx}+v_1\mathbf{q}_x+v_0\mathbf{q},\\ \beta_1v_{0,\tau_1}&=v_{0,x}-3(\mathbf{q}_x\mathcal{M}_0\mathbf{q}^*)_x,\ \beta_1v_{1,\tau_1}=v_{1,x}-3(\mathbf{q}\mathcal{M}_0\mathbf{q}^*)_x.\end{aligned} \tag{14}$$

This system admits the real version ($\mathcal{M}_0^\top=\mathcal{M}_0^*$, $\mathbf{q}^*=\mathbf{q}^\top$)

$$\alpha_3\mathbf{q}_{t_3}=\mathbf{q}_{xxx}+v_1\mathbf{q}_x+\tfrac{1}{2}v_{1,x}\mathbf{q},\quad \beta_1v_{1,\tau_1}=v_{1,x}-3(\mathbf{q}\mathcal{M}_0\mathbf{q}^\top)_x. \tag{15}$$

(3) $k=2$, $n=2$. (9) takes the form

$$[L_2, M_2]=0,\quad L_2=\beta_2\partial_{\tau_2}-D^2-u_0-\mathbf{q}\mathcal{M}_0D^{-1}\mathbf{r}^\top,\quad M_2=\alpha_2\partial_{t_2}-D^2-u_0. \tag{16}$$

Under the reduction $\alpha_2, \beta_2\in i\mathbb{R}$, $\mathbf{r}=\bar{\mathbf{q}}$, $\mathcal{M}_0=-\mathcal{M}_0^*$, $u_0=\bar{u}_0:=u$, the operators L_2 and M_2 in (16) become Hermitian and (16) takes the form

$$\alpha_2\mathbf{q}_{t_2}=\mathbf{q}_{xx}+u\mathbf{q},\qquad \alpha_2u_{t_2}=\beta_2u_{\tau_2}+2(\mathbf{q}\mathcal{M}_0\mathbf{q}^*)_x, \tag{17}$$

which is a (2+1)-dimensional vector generalization of the Yajima-Oikawa system.

(4) $k=2$, $n=3$. Now (9) becomes

$$\begin{aligned}L_2&=\beta_2\partial_{\tau_2}-D^2-2u_0-\mathbf{q}\mathcal{M}_0D^{-1}\mathbf{r}^\top,\\ M_3&=\alpha_3\partial_{t_3}-D^3-3u_0D-\tfrac{3}{2}\left(u_{0,x}+\beta_2D^{-1}\{u_{0,\tau_2}\}+\mathbf{q}\mathcal{M}_0\mathbf{r}^\top\right).\end{aligned} \tag{18}$$

With the additional reduction $\beta_2\in i\mathbb{R}$, $\alpha_3\in\mathbb{R}$, $u_0=\bar{u}_0:=u$, $\mathcal{M}_0=-\mathcal{M}_0^*$ and $\mathbf{r}=\bar{\mathbf{q}}$, this is equivalent to the following generalization of the Melnikov system [32],

$$\begin{aligned}\alpha_3\mathbf{q}_{t_3}&=\mathbf{q}_{xxx}+3u\mathbf{q}_x+\frac{3}{2}\left(u_x+\beta_2D^{-1}\{u_{\tau_2}\}+\mathbf{q}\mathcal{M}_0\mathbf{q}^*\right)\mathbf{q},\\ \alpha_3u_{t_3}&-\frac{1}{4}u_{xxx}-3uu_x+\frac{3}{4}\left(\mathbf{q}\mathcal{M}_0\mathbf{q}_x^*-\mathbf{q}_x\mathcal{M}_0\mathbf{q}^*\right)_x-\frac{3}{4}\beta_2\left(\mathbf{q}\mathcal{M}_0\mathbf{q}^*\right)_{\tau_2}\Big]_x=\frac{3}{4}\beta_2^2u_{\tau_2\tau_2}.\end{aligned} \tag{19}$$

Remark 2.1 *(12) and (17) are related by a linear change of the independent variables.*

Thus, for $k=1$ we have the DS-III hierarchy (its first members are DS-III ($k=1, n=2$) and a special (2+1)-dimensional extension of mKdV ($k=1, n=3$), see (12) and (14). For $k=2$, $k=3$, we have (2+1)-dimensional generalizations of the Yajima-Oikawa (in particular, it contains (17) and (19)) and the Melnikov hierarchy [15], respectively.

3. Lax representations for matrix generalizations of Davey-Stewartson systems

In this section we introduce an essential extension of the Lax pair (10) for the (2+1)-dimensional extended KP hierarchy to the case of two integro-differential operators. It leads to DS-I and DS-II systems. It is known that the Davey-Stewartson (DS-I) system is connected with the non-stationary Dirac operator (see [35]). By using a matrix differential representation for the DS system, involving the non-stationary Dirac operator, and the representation for DS-III in the (2+1)-dimensional k-cKP hierarchy, we obtain the following Lax integro-differential Lax pair,

$$L_1=\partial_y-\mathbf{q}\mathcal{M}_0D^{-1}\mathbf{r}^\top,$$

$$M_2 = \alpha_2\partial_{t_2} - c_1 D^2 - c_2\partial_y^2 + 2c_1 S_1 + 2c_2\mathbf{q}\mathcal{M}_0 D^{-1}\mathbf{r}_y^\top + 2c_2\mathbf{q}\mathcal{M}_0 D^{-1}\mathbf{r}^\top\partial_y, \tag{20}$$

where $\alpha_2, c_1, c_2 \in \mathbb{C}$. $\mathbf{q} = \mathbf{q}(x, y, t_2)$, $\mathbf{r} = \mathbf{r}(x, y, t_2)$ and $S_1 = S_1(x, y, t_2)$ are matrix functions with dimensions $N \times M$ and $N \times N$, respectively. $\mathcal{M}_0$ is a constant $M \times M$ matrix. The Lax equation $[L_1, M_2] = 0$ is equivalent to the following system,

$$\begin{aligned} \alpha_2\mathbf{q}_{t_2} &= c_1\mathbf{q}_{xx} + c_2\mathbf{q}_{yy} - 2c_1 S_1\mathbf{q} - 2c_2\mathbf{q}\mathcal{M}_0 S_2, \\ -\alpha_2\mathbf{r}_{t_2}^\top &= c_1\mathbf{r}_{xx}^\top + c_2\mathbf{r}_{yy}^\top - 2c_1\mathbf{r}^\top S_1 - 2c_2 S_2\mathcal{M}_0\mathbf{r}^\top, \\ S_{1y} &= (\mathbf{q}\mathcal{M}_0\mathbf{r}^\top)_x, \; S_{2x} = (\mathbf{r}^\top\mathbf{q})_y. \end{aligned} \tag{21}$$

After the reduction $c_1, c_2 \in \mathbb{R}$, $\alpha_2 \in i\mathbb{R}$; $\mathbf{r}^\top = \mathbf{q}^*$, $\mathcal{M}_0 = \mathcal{M}_0^*$, the operators L_1 and M_2 are skew-Hermitian and Hermitian, respectively, and (21) takes the form

$$\begin{aligned} \alpha_2\mathbf{q}_{t_2} &= c_1\mathbf{q}_{xx} + c_2\mathbf{q}_{yy} - 2c_1 S_1\mathbf{q} - 2c_2\mathbf{q}\mathcal{M}_0 S_2, \\ S_{1y} &= (\mathbf{q}\mathcal{M}_0\mathbf{q}^*)_x, \; S_{2x} = (\mathbf{q}^*\mathbf{q})_y. \end{aligned} \tag{22}$$

This has the following two interesting subcases:

(1) $c_2 = 0$. Then we have

$$\alpha_2\mathbf{q}_{t_2} = c_1\mathbf{q}_{xx} - 2c_1 S_1\mathbf{q}, \; S_{1y} = (\mathbf{q}\mathcal{M}_0\mathbf{q}^*)_x. \tag{23}$$

(2) $c_1 = 0$. Then (22) takes the form

$$\alpha_2\mathbf{q}_{t_2} = c_2\mathbf{q}_{yy} - 2c_2\mathbf{q}\mathcal{M}_0 S_2, \quad S_{2x} = (\mathbf{q}^*\mathbf{q})_y. \tag{24}$$

The systems (23) and (24) are two different matrix generalizations of the Davey-Stewartson equation (DS-III) [2, 30, 37]. In the vector case ($N = 1$), (23) has been obtained in [28] as a member of the DS-III hierarchy.

Remark 3.1 *If $N = 1$, the change of variables $\tilde{x} = x + y$, $\tau_1 = \beta_1 y$, $\tilde{\mathbf{q}}(\tilde{x}, \tau_1) = \mathbf{q}(x, y)$, $v_0(\tilde{x}, \tau_1) = -2S_1(x, y)$ maps (23) to the DS-III equation (12) (which we obtained from the (2+1)-dimensional k-cKP hierarchy).*

Let us consider (22) in the case where $u := \mathbf{q}$ and $\mathcal{M}_0 := \mu$ are scalars. Then (22) becomes

$$\alpha_2 u_{t_2} = c_1 u_{xx} + c_2 u_{yy} - 2c_1 S_1 u - 2\mu c_2 S_2 u, \; S_{1y} = \mu(|u|^2)_x, \; S_{2x} = (|u|^2)_y. \tag{25}$$

Setting $c_1 = c_2 = 1$ and $\mu = 1$, as a consequence of (25) we obtain

$$\alpha_2 u_{t_2} = u_{xx} + u_{yy} - 2Su, \; S_{xy} = (|u|^2)_{xx} + (|u|^2)_{yy}, \tag{26}$$

where $S = S_1 + S_2$. This is the well-known Davey-Stewartson system (DS-I) and (22) is therefore a matrix generalization.

Now we present integro-differential Lax pairs for generalizations of the second Davey-Stewartson equation (DS-II). By replacing the real variables x and y in the operators L_1, M_2 in (20) by a complex variable $z = x + iy$ and its conjugate $\bar{z}$, respectively, we obtain the following pair,

$$\begin{aligned} L_1 &= \partial_{\bar{z}} - \mathbf{q}\mathcal{M}_0 D_z^{-1}\mathbf{r}^\top, \\ M_2 &= \alpha_2\partial_{t_2} - c_1 D_{zz}^2 - c_2\partial_{\bar{z}\bar{z}}^2 + 2c_1 S_1 + 2c_2\mathbf{q}\mathcal{M}_0 D_z^{-1}\mathbf{r}_{\bar{z}}^\top + 2c_2\mathbf{q}\mathcal{M}_0 D_z^{-1}\mathbf{r}^\top\partial_{\bar{z}}, \end{aligned} \tag{27}$$

where $\mathbf{q} = \mathbf{q}(z, \bar{z}, t_2)$, $\mathbf{r} = \mathbf{r}(z, \bar{z}, t_2)$ and $S_1 = S_1(z, \bar{z}, t_2)$ are matrix functions with dimensions $N \times M$ and $N \times N$, respectively. $\mathcal{M}_0$ is a constant $M \times M$ matrix. The Lax equation $[L_1, M_2] = 0$ is equivalent to the system

$$\begin{aligned} \alpha_2 \mathbf{q}_{t_2} &= c_1 \mathbf{q}_{zz} + c_2 \mathbf{q}_{\bar{z}\bar{z}} - 2c_1 S_1 \mathbf{q} - 2c_2 \mathbf{q} \mathcal{M}_0 S_2, \\ -\alpha_2 \mathbf{r}^\top_{t_2} &= c_1 \mathbf{r}^\top_{zz} + c_2 \mathbf{r}^\top_{\bar{z}\bar{z}} - 2c_1 \mathbf{r}^\top S_1 - 2c_2 S_2 \mathcal{M}_0 \mathbf{r}^\top, \\ S_{1\bar{z}} &= (\mathbf{q}\mathcal{M}_0 \mathbf{r}^\top)_z, \quad S_{2z} = (\mathbf{r}^\top \mathbf{q})_{\bar{z}}. \end{aligned} \tag{28}$$

In terms of the real variables t_2, x and y, after setting $c_1 = c_2 = 1$, the system (28) becomes

$$\begin{aligned} 2\alpha_2 \mathbf{q}_{t_2} &= \mathbf{q}_{xx} - \mathbf{q}_{yy} - 4S_1 \mathbf{q} - 4\mathbf{q}\mathcal{M}_0 S_2, \\ -2\alpha_2 \mathbf{r}^\top_{t_2} &= \mathbf{r}^\top_{xx} - \mathbf{r}^\top_{yy} - 4\mathbf{r}^\top S_1 - 4S_2 \mathcal{M}_0 \mathbf{r}^\top, \\ S_{1x} + iS_{1y} &= (\mathbf{q}\mathcal{M}_0\mathbf{r}^\top)_x - i(\mathbf{q}\mathcal{M}_0\mathbf{r}^\top)_y, \quad S_{2x} - iS_{2y} = (\mathbf{r}^\top\mathbf{q})_x + i(\mathbf{r}^\top\mathbf{q})_y. \end{aligned} \tag{29}$$

If $N = M$ (so that $\mathbf{q}$ and $\mathbf{r}$ are square matrices), and $\alpha_2 \in i\mathbb{R}$, (29) admits the reduction $\mathcal{M}_0\mathbf{r}^\top = \bar{\mathbf{q}}$, $S_1 = \bar{S}_2$, and then takes the following form,

$$\begin{aligned} 2\alpha_2 \mathbf{q}_{t_2} &= \mathbf{q}_{xx} - \mathbf{q}_{yy} - 4S_1\mathbf{q} - 4\mathbf{q}\bar{S}_1, \\ S_{1x} + iS_{1y} &= (\mathbf{q}\bar{\mathbf{q}})_x - i(\mathbf{q}\bar{\mathbf{q}})_y. \end{aligned} \tag{30}$$

In the scalar case ($N = M = 1$), writing $u = \mathbf{q}$, we obtain the following consequence of (30),

$$2\alpha_2 u_{t_2} = u_{xx} - u_{yy} - 8\hat{S}u, \quad \hat{S}_{xx} + \hat{S}_{yy} = |u|^2_{xx} - |u|^2_{yy}, \tag{31}$$

where $\hat{S} = \mathrm{Re}(S_1)$. This is the second Davey-Stewartson system (DS-II).

4. Integro-differential Lax representations for a (2+1)-dimensional matrix generalization of the mKdV equation

In this section we generalize the Lax representation (13) for the (2+1)-dimensional mKdV equation to the case of two integro-differential operators. More precisely, we consider an extension of the operator M_3 in (13). As in the case of the operator M_2 in (20), by using a differential representation for the (2+1)-dimensional mKdV equation that involves the non-stationary Dirac operator [35], we obtain the following integro-differential Lax pair,

$$\begin{aligned} L_1 &= \partial_y - \mathbf{q}\mathcal{M}_0 D^{-1}\mathbf{r}^\top, \\ M_3 &= \alpha_3\partial_{t_3} + c_1 D^3 - c_2\partial_y^3 - 3c_1 v_1 D - 3c_1 v_3 + 3c_2\mathbf{q}_y\mathcal{M}_0 D^{-1}\mathbf{r}^\top_y \\ &\quad -3c_2\mathbf{q}\mathcal{M}_0\partial_y D^{-1}\mathbf{r}^\top\mathbf{q}\mathcal{M}_0 D^{-1}\mathbf{r}^\top + 3c_2\mathbf{q}\mathcal{M}_0 D^{-1}\{\mathbf{r}^\top\mathbf{q}\}_y\mathcal{M}_0 D^{-1}\mathbf{r}^\top \\ &\quad +3c_2\partial_y\mathbf{q}\mathcal{M}_0 D^{-1}\mathbf{r}^\top\partial_y, \end{aligned} \tag{32}$$

where $\mathbf{q} = \mathbf{q}(x, y, t_3)$, $\mathbf{r} = \mathbf{r}(x, y, t_3)$ and $v_1 = v_1(x, y, t_3)$, $v_3 = v_3(x, y, t_3)$ are $N \times M$, respectively $N \times N$, matrix functions. As in the case of the pair L_1, M_2 in (20), a change of variables shows that in the case $c_2 = 0$ and $N = 1$ the Lax pairs (32) and (13) are equivalent. The Lax equation $[L_1, M_3] = 0$ results in the system

$$\begin{aligned} &\alpha_3\mathbf{q}_{t_3} + c_1\mathbf{q}_{xxx} - c_2\mathbf{q}_{yyy} - 3c_1 v_1\mathbf{q}_x + 3c_2\mathbf{q}_y\mathcal{M}_0 v_2 + 3c_2\mathbf{q}\mathcal{M}_0 v_{2y} - 3c_1 v_3\mathbf{q} - 3c_2\mathbf{q}\mathcal{M}_0 v_4 \\ &-3c_2\mathbf{q}D^{-1}\left\{\mathcal{M}_0\mathbf{r}^\top\mathbf{q}\mathcal{M}_0 v_2 - \mathcal{M}_0 v_2\mathcal{M}_0\mathbf{r}^\top\mathbf{q}\right\} = 0, \\ &\alpha_3\mathbf{r}^\top_{t_3} + c_1\mathbf{r}^\top_{xxx} - c_2\mathbf{r}^\top_{yyy} - 3c_1\mathbf{r}^\top_x v_1 - 3c_1\mathbf{r}^\top v_{1x} + 3c_2 v_2\mathcal{M}_0\mathbf{r}^\top_y + 3c_1\mathbf{r}^\top v_3 + 3c_2 v_4\mathcal{M}_0\mathbf{r}^\top \\ &-3c_2 D^{-1}\left\{v_2\mathcal{M}_0\mathbf{r}^\top\mathbf{q} - \mathbf{r}^\top\mathbf{q}\mathcal{M}_0 v_2\right\}\mathcal{M}_0\mathbf{r}^\top = 0, \\ &v_{1y} = (\mathbf{q}\mathcal{M}_0\mathbf{r}^\top)_x,\ v_{2x} = (\mathbf{r}^\top\mathbf{q})_y,\ v_{3y} = (\mathbf{q}_x\mathcal{M}_0\mathbf{r}^\top)_x + [\mathbf{q}\mathcal{M}_0\mathbf{r}^\top, v_1],\ v_{4x} = (\mathbf{r}^\top_y\mathbf{q})_y. \end{aligned} \tag{33}$$

Let us consider some reductions of this system:

(1) $\alpha_3, c_1, c_2 \in \mathbb{R}$, $\mathbf{r}^\top = \mathbf{q}^*$, $\mathcal{M}_0 = \mathcal{M}_0^*$. The operators L_1 and M_3 are then skew-Hermitian and (33) takes the form

$$\alpha_3 \mathbf{q}_{t_3} + c_1 \mathbf{q}_{xxx} - c_2 \mathbf{q}_{yyy} - 3c_1 v_1 \mathbf{q}_x + 3c_2 \mathbf{q}_y \mathcal{M}_0 v_2 + 3c_2 \mathbf{q} \mathcal{M}_0 v_{2y} - 3c_1 v_3 \mathbf{q} - 3c_2 \mathbf{q} \mathcal{M}_0 v_4$$
$$-3c_2 \mathbf{q} D^{-1} \{\mathcal{M}_0 \mathbf{q}^* \mathbf{q} \mathcal{M}_0 v_2 - \mathcal{M}_0 v_2 \mathcal{M}_0 \mathbf{q}^* \mathbf{q}\} = 0,$$
$$v_{1y} = (\mathbf{q}\mathcal{M}_0 \mathbf{q}^*)_x,\ v_{2x} = (\mathbf{q}^* \mathbf{q})_y,\ v_{3y} = (\mathbf{q}_x \mathcal{M}_0 \mathbf{q}^*)_x + [\mathbf{q}\mathcal{M}_0 \mathbf{q}^*, v_1],\ v_{4x} = (\mathbf{q}_y^* \mathbf{q})_y. \quad (34)$$

In the scalar case ($N = M = 1$), setting $\mathbb{R} \ni \mu := \mathcal{M}_0$, $q(x,y,t_3) := \mathbf{q}(x,y,t_3)$, (34) reads

$$\alpha_3 q_{t_3} + c_1 q_{xxx} - c_2 q_{yyy} - 3c_1 \mu q_x \int |q|_x^2 dy + 3c_2 \mu q_y \int |q|_y^2 dx + 3c_2 \mu q \int (\bar{q} q_y)_y dx$$
$$-3c_1 \mu q \int (q_x q)_x dy = 0. \quad (35)$$

In the real case $\mathbf{q} = \bar{\mathbf{q}}$, (34) becomes

$$\alpha_3 \mathbf{q}_{t_3} + c_1 \mathbf{q}_{xxx} - c_2 \mathbf{q}_{yyy} - 3c_1 v_1 \mathbf{q}_x + 3c_2 \mathbf{q}_y \mathcal{M}_0 v_2 + 3c_2 \mathbf{q} \mathcal{M}_0 v_{2y} - 3c_1 v_3 \mathbf{q} - 3c_2 \mathbf{q} \mathcal{M}_0 v_4$$
$$-3c_2 \mathbf{q} D^{-1} \left\{\mathcal{M}_0 \mathbf{q}^\top \mathbf{q} \mathcal{M}_0 v_2 - \mathcal{M}_0 v_2 \mathcal{M}_0 \mathbf{q}^\top \mathbf{q}\right\} = 0,$$
$$v_{1y} = (\mathbf{q}\mathcal{M}_0 \mathbf{q}^\top)_x,\ v_{2x} = (\mathbf{q}^\top \mathbf{q})_y,\ v_{3y} = (\mathbf{q}_x \mathcal{M}_0 \mathbf{q}^\top)_x + [\mathbf{q}\mathcal{M}_0 \mathbf{q}^\top, v_1],\ v_{4x} = (\mathbf{q}_y^\top \mathbf{q})_y. \quad (36)$$

In the scalar case ($N = M = 1$), writing $\mathcal{M}_0 = \mu$ and $q = q(x,y,t_3) = \mathbf{q}(x,y,t_3)$, after setting $y = x$ and $c_1 - c_2 = 1$, (36) has the form

$$\alpha_3 q_{t_3} + q_{xxx} - 6\mu q^2 q_x = 0, \quad (37)$$

which is the mKdV equation. The systems (34) and (36) are therefore, respectively, complex and real, spatially two-dimensional matrix generalizations of it.

(2) $\mathcal{M}_0 \mathbf{r}^\top = \nu$ with a constant matrix ν. In terms of $u := \mathbf{q}\nu$, (33) takes the form

$$\alpha_3 u_{t_3} + c_1 u_{xxx} - c_2 u_{yyy} - 3c_1 D\left\{\left(\int u_x dy\right) u\right\} + 3c_2 \partial_y \left\{u \left(\int u_y dx\right)\right\}$$
$$-3c_1 \left(\int [u, v_1] dy\right) u - 3c_2 u \left(\int [u, v_2] dx\right) = 0,$$
$$\nu \left(c_1 \int [u, v_1] dy - c_2 \int [v_2, u] dx\right) = 0, \quad v_{1y} = u_x, \quad v_{2x} = u_y. \quad (38)$$

In the scalar case ($N = 1, M = 1$), this reduces to

$$\alpha_3 u_{t_3} + c_1 u_{xxx} - c_2 u_{yyy} - 3c_1 D\left\{\left(\int u_x dy\right) u\right\} + 3c_2 \partial_y \left\{u \left(\int u_y dx\right)\right\} = 0, \quad (39)$$

which is the Nizhnik equation [35]. The system (38) thus generalizes the latter to the matrix case.

5. Integro-differential Lax representation for a spatially two-dimensional matrix generalization of the Chen-Lee-Liu equation

In this section we apply a gauge transformation to the Lax pair (20) in order to obtain an integro-differential Lax pair for the (2+1)-dimensional Chen-Lee-Liu equation. The resulting Lax pair generalizes a corresponding Lax pair obtained from the (2+1)-dimensional k-cmKP hierarchy. Let $f = f(x, y, t_2)$ be an $M \times M$ matrix solution of $L_1\{f\} = 0$, where the operator L_1 has the form given in (20). Consider the following gauge transformation,

$$f^{-1}L_1 f = f^{-1}\Big(\partial_y - \mathbf{q}\mathcal{M}_0 D^{-1}\mathbf{r}^\top\Big) f = \partial_y + f^{-1}\mathbf{q}\mathcal{M}_0 D^{-1}D^{-1}\{\mathbf{r}^\top f\}D := \partial_y - \tilde{\mathbf{q}}\mathcal{M}_0 D^{-1}\tilde{\mathbf{r}}^\top D,$$

where $\tilde{\mathbf{q}} := f^{-1}\mathbf{q}$, $\tilde{\mathbf{r}}^\top := -D^{-1}\{\mathbf{r}^\top f\}$. It is also possible to apply a similar transformation to the operator M_2 in (20). In order to simplify notation, in the following we write $\mathbf{q}, \mathbf{r}$ instead of $\tilde{\mathbf{q}}, \tilde{\mathbf{r}}$. As a result of the gauge transformation, we obtain a Lax pair of the form

$$L_1 = \partial_y - \mathbf{q}\mathcal{M}_0 D^{-1}\mathbf{r}^\top D, \tag{40}$$

$$M_2 = \alpha_2\partial_{t_2} - c_1 D^2 - c_2\partial_y^2 + 2c_1 S_1 D + 2c_2\mathbf{q}\mathcal{M}_0 D^{-1}\partial_y \mathbf{r}^\top D, \tag{41}$$

where $\mathbf{q} = \mathbf{q}(x, y, t_2)$, $\mathbf{r} = \mathbf{r}(x, y, t_2)$ and $S_1 = S_1(x, y, t_2)$ are $N \times M$, respectively $N \times N$, matrix functions. $\mathcal{M}_0$ is a constant $M \times M$ matrix. The condition $[L_1, M_2] = 0$ is then equivalent to the following system,

$$\begin{aligned} &\alpha_2\mathbf{q}_{t_2} - c_1\mathbf{q}_{xx} - c_2\mathbf{q}_{yy} + 2c_1 S_1\mathbf{q}_x - 2c_2\mathbf{q}\mathcal{M}_0 S_2 + 2c_2\mathbf{q}\mathcal{M}_0(\mathbf{r}^\top\mathbf{q})_y = 0,\\ &\big(\alpha_2\mathbf{r}^\top_{t_2} + c_1\mathbf{r}^\top_{xx} + c_2\mathbf{r}^\top_{yy} + 2c_1\mathbf{r}^\top_x S_1 + 2c_2 S_2\mathcal{M}_0\mathbf{r}^\top\big)_x = 0,\\ &S_{1y} = (\mathbf{q}\mathcal{M}_0\mathbf{r}^\top)_x + [\mathbf{q}\mathcal{M}_0\mathbf{r}^\top, S_1],\ \ S_{2x} = (\mathbf{r}^\top_x\mathbf{q})_y. \end{aligned} \tag{42}$$

Let us consider some of its reductions:

(1) The system (42) contains three different (2+1)-dimensional matrix generalizations of the Chen-Lee-Liu system: a) $c_1 = 0$, b) $c_2 = 0$, c) $c_1 \neq 0$ and $c_2 \neq 0$.
In case b), M_2 becomes a differential operator. With the restriction $N = 1$, this case appeared in [28].

(2) $\alpha_2 \in i\mathbb{R}$, $c_1, c_2 \in \mathbb{R}$, $\mathcal{M}_0 = -\mathcal{M}_0^*$, $\mathbf{r}^\top = \mathbf{q}^*$, $S_1 = S_1^*$. Then L_1 and M_2, given by (40) and (41), are D-skew-Hermitian ($L_1^* = -DL_1D^{-1}$) and D-Hermitian ($M_2^* = DM_2D^{-1}$), respectively. (42) has the following form,

$$\begin{aligned} &\alpha_2\mathbf{q}_{t_2} - c_1\mathbf{q}_{xx} - c_2\mathbf{q}_{yy} + 2c_1 S_1\mathbf{q}_x - 2c_2\mathbf{q}\mathcal{M}_0 S_2 + 2c_2\mathbf{q}\mathcal{M}_0(\mathbf{q}^*\mathbf{q})_y = 0,\\ &S_{1y} = (\mathbf{q}\mathcal{M}_0\mathbf{q}^*)_x + [\mathbf{q}\mathcal{M}_0\mathbf{q}^*, S_1],\ \ S_{2x} = (\mathbf{q}^*_x\mathbf{q})_y. \end{aligned} \tag{43}$$

In the scalar case ($N = M = 1$), setting $c_1 = 1$, $c_2 = 0$, $y = x$, and writing $\mu = \mathcal{M}_0$, $q = q(x, y, t_2) = \mathbf{q}(x, y, t_2)$, (43) reduces to the Chen-Lee-Liu equation [38]

$$\alpha_2 q_{t_2} - q_{xx} + 2\mu|q|^2 q_x = 0.$$

(3) $\mathcal{M}_0\mathbf{r}^\top = \nu$ with a constant matrix ν. Then the operators (40) and (41) are purely differential and, in terms of $u := \mathbf{q}\nu$, (42) reads

$$\alpha_2 u_{t_2} - c_1 u_{xx} - c_2 u_{yy} + 2c_1 S_1 u_x + 2c_2 u u_y = 0, \qquad S_{1y} = u_x + [u, S_1]. \tag{44}$$

This is a spatially two-dimensional matrix generalization of the Burgers equation. A generalization of (44) to an arbitrary number of spatial dimensions has been considered in [39].

It is possible to apply a similar gauge transformation to the pair of operators (32). Corresponding Lax representations that lead to matrix versions of spatially two-dimensional higher Chen-Lee-Liu equations were investigated in [39].

6. Conclusions

In this paper we considered several members of a (2+1)-dimensional generalization of the k-cKP hierarchy (9). Originally, this substantial generalization of the k-cKP hierarchy had been proposed in [29, 30]. For some members of this hierarchy (e.g. (17) and (19)), solutions were obtained via the binary transformation dressing method [31, 32]. The (2+1)-dimensional extension of the k-cKP hierarchy has been rediscovered more recently in [33], also see [34, 40] for further investigations. The authors of [33] and [34] considered the (2+1)-dimensional k-cKP and (2+1)-dimensional k-cmKP hierarchies with two operators of the Lax pair having different orders of differentiation with respect to x. This excludes, for example, (2+1)-dimensional vector generalizations of the Yajima-Oikawa (17) and Drinfeld-Sokolov-Wilson equation [41–43]:

$$\mathbf{q}_{t_3} = \mathbf{q}_{xxx} + 3u\mathbf{q}_x + \frac{3}{2}u_x\mathbf{q}, \; u_{t_3} = u_{\tau_3} + 3(\mathbf{q}\mathcal{M}_0\mathbf{q}^\top)_x. \tag{45}$$

It is obtained from $[L_3, M_3] = 0$ with

$$L_3 = \partial_{\tau_3} - D^3 - 3uD - \frac{3}{2}u_x - \mathbf{q}\mathcal{M}_0 D^{-1}\mathbf{q}^\top, \quad M_3 = \partial_{t_3} - D^3 - 3uD - \frac{3}{2}u_x. \tag{46}$$

Remark 6.1 *Analogously to Remark 2.1, it can be shown that the (2+1)-dimensional vector generalization of the Drinfeld-Sokolov-Wilson equation (45) is equivalent to the real version of the (2+1)-dimensional multi-component mKdV equation (15) via a linear change of independent variables.*

The aim of our work was also to generalize Lax representations for members of the (2+1)-dimensional k-cKP hierarchy [29, 30, 33, 34], in order to obtain integrable equations that do not belong to the (2+1)-dimensional k-cKP hierarchy. In particular, we constructed Lax integro-differential representations for matrix generalizations of the Davey-Stewartson systems (DS-I,DS-II,DS-III), matrix generalizations of (2+1)-dimensional extensions of the mKdV, the Nizhnik [35] and the Chen-Lee-Liu [38] equations. Representations for some of those systems in the algebra of purely differential operators with matrix coefficients can be found in [44].

One of the problems left for further investigation are the dressing methods for the corresponding Lax representations. It was shown that for the Lax representations for the (2+1)-dimensional k-cKP hierarchy [29, 30, 34], one can use differential operators for the dressing. The most interesting systems obtained from the (2+1)-dimensional k-cKP hierarchy and in Sections 3-5 arise after a Hermitian conjugation reduction. This imposes nontrivial constraints on the dressing differential operator. It was shown for the k-cKP hierarchy [45, 46] that it is more suitable to use a binary transformation operator in this situation. The dressing method by binary transformations for evolution integro-differential operators that arise in the (2+1)-dimensional k-cKP case has been considered in [36]. We plan to apply this method to the systems in Sections 3-5 in a forthcoming paper. Another interesting question is the possibility of generalizations of other representations from the (2+1)-dimensional k-cKP hierarchy to the case of two integro-differential operators, e.g. (2+1)-dimensional extensions of the Yajima-Oikawa and the Melnikov systems.

7. Acknowledgments

The authors thank Professor Müller-Hoissen for fruitful discussions and useful advice in preparation of this paper. O. Chvartatskyi also thanks the organizers of the ISQS-20 Conference for the warm hospitality in Prague.

References

[1] Dickey L A 2003 *Soliton Equations and Hamiltonian Systems*, Advanced Series in Mathematical Physics, 2nd ed. World Scientific, River Edge, NJ, Vol. 26.

[2] Zakharov V E and Shabat A B 1974 *Funct. Anal. Appl.* **8** (3), 226
[3] Novikov S P, Manakov S V, Pitaevskij L P and Zakharov V E 1984 *Theory of solitons. The inverse scattering methods.* (Transl. from the Russian), Contemporary Soviet Mathematics. New York - London: Plenum Publishing Corporation. Consultants Bureau,
[4] Bullough R K and Caudrey P J (eds.), 1980 *Solitons*, Springer-Verlag, Berlin
[5] Marchenko V A 1988 *Nonlinear equations and operator algebras,* Dordrecht, Boston, Lancaster, Tokyo, Reidel
[6] Matveev V B 1979 *Lett. in Math. Phys.* **3** 213
[7] Matveev V B and Salle M A 1991 *Darboux transformations and solitons*, Berlin Heidelberg, Springer-Verlag.
[8] Date E, Jimbo M, Kashiwara M, and Miwa T 1981 *J. Phys. Soc. Jpn.* **50** 3806
[9] Date E, Jimbo M, Kashiwara M, and Miwa T 1982 *Publ. Res. Inst. Math. Sci.* **18** 1077
[10] Sato M and Sato Y 1983 *North-Holland Math. Stud.* **81** 259
[11] Jimbo M and Miwa T 1983 *Publ. Res. Inst. Math. Sci.* **19** 943
[12] Ohta Y, Satsuma J, Takahashi D and Tokihiro T 1988 *Prog. Theor. Phys. Suppl.* **94** 210
[13] Melnikov V K 1983 *Lett. Math. Phys.* **7** 129
[14] Melnikov V K 1985 *Preprint JINR P2-85-958*, Dubna (in Russian)
[15] Melnikov V K 1987 *Preprint JINR P2-87-136*, Dubna (in Russian)
[16] Melnikov V K 1987 *Commun. Math. Phys.* **112** 639
[17] Melnikov V K 1988 *Phys. Lett. A* **128** 488
[18] Sidorenko J and Strampp W 1991 *Inverse Problems* **7** L37
[19] Konopelchenko B G, Sidorenko J and Strampp W 1991 *Phys. Lett. A* **157** 17
[20] Cheng Y and Li Y S 1991 *Phys. Lett. A* **157** 22
[21] Cheng Y 1992 *J. Math. Phys.* **33** 3774
[22] Cheng Y and Li Y S 1992 *J. Phys. A.* **25** 419
[23] Sidorenko J and Strampp W 1993 *J. Math. Phys.* **34** (4) 1429
[24] Oevel W 1993 *Physica A* **195** 533
[25] Oevel W and Strampp W 1996 *J. Math. Phys.* **37** 6213
[26] Aratyn H, Nissimov E, and Pacheva S 1997 *Int. J. Mod. Phys. A* **12** 1265
[27] Kundu A, Strampp W and Oevel W 1995 *J. Math. Phys.* **36** 2972
[28] Oevel W and Carillo S 1998 *J. Math. Anal. Appl.* **217** 161
[29] Mytropolsky Yu O, Samoilenko V H and Sidorenko Yu M 1999 *Proceedings of NSA of Ukraine* **8** 19
[30] Samoilenko A M, Samoilenko V G and Sidorenko Yu M 1999 *Ukr. Math. Journ.* **51** (1) 86
[31] Berkela Yu Yu and Sydorenko Yu M 2002 *Mat. Studii* **25** (1) 38
[32] Berkela Yu Yu "Integration of nonlinear evolution systems with nonlocal constraints" Ph.D. thesis, Ivan Franko National University of Lviv (in Ukrainian)
[33] Liu X J, Zeng Y B and Lin R 2008 *Phys. Lett.* **372** 3819
[34] Liu X J, Lin R, Jin B and Zeng Y B 2009 *J. Math. Phys.* **50** 053506
[35] Nizhnik L P 1980 *Dokl. Akad. Nauk SSSR* **254** 332
[36] Sydorenko Yu and Chvartatskyi O 2009 *Visn. Kyiv. Univ. Ser: mech.-mat.* **22** 32 (in Ukrainian)
[37] Fokas A S 1994 *Inverse Problems* **10** L19
[38] Chen H H, Lee Y C, Liu C S 1979 *Physica Scr.* **20** 490
[39] Sydorenko Yu and Chvartatskyi O 2012 *Carpathian Mathematical publications* **4** (1) 124 (in Ukrainian)
[40] Huang Y,Yao Y Q and Zeng Y B 2012 *Commun. Theor. Phys.* **57** (4) 515
[41] Drinfeld V G and Sokolov V V 1981 *Sov. Math. Dokl.* **23** 457
[42] Wilson G 1982 *Phys. Lett. A* **89** 332
[43] Hirota R, Grammaticos B and Ramani A 1986 *J. Math. Phys* **27** (6) 1499
[44] Athorne C, Fordy A P 1987 *J. Math. Phys.* **28** 2018
[45] Oevel W and Schief W 1994 *Reviews in Mathematical Physics* **6** 1301
[46] Berkela Yu Yu and Sydorenko Yu M 2006 *Mat. Studii* **25** (1) 38 (in Ukrainian)

Kinematics of semiclassical spin and spin fiber bundle associated with so(n) Lie-Poisson manifold

A A Deriglazov1

1 Dept. de Matematica, ICE, Universidade Federal de Juiz de Fora, MG, Brazil

E-mail: alexei.deriglazov@ufjf.edu.br

Abstract. We describe geometric construction underlying the Lagrangian actions for non-Grassmann spinning particles proposed in our recent works. If we discard the spatial variables (the case of frozen spin), the problem reduces to formulation of a variational problem for Hamiltonian system on a manifold with so(n) Lie-Poisson bracket. To achieve this, we identify dynamical variables of the problem with coordinates of the base of a properly constructed fiber bundle. In turn, the fiber bundle is embedded as a surface into the phase space equipped with canonical Poisson bracket. This allows us to formulate the variational problem using the standard methods of Dirac theory for constrained systems.

Consider $\frac{1}{2}n(n-1)$-dimensional Euclidean space equipped with the coordinates $J^{ij} = -J^{ji}$, $i, j = 1, 2, \ldots, n$ and with the Lie-Poisson bracket

$$\mathbb{R}^{\frac{n(n-1)}{2}} = \left\{ J^{ij}, \ \{J^{ij}, J^{kl}\}_{LPB} = 2(\delta^{ik}J^{jl} - \delta^{il}J^{jk} - \delta^{jk}J^{il} + \delta^{jl}J^{ik}) \right\}. \tag{1}$$

This is the Lie-Poisson manifold associated with $so(n)$ algebra [1, 2, 3]. The construction presented below works for the case of Minkowski space with $so(m,n)$ algebra as well.

We discuss the Hamiltonian flow

$$\dot{J} = \{J, H_0\}_{LPB}, \tag{2}$$

generated by given Hamiltonian $H_0(J)$. Our aim is to formulate variational problem for the Hamiltonian flow on the submanifold $\mathbb{S}$ which will be specified below. We call $\mathbb{S}$ spin surface, as the canonical quantization of the submanifold gives quantum mechanics of the spin one-half particle. In [8] it has been demonstrated that $SO(3)$ spin surface leads to a reasonable model of non-relativistic spin. $SO(1,3)$ spin surface can be used to construct variational problem for unified description of both the Frenkel [6] and BMT [7] theories of relativistic spin [9]. $SO(2,3)$ spin surface implies the model of Dirac electron [11, 10], and represents an example of pseudoclassical mechanics [12].

Roughly speaking, our procedure is as follows. The formulation of a variational problem in closed form is known for the equations constructed on the phase space equipped with canonical Poisson bracket, $\{\omega_i, \pi_j\} = \delta_{ij}$. In this case the Hamiltonian equations can be obtained by variation of the action $\int \pi\dot{\omega} - H(\omega, \pi)$, where $H(\omega, \pi)$ stands for Hamiltonian of the system. To apply this prescription, we reformulate our problem (1), (2) in terms of the properly constructed phase space with constraints.

CONSTRUCTION 1.

1.1. Take $2n$-dimensional phase space equipped with the Poisson bracket

$$\mathbb{R}^{2n} = \{ \ \omega^i, \pi^j, \ \{\omega^i, \pi^j\}_{PB} = \delta^{ij} \ \}. \tag{3}$$

1.2. Define the map from the phase to angular-momentum space (1)

$$\begin{aligned} f:\ \mathbb{R}^{2n} \ &\rightarrow \ \mathbb{R}^{\frac{n(n-1)}{2}}, \\ f:(\omega^i, \pi^j) \ &\rightarrow \ J^{ij} = 2(\omega^i\pi^j - \omega^j\pi^i). \end{aligned} \tag{4}$$

We have, for $n > 2$,

$$\text{rank}\frac{\partial(J^{ij})}{\partial(\omega^k, \pi^l)} = 2n-3, \tag{5}$$

so an image of the map is $(2n-3)$-dimensional surface $\mathbb{M}$

$$f(\mathbb{R}^{2n}) = \mathbb{M}^{2n-3} \in \mathbb{R}^{\frac{n(n-1)}{2}}. \tag{6}$$

Poisson bracket of the functions $J^{ij}(\omega,\pi)$ coincides with the Lie-Poisson bracket (1). More generally, for any functions $A(J)$, $B(J)$,

$$\{A(J), B(J)\}_{PB} = \{A(J), B(J)\}_{LPB}|_{J\rightarrow J(\omega,\pi)}\,. \tag{7}$$

1.3. Look for the surface $\mathbb{T} = \{ \ \omega, \pi \mid T_a(\omega^i, \pi^j) = 0 \ \} \in \mathbb{R}^{2n}$ which is invariant under action of $SO(n)$, that is

$$\{T_a, J^{ij}\}_{PB} = 0. \tag{8}$$

This is essentially unique surface of $2n-3$ dimensions

$$\mathbb{T}^{2n-3} = \{T_3 = \pi^2 + a_3 = 0, \ T_4 = \omega^2 + a_4 = 0, \ T_5 = \omega\pi = 0\}. \tag{9}$$

where $a_3, a_4 \in \mathbb{R}$, and it has been denoted $\pi^2 = \pi^i\pi^i$, and so on.

COMMENTS. A. Any trajectory of $H_0(J)$ which starts on $\mathbb{M}^{2n-3}$ lies entirely on $\mathbb{M}^{2n-3}$ (the proof is similar to those of Proposition 3 below).

B. $SO(3)$ is the exceptional case, when $\mathbb{M} = \mathbb{R}^{\frac{n(n-1)}{2}}$, and the vector representation of $SO(3)$ coincides with the adjoint one. Besides the surface $\mathbb{T}^{2n-3}$ can be identified with the group manifold $SO(3)$, see [13].

C. The invariance condition (8) guarantees the validity of important Propositions 2, 3, see below.

D. Casimir operators of $SO(n)$ group are scalar functions of the generators, $C(J^{ij})$. On the surface (9) they will have fixed values determined by the constants a_3 and a_4: $C(J^{ij}) = C(\omega^2, \pi^2, \omega\pi) = C(a_3, a_4)$. In particular, the first Casimir operator is $J^2 = 8a_3a_4$, see Eq. (22) below.

Denote $\mathbb{S}$ image of $\mathbb{T}^{2n-3}$ under the map f (this is called the spin surface, see Figure 1)

$$\mathbb{S} = f(\mathbb{T}^{2n-3}) \in \mathbb{R}^{\frac{n(n-1)}{2}}. \tag{10}$$

Denote $\mathbb{F}_J \in \mathbb{T}^{2n-3}$ preimage of a point $J \in \mathbb{S}$, $\mathbb{F}_J = f^{-1}(J)$.

The manifold $\mathbb{T}^{2n-3}$ has natural structure of fiber bundle

$$\mathbb{T}^{2n-3} = (\mathbb{S}, \mathbb{F}, f), \tag{11}$$

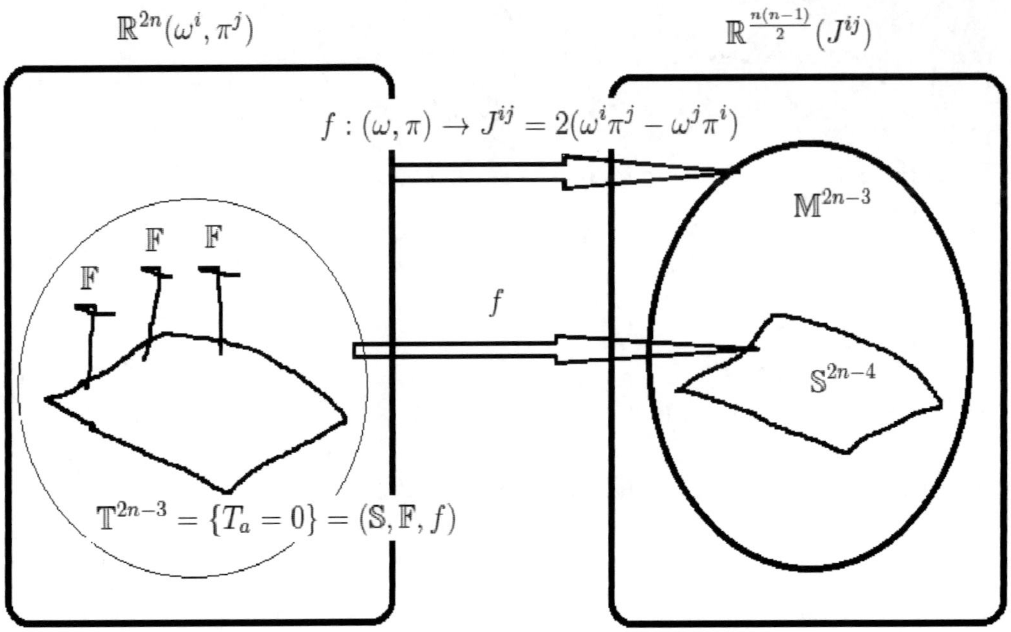

Figure 1. Identification of spin surface $\mathbb{S}^{2n-4}$ with base of the fiber bundle $\mathbb{T}^{2n-3}$.

with the base $\mathbb{S}$, the projection map f, the standard fiber $\mathbb{F}$ and structure group of the fiber $SO(2)$. The standard fiber can be described as follows. Let us identify the vector $(\omega, \pi) \in \mathbb{T}^{2n-3}$ with the pair of orthogonal vectors of $\mathbb{R}^n$

$$(\omega, \pi) \sim \vec{\omega} \perp \vec{\pi}, \quad \vec{\pi}^2 = -a_4, \quad \vec{\omega}^2 = -a_3. \tag{12}$$

Then $\mathbb{F}$ is composed by all the pairs obtained from $(\vec{\omega}, \vec{\pi})$ by rotations in the plane of these vectors. So the structure group is composed of rotations

$$\vec{\omega}' = \vec{\omega} \cos\beta + \sqrt{\frac{a_4}{a_3}} \vec{\pi} \sin\beta, \qquad \vec{\pi}' = -\sqrt{\frac{a_3}{a_4}} \vec{\omega} \sin\beta + \vec{\pi} \cos\beta. \tag{13}$$

By construction, $(\omega', \pi') \in \mathbb{T}^{2n-3}$. Infinitesimal form of the symmetry is

$$\delta\vec{\omega} = \sqrt{\frac{a_4}{a_3}} \beta \vec{\pi}, \qquad \delta\vec{\pi} = -\sqrt{\frac{a_3}{a_4}} \beta \vec{\omega}. \tag{14}$$

As it must be, the transformations leave inert the points of base, $\delta J^{ij} = 0$. In the dynamical realization, the structure group acts independently at each instance of time and turn into the local (gauge) symmetry of a Lagrangian. The spin-plane symmetry determines physical sector of the theory, and hence play the fundamental role in our construction. Indeed, according to Eq. (4), we consider the spin J^{ij} as angular-momentum of an "inner-space particle" $\vec{\omega}$. The crucial difference with the usual (spacial) angular momentum is the presence of the plane-spin symmetry, which acts on the basic variables ω, π, while leaves invariant the spin variables J. According to the general theory [4, 5], the gauge non-invariant coordinates ω of the inner-space are not physical (observable) quantities. The only observable quantities are the gauge-invariant variables J. So our geometric construction realizes, in a systematic form, the oldest idea about spin as the "intrinsic angular momentum".

Local coordinates on $\mathbb{M}^{2n-3}$ and equations of this surface. They can be obtained solving Eq. (4). Namely, the subset J' of $2n-3$ independent functions among $J(\omega,\pi)$, represents the local coordinates.

Let us discuss this in some details. In accordance with the rank condition (5), we can separate J^{ij} on two groups, $J=(J',J'')$, in such a way that the number of J' is equal to $2n-3$, and

$$\mathrm{rank}\frac{\partial J'}{\partial(\omega,\pi)}=2n-3. \tag{15}$$

Then equations for J' from the system (4) can be resolved with respect some of $2n-3$ variables among ω,π. Substitute these expressions into the remaining equations from (4). By construction, the result does not depend on ω and π, and we obtain the expressions for J'' through J'

$$J''=g(J')\equiv J''(J'). \tag{16}$$

In the result, the image $f(\mathbb{R}^{2n})$ is the surface with equations (16) and with local coordinates J'

$$\mathbb{M}^{2n-3}=\left\{J=(J',J'')\;\Big|\;J''=J''(J')\right\}. \tag{17}$$

So, points of $\mathbb{R}^{2n}$ are mapped into

$$f(\omega,\pi)=(J',J''(J'))\in\mathbb{M}^{2n-3}. \tag{18}$$

For example, for $SO(2,3)$ case, the local coordinates are $J^{5\mu}$, J^{0i} while as the equations of the surface we can take [12]

$$\epsilon^{\mu\nu\alpha\beta}J^5{}_\nu J_{\alpha\beta}=0,\Leftrightarrow J^{ij}=(J^{50})^{-1}(J^{5i}J^{0j}-J^{5j}J^{0i}). \tag{19}$$

Among the four relativistic-covariant equations on the l.h.s., there are only three independent.

Local coordinates on $\mathbb{T}^{2n-3}$ adjusted with the structure of fibration. Since

$$\mathrm{rank}\frac{\partial(J',\omega^n,T_4,T_5)}{\partial(\omega^k,\pi^l)}=2n, \tag{20}$$

we can make the change of coordinates on $\mathbb{R}^{2n}$

$$(\omega^i,\pi^j)\;\leftrightarrow\;(J',\omega^n,T_4,T_5). \tag{21}$$

In the new coordinates the function T_3 does not depend on ω^n. Indeed, $J^{ij}J_{ij}$ can be identically rewritten through T_a as follows

$$J^2=8\left[(T_4-a_4)T_3-T_5^2-a_3T_4+a_3a_4\right], \tag{22}$$

then

$$T_3=\frac{J^2+8T_5^2+8a_3T_4-8a_3a_4}{8(T_4-a_4)}. \tag{23}$$

On the other hand, substitute the new coordinates into the expression $(J^{ij}(\omega,\pi))^2$. By construction, this gives $J^2=(J')^2+(J''(J'))^2$. Using this in Eq. (23), we obtain

$$T_3(\omega,\pi)|_{(J',\omega^n,T_4,T_5)}=T_3(J',T_4,T_5). \tag{24}$$

Hence in the new coordinates the surface looks as

$$\mathbb{T}^{2n-3} = \left\{ J', \omega^n, T_4, T_5 \;\middle|\; T_4 = T_5 = 0,\ T_3(J') = 0 \right\}. \tag{25}$$

PROPOSITION 1. $\dim \mathbb{S} = 2n - 4$, (then $\dim \mathbb{F} = 1$).

Indeed, take restriction of the map (18) on $\mathbb{T}^{2n-3}$. Since any point on the surface obeys the condition $T_3(J') = 0$, we have

$$f|_{\mathbb{T}} : (\omega, \pi) \to (J', J''(J')), \quad \text{where} \quad T_3(J') = 0. \tag{26}$$

Hence $T_3(J') = 0$ is equation of the base in space $\mathbb{M}^{2n-3}$, and $\dim \mathbb{S} = 2n - 4$.

Let as take some $2n-4$ variables $\tilde{J}$ among J' which form the coordinate system of the base $\mathbb{S}^{2n-4}$. Then (21) and (25) imply that $(\tilde{J}, \omega^n)$ can be taken as local coordinates of the fibration $\mathbb{T}^{2n-3}$.

CONSTRUCTION 2.

2.1. Let $H_0(J)$ is some Hamiltonian on $\mathbb{R}^{\frac{n(n-1)}{2}}$. The map f can be used to induce the Hamiltonian $H_0(\omega, \pi)$ on the phase space $\mathbb{R}^{2n}$

$$H_0(\omega, \pi) \equiv H_0(J(\omega, \pi)). \tag{27}$$

Let us confirm that Hamiltonian flows of $H_0(\omega, \pi)$ and $H_0(J)$ are adjusted with the surfaces $\mathbb{T}^{2n-3}$ and $\mathbb{S}$.

PROPOSITION 2. Any trajectory of $H_0(\omega, \pi)$ which starts on $\mathbb{T}^{2n-3}$ lies entirely on $\mathbb{T}^{2n-3}$.

Indeed, let $T_a(\omega(\tau_0), \pi(\tau_0)) = 0$ for some trajectory of $H_0(\omega, \pi)$. We have

$$\dot{T}_a(\omega(\tau), \pi(\tau)) = \{T_a, H_0(J(\omega, \pi))\}_{PB} = \{T_a, J\}_{PB} \frac{\partial H_0}{\partial J} = 0, \tag{28}$$

due to the invariance condition $\{T_a, J\} = 0$. Hence $T_a(\tau) = T_a(\tau_0) = 0$ for any τ, that is $(\omega(\tau), \pi(\tau))$ belong to $\mathbb{T}^{2n-3}$ at each τ.

PROPOSITION 3. Any trajectory of $H_0(J)$ which starts on $\mathbb{S}$ lies entirely on $\mathbb{S}$.

Indeed, the problem

$$\dot{J} = \{J, H_0(J)\}_{LPB} = 0, \qquad J(\tau_0) = J_0 \in \mathbb{S}, \tag{29}$$

has unique solution, we denote it $J(\tau)$. Take any point (ω_0, π_0) which belong to preimage of J_0, $f(\omega_0, \pi_0) = J_0$. Construct the (unique) solution to the problem $\dot{\omega} = \{\omega, H_0(\omega, \pi)\}_{PB} = 0$, $\dot{\pi} = \{\pi, H_0(\omega, \pi)\}_{PB} = 0$, $(\omega(\tau_0), \pi(\tau_0)) = (\omega_0, \pi_0)$. According the Proposition 2, it lies in $\mathbb{T}^{2n-3}$, then $f(\omega(\tau), \pi(\tau))$ lies in $\mathbb{S}$. Besides, this obeys the problem (29). Since solution to the problem is unique, we conclude $J(\tau) = f(\omega(\tau), \pi(\tau)) \in \mathbb{S}$ for each τ.

2.2. Consider variational problem on the extended phase space $(\omega^i, \pi^j, e_a, \pi_a, \lambda_{ea})$, $a = 3, 4$, with the functional

$$S_H = \int d\tau \ \pi\dot{\omega} - \left[H_0(\omega, \pi) + \frac{e_3}{2} T_3 + \frac{e_4}{2} T_4 + \pi_{ea}(\lambda_{ea} - \dot{e}_a) \right]. \tag{30}$$

Variation of the functional leads to the equations

$$\pi_{ea} = 0, \quad \dot{e}_a = \lambda_{ea}, \tag{31}$$

$$\pi^2 + a_3 = 0, \quad \omega^2 + a_4 = 0, \tag{32}$$

$$\dot{\omega}^i = \frac{\partial H_0}{\partial \pi^i} - e_3 \pi^i, \quad \dot{\pi}^i = -\frac{\partial H_0}{\partial \omega^i} + e_4 \omega^i. \tag{33}$$

Eq. (31) has been used in obtaining (32). Compute derivative of the first equation from (32)

$$0 = (\pi^2 + a_3)^{\cdot} = 2(-\pi^i \frac{\partial H_0}{\partial \omega^i} + e_4(\omega\pi)) = 2e_4(\omega\pi) \Rightarrow (\omega\pi) = 0, \tag{34}$$

where we have used $\pi^i \frac{\partial H_0(J(\omega,\pi))}{\partial \omega^i} = 4\frac{\partial H_0}{\partial J^{ij}}\pi^i\pi^j = 0$. We continue the process, obtaining the following algebraic consequences of the system (31)-(33)

$$(\omega\pi) = 0, \quad e_4 = \frac{a_3}{a_4} e_3, \quad \lambda_{e4} = \frac{a_3}{a_4}\lambda_{e3}. \tag{35}$$

The auxiliary variables λ_{e4}, e_4, π_{e4} and π_{e3} are fixed by the algebraic equations in terms of λ_{e3}. For the remaining variables we have the differential equations

$$\dot{e}_3 = \lambda_{e3}, \tag{36}$$

$$\dot{\omega}^i = \frac{\partial H_0}{\partial \pi^i} - e_3 \pi^i, \quad \dot{\pi}^i = -\frac{\partial H_0}{\partial \omega^i} + \frac{a_3}{a_4} e_3 \omega^i, \tag{37}$$

as well as the algebraic constraints

$$T_3 = \pi^2 + a_3 = 0, \quad T_4 = \omega^2 + a_4 = 0, \quad T_5 = (\omega\pi) = 0. \tag{38}$$

Equations (38) and (9) imply the following

PROPOSITION 4. All the trajectories $(\omega(\tau), \pi(\tau))$ of the problem (30) live on the fiber bundle $\mathbb{T}^{2n-3}$.

Let us consider the projection map (4), $J^{ij}(\tau) = f(\omega(\tau), \pi(\tau))$. According the Proposition 4 and Eq. (10), $J(\tau)$ lies on $\mathbb{S}$. Besides, $J(\tau)$ represents a solution to the problem (2)

$$\begin{gathered}
\dot{J}^{ij} = \frac{\partial J^{ij}}{\partial \omega^a}\dot{\omega}^a + \frac{\partial J^{ij}}{\partial \pi_a}\dot{\pi}_a = \\
\frac{\partial J^{ij}}{\partial \omega^a}\left(\frac{\partial H_0}{\partial \pi_a} - e_3\pi^a\right) + \frac{\partial J^{ij}}{\partial \pi_a}\left(-\frac{\partial H_0}{\partial \omega^a} + \frac{a_3}{a_4}e_3\omega^a\right) = \\
\frac{\partial J^{ij}}{\partial \omega^a}\frac{\partial H_0}{\partial \pi_a} - \frac{\partial J^{ij}}{\partial \pi_a}\frac{\partial H_0}{\partial \omega^a} = \left(\frac{\partial J^{ij}}{\partial \omega^a}\frac{\partial J^{kl}}{\partial \pi^a} - \frac{\partial J^{ij}}{\partial \pi_a}\frac{\partial J^{kl}}{\partial \omega^a}\right)\frac{\partial H_0}{\partial J^{kl}} = \\
\{J^{ij}, J^{kl}\}_{PB}\frac{\partial H_0}{\partial J^{kl}} = \{J^{ij}, J^{kl}\}_{LPB}\frac{\partial H_0}{\partial J^{kl}} = \{J^{ij}, H_0(J)\}_{LPB}.
\end{gathered} \tag{39}$$

Hence we have obtained

PROPOSITION 5. Any trajectory of the Hamiltonian flow of $H_0(J)$ on $\mathbb{S}$, $J(\tau)|_{\mathbb{S}}$, is a projection of some trajectory $(\omega(\tau), \pi(\tau))$ of the variational problem (30), $J(\tau)|_{\mathbb{S}} = f(\omega(\tau), \pi(\tau))$.

In this sense, trajectories of the Lie-Poisson system (2) lying on $\mathbb{S} \in \mathbb{R}^{\frac{n(n-1)}{2}}$, are obtained starting from the variational problem (30) formulated on $\mathbb{R}^{2n}$.

For the physically interesting cases $SO(3)$, $SO(1,3)$ and $SO(2,3)$, the Lagrangian actions can be restored from the Hamiltonian one, (30), see [8-13].

Acknowledgments
This work has been supported by the Brazilian foundation FAPEMIG.

References

[1] Perelomov A M 1990 *Integrable systems of classical mechanics and Lie algebras* (Birkhauser Verlag)
[2] Arnold V I 1989 *Mathematical Methods of Classical Mechanics* (Springer, New York, NY)
[3] Nersessian A 2006 *Lect. Notes Phys.* **698** 139
[4] Dirac P A M 1964 *Lectures on Quantum Mechanics* (Yeshiva University, New York)
[5] Deriglazov A A 2010 *Classical mechanics, Hamiltonian and Lagrangian formalism* (Springer-Verlag, ISBN 978-3-642-14036-5)
[6] Frenkel J 1926 *Z. fur Physik* **37** 243
[7] Bargmann V, Michel L and Telegdi V L 1959 *Phys. Rev. Lett.* **2** 435
[8] Deriglazov A A 2010 *Mod. Phys. Lett.* A **25** 2769
[9] Deriglazov A A 2012 Variational problem for the Frenkel and the Bargmann-Michel-Telegdi (BMT) equations (Preprint arXiv: 1204.2494)
[10] Deriglazov A A 2012 *Annals of Physics* **327** 398
[11] Deriglazov A A 2012 *Phys. Lett.* A **376** 309
[12] Deriglazov A A 2012 Classical-mechanical models without observable trajectories and the Dirac electron (Preprint arXiv: 1203.5697)
[13] Deriglazov A A, Rizzuti B F and Zamudio G P 2012 *Spinning particles. Possibility of space-time interpretation for the inner space of spin* (LAP LAMBERT Academic Publishing, ISBN 978-3-659-14374-8)

Journal of Physics: Conference Series **411** (2013) 012012 doi:10.1088/1742-6596/411/1/012012

Invariant Differential Operators for Non-Compact Lie Groups: Euclidean Jordan Groups or Conformal Lie Groups

V.K. Dobrev

Institute of Nuclear Research and Nuclear Energy,
Bulgarian Academy of Sciences,
72 Tsarigradsko Chaussee, 1784 Sofia, Bulgaria

E-mail: dobrev@inrne.bas.bg

Abstract. In the present paper we continue the project of systematic construction of invariant differential operators for non-compact semisimple Lie groups. Our starting points is the class of algebras, which we call 'conformal Lie algebras' (CLA), which have very similar properties to the conformal algebras of Minkowski space-time, though our aim is to go beyond this class in a natural way. For this we introduce the new notion of *parabolic relation* between two non-compact semisimple Lie algebras $\mathcal{G}$ and $\mathcal{G}'$ that have the same complexification and possess maximal parabolic subalgebras with the same complexification.

1. Introduction

Invariant differential operators play very important role in the description of physical symmetries - starting from the early occurrences in the Maxwell, d'Allembert, Dirac, equations, (for more examples cf., e.g., [1]), to the latest applications of (super-)differential operators in conformal field theory, supergravity and string theory (for reviews, cf. e.g., [2],[3]. Thus, it is important for the applications in physics to study systematically such operators.

In a recent paper [4] we started the systematic explicit construction of invariant differential operators. We gave an explicit description of the building blocks, namely, the *parabolic subgroups and subalgebras* from which the necessary representations are induced. Thus we have set the stage for study of different non-compact groups.

Since the study and description of detailed classification should be done group by group we had to decide which groups to study first. A natural choice would be non-compact groups that have *discrete series* of representations. By the Harish-Chandra criterion [5] these are groups where holds:

$$\operatorname{rank} G = \operatorname{rank} K, \tag{1}$$

where K is the *maximal compact subgroup* of the non-compact group G. Another formulation is to say that the Lie algebra $\mathcal{G}$ of G has a compact Cartan subalgebra.

Example: The groups $SO(p,q)$ have discrete series, *except* when both p,q are *odd* numbers. $\diamondsuit$

This class is still rather big, thus, we decided to start with a subclass, namely, the class of *Hermitian symmetric spaces.* The practical criterion is that in these cases, the *maximal compact subalgebra* $\mathcal{K}$ is of the form:

$$\mathcal{K} = so(2) \oplus \mathcal{K}' . \tag{2}$$

The Lie algebras from this class are:

$$so(n,2), \quad sp(n,R), \quad su(m,n), \quad so^*(2n), \quad E_{6(-14)}, \quad E_{7(-25)} \tag{3}$$

These groups/algebras have *highest/lowest weight representations*, and relatedly *holomorphic discrete series representations* [6].

The most widely used of these algebras are the *conformal algebras* $so(n,2)$ in n-dimensional Minkowski space-time. In that case, there is a maximal *Bruhat decomposition* [7]:

$$\begin{aligned} so(n,2) &= \mathcal{P} \oplus \tilde{\mathcal{N}} = \mathcal{M} \oplus \mathcal{A} \oplus \mathcal{N} \oplus \tilde{\mathcal{N}} , \\ \mathcal{M} &= so(n-1,1) , \quad \dim \mathcal{A} = 1, \quad \dim \mathcal{N} = \dim \tilde{\mathcal{N}} = n \end{aligned} \tag{4}$$

that has direct physical meaning, namely, $so(n-1,1)$ is the *Lorentz algebra* of n-dimensional Minkowski space-time, the subalgebra $\mathcal{A} = so(1,1)$ represents the *dilatations*, the conjugated subalgebras $\mathcal{N}$, $\tilde{\mathcal{N}}$ are the algebras of *translations*, and *special conformal transformations*, both being isomorphic to n-dimensional Minkowski space-time.

The subalgebra $\mathcal{P} = \mathcal{M} \oplus \mathcal{A} \oplus \mathcal{N}$ ($\cong \mathcal{M} \oplus \mathcal{A} \oplus \tilde{\mathcal{N}}$) is a *maximal parabolic subalgebra.*

There are other special features which are important. In particular, the complexification of the maximal compact subgroup is isomorphic to the complexification of the first two factors of the Bruhat decomposition:

$$\mathcal{K}^{\mathbb{C}} = so(n,\mathbb{C}) \oplus so(2,\mathbb{C}) \cong so(n-1,1)^{\mathbb{C}} \oplus so(1,1)^{\mathbb{C}} = \mathcal{M}^{\mathbb{C}} \oplus \mathcal{A}^{\mathbb{C}} . \tag{5}$$

In particular, the coincidence of the complexification of the semi-simple subalgebras:

$$\mathcal{K}'^{\mathbb{C}} = \mathcal{M}^{\mathbb{C}} \tag{6}$$

means that the sets of finite-dimensional (nonunitary) representations of $\mathcal{M}$ are in 1-to-1 correspondence with the finite-dimensional (unitary) representations of $\mathcal{K}'$.

It turns out that some of the hermitian-symmetric algebras share the above-mentioned special properties of $so(n,2)$. This subclass consists of:

$$so(n,2), \quad sp(n,\mathbb{R}), \quad su(n,n), \quad so^*(4n), \quad E_{7(-25)} \tag{7}$$

the corresponding analogs of Minkowski space-time V being:

$$\mathbb{R}^{n-1,1}, \quad \mathrm{Sym}(n,\mathbb{R}), \quad \mathrm{Herm}(n,\mathbb{C}), \quad \mathrm{Herm}(n,\mathbb{Q}), \quad \mathrm{Herm}(3,\mathbb{O}) \tag{8}$$

where we use standard notation $\mathbb{R}, \mathbb{C}, \mathbb{Q}, \mathbb{O}$ for the four division algebras (real, complex, quaternion, octonion).

In view of applications to physics, we proposed to call these algebras '*conformal Lie algebras*', (or groups) [8].

They were used by other authors under different names. First, the corresponding groups are also called '*Hermitian symmetric spaces of tube type*' [9]. The same class was identified from different considerations in [10] who called them '*conformal groups of simple Jordan algebras*'. In fact, the relation between Jordan algebras and division algebras was known long time ago. Our class was identified from still different considerations also in [11] where they were called '*simple space-time symmetries generalizing conformal symmetry*'.

We have started the study of the above class in the framework of the present approach in the cases: $so(n,2)$, $su(n,n)$, $sp(n,\mathbb{R})$, $E_{7(-25)}$, in [24], [12], [13], [8], resp., and we have considered also the algebra $E_{6(-14)}$, [14].

Lately, in [15], which we review here, we discovered an efficient way to extend our considerations beyond this class introducing the notion of 'parabolically related non-compact semisimple Lie algebras'.

• *Definition:* Let $\mathcal{G},\mathcal{G}'$ be two non-compact semisimple Lie algebras with the same complexification $\mathcal{G}^{\mathbb{C}} \cong \mathcal{G}'^{\mathbb{C}}$. We call them *parabolically related* if they have parabolic subalgebras $\mathcal{P} = \mathcal{M}\oplus\mathcal{A}\oplus\mathcal{N}$, $\mathcal{P}' = \mathcal{M}'\oplus\mathcal{A}'\oplus\mathcal{N}'$, such that: $\mathcal{M}^{\mathbb{C}} \cong \mathcal{M}'^{\mathbb{C}}$ $(\Rightarrow \mathcal{P}^{\mathbb{C}} \cong \mathcal{P}'^{\mathbb{C}})$. $\diamondsuit$

Certainly, there are many such parabolic relationships for any given algebra $\mathcal{G}$. Furthermore, two algebras $\mathcal{G},\mathcal{G}'$ may be parabolically related with different parabolic subalgebras.

For example, the exceptional Lie algebras $E_{6(6)}$ and $E_{6(2)}$ are parabolically related (and also to $E_{6(-14)}$) with maximal parabolics such that $\mathcal{M}^{\mathbb{C}} \cong sl(6,\mathbb{C})$. But these two algebras are related also by another pair of maximal parabolics such that $\tilde{\mathcal{M}}^{\mathbb{C}} \cong sl(3,\mathbb{C})\oplus sl(3,\mathbb{C})\oplus sl(2,\mathbb{C})$, see Section 7.

Another interesting example are the algebras $so^*(2m)$ and $so(p,q)$ which have a series of maximal parabolics with $\mathcal{M}$-factors [4]:

$$\begin{aligned} \mathcal{M}_j &= su^*(2j)\oplus so^*(2m-4j)\ , \qquad j\leq [\tfrac{m}{2}]\ , \\ \mathcal{M}'_j &= sl(2j,\mathbb{R})\oplus so(p-2j,q-2j)\ , \qquad j\leq [\tfrac{q}{2}]\leq [\tfrac{p}{2}]\ , \end{aligned} \tag{9}$$

whose complexifications coincide for $p+q=2m$

$$\begin{aligned} (\mathcal{M}_j)^{\mathbb{C}} &= (\mathcal{M}'_j)^{\mathbb{C}} = sl(2j,\mathbb{C})\oplus so(2m-4j,\mathbb{C})\ , \\ & j\leq [\tfrac{q}{2}]\leq [\tfrac{m}{2}] = [\tfrac{p+q}{4}]\ . \end{aligned} \tag{10}$$

As we know only for $m = 2n$, i.e., for $so^*(4n)$ is fulfilled relation (6), with $\mathcal{M} = \mathcal{M}_n = su^*(2n)$ from (9), (recalling that $\mathcal{K}' \cong su(2n)$). Obviously, $so(p,q)$ is parabolically related to $so^*(4n)$ with this $\mathcal{M}$-factor only when $p=q=2n$, i.e., $\mathcal{G}' = so(2n,2n)$ with $\mathcal{M}'_n = sl(2n,\mathbb{R})$ (which is outside the range of (10)).

We summarize the algebras parabolically related to conformal Lie algebras with maximal parabolics fulfilling (6) in the table below, where we have included also the algebra $E_{6(-14)}$; we display only the semisimple part $\mathcal{K}'$ of $\mathcal{K}$; $sl(n,\mathbb{C})_{\mathbb{R}}$ denotes $sl(n,\mathbb{C})$ as a real Lie algebra, (thus, $(sl(n,\mathbb{C})_{\mathbb{R}})^{\mathbb{C}} = sl(n,\mathbb{C})\oplus sl(n,\mathbb{C})$); e_6 denotes the compact real form of E_6; and we have imposed restrictions to avoid coincidences or degeneracies due to well known isomorphisms: $so(1,2)\cong sp(1,\mathbb{R})\cong su(1,1)$, $so(2,2)\cong so(1,2)\oplus so(1,2)$, $su(2,2)\cong so(4,2)$, $sp(2,\mathbb{R})\cong so(3,2)$, $so^*(4)\cong so(3)\oplus so(2,1)$, $so^*(8)\cong so(6,2)$.

Journal of Physics: Conference Series **411** (2013) 012012 doi:10.1088/1742-6596/411/1/012012

Table of conformal Lie algebras (CLA) $\mathcal{G}$ with $\mathcal{M}$-factor fulfilling (6)
and the corresponding parabolically related algebras $\mathcal{G}'$

$\mathcal{G}$	$\mathcal{K}'$	$\mathcal{M}$ $\dim V$	$\mathcal{G}'$	$\mathcal{M}'$
$so(n,2)$ $n \geq 3$	$so(n)$	$so(n-1,1)$ n	$so(p,q)$, $p+q=$ $=n+2$	$so(p-1,q-1)$
$su(2k,2k)$ $k \geq 2$	$su(2k) \oplus su(2k)$	$sl(2k,\mathbb{C})_{\mathbb{R}}$ $(2k)^2$	$su^*(4k)$ $sl(4k,\mathbb{R})$	$su^*(2k) \oplus su^*(2k)$ $sl(2k,\mathbb{R}) \oplus sl(2k,\mathbb{R})$
$sp(2r,\mathbb{R})$ rank $= 2r \geq 4$	$su(2r)$	$sl(2r,\mathbb{R})$ $r(2r+1)$	$sp(r,r)$	$su^*(2r)$
$so^*(4n)$ $n \geq 3$	$su(2n)$	$su^*(2n)$ $n(2n-1)$	$so(2n,2n)$	$sl(2n,\mathbb{R})$
$E_{7(-25)}$	e_6	$E_{6(-26)}$ 27	$E_{7(7)}$	$E_{6(6)}$
below not CLA !				
$E_{6(-14)}$	$so(10)$	$su(5,1)$ 21	$E_{6(6)}$ $E_{6(2)}$	$sl(6,\mathbb{R})$ $su(3,3)$

2. Preliminaries

Let G be a semisimple non-compact Lie group, and K a maximal compact subgroup of G. Then we have an *Iwasawa decomposition* $G = KA_0N_0$, where A_0 is Abelian simply connected vector subgroup of G, N_0 is a nilpotent simply connected subgroup of G preserved by the action of A_0. Further, let M_0 be the centralizer of A_0 in K. Then the subgroup $P_0 = M_0A_0N_0$ is a *minimal parabolic subgroup* of G. A *parabolic subgroup* $P = M'A'N'$ is any subgroup of G which contains a minimal parabolic subgroup.

Further, let $\mathcal{G}, \mathcal{K}, \mathcal{P}, \mathcal{M}, \mathcal{A}, \mathcal{N}$ denote the Lie algebras of G, K, P, M, A, N, resp.

For our purposes we need to restrict to *maximal parabolic subgroups* $P = MAN$, i.e. rank $A = 1$, resp. to *maximal parabolic subalgebras* $\mathcal{P} = \mathcal{M} \oplus \mathcal{A} \oplus \mathcal{N}$ with dim $\mathcal{A} = 1$.

Let ν be a (non-unitary) character of A, $\nu \in \mathcal{A}^*$, parameterized by a real number d, called the *conformal weight* or energy.

Further, let μ fix a discrete series representation D^μ of M on the Hilbert space V_μ, or a finite-dimensional (non-unitary) representation of M.

We call the induced representation $\chi = \mathrm{Ind}_P^G(\mu \otimes \nu \otimes 1)$ an *elementary representation* of G [16]. (These are called *generalized principal series representations* (or *limits thereof*) in, e.g., [6].) Their spaces of functions are:

$$\mathcal{C}_\chi = \{\mathcal{F} \in C^\infty(G, V_\mu) \mid \mathcal{F}(gman) = e^{-\nu(H)} \cdot D^\mu(m^{-1})\,\mathcal{F}(g)\} \tag{11}$$

where $a = \exp(H) \in A'$, $H \in \mathcal{A}'$, $m \in M'$, $n \in N'$. The representation action is the *left regular action*:

$$(\mathcal{T}^\chi(g)\mathcal{F})(g') = \mathcal{F}(g^{-1}g') , \quad g, g' \in G . \tag{12}$$

• An important ingredient in our considerations are the *highest/lowest weight representations* of $\mathcal{G}^{\mathbb{C}}$. These can be realized as (factor-modules of) Verma modules V^Λ over $\mathcal{G}^{\mathbb{C}}$, where $\Lambda \in (\mathcal{H}^{\mathbb{C}})^*$, $\mathcal{H}^{\mathbb{C}}$ is a Cartan subalgebra of $\mathcal{G}^{\mathbb{C}}$, weight $\Lambda = \Lambda(\chi)$ is determined uniquely from χ [17].

Actually, since our ERs may be induced from finite-dimensional representations of $\mathcal{M}$ (or their limits) the Verma modules are always reducible. Thus, it is more convenient to use *generalized Verma modules* $\tilde{V}^\Lambda$ such that the role of the highest/lowest weight vector v_0 is taken by the (finite-dimensional) space $V_\mu v_0$. For the generalized Verma modules (GVMs) the reducibility is controlled only by the value of the conformal weight d. Relatedly, for the intertwining differential operators only the reducibility w.r.t. non-compact roots is essential.

• Another important ingredient of our approach is as follows. We group the (reducible) ERs with the same Casimirs in sets called *multiplets* [18],[19]. A multiplet is depicted as a connected graph, the *vertices* of which correspond to the reducible ERs and the *lines (arrows)* between the vertices correspond to intertwining operators. The explicit parametrization of the multiplets and of their ERs is important for understanding of the situation.

In the case of a conformal Lie algebra $\mathcal{G}$ (cf. (7)) the number of ERs in a multiplet is equal to [15]:

$$|W(\mathcal{G}^{\mathbb{C}}, \mathcal{H}^{\mathbb{C}})| \,/\, |W(\mathcal{M}^{\mathbb{C}}, \mathcal{H}_m^{\mathbb{C}})| \tag{13}$$

where $\mathcal{H}^{\mathbb{C}}$, $\mathcal{H}_m^{\mathbb{C}}$ are Cartan subalgebras of $\mathcal{G}^{\mathbb{C}}$, $\mathcal{M}^{\mathbb{C}}$, resp.

In fact, the multiplets contain explicitly all the data necessary to construct the intertwining differential operators. Actually, the data for each intertwining differential operator consists of the pair (β, m), where β is a (non-compact) positive root of $\mathcal{G}^{\mathbb{C}}$, $m \in \mathbb{N}$, such that the *BGG Verma module reducibility condition* (for highest weight modules) is fulfilled:

$$(\Lambda + \rho, \beta^\vee) = m , \quad \beta^\vee \equiv 2\beta/(\beta, \beta) \tag{14}$$

ρ is half the sum of the positive roots of $\mathcal{G}^{\mathbb{C}}$.When the above holds then the Verma module with shifted weight $V^{\Lambda-m\beta}$ (or $\tilde{V}^{\Lambda-m\beta}$ for GVM and β non-compact) is embedded in the Verma module V^{Λ} (or $\tilde{V}^{\Lambda}$). This embedding is realized by a singular vector v_s determined by a polynomial $\mathcal{P}_{m,\beta}(\mathcal{G}^{-})$ in the universal enveloping algebra $(U(\mathcal{G}_{-}))\, v_0$, $\mathcal{G}^{-}$ is the subalgebra of $\mathcal{G}^{\mathbb{C}}$ generated by the negative root generators [20]. More explicitly, $v^s_{m,\beta} = \mathcal{P}_{m,\beta}\, v_0$ (or $v^s_{m,\beta} = \mathcal{P}_{m,\beta}\, V_\mu\, v_0$ for GVMs) [17]. Then there exists [17] an *intertwining differential operator*

$$\mathcal{D}_{m,\beta} \ : \ \mathcal{C}_{\chi(\Lambda)} \ \longrightarrow \ \mathcal{C}_{\chi(\Lambda-m\beta)} \tag{15}$$

given explicitly by:

$$\mathcal{D}_{m,\beta} \ = \ \mathcal{P}_{m,\beta}(\widehat{\mathcal{G}^{-}}) \tag{16}$$

where $\widehat{\mathcal{G}^{-}}$ denotes the *right action* on the functions $\mathcal{F}$.

In most of these situations the invariant operator $\mathcal{D}_{m,\beta}$ has a non-trivial invariant kernel in which a subrepresentation of $\mathcal{G}$ is realized. Thus, studying the equations with trivial RHS:

$$\mathcal{D}_{m,\beta}\, f \ = \ 0 \ , \qquad f \in \mathcal{C}_{\chi(\Lambda)} \ , \tag{17}$$

is also very important. For example, in many physical applications in the case of first order differential operators, i.e., for $m = m_\beta = 1$, these equations are called *conservation laws*, and the elements $f \in \ker \mathcal{D}_{m,\beta}$ are called *conserved currents*.

The above construction works also for the *subsingular vectors* v_{ssv} of Verma modules. Such a vector is also expressed by a polynomial $\mathcal{P}_{ssv}(\mathcal{G}^{-})$ in the universal enveloping algebra: $v^s_{ssv} = \mathcal{P}_{ssv}(\mathcal{G}^{-})\, v_0$. Thus, there exists a *conditionally invariant differential operator* given explicitly by: $\mathcal{D}_{ssv} = \mathcal{P}_{ssv}(\widehat{\mathcal{G}^{-}})$, and a *conditionally invariant differential equation*, for many more details, see [21]. (Note that these operators (equations) are not of first order.)

Below in our exposition we shall use the so-called Dynkin labels:

$$m_i \ \equiv \ (\Lambda + \rho, \alpha_i^\vee) \ , \quad i = 1, \ldots, n, \tag{18}$$

where $\Lambda = \Lambda(\chi)$, ρ is half the sum of the positive roots of $\mathcal{G}^{\mathbb{C}}$.

We shall use also the so-called Harish-Chandra parameters [22]:

$$m_\beta \equiv (\Lambda + \rho, \beta) \ , \tag{19}$$

where β is any positive root of $\mathcal{G}^{\mathbb{C}}$. These parameters are redundant, since they are expressed in terms of the Dynkin labels, however, some statements are best formulated in their terms. (Clearly, both the Dynkin labels and Harish-Chandra parameters have their origin in the BGG reducibility condition.)

3. Conformal algebras $so(n,2)$ and parabolically related

Let $\mathcal{G} = so(n,2)$, $n > 2$. We label the signature of the ERs of $\mathcal{G}$ as follows:

$$\begin{aligned}\chi \ &= \ \{ n_1 , \ldots , n_{\tilde h} ; c \} \ , \quad n_j \in \mathbb{Z}/2 \ , \quad c = d - \tfrac{n}{2} \ , \quad \tilde h \equiv [\tfrac{n}{2}], \\ & |n_1| < n_2 < \cdots < n_{\tilde h} \ , \qquad n \text{ even} \ , \\ & 0 < n_1 < n_2 < \cdots < n_{\tilde h} \ , \qquad n \text{ odd} \ ,\end{aligned} \tag{20}$$

where the last entry of χ labels the characters of $\mathcal{A}$, and the first $\tilde h$ entries are labels of the finite-dimensional nonunitary irreps of $\mathcal{M} \cong so(n-1,1)$.

The reason to use the parameter c instead of d is that the parametrization of the ERs in the multiplets is given in a simple intuitive way:

$$\begin{aligned}\chi_1^\pm \ &= \ \{\epsilon n_1 , \ldots , n_{\tilde h} ; \pm n_{\tilde h +1}\} \ , \quad n_{\tilde h} < n_{\tilde h+1} \ , \\ \chi_2^\pm \ &= \ \{\epsilon n_1 , \ldots , n_{\tilde h -1} , n_{\tilde h+1} ; \pm n_{\tilde h}\} \\ \chi_3^\pm \ &= \ \{\epsilon n_1, \ldots , n_{\tilde h -2}, n_{\tilde h}, n_{\tilde h+1} ; \pm n_{\tilde h -1}\} \\ & \cdots \\ \chi_{\tilde h}^\pm \ &= \ \{\epsilon n_1 , n_3 , \ldots , n_{\tilde h} , n_{\tilde h+1} ; \pm n_2\} \\ \chi_{\tilde h+1}^\pm \ &= \ \{\epsilon n_2 , \ldots , n_{\tilde h} , n_{\tilde h+1} ; \pm n_1\} \\ \epsilon \ &= \ \begin{cases} \pm, & n \ \mathit{even} \\ 1, & n \ \mathit{odd} \end{cases}\end{aligned} \tag{21}$$

($\epsilon = \pm$ is correlated with $\chi^\pm$).

Further, we denote by $\tilde{\mathcal{C}}_i^\pm$ the representation space with signature $\chi_i^\pm$.

The number of ERs in the multiplets according to (13) is equal to:

$$|W(\mathcal{G}^\mathbb{C},\mathcal{H}^\mathbb{C})| \,/\, |W(\mathcal{M}^\mathbb{C},\mathcal{H}_m^\mathbb{C})| \ = \ 2(1+\tilde h) \tag{22}$$

All the above is valid for $so(p,q)$ for $p+q = n+2$, (with $\mathcal{M} \cong so(p-1,q-1)$), cf. [23] (for $q=1$), [15].

The multiplets look a little differently for n even and n odd.

We give two diagrams for n even. The first shows the multiplets as they appeared for $so(n+1,1)$ [23] and $so(n,2)$ [24]. The next picture is for the general case $so(p,q)$ with $p+q = n+2$ [15].

Then we give the two diagrams for n odd.

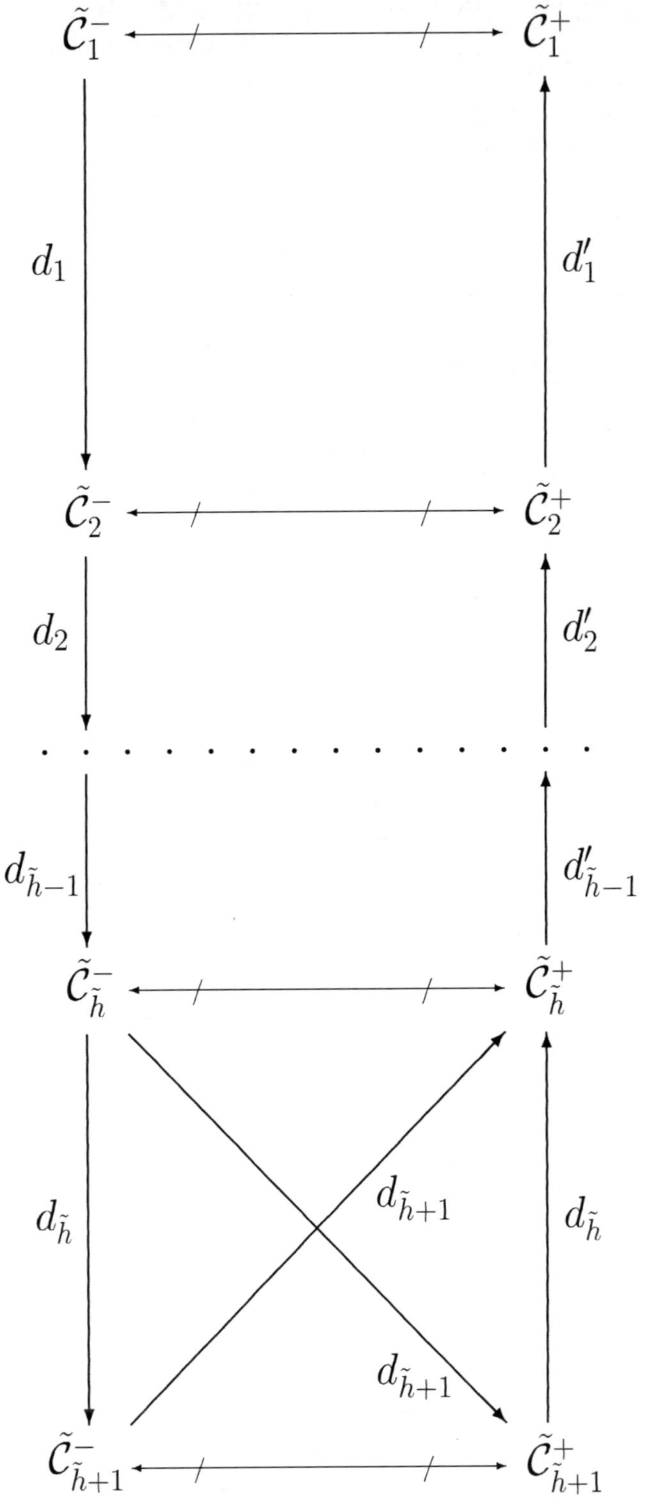

$SO(n,2)$ and $SO(n+1,1)$ for $n \geq 4$ even, $\tilde{h} = \frac{n}{2}$
(arrows are differential operators, dashed arrows are integral operators)

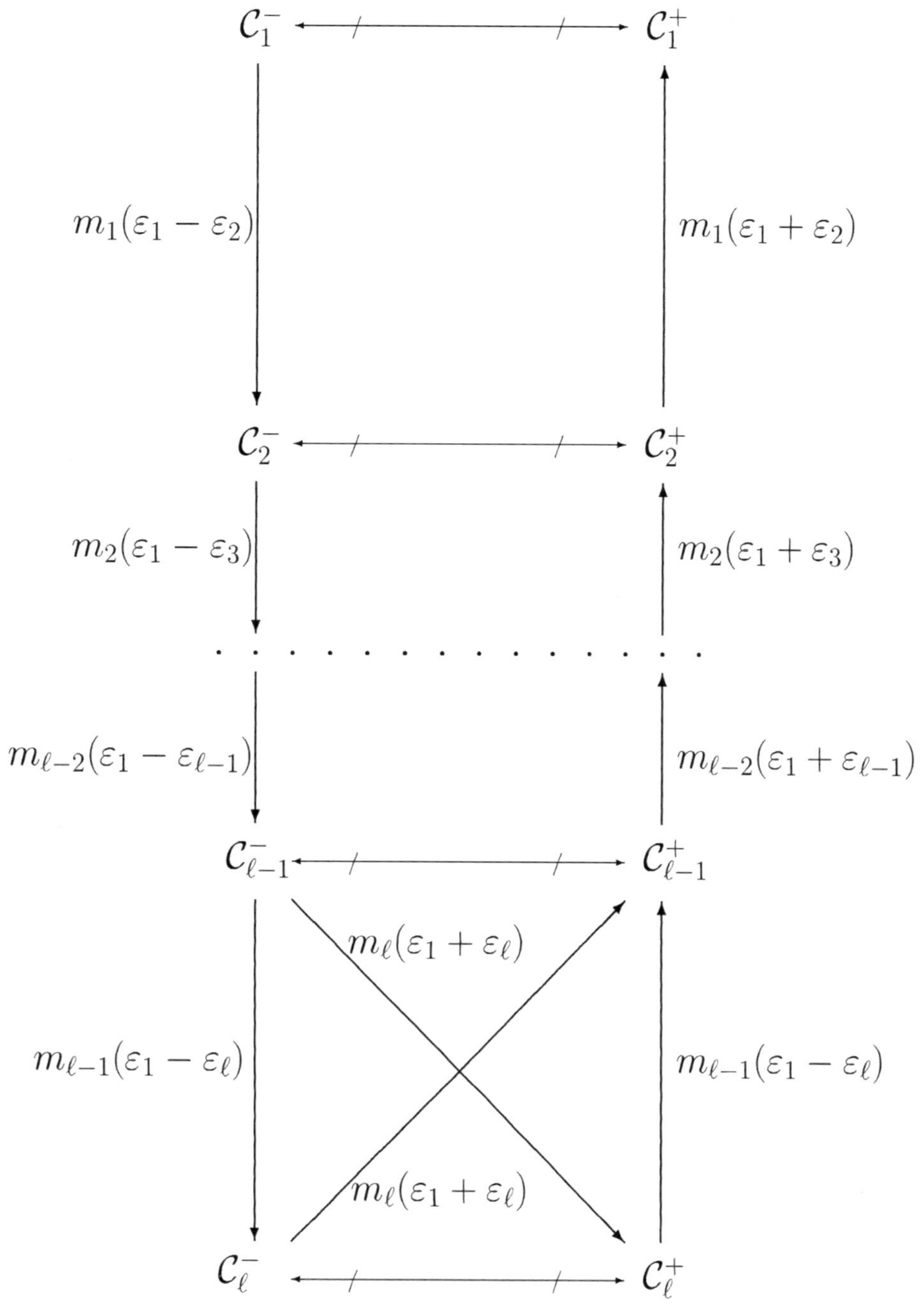

$SO(p,q)$ for $p+q=n+2=2\ell=2(\tilde{h}+1)\geq 6$ with
maximal parabolic subgroup $P=MAN$, where $M=SO(p-1,q-1)$
(arrows are differential operators, dashed arrows are integral operators)
$\varepsilon_1 \pm \varepsilon_k$ are the non-compact roots, m_j are Harish-Chandra parameters

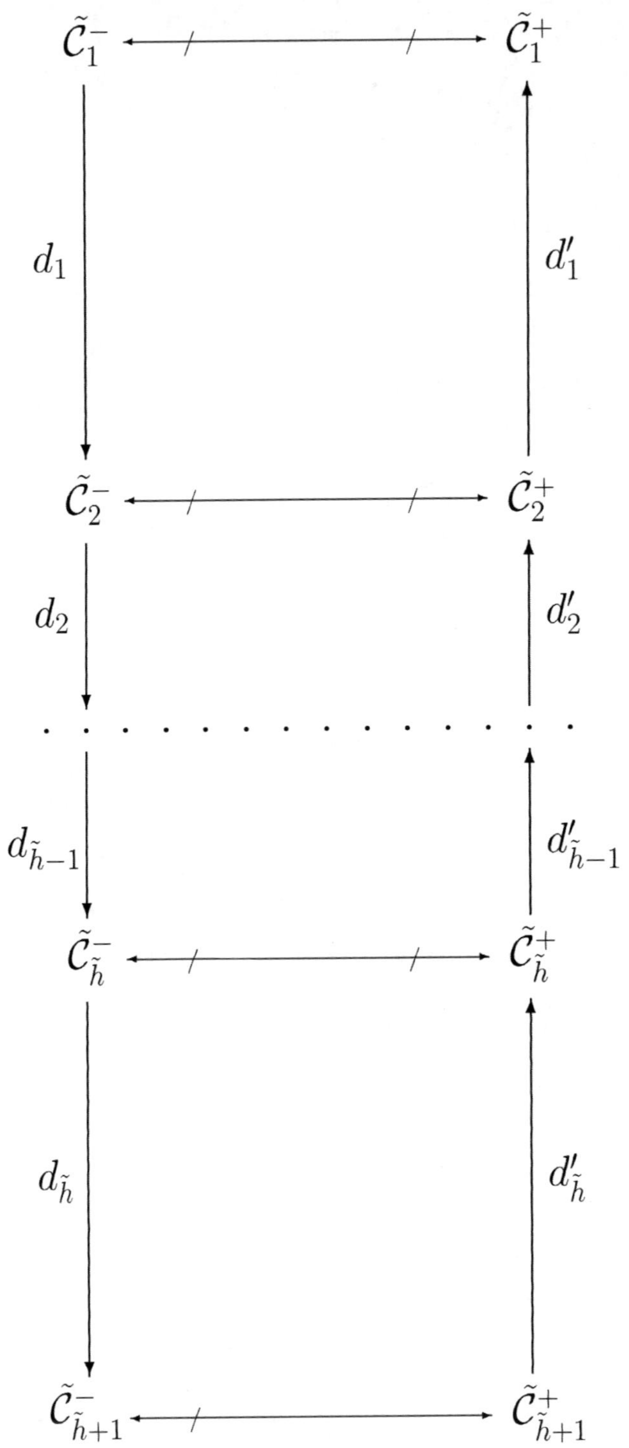

$SO(n,2)$ and $SO(n+1,1)$ for $n \geq 3$ odd, $\tilde{h} = \frac{1}{2}(n-1)$
(arrows are differential operators, dashed arrows are integral operators)

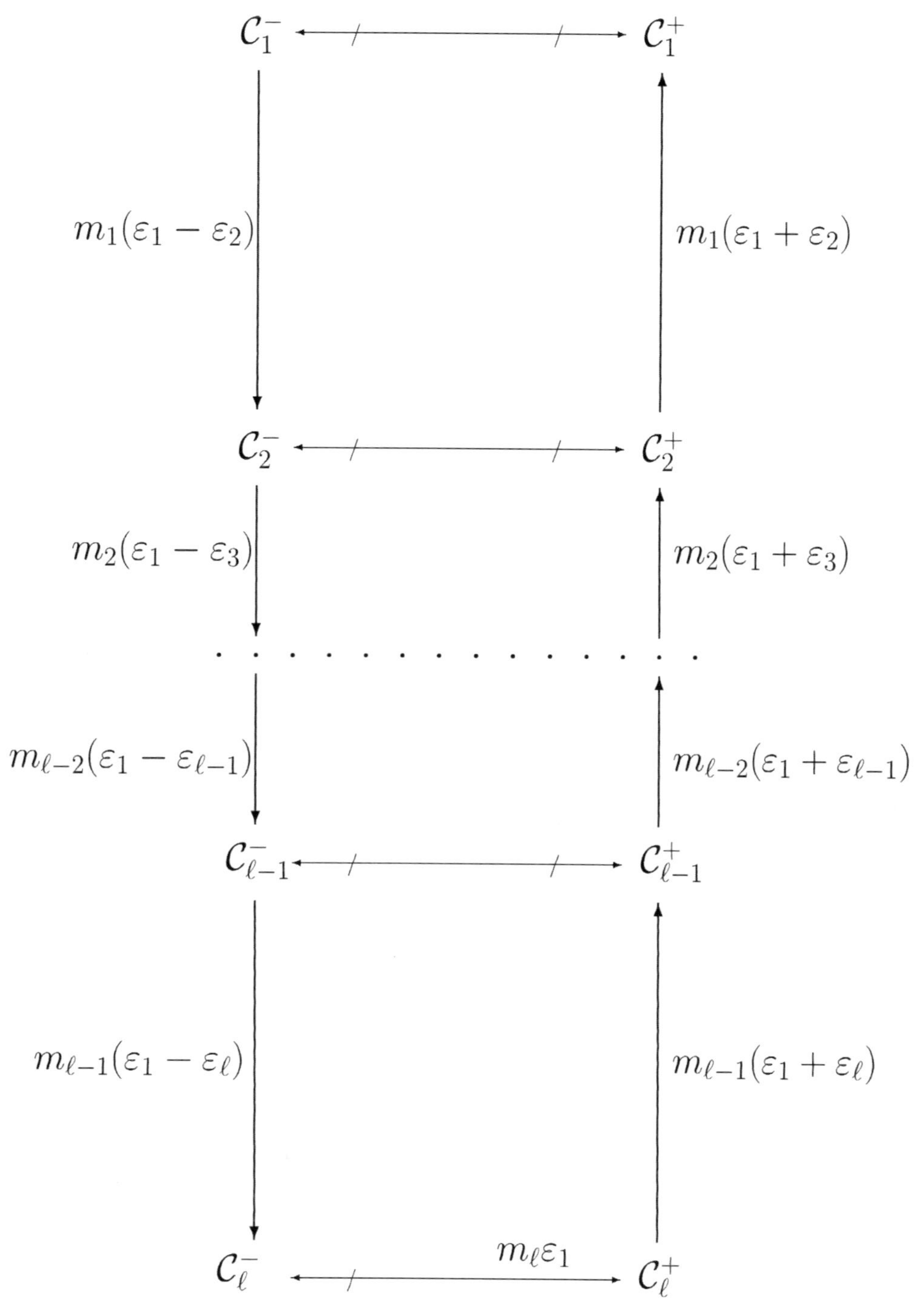

$SO(p,q)$ for $p+q = n+2 = 2\ell+1 = 2\tilde{h}+3 \geq 5$ with
maximal parabolic subgroup $P = MAN$, where $M = SO(p-1, q-1)$
(arrows are differential operators, dashed arrows are integral operators)
$\varepsilon_1 \pm \varepsilon_k$, ε_1 are the non-compact roots, m_j are Harish-Chandra parameters

The ERs in the multiplets are related by *intertwining integral and differential operators.* The *integral operators* were introduced by Knapp and Stein [25]. They correspond to elements of the restricted Weyl group of $\mathcal{G}$. These operators intertwine the pairs $\tilde{C}_i^\pm$

$$G_i^\pm \ : \ \tilde{C}_i^\mp \longrightarrow \tilde{C}_i^\pm \ , \quad i \ = \ 1,\ldots,1+\tilde{h} \tag{23}$$

The *intertwining differential operators* correspond to non-compact positive roots of the root system of $so(n+2,\mathbb{C})$, cf. [17]. [In the current context, compact roots of $so(n+2,\mathbb{C})$ are those that are roots also of the subalgebra $so(n,\mathbb{C})$, the rest of the roots are non-compact.] The degrees of these intertwining differential operators are given just by the differences of the c entries:

$$\begin{aligned} &\deg d_i = \deg d_i' = n_{\tilde{h}+2-i} - n_{\tilde{h}+1-i}\,, \qquad && i = 1,\ldots,\tilde{h}\,, \quad \forall n\,, \\ &\deg d_{\tilde{h}+1} = n_2 + n_1\,, \quad n \text{ even} && \end{aligned} \tag{24}$$

Matters are arranged so that in every multiplet only the ER with signature χ_1^- contains a *finite-dimensional nonunitary subrepresentation* in a subspace $\mathcal{E}$. The latter corresponds to the finite-dimensional unitary irrep of $so(n+2)$ with signature $\{n_1\,,\ldots,\,n_{\tilde{h}}\,,\,n_{\tilde{h}+1}\}$. The subspace $\mathcal{E}$ is annihilated by the operator G_1^+, and is the image of the operator G_1^-.

Although the diagrams are valid for arbitrary $so(p,q)$ $(p+q \geq 5)$ the contents is very different. We comment only on the ER with signature χ_1^+. In all cases it contains an UIR of $so(p,q)$ realized on an invariant subspace $\mathcal{D}$ of the ER χ_1^+. That subspace is annihilated by the operator G_1^-, and is the image of the operator G_1^+. (Other ERs contain more UIRs.)

If $pq \in 2\mathbb{N}$ the mentioned UIR is a discrete series representation. (Other ERs contain more discrete series UIRs.)

And if $q = 2$ the invariant subspace $\mathcal{D}$ is the direct sum of two subspaces $\mathcal{D} = \mathcal{D}^+ \oplus \mathcal{D}^-$, in which are realized a *holomorphic discrete series representation* and its conjugate *anti-holomorphic discrete series representation*, resp. Note that the corresponding *lowest weight GVM* is infinitesimally equivalent only to the holomorphic discrete series, while the conjugate *highest weight GVM* is infinitesimally equivalent to the anti-holomorphic discrete series.

Note that the $\deg d_i$, $\deg d_i'$, are Harish-Chandra parameters corresponding to the non-compact positive roots of $so(n+2,\mathbb{C})$. From these, only $\deg d_1$ corresponds to a simple root, i.e., is a Dynkin label.

Above we restricted to $n > 2$. The case $n = 2$ is reduced to $n = 1$ since $so(2,2) \cong so(1,2) \oplus so(1,2)$. The case $so(1,2)$ is special and must be treated separately. But in fact, it is contained in what we presented already. In that case the multiplets contain only *two ERs* which may be depicted by the *top pair* $\chi_1^\pm$ in all pictures that we presented. And they have the properties that we described. That case was the first given already in 1946-7 independently by Gel'fand et al [26] and Bargmann [27].

4. The Lie algebra $su(n,n)$ and parabolically related

Let $\mathcal{G} = su(n,n)$, $n \geq 2$. The maximal compact subgroup is $\mathcal{K} \cong u(1) \oplus su(n) \oplus su(n)$, while $\mathcal{M} = sl(n,\mathbb{C})_{\mathbb{R}}$.

The signature of the ERs of $\mathcal{G}$ is [12]:

$$\chi \ = \ \{\,n_1\,,\ldots,\,n_{n-1}\,,\,n_{n+1}\,\ldots,\,n_{2n-1}\,;\,c\,\}\ , \qquad n_j \in \mathbb{N}\ , \quad c = d - n \tag{25}$$

The Knapp–Stein restricted Weyl reflection is given by:

$$\begin{aligned} G_{KS} &: \mathcal{C}_\chi \longrightarrow \mathcal{C}_{\chi'} , \\ \chi' &= \{(n_1, \ldots, n_{n-1}, n_{n+1}, \ldots, n_{2n-1})^*; -c\} , \\ (n_1, \ldots, n_{n-1}, n_{n+1}, \ldots, n_{2n-1})^* &\doteq (n_{n+1}, \ldots, n_{2n-1}, n_1, \ldots, n_{n-1}) \end{aligned} \tag{26}$$

Further, we use the root system of the complex algebra $sl(2n, \mathbb{C})$. The positive roots α_{ij} in terms of the simple roots α_i are:

$$\begin{aligned} &\alpha_{ij} = \alpha_i + \alpha_{i+1} + \cdots + \alpha_j , \qquad 1 \leq i < j \leq 2n-1 , \\ &\alpha_{ii} \equiv \alpha_i , \quad 1 \leq i \leq 2n-1 \end{aligned} \tag{27}$$

from which the non-compact are:

$$\alpha_{ij} , \qquad 1 \leq i \leq n , \qquad n \leq j \leq 2n-1 \tag{28}$$

The correspondence between the signatures χ and the highest weight Λ is through the Dynkin labels:

$$\begin{aligned} n_i = m_i &\equiv (\Lambda + \rho, \alpha_i^\vee) = (\Lambda + \rho, \alpha_i) , \qquad i = 1, \ldots, 2n-1, \\ c = -\tfrac{1}{2}(m_{\tilde\alpha} + m_n) &= -\tfrac{1}{2}(m_1 + \cdots + m_{n-1} + 2m_n + m_{n+1} + \cdots + m_{2n-1}) , \end{aligned} \tag{29}$$

$\Lambda = \Lambda(\chi)$, $\tilde\alpha = \alpha_1 + \cdots + \alpha_{2n-1}$ is the highest root.

The number of ERs in the corresponding multiplets by (13) is equal to:

$$|W(\mathcal{G}^{\mathbb{C}}, \mathcal{H}^{\mathbb{C}})| \,/\, |W(\mathcal{M}^{\mathbb{C}}, \mathcal{H}_m^{\mathbb{C}})| = \binom{2n}{n} \tag{30}$$

In our diagrams we need also the Harish-Chandra parameters for the non-compact roots using the following notation:

$$m_{ij} \equiv m_{\alpha_{ij}} = m_i + \cdots + m_j , \quad i < j \tag{31}$$

We use the following conventions. Each intertwining differential operator is represented by an arrow accompanied by a symbol $i_{j...k}$ encoding the root $\beta_{j...k}$ and the number $m_{\beta_{j...k}}$ which is involved in the BGG criterion. This notation is used to save space, but it can be used due to the fact that only intertwining differential operators which are non-composite are displayed, and that the data β, m_β, which is involved in the embedding $V^\Lambda \longrightarrow V^{\Lambda - m_\beta,\beta}$ turns out to involve only the m_i corresponding to simple roots, i.e., for each β, m_β there exists $i = i(\beta, m_\beta, \Lambda) \in \{1, \ldots, r\}$, $(r = \mathrm{rank}\,\mathcal{G})$, such that $m_\beta = m_i$. Hence the data $\beta_{j...k}, m_{\beta_{j...k}}$ is represented by $i_{j...k}$ on the arrows.

Below we give the diagrams for $su(n,n)$ for $n = 2, 3, 4$. For $n = 2k$ these are diagrams also for the parabolically related $su^*(4k)$ and $sl(4k, \mathbb{R})$. We present these introducing a new diagram look which can handle the more complicated cases that follow further. In this new look only the invariant differential operators are presented explicitly. The integral Knapp-Stein operators, more precisely the restricted Weyl reflection action is understood by a symmetry of the picture, either w.r.t. a central point, or w.r.t. middle line.

We give first the diagram for $n = 2$ where the Knapp-Stein symmetry is w.r.t. to the bullet in the middle of the figure. The diagram describes also $su^*(4)$ and $sl(4, \mathbb{R})$. And since we have the isomorphisms $su^*(4) \cong so(5,1)$, $su(2,2) \cong so(4,2)$, $sl(4,\mathbb{R}) \cong so(3,3)$, we repeat in the second picture the diagram as it was given originally in the cases $so(p,q)$ for $p+q=6$: for $q=1$ in [28], and for $q=2$ in [29].

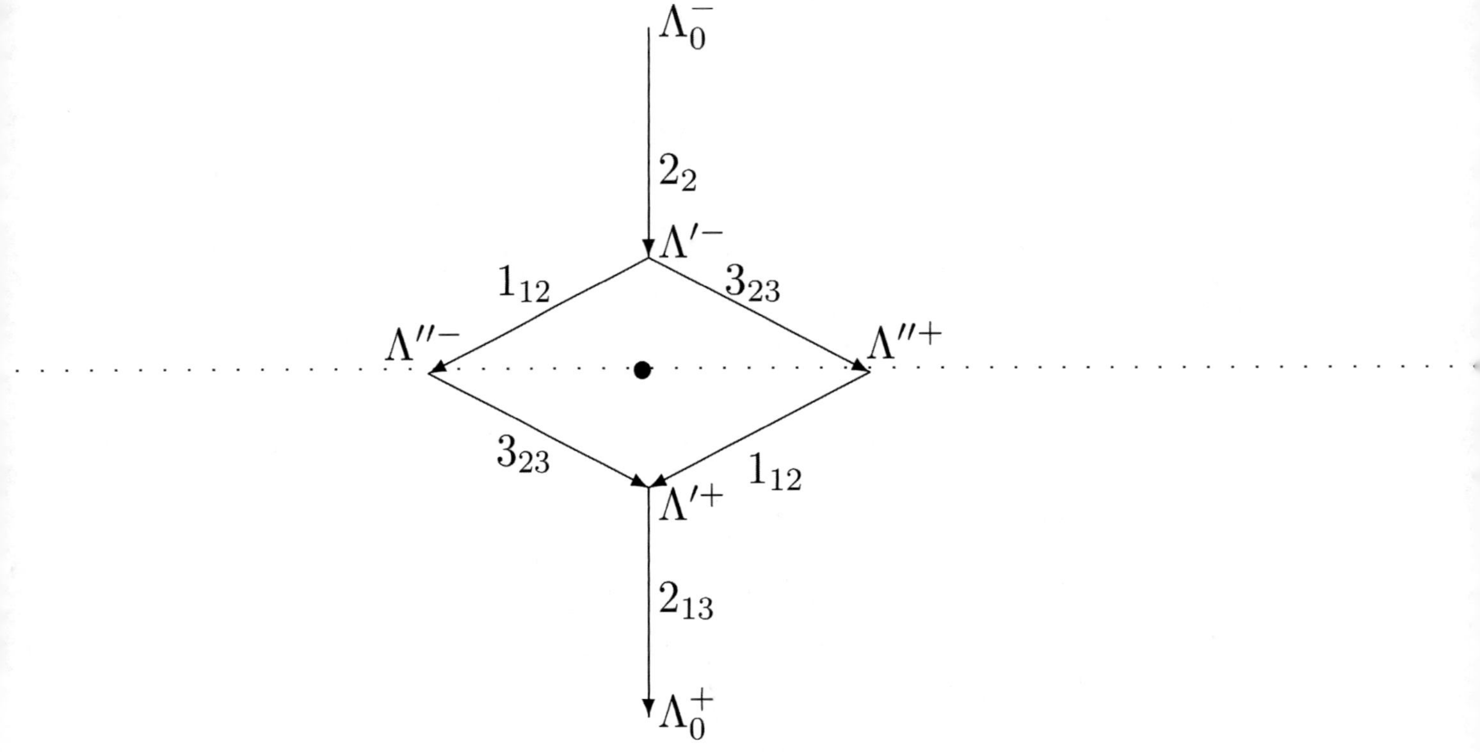

Main multiplets for $su(2,2) \cong so(4,2)$, $su^*(4) \cong so(5,1)$, and $sl(4,\mathbb{R}) \cong so(3,3)$

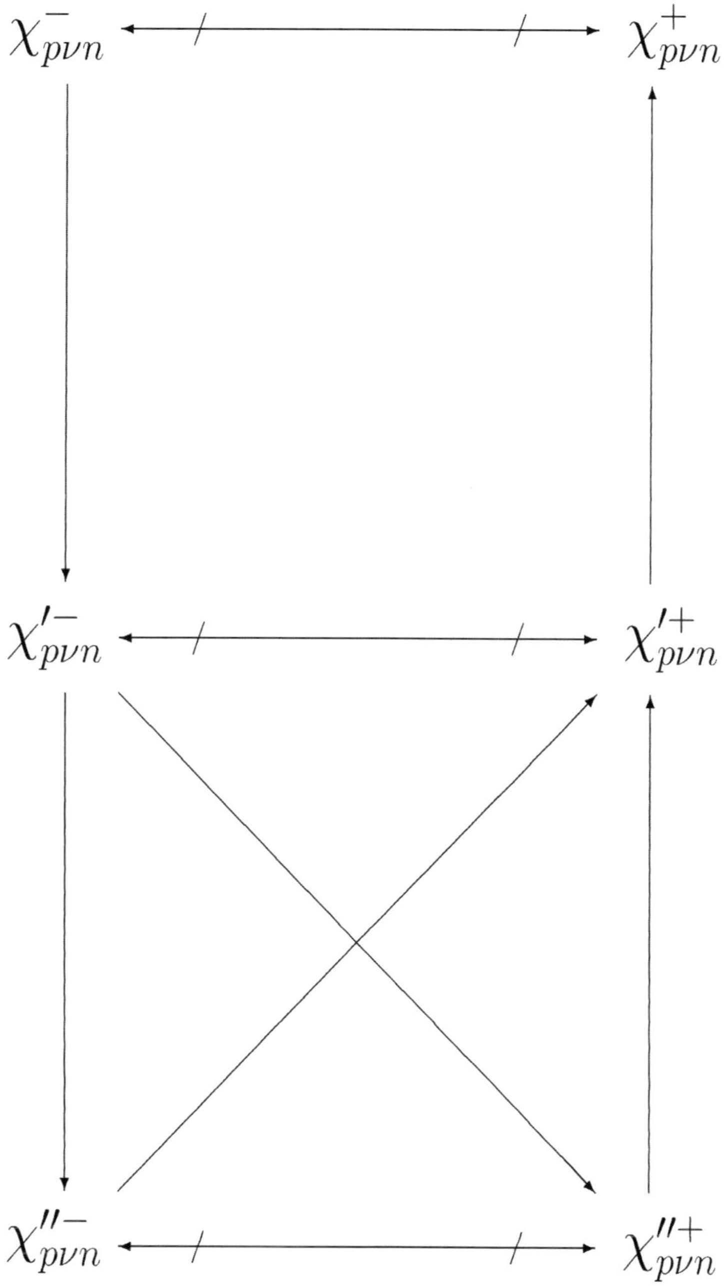

Sextet of partially equivalent ERs and intertwining operators
given first for $so(5,1)$, then for $so(4,2)$ (valid also for $so(3,3)$)
(arrows are differential operators, dashed arrows are integral operators)

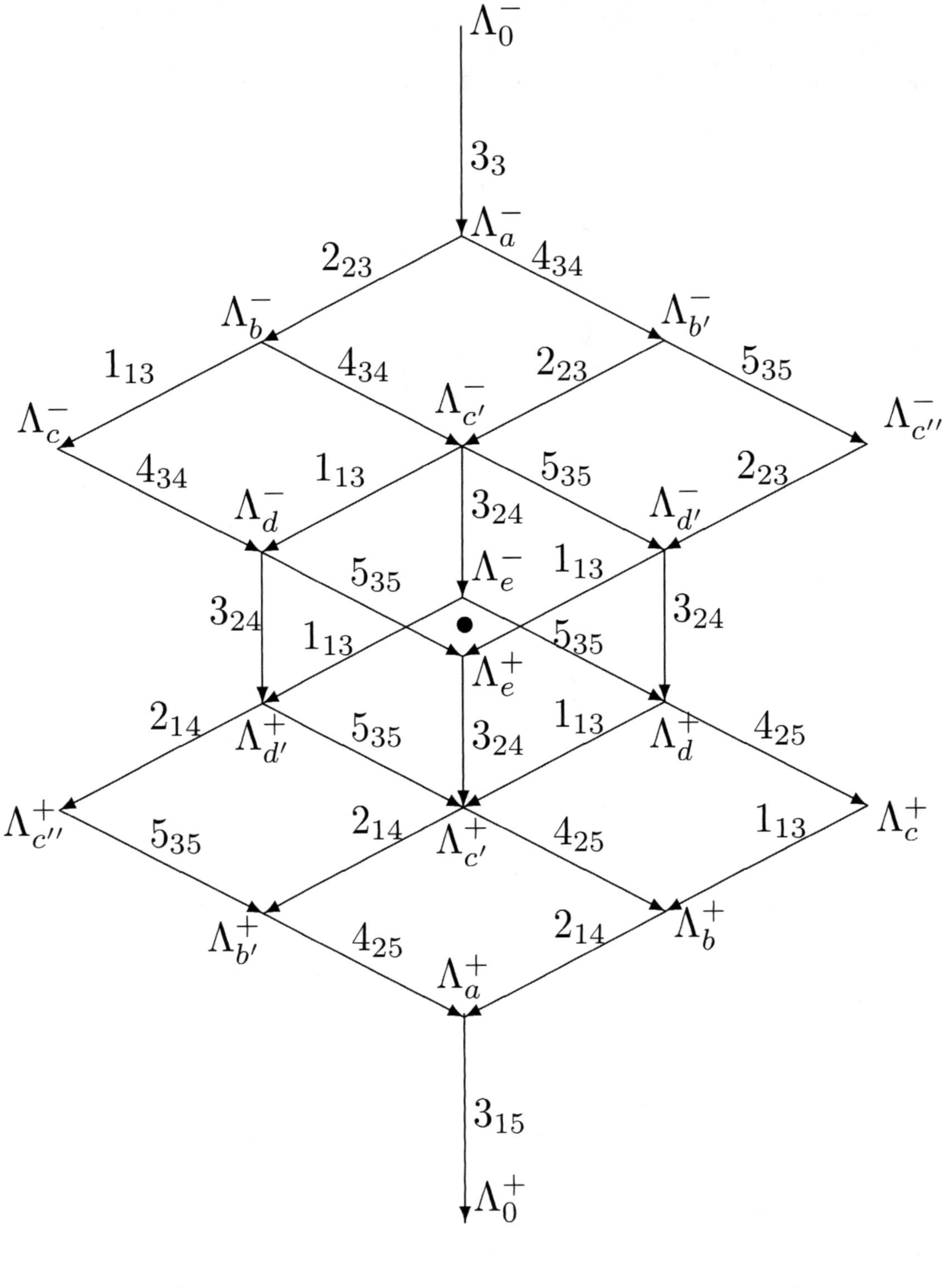

Main multiplets for $su(3,3)$

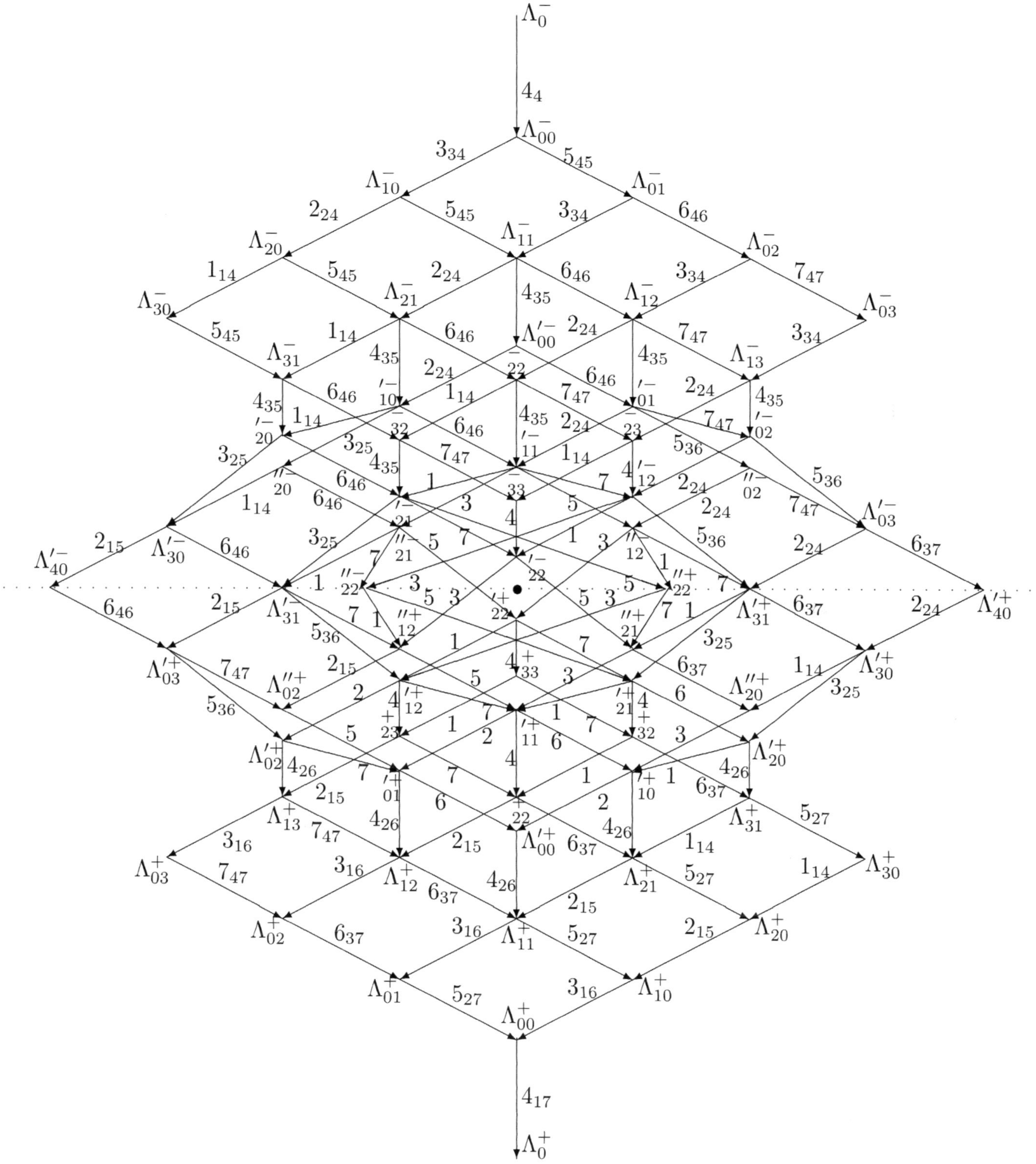

Main multiplets for $su(4,4)$, $su^*(8)$ and $sl(8,\mathbb{R})$

5. The Lie algebras $sp(n,\mathbb{R})$ and $sp(\frac{n}{2},\frac{n}{2})$ (n–even)

Let $n \geq 2$. Let $\mathcal{G} = sp(n,\mathbb{R})$, the rank n split real form of $sp(n,\mathbb{C}) = \mathcal{G}^{\mathbb{C}}$. The maximal compact subgroup is $\mathcal{K} \cong u(1) \oplus su(n)$, while $\mathcal{M} = sl(n,\mathbb{R})$.

The signature of the ERs of $\mathcal{G}$ is [13]:

$$\chi = \{ n_1 , \ldots , n_{n-1} ; c \} , \qquad n_j \in \mathbb{N} . \tag{32}$$

The Knapp-Stein restricted Weyl reflection acts as follows:

$$\begin{aligned} G_{KS} &: \mathcal{C}_\chi \longrightarrow \mathcal{C}_{\chi'} , \\ \chi' &= \{ (n_1, \ldots, n_{n-1})^* ; -c \} , \\ (n_1, \ldots, n_{n-1})^* &\doteq (n_{n-1}, \ldots, n_1) \end{aligned} \tag{33}$$

In terms of an orthonormal basis ϵ_i, $i = 1, \ldots, n$, the positive roots are:

$$\Delta^+ = \{ \epsilon_i \pm \epsilon_j, \ 1 \leq i < j \leq n; \qquad 2\epsilon_i, \ 1 \leq i \leq n \} \tag{34}$$

the simple roots are:

$$\pi = \{ \alpha_i = \epsilon_i - \epsilon_{i+1}, \ 1 \leq i \leq n-1, \qquad \alpha_n = 2\epsilon_n \} \tag{35}$$

the non-compact roots:

$$\beta_{ij} \equiv \epsilon_i + \epsilon_j, \ , \ 1 \leq i \leq j \leq n , \tag{36}$$

the Harish-Chandra parameters: $m_\beta \equiv (\Lambda + \rho, \beta)$ for the noncompact roots are:

$$\begin{aligned} m_{\beta_{ij}} &= \Big(\sum_{s=i}^{n} + \sum_{s=j}^{n} \Big) m_s , \qquad i < j , \\ m_{\beta_{ii}} &= \sum_{s=i}^{n} m_s \end{aligned} \tag{37}$$

The correspondence between the signatures χ and the highest weight Λ is:

$$n_i = m_i , \quad c = -\tfrac{1}{2}(m_{\tilde{\alpha}} + m_n) = -\tfrac{1}{2}(m_1 + \cdots + m_{n-1} + 2m_n) \tag{38}$$

where $\tilde{\alpha} = \beta_{11}$ is the highest root.

The number of ERs in the corresponding multiplets by (13) is:

$$|W(\mathcal{G}^{\mathbb{C}}, \mathcal{H}^{\mathbb{C}})| \, / \, |W(\mathcal{M}^{\mathbb{C}}, \mathcal{H}^{\mathbb{C}}_m)| = 2^n \tag{39}$$

Below we give pictorially the multiplets for $sp(n,\mathbb{R})$ for $n = 2,3,4,5,6$. For $n = 2r$ these are also multiplets for $sp(r,r)$, $r = 1,2,3$.

Recalling the isomorphisms $sp(1,1) \cong so(4,1)$, $sp(2) \cong so(3,2)$, we repeat the $n = 2$ diagram as it was given originally in the cases $so(p,q)$ with $p + q = 5$: for $q = 1$ in [16] and for $q = 2$ in [30].

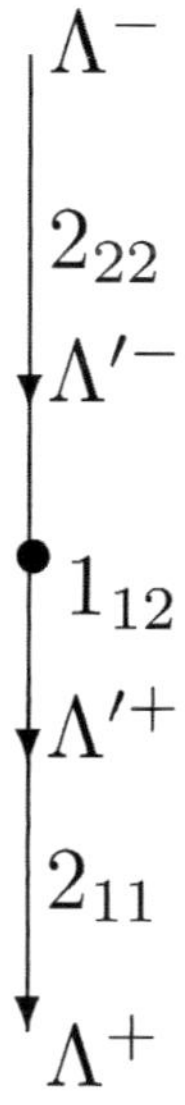

Main multiplets for $Sp(2, I\!\!R)$ and $Sp(1,1)$

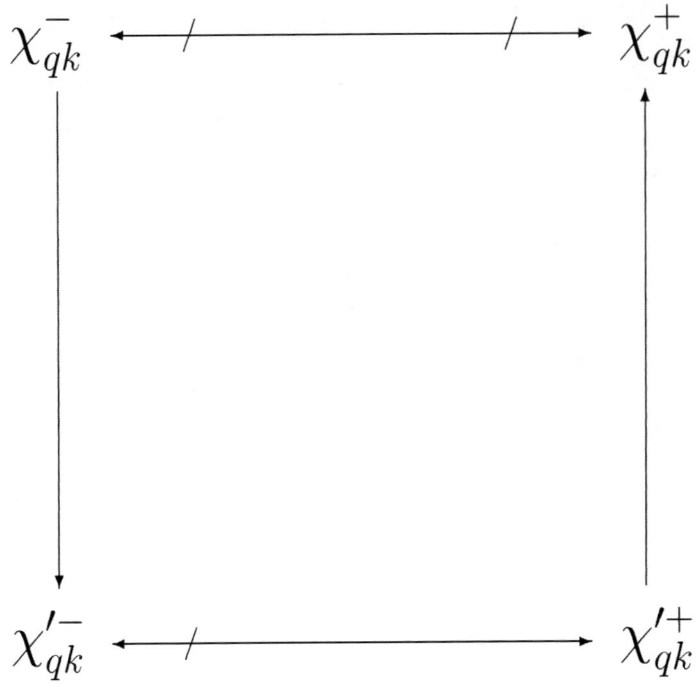

Quartet of partially equivalent ERs and intertwining operators given first for $so(4,1) \cong sp(1,1)$, then for $so(3,2) \cong sp(2,\mathbb{R})$ (arrows are differential operators, dashed arrows are integral operators)

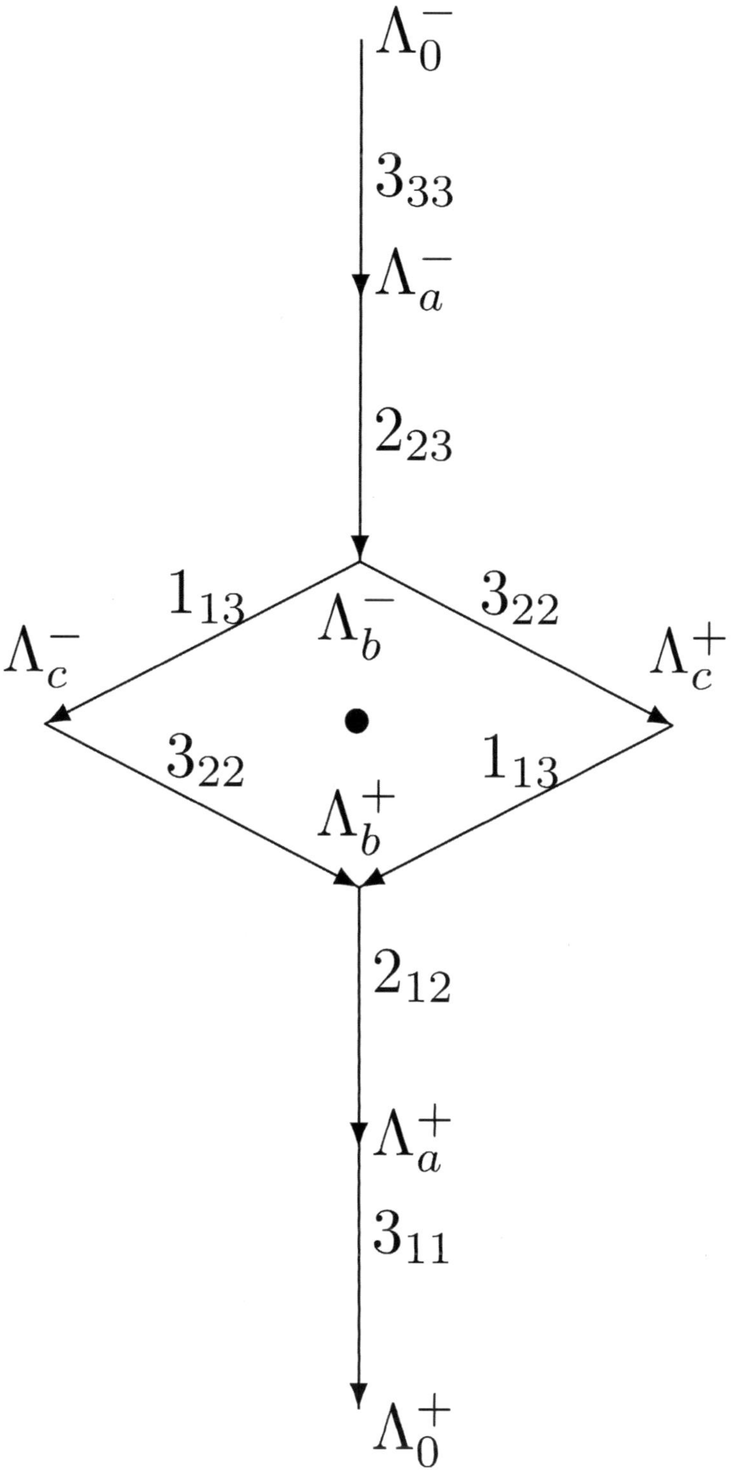

Main multiplets for $Sp(3, \mathbb{R})$

Journal of Physics: Conference Series **411** (2013) 012012 doi:10.1088/1742-6596/411/1/012012

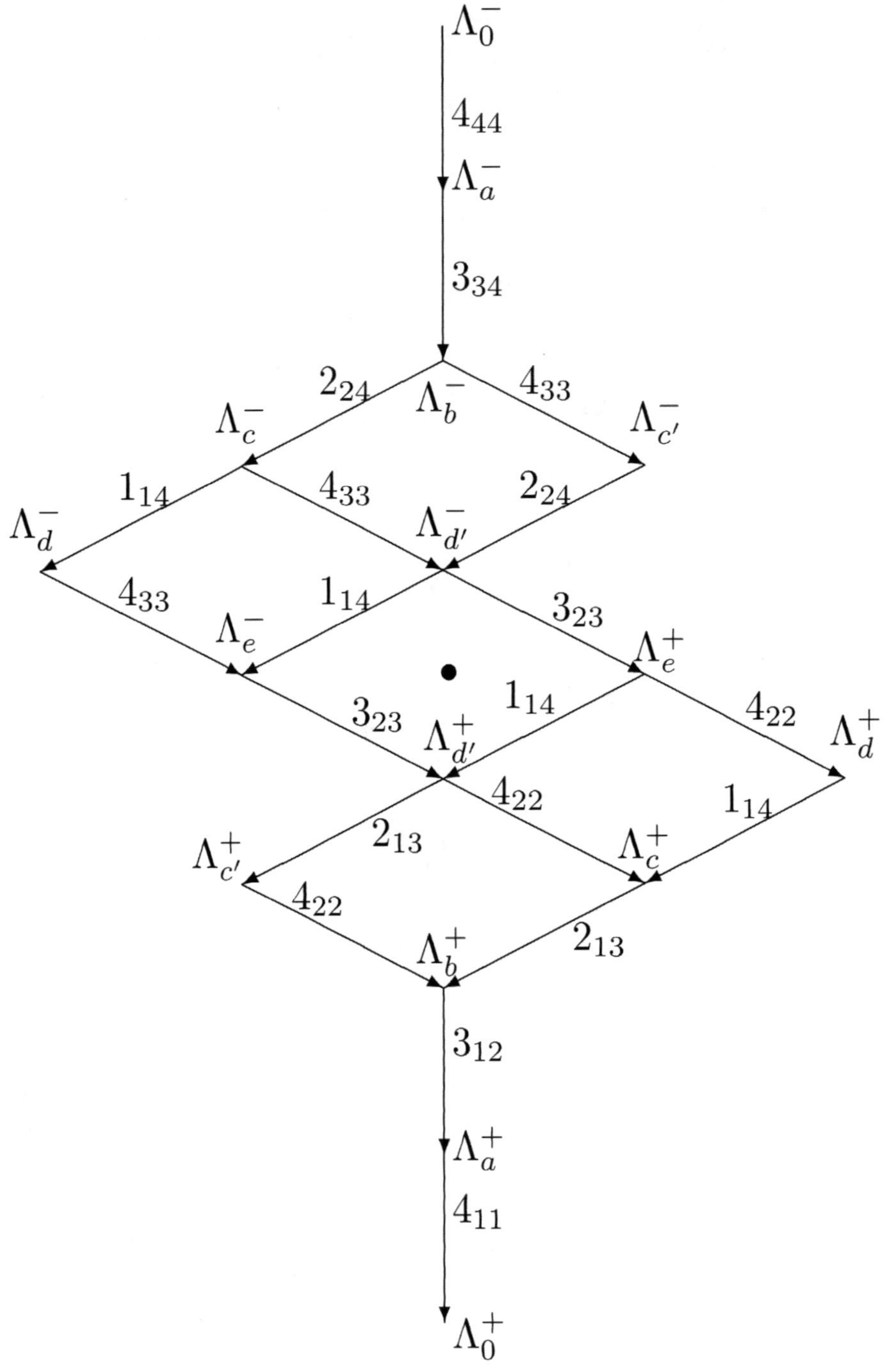

Main multiplets for $Sp(4, I\!R)$ and $Sp(2, 2)$

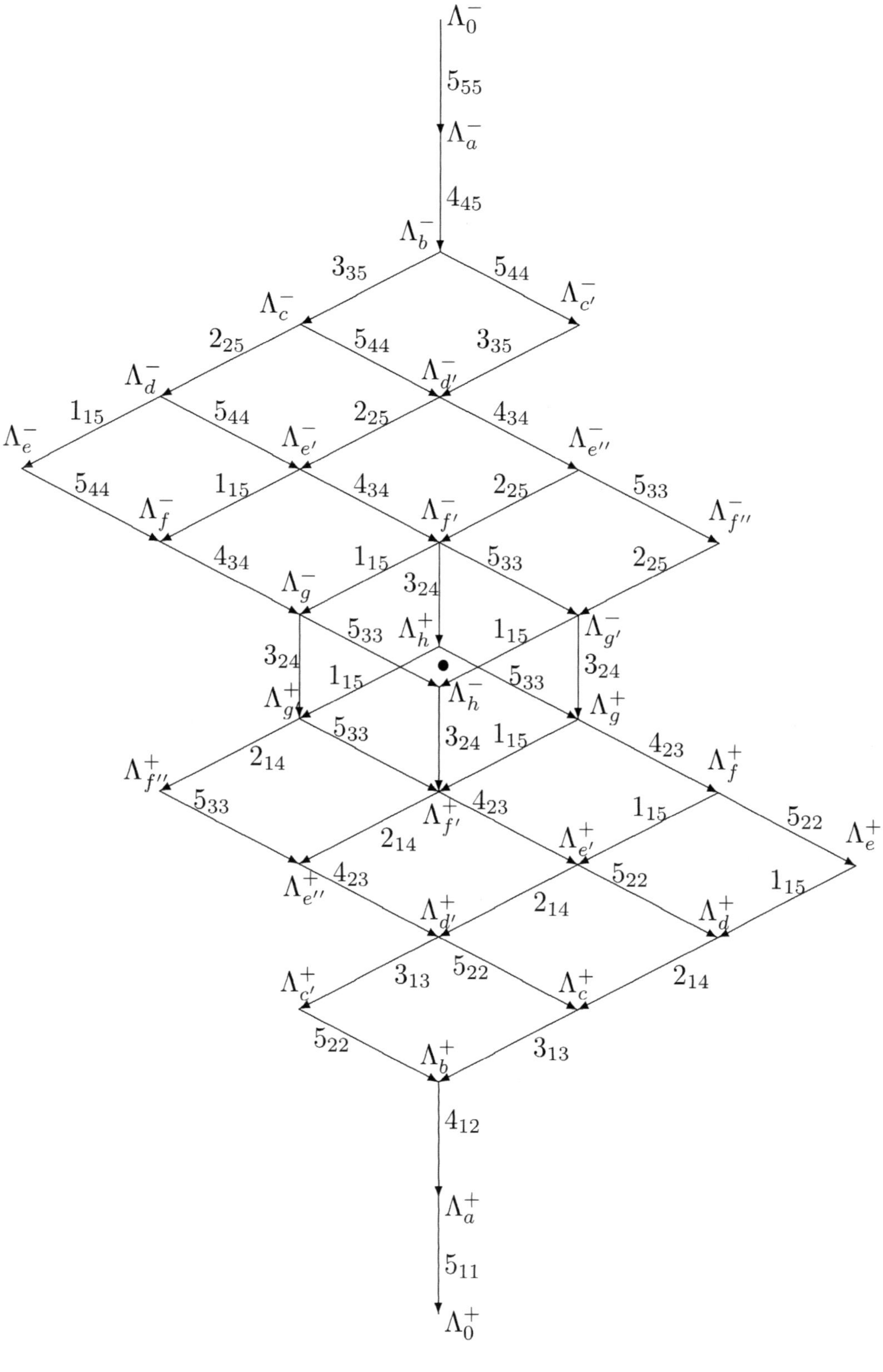

Main multiplets for $Sp(5, I\!R)$

Journal of Physics: Conference Series **411** (2013) 012012 doi:10.1088/1742-6596/411/1/012012

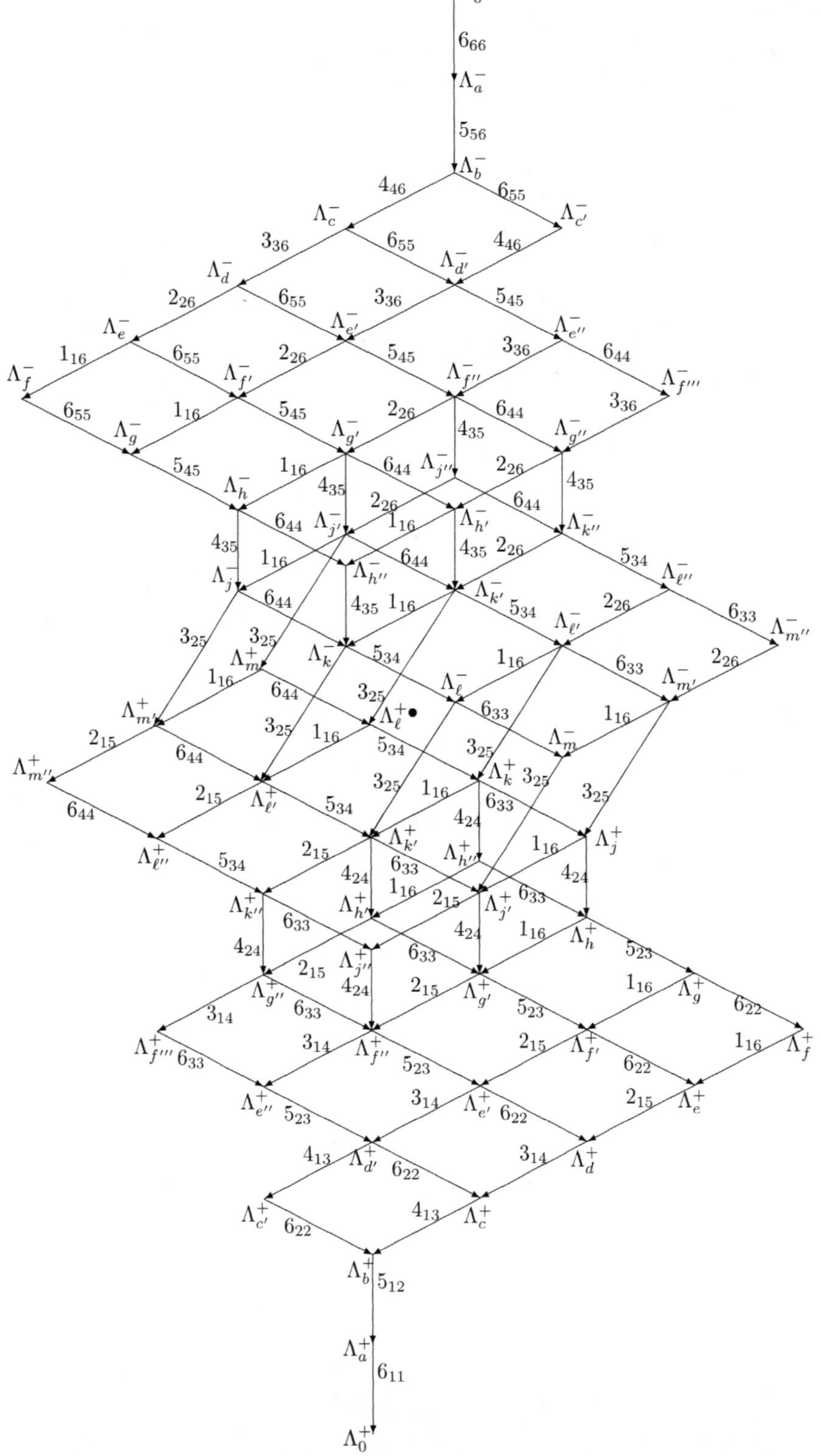

Main multiplets for $Sp(6,\mathbb{R})$ and $Sp(3,3)$

Journal of Physics: Conference Series **411** (2013) 012012 doi:10.1088/1742-6596/411/1/012012

6. The Lie algebras $E_{7(-25)}$ and $E_{7(7)}$

Let $\mathcal{G} = E_{7(-25)}$. The maximal compact subgroup is $\mathcal{K} \cong e_6 \oplus so(2)$. We work with maximal parabolic $\mathcal{P} = \mathcal{M} \oplus \mathcal{M} \oplus \mathcal{N}$ with $\mathcal{M} \cong E_{6(-26)}$.

The signatures of the ERs of $\mathcal{G}$ are [8]:

$$\chi = \{n_1, \ldots, n_6; c\}, \qquad n_j \in \mathbb{N}, \tag{40}$$

expressed through the Dynkin labels:

$$n_i = m_i, \quad c = -\tfrac{1}{2}(m_{\tilde{\alpha}} + m_7) = -\tfrac{1}{2}(2m_1 + 2m_2 + 3m_3 + 4m_4 + 3m_5 + 2m_6 + 2m_7) \tag{41}$$

The same holds for the parabolically related exceptional Lie algebra $E_{7(7)}$ (with $\mathcal{M}$-factor $E_{6(6)}$).

The noncompact roots of the complex algebra E_7 are:

$$\begin{aligned}
&\alpha_7, \quad \alpha_{17}, \quad \ldots, \alpha_{67}, \\
&\alpha_{1,37}, \ \alpha_{2,47}, \ \alpha_{17,4}, \ \alpha_{27,4}, \\
&\alpha_{17,34}, \ \alpha_{17,35}, \ \alpha_{17,36}, \ \alpha_{17,45}, \ \alpha_{17,46}, \\
&\alpha_{27,45}, \ \alpha_{27,46}, \\
&\alpha_{17,25,4}, \ \alpha_{17,26,4}, \ \alpha_{17,35,4}, \ \alpha_{17,36,4}, \\
&\alpha_{17,26,45}, \ \alpha_{17,36,45}, \\
&\alpha_{17,26,35,4}, \ \alpha_{17,26,45,4}, \\
&\alpha_{17,16,35,4} = \tilde{\alpha},
\end{aligned} \tag{42}$$

using compact notation:

$$\begin{aligned}
\alpha_{ij} &= \alpha_i + \alpha_{i+1} + \cdots + \alpha_j, \ i < j, \\
\alpha_{ij,k} &= \alpha_{k,ij} = \alpha_i + \alpha_{i+1} + \cdots + \alpha_j + \alpha_k, \qquad i < j, \qquad \text{etc.}
\end{aligned} \tag{43}$$

The multiplets of the main type are in 1-to-1 correspondence with the finite-dimensional irreps of E_7, i.e., they will be labelled by the seven positive Dynkin labels $m_i \in \mathbb{N}$. The number of ERs in the corresponding multiplets by (13) is equal to:

$$|W(\mathcal{G}^{\mathbb{C}}, \mathcal{H}^{\mathbb{C}})| \,/\, |W(\mathcal{M}^{\mathbb{C}}, \mathcal{H}_m^{\mathbb{C}})| = 56 \tag{44}$$

The Knapp-Stein operators $G^{\pm}_{\chi}$ act pictorially as reflection w.r.t. the bullet intertwining each $\mathcal{T}^-_{\chi}$ member with the corresponding $\mathcal{T}^+_{\chi}$ member.

Journal of Physics: Conference Series **411** (2013) 012012 doi:10.1088/1742-6596/411/1/012012

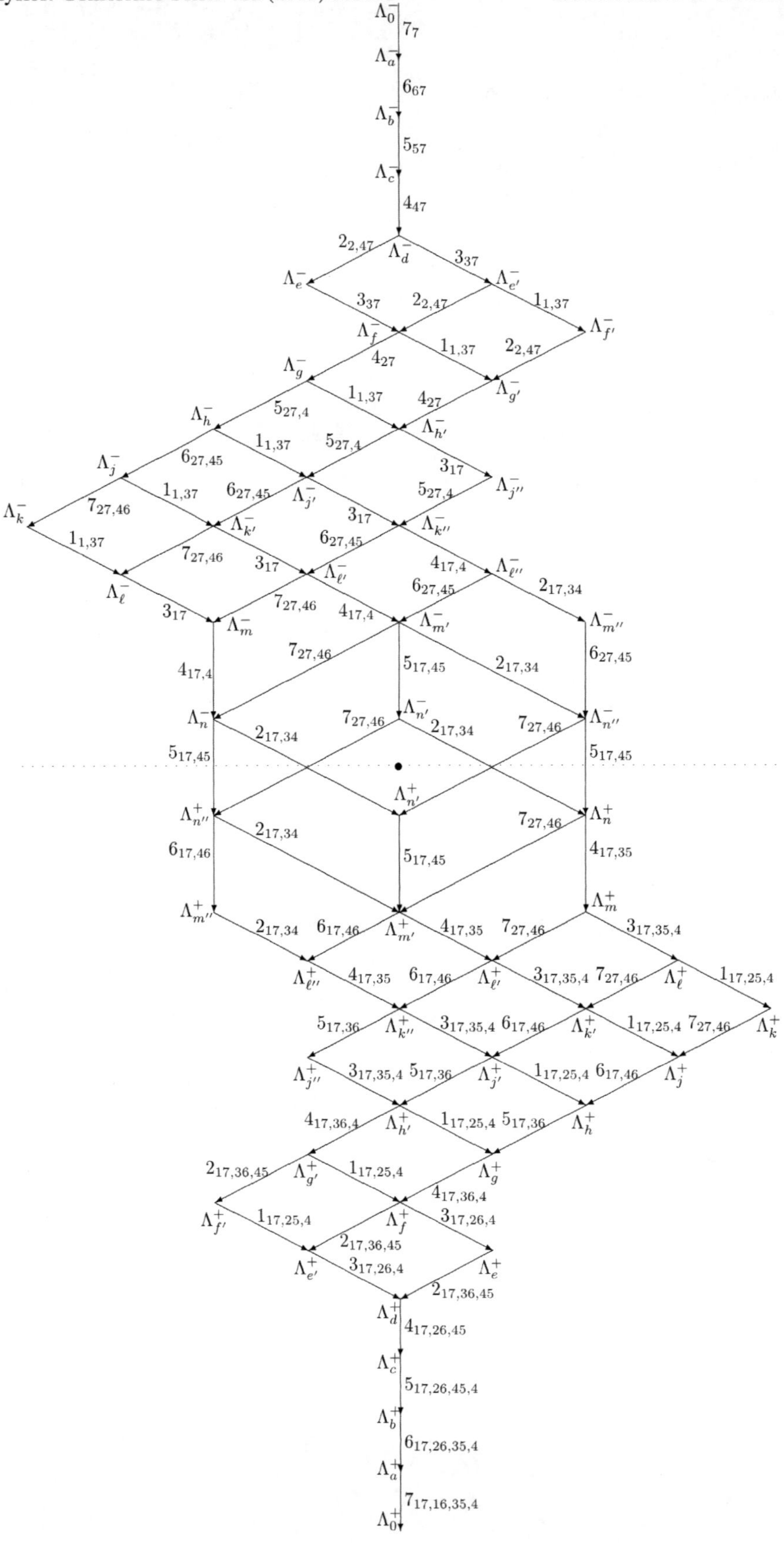

Main Type for $E_{7(-25)}$ and $E_{7(7)}$

Journal of Physics: Conference Series **411** (2013) 012012 doi:10.1088/1742-6596/411/1/012012

7. The Lie algebras $E_{6(-14)}$, $E_{6(6)}$ and $E_{6(2)}$

Let $\mathcal{G} = E_{6(-14)}$. The maximal compact subalgebra is $\mathcal{K} \cong so(10) \oplus so(2)$, while $\mathcal{M} \cong su(5,1)$.

The signature of the ERs of $\mathcal{G}$ is [14]:

$$\chi = \{\, n_1\,, n_3\,, n_4\,, n_5\,, n_6\,;\, c\} \;, \quad c = d - \tfrac{11}{2}\;, \tag{45}$$

expressed through the Dynkin labels as:

$$n_i = m_i\;, \quad -c \;=\; \tfrac{1}{2} m_{\tilde\alpha} \;=\; \tfrac{1}{2}(m_1 + 2m_2 + 2m_3 + 3m_4 + 2m_5 + m_6) \tag{46}$$

The same holds for the parabolically related exceptional Lie algebras $E_{6(6)}$ and $E_{6(2)}$ with $\mathcal{M}$-factors $sl(6,\mathbb{R})$ and $su(3,3)$, resp.

Further, we need the noncompact roots of the complex algebra E_6 :

$$\begin{aligned}
&\alpha_2\,,\ \alpha_{14}\,,\ \alpha_{15}\,,\ \alpha_{16}\,,\ \alpha_{24}\,,\ \alpha_{25}\,,\ \alpha_{26} \\
&\alpha_{2,4}\,,\ \alpha_{2,45}\,,\ \alpha_{2,46}\,,\ \alpha_{25,4}\,,\ \alpha_{15,4}\,,\ \alpha_{26,4} \\
&\alpha_{16,4}\,,\ \alpha_{15,34}\,,\ \alpha_{26,45}\,,\ \alpha_{16,34}\,,\ \alpha_{16,45} \\
&\alpha_{16,35}\,,\ \alpha_{16,35,4}\,,\ \alpha_{16,25,4} \;=\; \tilde\alpha
\end{aligned} \tag{47}$$

The multiplets of the main type are in 1-to-1 correspondence with the finite-dimensional irreps of $\mathcal{G}$, i.e., they will be labelled by the six positive Dynkin labels $m_i \in \mathbb{N}$. It turns out that each such multiplet contains 70 ERs/GVMs - see the figure below.

The Knapp-Stein operators $G^{\pm}_{\chi}$ act pictorially as reflection w.r.t. the dotted line separating the $\mathcal{T}^-_{\chi}$ members from the $\mathcal{T}^+_{\chi}$ members.

Note that there are five cases when the embeddings correspond to the highest root $\tilde\alpha$: $V^{\Lambda^-} \longrightarrow V^{\Lambda^+}$, $\Lambda^+ \;=\; \Lambda^- - m_{\tilde\alpha}\,\tilde\alpha$. In these five cases the weights are denoted as: $\Lambda^{\pm}_{k''}$, $\Lambda^{\pm}_{k'}$, $\Lambda^{\pm}_{\tilde k}$, $\Lambda^{\pm}_{k}$, $\Lambda^{\pm}_{k^o}$, then: $m_{\tilde\alpha} \;=\; m_1, m_3, m_4, m_5, m_6$, resp. Thus, their action coincides with the action of the Knapp-Stein operators G^{+}_{χ} which in the above five cases degenerate to differential operators as we discussed for $so(3,2)$.

Note that the figure has the standard E_6 symmetry, namely, conjugation exchanging indices $1 \longleftrightarrow 6$, $3 \longleftrightarrow 5$.

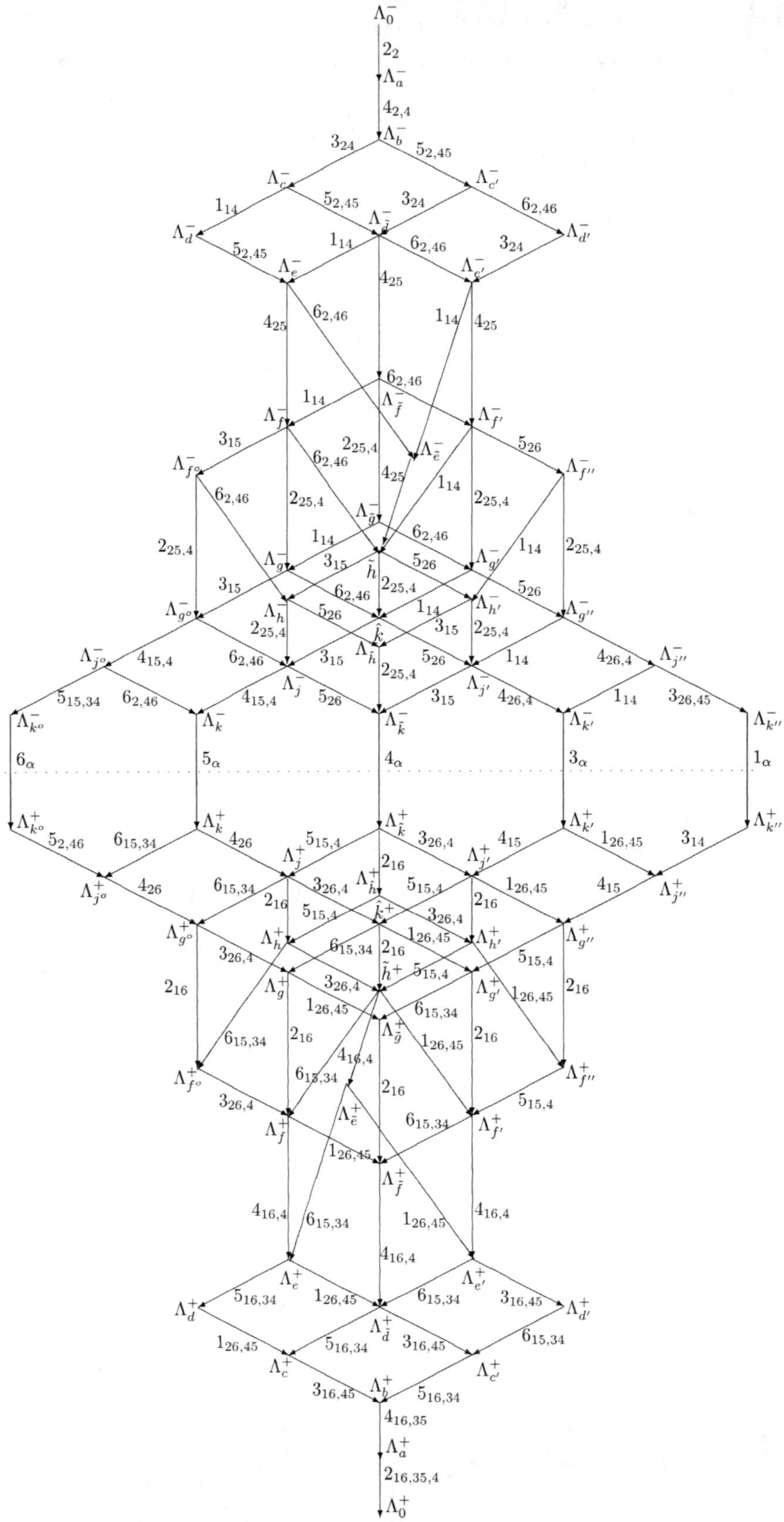

Main Type for $E_{6(-14)}$, $E_{6(6)}$ and $E_{6(2)}$

Acknowledgments

This work was supported in part by the Bulgarian National Science Fund, grant DO 02-257.

References

[1] J.F. Cornwell, *Group Theory in Physics, Vol. III*, (Academic Press, London & San Diego, 1989).
[2] J. Maldacena, Large N Field Theories, String Theory and Gravity, in: Lectures on Quantum Gravity, Series of the Centro De Estudios Cientficos, eds. A Gomberoff and D. Marolf, 2005, 91-150.
[3] J. Terning, *Modern Supersymmetry: Dynamics and Duality*, International Series of Monographs on Physics # 132, (Oxford University Press, 2005).
[4] V.K. Dobrev, Rev. Math. Phys. **20** (2008) 407-449.
[5] Harish-Chandra, Ann. Math. **116** (1966) 1-111.
[6] A.W. Knapp, *Representation Theory of Semisimple Groups (An Overview Based on Examples)*, (Princeton Univ. Press, 1986).
[7] F. Bruhat, "Sur les representations induites des groups de Lie", Bull. Soc. Math. France, **84** (1956) 97-205.
[8] V.K. Dobrev, J. Phys. **A42** (2009) 285203.
[9] J. Faraut and A. Korányi, *Analysis on Symmetric Cones*, Oxford Mathematical Monographs, (Clarendon Press, Oxford, 1994).
[10] M. Gunaydin, Mod. Phys. Lett. **A8** (1993) 1407-1416.
[11] G. Mack and M. de Riese, J. Math. Phys. **48** (2007) 052304.
[12] V.K. Dobrev, Invariant differential operators for non-compact Lie groups: the main $su(n,n)$ cases, Plenary talk at SYMPHYS XV, (Dubna, 12-16.7.2011), to appear in the Proceedings, eds. G. Pogosyan et al.
[13] V.K. Dobrev, Invariant differential operators for non-compact Lie groups: the $sp(n,r)$ case, arXiv:1205.5521v1 [hep-th], CERN-PH-TH/2012-143.
[14] V.K. Dobrev, arXiv:0812.2655 [math-ph], Proceedings, eds. B. Dragovich and Z. Rakic, (Institute of Physics, Belgrade, 2009) pp. 95-124.
[15] V.K. Dobrev, Invariant Differential Operators for Non-Compact Lie Algebras Parabolically Related to Conformal Lie Algebras, arXiv:1208.0409 [hep-th], CERN-PH-TH/2012-215.
[16] V.K. Dobrev, G. Mack, V.B. Petkova, S.G. Petrova and I.T. Todorov, *Harmonic Analysis on the n-Dimensional Lorentz Group and Its Applications to Conformal Quantum Field Theory*, Lecture Notes in Physics, Vol. 63 (Springer, Berlin, 1977).
[17] V.K. Dobrev, Rept. Math. Phys. **25** (1988) 159-181; first as ICTP Trieste preprint IC/86/393 (1986).
[18] V.K. Dobrev, Lett. Math. Phys. **9**, 205-211 (1985).
[19] V.K. Dobrev, J. Math. Phys. **26** (1985) 235-251.
[20] J. Dixmier, *Enveloping Algebras*, (North Holland, New York, 1977).
[21] V.K. Dobrev, J. Phys. **A28** (1995) 7135-7155; in: Proceedings, Symposium "Symmetries in Science IX", Bregenz, Austria, (August 1996), eds. B. Gruber and M. Ramek, (Plenum Press, New York and London, 1997) pp. 47-80.
[22] Harish-Chandra, "Representations of semisimple Lie groups: IV,V", Am. J. Math. **77** (1955) 743-777, **78** (1956) 1-41.
[23] V.K. Dobrev, Suppl. Rendiconti Circolo Matematici di Palermo, Serie II, Numero 43 (1996) 15-56.
[24] V.K. Dobrev, J. Phys. **A41** (2008) 425206; arXiv:0712.4375.
[25] A.W. Knapp and E.M. Stein, Ann. Math. **93** (1971) 489-578; II : Inv. Math. **60** (1980) 9-84.
[26] I.M. Gelfand and M.A. Naimark, Acad. Sci. USSR. J. Phys. **10** (1946) 93-94.
[27] V. Bargmann, Annals Math. **48**, (1947) 568-640.
[28] V.K. Dobrev and V.B. Petkova, Rept. Math. Phys. **13** (1978) 233-277.
[29] V.B. Petkova and G.M. Sotkov, Lett. Math. Phys. **8** (1984) 217-226, Erratum-ibid. 9 (1985) 83.
[30] V.K. Dobrev, J. Phys. **A39** (2006) 5995-6020.

XXth International Conference on Integrable Systems and Quantum Symmetries (ISQS-20) IOP Publishing
Journal of Physics: Conference Series **411** (2013) 012013 doi:10.1088/1742-6596/411/1/012013

Introduction to Khovanov Homologies. II. Reduced Jones superpolynomials

V. Dolotin and A. Morozov
ITEP, Moscow, Russia

Abstract. A second part of detailed elementary introduction into Khovanov homologies. This part is devoted to *reduced* Jones superpolynomials. The story is still about a hypercube of resolutions of a link diagram. Each resolution is a collection of non-intersecting cycles, and one associates a 2-dimensional vector space with each cycle. *Reduced* superpolynomial arises when for all cycles, containing a "marked" edge of the link diagram, the vector space is reduced to 1-dimensional. The rest remains the same. Edges of the hypercube are associated with cut-and-join operators, acting on the cycles. Superpartners of these operators can be combined into differentials of a complex, and superpolynomial is the Poincare polynomial of this complex. HOMFLY polynomials are practically the same in reduced and unreduced case, but superpolynomials are essentially different, already in the simplest examples of trefoil and figure-eight knot.

1. Introduction

This text is a continuation of [1], and is formulated not independently, but rather as a set of comments to that one. We assume that the reader is familiar with [1], use the same terminology and constructions, without going into lengthy explanations. All the references to relevant original works can be also found in [1]. References to formulas from [1] are given in the form (I.x) – in the present paper we refer to the version 1 (v1) of [1].

For relevant foundations of knot theory see [2]-[4]. For basic references on Khovanov-Rozansky homologies and superpolynomials see [5]-[21] and [22]-[41] respectively.

2. From unreduced to reduced Jones superpolynomial

2.1. Ordinary Jones

The unreduced Jones polynomial in the form (I.13) is obviously divisible by $D = q + q^{-1}$:

$$J^{\Gamma_c}(q) = (-)^{n_\circ} q^{n_\bullet - 2n_\circ} \sum_{\text{resolutions } r \text{ of } \Gamma} (-q)^{|r - r_c|} D^{\nu_r} = D \cdot \underline{J}^{\Gamma_c}(q) \tag{1}$$

The ratio $\underline{J}^{\Gamma}(q)$ is called *reduced* Jones polynomial. In what follows we underline the variables (times and their superpartners), which are being reduced, as well as the objects (differential, cohomologies, knot polynomials) obtained *in result* of the reduction – hopefully, this does not cause confusion.

To obtain $\underline{J}^{\Gamma}(q)$ from (I.13) one should say that one of the cycles at each vertex of the resolutions hypercube contributes not D, but just 1. For this purpose one can mark one *edge*

$E = e$ of Γ – in each resolution there will be exactly one cycle, containing e,– and let *this* cycle contribute 1 instead of D.

Since there is always an item in the sum (1) with $\nu_r = 1$, there is only one common power of D – and thus only one edge can be marked in Γ in above construction. Of course, the answer does not depend on the choice of this single e.

2.2. Reduced superpolynomial

Now it is clear, what should be done in Khovanov's T-deformation: one should reduce the vector spaces V in (I.33), associated with all the cycles, which contain the marked edge e, from two- to one-dimensional. This means that the corresponding θ-variables in (I.46) should be nullified.

It is important here that the maps Q in (I.40) diminish the q-grading by one: because of this θ-derivatives are always multiplied by θ – what makes such reduction self-consistent. This would not be true if we tried to nullify η-variables instead.

In other words, nullification of underlined elements v_- in the following formulas is self-consistent, while it would not be like that if one attempts to nullify v_+:

$$
\begin{array}{lccc}
\underline{V} \to \underline{V} \otimes V & & & \\
 & v_+ \longrightarrow v_+ \otimes v_- + \underline{v_-} \otimes v_+ & \Longrightarrow & v_+ \longrightarrow v_+ \otimes v_-, \\
 & \underline{v_-} \longrightarrow \underline{v_-} \otimes v_- & \Longrightarrow & 0 \longrightarrow 0 \\
\underline{V} \otimes V \longrightarrow \underline{V} & & & \\
 & \underline{v_-} \otimes v_- \longrightarrow 0 & \Longrightarrow & 0 \longrightarrow 0, \\
 & \underline{v_-} \otimes v_+ \longrightarrow \underline{v_-} & \Longrightarrow & 0 \longrightarrow 0, \\
 & v_+ \otimes v_- \longrightarrow \underline{v_-} & \Longrightarrow & v_+ \otimes v_- \longrightarrow 0, \\
 & v_+ \otimes v_+ \longrightarrow v_+ & \Longrightarrow & v_+ \otimes v_+ \longrightarrow v_+
\end{array}
$$

3. Example: Trefoil 3_1 in a 2-strand realization

3.1. Hypercube, Jones and the cut-and-join operator

Extended Jones polynomial (I.7.4.1) is

$$
\mathcal{J}^{\bullet\bullet\bullet} = \underline{p_3}p_3' + t(\underline{p_6''} + \underline{p_6'} + \underline{p_6}) + t^2(\underline{p_4}p_2 + p_4'\underline{p_2'} + \underline{p_4''}p_2'') + t^3 p_2\underline{p_2'}p_2'' \qquad (2)
$$

Underlined are the terms, eliminated by reduction – and there is exactly one such factor in each item.

This Jones polynomial and the cut an join operator below are build with the help of the hypercube quiver:

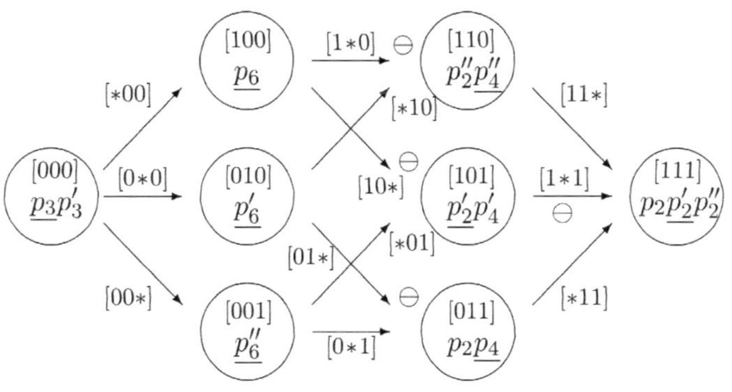

Dimension three of the hypercube is the number of vertices in the knot diagram, these vertices correspond to the three positions in the triple sequences $[\cdot\ \cdot\ \cdot]$.

Vertices of the hypercube are in one-to-one correspondence with resolutions of the knot diagram into sets of non-intersecting cycles. If one of the two possible resolutions of the crossing is chosen in a given vertex of the knot diagram, we put 0 or 1 into the binary label $[\cdot\ \cdot\ \cdot]$ of the vertex. Cycles are labeled by p_k, where k is the number of edges of the knot diagram in the cycle, and different cycles of the same length are labeled by different number of primes. The sum of labels k in each hypercube vertex is equal to the number of edges in the knot diagram (in the present case it is 6). Underlined are the cycles, which contain the marked edge of the knot diagram, there is exactly one underlined p at each hypercube vertex.

Edges of the hypercube are in one-two-one correspondence with the flips between two possible resolutions at some vertex of the knot diagram. Thus they connect two hypecube vertices differing by a single entry in the binary labels – it stands at the position of the vertex where the flip takes place and we put $*$ at this position to label the edges. A sign is attributed to the edge, to which a factor -1 is contributed by all unities in the binary label, which stand to the left of the $*$, see (I.45). Hypercube edges to which minus sign is attributed in this way are labeled by $\ominus$.

Hypercubes are the same for all knot diagrams of the same topology, and do not depend on the coloring of the vertices of the knot diagram. This affects only the choice "initial" hypercube vertex, when 0 is attributed to all black vertices (where $\mathcal{R}$ stands in Turaev-Reshetikhin representation of HOMFLY polynomial) and 1 – to all white vertices (where $\mathcal{R}^{-1}$ stands).

Extended Jones polynomial is just a sum of the entries over all hypercube vertices, weighted with t in the power, equal to the number of arrows, containing the vertex with "initial" one.

Similarly the cut-and-join operator is a sum over all edges, with $\oplus$ signs marking the terms, where differentials from associated BRST operator will act with the negative signs, see below.

Thus the cut-and-join operator, *dual* to (2), is $K^{\bullet\bullet\bullet} = \sum_{i=1}^{3} K_i$ with

$$K_1 = (\underline{p_6} + \underline{p_6'} + \underline{p_6''})\frac{\partial^2}{\partial \underline{p_3}\partial p_3'},$$

$$K_2 = p_2\underline{p_4}\left(\frac{\partial}{\partial \underline{p_6''}} \ominus \frac{\partial}{\partial \underline{p_6'}}\right) + \underline{p_2'}p_4\left(\frac{\partial}{\partial \underline{p_6''}} \ominus \frac{\partial}{\partial \underline{p_6}}\right) + p_2''\underline{p_4''}\left(\frac{\partial}{\partial \underline{p_6'}} \ominus \frac{\partial}{\partial \underline{p_6}}\right),$$

$$K_3 = \underline{p_2'}p_2''\frac{\partial}{\partial \underline{p_4}} \ominus p_2p_2''\frac{\partial}{\partial p_4'} + p_2\underline{p_2'}\frac{\partial}{\partial \underline{p_4''}} \tag{3}$$

Note that whenever all derivatives over underlined variables are multiplied by underlined variables – this will make reduction procedure consistent for the differentials. Note also that there are terms in the cut-and-join operator, where no variables are underlined: reduction touches all the vertices, but not all the edges of the hypercube.

3.2. Differentials and cohomologies

Differentials are naturally written in [1] through Grassmannian variables $\vartheta_k = (\theta_k, \eta_k)$, and reduction implies that the underlined variables $\underline{\theta_k}$ and derivatives over them are eliminated, leaving only the corresponding η_k (which we do not underline after that). However, introduction of Grassmannian variables requires accurate work with the signs. We fix the basis vectors in the space of Grassmannian variables lexicographically: θ_2' stands before η_2', before θ_2'', before θ_3'..., i.e. the basis vector is $\eta_2''\theta_3$ rather than $\theta_3\eta_2''$ or $\theta_3\eta_3'$ rather than $\eta_3'\theta_3$. Note that the sign $\ominus$ refers to the sign in the transformation matrix, the sign of the corresponding term in the differential depends on the way the variables are ordered in this term – and does not need to coincide with $\ominus$ in bosonic form of the cut-and-join operators. The second derivative over Grassmann variables is defined in inverse order, as $\frac{\partial^2}{\partial\theta\partial\eta} = \frac{\partial}{\partial\eta}\frac{\partial}{\partial\theta}$, so that $\frac{\partial^2}{\partial\theta\partial\eta}\theta\eta = +1$.

For example, the differential

$$d_1 = (\eta_6 + \eta_6' + \eta_6'')\frac{\partial^2}{\partial\eta_3\partial\eta_3'} + (\theta_6 + \theta_6' + \theta_6'')\left(\frac{\partial^2}{\partial\eta_3\partial\theta_3'} + \frac{\partial^2}{\partial\theta_3\partial\eta_3'}\right) \tag{4}$$

acts on the basis vectors of the space $\mathcal{C}_1$ as follows:

$$\begin{array}{c|cccc} \mathcal{C}_1 & \theta_3\theta_3' & \theta_3\eta_3' & \eta_3\theta_3' & \eta_3\eta_3' \\ d_1 \downarrow & & & & \\ d\mathcal{C}_1 & 0 & \theta_6+\theta_6'+\theta_6'' & \theta_6+\theta_6'+\theta_6'' & \eta_6+\eta_6'+\eta_6'' \end{array} \tag{5}$$

i.e. provides basis vectors of $\mathcal{C}_2$ with all signs positive.

If, however, we take d_3 the situation will be different. The naive superpartner to K_3 would be

$$\theta_2\theta_2'\frac{\partial}{\partial\theta_4''} \ominus \theta_2\theta_2''\frac{\partial}{\partial\theta_4'} + \theta_2'\theta_2''\frac{\partial}{\partial\theta_4} + (\theta_2\eta_2' + \eta_2\theta_2')\frac{\partial}{\partial\eta_4''} \ominus (\theta_2\eta_2'' + \eta_2\theta_2'')\frac{\partial}{\partial\eta_4'} + (\theta_2'\eta_2'' + \eta_2'\theta_2'')\frac{\partial}{\partial\eta_4} \tag{6}$$

However such operator converts a basis vector $\theta_2\theta_4$ into $-\theta_2\theta_2'\theta_2''$, which is minus the basis vector. At the same time another basis vector $\theta_2'\theta_4'$ is converted into $\oplus\theta_2\theta_2'\theta_2''$, i.e. would be correct if we substitute $-$ instead of $\ominus$. This means that the signs should be changed appropriately:

$$d_3 = -\left(\theta_2\theta_2'\frac{\partial}{\partial\theta_4''}\hat{+}\theta_2\theta_2''\frac{\partial}{\partial\theta_4'} + \theta_2'\theta_2''\frac{\partial}{\partial\theta_4} + (\theta_2\eta_2' + \eta_2\theta_2')\frac{\partial}{\partial\eta_4''}\hat{+}(\theta_2\eta_2'' + \eta_2\theta_2'')\frac{\partial}{\partial\eta_4'} + (\theta_2'\eta_2'' + \eta_2'\theta_2'')\frac{\partial}{\partial\eta_4}\right) \tag{7}$$

and it turns out that on the place of $\ominus$ in this case one should substitute the same signs as everywhere else: we put hats over the two terms with $\ominus$ which would naively have different signs. In eqs.(13) and (15) below we shall see that the hatted signs are absolutely necessary to guarantee the nilpotency property $d_3d_2 = 0$, in particular that $\mathrm{Im}(d_2) \subset \mathrm{Ker}(d_3)$.

After this comment we can return to our main line. The differential d_1 is reduced as follows:

$$d_1 = (\eta_6'' + \eta_6' + \eta_6)\frac{\partial^2}{\partial\eta_3\partial\eta_3'} + (\underline{\theta_6''} + \underline{\theta_6'} + \underline{\theta_6})\left(\frac{\partial^2}{\partial\eta_3\partial\theta_3'} + \frac{\partial^2}{\partial\underline{\theta_3}\partial\eta_3'}\right) \Longrightarrow \underline{d_1} = (\eta_6'' + \eta_6' + \eta_6)\frac{\partial^2}{\partial\eta_3\partial\eta_3'} \tag{8}$$

Underlined here are the variables which should be nullified in the reduction. Obviously the only derivative w.r.t. an underlined variables ($\underline{\theta_3}$) comes multiplied by the underlined variables.

The kernel and cohomology of d_1 are changed by the reductions as follows:

$$\begin{aligned} \mathrm{Ker}(d_1) = \left\{\underline{\theta_3}\theta_3,\ \eta_3\theta_3' - \underline{\theta_3}\eta_3'\right\} &\Longrightarrow \mathrm{Ker}(\underline{d_1}) = \left\{\eta_3\theta_3'\right\}, \\ \dim_q(H_0) = \dim_q\mathrm{Ker}(d_1) = q^{-2} + 1 &\Longrightarrow \dim_q(\underline{H_0}) = \dim_q\mathrm{Ker}(\underline{d_1}) = 1 \end{aligned} \tag{9}$$

and the image of d_1 is

$$\mathrm{Im}(d_1) = \left\{\underline{\theta_6''} + \underline{\theta_6'} + \underline{\theta_6},\ \eta_6'' + \eta_6' + \eta_6\right\} \Longrightarrow \mathrm{Im}(\underline{d_1}) = \left\{\eta_6'' + \eta_6' + \eta_6\right\} \tag{10}$$

Similarly, for d_2:

$$\begin{aligned} d_2 = \theta_2\underline{\theta_4}\left(\frac{\partial}{\partial\underline{\theta_6''}} - \frac{\partial}{\partial\underline{\theta_6'}}\right) + \underline{\theta_2'}\theta_4'\left(\frac{\partial}{\partial\underline{\theta_6''}} - \frac{\partial}{\partial\underline{\theta_6}}\right) + \theta_2''\underline{\theta_4''}\left(\frac{\partial}{\partial\underline{\theta_6'}} - \frac{\partial}{\partial\underline{\theta_6}}\right) + \\ +(\underline{\eta_2\theta_4} + \theta_2\eta_4)\left(\frac{\partial}{\partial\eta_6''} - \frac{\partial}{\partial\eta_6'}\right) + (\eta_2'\theta_4' + \underline{\theta_2'}\eta_4')\left(\frac{\partial}{\partial\eta_6''} - \frac{\partial}{\partial\eta_6}\right) + (\eta_2''\underline{\theta_4''} + \theta_2''\eta_4'')\left(\frac{\partial}{\partial\eta_6'} - \frac{\partial}{\partial\eta_6}\right) \\ \Longrightarrow \quad \underline{d_2} = \theta_2\eta_4\left(\frac{\partial}{\partial\eta_6''} - \frac{\partial}{\partial\eta_6'}\right) + \eta_2'\theta_4'\left(\frac{\partial}{\partial\eta_6''} - \frac{\partial}{\partial\eta_6}\right) + \theta_2''\eta_4''\left(\frac{\partial}{\partial\eta_6'} - \frac{\partial}{\partial\eta_6}\right) \end{aligned} \tag{11}$$

so that

$$\mathrm{Ker}(d_2) = \Big\{\eta_6'' + \eta_6' + \eta_6,\ \underline{\theta_6''} + \underline{\theta_6'} + \underline{\theta_6}\Big\} \ \Longrightarrow\ \mathrm{Ker}(\underline{d_2}) = \Big\{\eta_6'' + \eta_6' + \eta_6\Big\},$$
$$H_1 = \mathrm{Ker}(d_2)/\mathrm{Im}(d_1) = \emptyset \ \Longrightarrow\ \underline{H_1} = \mathrm{Ker}(\underline{d_2})/\mathrm{Im}(\underline{d_1}) = \emptyset \tag{12}$$

and

$$\mathrm{Im}(d_2) = \Big\{\theta_2\underline{\theta_4} + \underline{\theta_2'}\theta_4',\ \theta_2''\underline{\theta_4''} + \underline{\theta_2'}\theta_4',\ (\eta_2\underline{\theta_4} + \theta_2\eta_4) + (\eta_2'\theta_4' + \underline{\theta_2'}\eta_4'),\ (\eta_2''\underline{\theta_4''} + \theta_2''\eta_4'') + (\eta_2'\theta_4' + \underline{\theta_2'}\eta_4')\Big\}$$
$$\Longrightarrow\quad \mathrm{Im}(\underline{d_2}) = \Big\{\theta_2\eta_4 + \eta_2'\theta_4',\ \theta_2''\eta_4'' + \eta_2'\theta_4'\Big\} \tag{13}$$

Finally, for d_3:

$$-d_3 = \theta_2\underline{\theta_2'}\frac{\partial}{\partial\theta_4''}\hat{+}\theta_2\theta_2''\frac{\partial}{\partial\theta_4'} + \underline{\theta_2'}\theta_2''\frac{\partial}{\partial\theta_4} + (\theta_2\eta_2' + \eta_2\underline{\theta_2'})\frac{\partial}{\partial\eta_4''}\hat{+}(\theta_2\eta_2'' + \eta_2\theta_2'')\frac{\partial}{\partial\eta_4'} + (\underline{\theta_2'}\eta_2'' + \eta_2'\theta_2'')\frac{\partial}{\partial\eta_4}$$
$$\Longrightarrow\quad -\underline{d_3} = \hat{+}\theta_2\theta_2''\frac{\partial}{\partial\theta_4'} + \theta_2\eta_2'\frac{\partial}{\partial\eta_4''}\hat{+}(\theta_2\eta_2'' + \eta_2\theta_2'')\frac{\partial}{\partial\eta_4'} + \theta_2''\eta_2'\frac{\partial}{\partial\eta_4} \tag{14}$$

so that

$$\mathrm{Ker}(d_3) = \Big\{\theta_2\underline{\theta_4}\hat{+}\underline{\theta_2'}\theta_4',\ \theta_2''\underline{\theta_4}''\hat{+}\underline{\theta_2'}\theta_4',\ \theta_2\eta_4\hat{+}\eta_2'\theta_4' - \eta_2''\underline{\theta_4''},\ \theta_2''\eta_4''\hat{+}\eta_2'\theta_4' - \eta_2\underline{\theta_4},\ \eta_2\underline{\theta_4}\hat{+}\underline{\theta_2'}\eta_4' + \eta_2''\underline{\theta_4''}\Big\}$$
$$\boxed{\Longrightarrow}\quad \mathrm{Ker}(\underline{d_3}) = \Big\{\theta_2\eta_4\hat{+}\eta_2'\theta_4',\ \theta_2''\eta_4''\hat{+}\eta_2'\theta_4',\ \eta_2\eta_4\hat{+}\eta_2'\eta_4' + \eta_2''\eta_4''\Big\},$$
$$H_2 = \mathrm{Ker}(d_3)/\mathrm{Im}(d_2) = \Big\{\eta_2\underline{\theta_4}\hat{+}\underline{\theta_2'}\eta_4' + \eta_2''\underline{\theta_4''}\Big\}$$
$$\boxed{\Longrightarrow}\quad \underline{H_2} = \mathrm{Ker}(\underline{d_3})/\mathrm{Im}(\underline{d_2}) = \Big\{\eta_2\eta_4 + \eta_2'\eta_4' + \eta_2''\eta_4''\Big\},$$
$$\dim_q(H_2) = q^0 = 1, \qquad \dim_q(\underline{H_2}) = q^2, \tag{15}$$

and

$$\mathrm{Im}(d_3) = \Big\{\theta_2\underline{\theta_2'}\theta_2'',\ \eta_2\underline{\theta_2'}\theta_2'',\ \theta_2\eta_2'\theta_2'',\ \theta_2\underline{\theta_2'}\eta_2'',\ \theta_2\eta_2'\eta_2'',\ \eta_2\eta_2'\theta_2'',\ \eta_2\underline{\theta_2'}\eta_2''\Big\}$$
$$\Longrightarrow\quad \mathrm{Im}(\underline{d_3}) = \Big\{\theta_2\eta_2'\theta_2'',\ \eta_2\eta_2'\theta_2'',\ \theta_2\eta_2'\eta_2''\Big\} \tag{16}$$

i.e.

$$\underline{H_3} = \mathrm{Coim}(\underline{d_3}) = \Big\{\eta_2\eta_2'\eta_2''\Big\}, \qquad \dim_{\mathrm{q}}(\underline{H_3}) = q^3 \tag{17}$$

and this time this coincides with

$$H_3 = \mathrm{Coim}(\underline{d_3}) = \Big\{\eta_2\eta_2'\eta_2''\Big\}, \qquad \dim_{\mathrm{q}}(H_3) = q^3 \tag{18}$$

In fact the boxed arrow above is not naive. Note that in variance with the cases of d_1 and d_2 the dimensions of kernels and images are not just divided by two when the reduction is performed: it is enough to say that for d_3 these dimensions are odd. Here we encounter a more serious reshuffling. It is implied by the fact that

$$d_3\Big(\eta_2\eta_4 + \eta_2'\eta_4' + \eta_2''\eta_4''\Big) = 2\eta_2\underline{\theta_2'}\eta_2'' \tag{19}$$

In the image of d_3 this fact is reflected just by nullification of $\eta_2\underline{\theta_2'}\eta_2'' \in \mathrm{Im}(d_3)$, which thus drops away from $\mathrm{Im}(\underline{d_3})$. However, in the kernels this looks slightly more involved: this combination of the q-grading level 2 did not belong to $\mathrm{Ker}(d_3)$, but it appears $\mathrm{Ker}(\underline{d_3})$ and starts contributing to $\underline{H_2}$. Instead, the terms of q-grading 0, which contributed to H_2, are nullified and disappear from $\underline{H_2}$. This has a drastic impact on the form of the two superpolynomial, unreduced and reduced: the powers of q in front of T^2 will be different in these two cases.

3.3. The Jones superpolynomial

In result unreduced Jones superpolynomial is

$$P^{3_1}(q|T) = q^3 \sum_{j=0}^{3} (qT)^j \dim_q(H_j) =$$
$$= q^3\Big((q^{-2}+1)\cdot(qT)^0 + 0\cdot(qT)^1 + q^0(qT)^2 + q^3(qT)^3\Big) = q + q^3 + q^5T^2 + q^9T^3 \tag{20}$$

while the reduced one is

$$\underline{P}^{3_1}(q|T) = q^{-1}\cdot q^3 \sum_{j=0}^{3} (qT)^j \dim_q(\underline{H_j}) =$$
$$= q^2\Big(1\cdot(qT)^0 + 0\cdot(qT)^1 + q^2(qT)^2 + q^3(qT)^3\Big) = \boxed{q^2 + q^6T^2 + q^8T^3} \tag{21}$$

and

$$\boxed{P^{3_1} = (q+q^{-1})\underline{P}^{3_1} \;-\; q^7(1+T)\,T^2} \tag{22}$$

The extra factor q^{-1} in the definition of $\underline{P}^{3_1}(q|T)$ compensates for non-trivial q-grading power of the reduced 1-dimensional space (we keep v_+ with the grading q as its basis element, while it should be rather shifted to 1). Minus sign in front of the last "correction term" shows that the *unreduced* superpolynomial is "more refined": the T-deformation of reduced Jones, after multiplication by D/q, can be further reduced – to provide a "smaller" superpolynomial, which is a deformation of the unreduced Jones.

4. Reduced Jones superpolynomial for the figure-eight knot

4.1. Braid

The trefoil can be made not only from the 2-strand braid with 3 crossings, but also from the 3-strand braid with 4 crossings. Remarkably, in Khovanov formalism a similar representation for the non-torus figure-eight knot 4_1 differs only by the change of coloring.

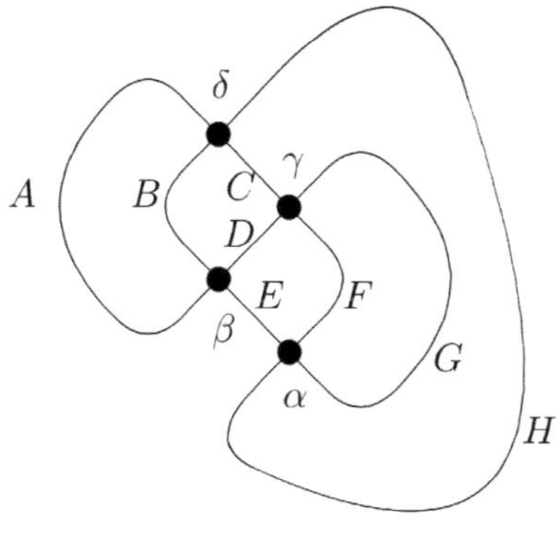

3_1 knot (trefoil)

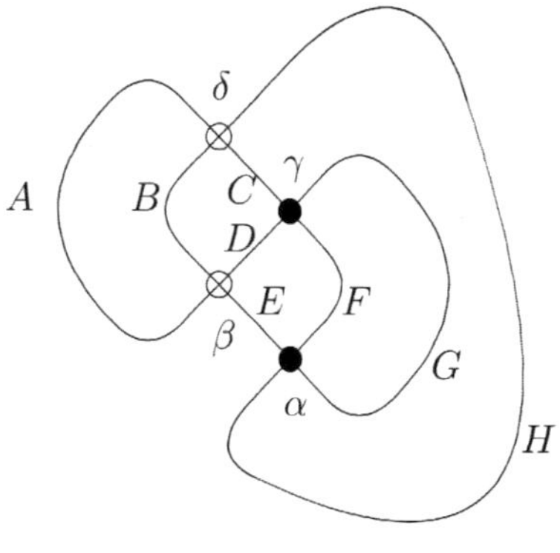

4_1 knot (figure eight)

The graph has eight edges and four vertices, labeled by capital latin and small green letters respectively. The hypercube is 4-dimensional and has 16 vertices: the graph possesses 16 resolutions. These involve $2+4+1+4+4+5 = 20$ different cycles of the lengths $2, 3, 4, 5, 6, 8$.

Journal of Physics: Conference Series **411** (2013) 012013 doi:10.1088/1742-6596/411/1/012013

Drawing a four-dimensional hypecube is not very informative, therefore we substitute a picture by a table. The resolutions (hypecube vertices) are separated by double horizontal lines into sets with the same $|r - r_c|$ from $r_c = I$, associated with the 3_1 knot. We also explicitly list the vertices, where flips are made to obtain the resolution from I.

resolution	r	flips at	cycles			
I	[0000]		$p_2 = AB$ $p'_2 = FG$	$p_4 = CDEH$		
II	[1000]	α	$p_2 = AB$		$p_6 = CDEFGH$	
III	[0100]	β	$p'_2 = FG$		$p'_6 = ABDCHE$	
IV	[0010]	γ	$p_2 = AB$		$p''_6 = CGFDEH$	
V	[0001]	δ	$p'_2 = FG$		$p'''_6 = ABCDEH$	
VI	[1100]	α, β				$p'_8 = ABDCHGFE$
VII	[1010]	α, γ	$p_2 = AB$	$p_3 = DEF$ $p''_3 = CGH$		
VIII	[1001]	α, δ				$p_8 = ABCDEFGH$
IX	[0110]	β, γ				$p''_8 = ABDFGCHE$
X	[0101]	β, δ	$p'_2 = FG$	$p'_3 = AEH$ $p'''_3 = BCD$		
XI	[0011]	γ, δ				$p'''_8 = ABCGFDEH$
XII	[1110]	α, β, γ		$p''_3 = CGH$	$p''_5 = ABDFE$	
XIII	[1101]	α, β, δ		$p'''_3 = BCD$	$p'''_5 = AHGFE$	
XIV	[1011]	α, γ, δ		$p_3 = DEF$	$p_5 = ABCGH$	
XV	[0111]	β, γ, δ		$p'_3 = AEH$	$p'_5 = BCGFD$	
XVI	[1111]	$\alpha, \beta, \gamma, \delta$				$p''''_8 = AHGCBDFE$

For 4_1 the starting resolution is different: $r_c = X$, and the sets are different:

$$\begin{array}{cc} |r - r_c| & \text{resolutions } r \\ 0 & X \\ 1 & III,\ V,\ XIII,\ XV \\ 2 & I,\ VI,\ VIII,\ IX,\ XI,\ XVI \\ 3 & II,\ IV,\ XII,\ XIV \\ 4 & VII \end{array} \tag{23}$$

4.2. Jones polynomials
From these tables we immediately read the extended and unreduced Jones polynomials:

$$\mathcal{J}^{3_1} = p_2p_2'p_4 + t\Big(p_2p_6 + p_2'p_6' + p_2p_6'' + p_2'p_6'''\Big) +$$
$$+t^2\Big(p_2p_3p_3'' + p_2'p_3'p_3''' + p_8 + p_8' + p_8'' + p_8'''\Big) + t^3\Big(p_3p_5 + p_3'p_5' + p_3''p_5'' + p_3'''p_5'''\Big) + t^4p_8'''' \tag{24}$$

We remind that *extended* knot polynomials are not topological invariants, in particular this $\mathcal{J}^{3_1}$, defined through the 3-strand braid, is drastically different from $\mathcal{J}^{\bullet\bullet\bullet}$ in (2), defined for the 2-strand braid. However, the ordinary Jones polynomials, obtained after substitution (I.19) of $p_k = D$ and $t = -q$ are, of course, the same:

$$J^{\bullet\bullet\bullet}(q) = J^{3_1}(q) = q^4\Big(D^3 - 4qD^2 + q^2(2D^3 + 4D) - 4q^3D^2 + q^4D\Big) =$$
$$= D\cdot q^4\Big(D^2 - 4qD + q^2((2D^2+4) - 4q^3D + q^4\Big) = q + q^3 + q^5 - q^9 = D\cdot(q^2 + q^6 - q^8) \tag{25}$$

Similarly,

$$\mathcal{J}^{4_1} = p_2'p_3'p_3''' + t\Big(p_2'(p_6' + p_6''') + p_3'p_5' + p_3'''p_5'''\Big) +$$
$$+t^2\Big(p_2p_2'p_4 + p_8 + p_8' + p_8'' + p_8''' + p_8''''\Big) + t^3\Big(p_2(p_6 + p_6'') + p_3p_5 + p_3''p_5''\Big) + t^4p_2p_3p_3'' \tag{26}$$

and this implies

$$J^{4_1}(q) = \frac{q^2}{(-q^2)^2}\Big(D^3 - 4qD^2 + q^2(D^3 + 5D) - 4q^3D^2 + q^4D^3\Big) =$$
$$= D\cdot q^{-2}\Big(D^2 - 4qD + q^2(D^2+5) - 4q^3D + q^4D^2\Big) = q^5 + q^{-5} = D\left(q^{-4} - q^{-2} + 1 - q^2 + q^4\right) \tag{27}$$

4.3. Cut-and-join operators for the 3_1 knot
To construct superpolynomials, unreduced and reduced, we need also the differentials. According to the general procedure, outlined in secs.6 and 7 of [1], from (26) we read the bosonic cut-and-

join operators:

$$K_1 = (p_6 + p_6'')\frac{\partial^2}{\partial p_2'\partial p_4} + (p_6' + p_6''')\frac{\partial^2}{\partial p_2\partial p_4}$$

$$\begin{aligned}K_2 = &\underbrace{p_3p_3''\left(\frac{\partial}{\partial p_6''} \ominus \frac{\partial}{\partial p_6}\right)}_{\alpha\gamma} + \underbrace{p_3'p_3'''\left(\frac{\partial}{\partial p_6'''} \ominus \frac{\partial}{\partial p_6'}\right)}_{\beta\delta} + \underbrace{p_8\left(\frac{\partial^2}{\partial p_2'\partial p_6'''} \ominus \frac{\partial^2}{\partial p_2\partial p_6}\right)}_{\alpha\delta} + \\ &+ \underbrace{p_8'\left(\frac{\partial^2}{\partial p_2'\partial p_6'} \ominus \frac{\partial^2}{\partial p_2\partial p_6}\right)}_{\alpha\beta} + \underbrace{p_8''\left(\frac{\partial^2}{\partial p_2\partial p_6''} \ominus \frac{\partial^2}{\partial p_2'\partial p_6'}\right)}_{\beta\gamma} + \underbrace{p_8'''\left(\frac{\partial^2}{\partial p_2'\partial p_6'''} \ominus \frac{\partial^2}{\partial p_2\partial p_6''}\right)}_{\gamma\delta}\end{aligned}$$

$$\begin{aligned}K_3 = &\underbrace{p_3''p_5''\left(\frac{\partial}{\partial p_8'} + \frac{\partial}{\partial p_8''}\right) \ominus p_5''\frac{\partial^2}{\partial p_2\partial p_3}}_{\alpha\beta\gamma} + \underbrace{p_3'''p_5'''\left(\frac{\partial}{\partial p_8'} \ominus \frac{\partial}{\partial p_8}\right) + p_5'''\frac{\partial^2}{\partial p_2'\partial p_3'}}_{\alpha\beta\delta} + \\ &+ \underbrace{p_3p_5\left(\frac{\partial}{\partial p_8'''} \ominus \frac{\partial}{\partial p_8}\right) + p_5\frac{\partial^2}{\partial p_2\partial p_3''}}_{\alpha\gamma\delta} + \underbrace{p_3'p_5'\left(\frac{\partial}{\partial p_8''} + \frac{\partial}{\partial p_8'''}\right) \ominus p_5'\frac{\partial^2}{\partial p_2'\partial p_3'''}}_{\beta\gamma\delta}\end{aligned}$$

$$K_4 = p_8''''\left(\ominus\frac{\partial^2}{\partial p_3\partial p_5} + \frac{\partial^2}{\partial p_3'\partial p_5'} \ominus \frac{\partial^2}{\partial p_3''\partial p_5''} + \frac{\partial^2}{\partial p_3'''\partial p_5'''}\right) \tag{28}$$

Each item in the cut-and-join operator is associated with an edge of the hypercube, i.e. with a flip, made at exactly one vertex of original graph. By $\ominus$ in this formula we denote the terms where the sign factor in the super-analogue of the cut-and-join operator will be negative, $\epsilon^a_{bc} = -1$, according to the rule (I.45). There are 12 (out of 32) edges of this type.

4.4. Cut-and-join operators for the 4_1 knot

Similarly, for the 4_1 knot we have:

$$K_1 = (p_6''' \ominus p_6')\frac{\partial^2}{\partial p_3'\partial p_3'''} \ominus p_5'\frac{\partial^2}{\partial p_2'\partial p_3'''} + p_5'''\frac{\partial}{\partial p_2'\partial p_3'}$$

$$K_2 = \underbrace{p_2p_4\frac{\partial}{\partial p_6'} + p_2p_4\frac{\partial}{\partial p_6'''}}_{I} + \underbrace{p_8\left(\frac{\partial^2}{\partial p_2'\partial p_6'''} \ominus \frac{\partial^2}{\partial p_3'''\partial p_5'''}\right)}_{\alpha\delta} + \underbrace{p_8'\left(\frac{\partial^2}{\partial p_2'\partial p_6'} + \frac{\partial^2}{\partial p_3'''\partial p_5'''}\right)}_{\alpha\beta} +$$
$$+ \underbrace{p_8''\left(+\frac{\partial^2}{\partial p_3'\partial p_5'}\right) \ominus \frac{\partial^2}{\partial p_2'\partial p_6'}}_{\beta\gamma} + \underbrace{p_8'''\left(\frac{\partial^2}{\partial p_2'\partial p_6'''} + \frac{\partial^2}{\partial p_3'\partial p_5'}\right)}_{\gamma\delta} + \underbrace{p_8''''\left(\frac{\partial^2}{\partial p_3'\partial p_5'} + \frac{\partial^2}{\partial p_3'''\partial p_5'''}\right)}_{\alpha\beta\gamma\delta}$$

$$K_3 = \underbrace{p_6\frac{\partial^2}{\partial p_2'\partial p_4} \ominus p_2p_6\left(\frac{\partial}{\partial p_8} + \frac{\partial}{\partial p_8'}\right)}_{\alpha} + \underbrace{p_6''\frac{\partial^2}{\partial p_2'\partial p_4} + p_2p_6''\left(\frac{\partial}{\partial p_8''} \ominus \frac{\partial}{\partial p_8'''}\right)}_{\gamma} +$$
$$+ \underbrace{p_3''p_5''\left(\frac{\partial}{\partial p_8'} + \frac{\partial}{\partial p_8''} \ominus \frac{\partial}{\partial p_8''''}\right)}_{\alpha\beta\gamma} + \underbrace{p_3p_5\left(\frac{\partial}{\partial p_8'''} \ominus \frac{\partial}{\partial p_8} \ominus \frac{\partial}{\partial p_8''''}\right)}_{\alpha\gamma\delta}$$

$$K_4 = p_3p_3''\left(\frac{\partial}{\partial p_6''} \ominus \frac{\partial}{\partial p_6}\right) + p_2p_3''\frac{\partial}{\partial p_5} \ominus p_2p_3\frac{\partial}{\partial p_5''} \qquad (29)$$

Next we convert these bosonic operators into the BRST one, depending on Grassmannian variables with appropriately chosen signs in front of the different items to guarantee the nilpotency. This provides the collection of differentials.

We do not write the lengthy expressions for unreduced differentials, and proceed directly to reduced ones, which in the case of 4-vertex graphs are still slightly more concise.

4.5. Reductions

This time not all the edges are the same, and it looks like at least two different reductions could be performed, for the marked edges A (or B, F, G) and H (or C, D, E).

The following table shows by crosses the time-variables which are reduced (their θ-components eliminated, only η-components left) when different edges are marked. Surviving variables correspond to empty spaces in the table. The numbers of eliminated and survived variables are shown in the last two lines respectively. Clearly, the *a priori* (topologically) equivalent reductions imply nullification of very different θ-variables (all the 20 η's are always present), thus the reduced differentials are different and it is not technically trivial that the resulting cohomologies are the same.

It is even less trivial that the two *a priori* different reductions, which even leave different numbers of p-variables (8 and 6) in the differentials, give rise to the same reduced superpolynomials.

	A	B	C	D	E	F	G	H
$p_2 = AB$	x	x						
$p'_2 = FG$						x	x	
$p_3 = DEF$				x	x	x		
$p'_3 = AEH$	x				x			x
$p''_3 = CGH$			x				x	x
$p'''_3 = BBCD$		x	x	x				
$p_4 = CDEH$			x	x	x			x
$p_5 = ABCGH$	x	x	x				x	x
$p'_5 = BCGFD$		x	x	x		x	x	
$p''_5 = ABDFE$	x	x		x	x	x		
$p'''_5 = AHGFE$	x				x	x	x	x
$p_6 = CDEFGH$			x	x	x	x	x	x
$p'_6 = ABDCHE$	x	x	x	x	x			x
$p''_6 = CGFDEH$			x	x	x	x	x	x
$p'''_6 = ABHEDC$	x	x	x	x	x			x
all five p_8	x	x	x	x	x	x	x	x
	12	12	14	14	14	12	12	14
	8	8	6	6	6	8	8	6

Before proceeding further, we should check the consistency of our reductions – that derivatives w.r.t. reduced variables are always multiplied by reduced variables and thus do not contribute. Let us show how this works for the first operator K_1 in the more involved case of 4_1. The factor $(p'_6 + p'''_6)$ is reduced (θ'_6 and θ'''_6 vanish) in all the reductions, except for $F$and G – but in those cases neither p'_3 nor p'''_3 is vanishing: this makes all the reductions of the first item in K_1 consistent. Likewise, in the second item p'_5 is reduced in all cases except for A, E and H, but neither p'_2 nor p'''_3 are reduced. The same happens in the third item: when p'''_5 remains unreduced (in B, C and D reductions), neither p'_2 nor p'_3 are reduced. In a similar way for both 4_1 and 3_1 one can check self-consistency of all eight reductions of all the four operators K – and of their super-counterparts – the differentials d. All this will be transparently seen in explicit calculations below – but those we do only for one of the eight: for A-reduction.

5. A-reduction of the cut-and-join operator for the 3_1 knot

First of all, we write the cut-and-join operators (29), underlying the reduced variables p: this means that the corresponding Grassmannian variable ϑ contains only η-, but not the θ-component.

5.1. Reduction of bosonic operator (29)

$$\underline{K_1} = (p_6 + p_6'')\frac{\partial^2}{\partial p_2'\partial p_4} + (\underline{p_6'} + \underline{p_6'''})\frac{\partial^2}{\partial \underline{p_2}\partial p_4}$$

$$\underline{K_2} = p_3p_3''\left(\frac{\partial}{\partial p_6''} \ominus \frac{\partial}{\partial p_6}\right) + \underline{p_3'}p_3'''\left(\frac{\partial}{\partial \underline{p_6'''}} \ominus \frac{\partial}{\partial \underline{p_6'}}\right) + \underline{p_8}\left(\frac{\partial^2}{\partial p_2'\partial \underline{p_6'''}} \ominus \frac{\partial^2}{\partial \underline{p_2}\partial p_6}\right) +$$
$$+\underline{p_8'}\left(\frac{\partial^2}{\partial p_2'\partial \underline{p_6'}} \ominus \frac{\partial^2}{\partial \underline{p_2}\partial p_6}\right) + \underline{p_8''}\left(\frac{\partial^2}{\partial \underline{p_2}\partial p_6''} \ominus \frac{\partial^2}{\partial p_2'\partial \underline{p_6'}}\right) + \underline{p_8'''}\left(\frac{\partial^2}{\partial p_2'\partial \underline{p_6'''}} \ominus \frac{\partial^2}{\partial \underline{p_2}\partial p_6''}\right)$$

$$\underline{K_3} = p_3''\underline{p_5''}\left(\frac{\partial}{\partial \underline{p_8'}} + \frac{\partial}{\partial \underline{p_8''}}\right) \ominus \underline{p_5''}\frac{\partial^2}{\partial \underline{p_2}\partial p_3} + p_3'''\underline{p_5'''}\left(\frac{\partial}{\partial \underline{p_8'}} \ominus \frac{\partial}{\partial \underline{p_8}}\right) + \underline{p_5'''}\frac{\partial^2}{\partial p_2'\partial \underline{p_3'}} +$$
$$+p_3\underline{p_5}\left(\frac{\partial}{\partial \underline{p_8'''}} \ominus \frac{\partial}{\partial \underline{p_8}}\right) + \underline{p_5}\frac{\partial^2}{\partial \underline{p_2}\partial p_3''} + \underline{p_3'}p_5'\left(\frac{\partial}{\partial \underline{p_8''}} + \frac{\partial}{\partial \underline{p_8'''}}\right) \ominus p_5'\frac{\partial^2}{\partial p_2'\partial p_3'''}$$

$$\underline{K_4} = \underline{p_8''''}\left(\ominus\frac{\partial^2}{\partial p_3\partial \underline{p_5}} + \frac{\partial^2}{\partial \underline{p_3'}\partial p_5'} \ominus \frac{\partial^2}{\partial p_3''\partial \underline{p_5''}} + \frac{\partial^2}{\partial p_3'''\partial \underline{p_5'''}}\right) \tag{30}$$

These formulas allow us to write the A-reduced differentials in the 3_1 case and calculate their cohomologies.

5.2. Differential $\underline{d_1}$ and the cohomology $\underline{H_0}$
From

$$\underline{K_1} = (p_6 + p_6'')\frac{\partial^2}{\partial p_2'\partial p_4} + (\underline{p_6'} + \underline{p_6'''})\frac{\partial^2}{\partial \underline{p_2}\partial p_4} \tag{31}$$

we obtain:

$$\underline{d_1} = -(\theta_6 + \theta_6'')\left(\frac{\partial^2}{\partial\theta_2'\partial\eta_4} + \frac{\partial^2}{\partial\eta_2'\partial\theta_4}\right) - (\eta_6 + \eta_6'')\frac{\partial^2}{\partial\eta_2'\partial\eta_4} + (\eta_6' + \eta_6''')\frac{\partial^2}{\partial\eta_2\partial\eta_4} \tag{32}$$

Note non-trivial signs in this formula: they are such that the $\underline{d_1}$-images of all basis vectors of the 4-dimensional space $\underline{\mathcal{C}_1} = \underline{V_2} \otimes V_2' \otimes V_4$ (its quantum dimension is $\dim_q(\underline{\mathcal{C}_1}) = qD^2$) are expressed through the basis vectors of $\underline{\mathcal{C}_2}$ with non-negative coefficients (we remind our convention $\frac{\partial^2}{\partial\theta\partial\eta}(\theta\eta) = +1$):

$\underline{d_1} \downarrow$	$\underline{\mathcal{C}_0}$	$\eta_2\theta_2'\theta_4$	$\eta_2\theta_2'\eta_4$	$\eta_2\eta_2'\theta_4$	$\eta_2\eta_2'\eta_4$
	$\underline{d_1}\underline{\mathcal{C}_0}$	0	$\eta_2(\theta_6 + \theta_6'') + \theta_2'(\eta_6' + \eta_6''')$	$\eta_2(\theta_6 + \theta_6'')$	$\eta_2(\eta_6 + \eta_6'') + \eta_2'(\eta_6' + \eta_6''')$

Clearly,

$$\mathrm{Ker}(\underline{d_1}) = \{\eta_2\theta_2'\theta_4\}, \quad \dim_q(\underline{H_0}) = \dim_q\mathrm{Ker}(\underline{d_1}) = q^{-1} \tag{33}$$

The image

$$\mathrm{Im}(\underline{d_1}) = \Big\{\eta_2(\theta_6+\theta_6''),\ \theta_2'(\eta_6'+\eta_6'''),\ \eta_2(\eta_6+\eta_6'')+\eta_2'(\eta_6'+\eta_6''')\Big\}, \qquad \dim_q \mathrm{Im}(\underline{d_1}) = 2+q^2 \tag{34}$$

Note that differential itself is not really needed in this calculation: what we need are the vector spaces with basises and bosonic cut-and-join operator with the $\ominus$ labels – this is enough to construct the kernel and image of the differential, without knowing its explicit form (the sign assignments).

5.3. *Differential $\underline{d_2}$ and the cohomology $\underline{H_1}$*

This time we begin with

$$\underline{K_2} = p_3p_3''\left(\frac{\partial}{\partial p_6''}\ominus\frac{\partial}{\partial p_6}\right) + \underline{p_3'}p_3'''\left(\frac{\partial}{\partial \underline{p_6'''}}\ominus\frac{\partial}{\partial \underline{p_6'}}\right) + \underline{p_8}\left(\frac{\partial^2}{\partial p_2'\partial \underline{p_6'''}}\ominus\frac{\partial^2}{\partial \underline{p_2}\partial p_6}\right) +$$
$$+\underline{p_8'}\left(\frac{\partial^2}{\partial p_2'\partial \underline{p_6'}}\ominus\frac{\partial^2}{\partial \underline{p_2}\partial p_6}\right) + \underline{p_8''}\left(\frac{\partial^2}{\partial \underline{p_2}\partial p_6''}\ominus\frac{\partial^2}{\partial p_2'\partial \underline{p_6'}}\right) + \underline{p_8'''}\left(\frac{\partial^2}{\partial p_2'\partial \underline{p_6'''}}\ominus\frac{\partial^2}{\partial \underline{p_2}\partial p_6''}\right) \tag{35}$$

The signs in the differential $\underline{d_2}$ are now adjusted in two steps: first, we choose the signs so that basis vectors are mapped into basis vectors with non-negative coefficients, second we reverse the signs in the terms, marked by the $\ominus$ signs. As usual these terms will be marked by a hat. In this way we get:

$$\underline{d_2} = -\theta_3\theta_3''\left(\frac{\partial}{\partial \theta_6''}\,\hat{-}\,\frac{\partial}{\partial \theta_6}\right) - (\theta_3\eta_3''+\eta_3\theta_3'')\left(\frac{\partial}{\partial \eta_6''}\,\hat{-}\,\frac{\partial}{\partial \eta_6}\right) - \eta_3'\theta_3'''\left(\frac{\partial}{\partial \eta_6'''}\,\hat{-}\,\frac{\partial}{\partial \eta_6'}\right) +$$
$$+\eta_8\left(\frac{\partial^2}{\partial \eta_2'\partial \eta_6'''}\,\hat{-}\,\frac{\partial^2}{\partial \eta_2\partial \eta_6}\right) + \eta_8'\left(\frac{\partial^2}{\partial \eta_2'\partial \eta_6'}\,\hat{-}\,\frac{\partial^2}{\partial \eta_2\partial \eta_6}\right) +$$
$$+\eta_8''\left(\frac{\partial^2}{\partial \eta_2\partial \eta_6''}\,\hat{-}\,\frac{\partial^2}{\partial \eta_2'\partial \eta_6'}\right) + \eta_8'''\left(\frac{\partial^2}{\partial \eta_2'\partial \eta_6'''}\,\hat{-}\,\frac{\partial^2}{\partial \eta_2\partial \eta_6''}\right) \tag{36}$$

which acts on the space $\mathcal{C}_2$ as follows:

	$\underline{\mathcal{C}_1}$	$\eta_2\theta_6$	$\eta_2\eta_6$	$\theta_2'\eta_6'$	$\eta_2'\eta_6'$	$\eta_2\theta_6''$	$\eta_2\eta_6''$	$\theta_2'\eta_6'''$	$\eta_2'\eta_6'''$
$\underline{d_2}\downarrow$	$\underline{d_2\mathcal{C}_1}$	$\hat{-}\eta_2\theta_3\theta_3''$	$\hat{-}\eta_2(\theta_3\eta_3''+\eta_3\theta_3'')$ $\hat{-}\eta_8\hat{-}\eta_8'$	$\hat{-}\theta_2'\eta_3'\theta_3'''$	$\hat{-}\eta_2'\eta_3\theta_3'''$ $+\eta_8'\hat{-}\eta_8''$	$\eta_2\theta_3\theta_3''$	$\eta_2(\theta_3\eta_3''+\eta_3\theta_3'')$ $+\eta_8''\hat{-}\eta_8'''$	$\theta_2'\eta_3'\theta_3'''$	$\eta_2'\eta_3'\theta_3'''$ $+\eta_8+\eta_8'''$

Note that all the minus signs in the target space appear with hats: this is the rule to ascribe signs to the items in the differential.

It follows that

$$\mathrm{Ker}(\underline{d_2}) = \Big\{\eta_2(\theta_6+\theta_6''),\ \theta_2'(\eta_6'+\eta_6'''),\ \eta_2(\eta_6+\eta_6'')+\eta_2'(\eta_6'+\eta_6''')\Big\}, \qquad \dim_q\mathrm{Ker}(\underline{d_2}) = 2+q^2$$
$$\underline{H_1} = \mathrm{Ker}(\underline{d_2})/\mathrm{Im}(\underline{d_1}) = \emptyset \tag{37}$$

and

$$\mathrm{Im}(\underline{d_2}) = \Big\{\eta_2\theta_3\theta_3'',\ \theta_2'\eta_3'\theta_3'',\ \eta_2(\theta_3\eta_3''+\eta_3\theta_3'')+\eta_8+\eta_8',\ \eta_2'\eta_3'\theta_3'''+\eta_8+\eta_8''',\ \eta_8+\eta_8'-\eta_8''+\eta_8'''\Big\},$$
$$\dim_q\mathrm{Im}(\underline{d_2}) = 2q^{-1}+3q \tag{38}$$

5.4. Differential $\underline{d_3}$ and the cohomology $\underline{H_2}$

Starting from

$$\underline{K_3} = p_3''\underline{p_5''}\left(\frac{\partial}{\partial \underline{p_8'}} + \frac{\partial}{\partial \underline{p_8''}}\right) \ominus \underline{p_5''}\frac{\partial^2}{\partial \underline{p_2}\partial p_3} + p_3'''\underline{p_5'''}\left(\frac{\partial}{\partial \underline{p_8'}} \ominus \frac{\partial}{\partial \underline{p_8}}\right) + \underline{p_5'''}\frac{\partial^2}{\partial p_2'\partial \underline{p_3'}} +$$
$$+p_3\underline{p_5}\left(\frac{\partial}{\partial \underline{p_8'''}} \ominus \frac{\partial}{\partial \underline{p_8}}\right) + \underline{p_5}\frac{\partial^2}{\partial \underline{p_2}\partial p_3''} + \underline{p_3'}p_5'\left(\frac{\partial}{\partial \underline{p_8''}} + \frac{\partial}{\partial \underline{p_8'''}}\right) \ominus p_5'\frac{\partial^2}{\partial p_2'\partial p_3'''} \tag{39}$$

we obtain by the same procedure:

$$\underline{d_3} = \theta_3''\eta_5''\left(\frac{\partial}{\partial \eta_8'} + \frac{\partial}{\partial \eta_8''}\right) \hat{+}\, \eta_5''\frac{\partial^2}{\partial \eta_2\partial \eta_3} + \theta_3'''\eta_5'''\left(\frac{\partial}{\partial \eta_8'} \hat{-} \frac{\partial}{\partial \eta_8}\right) - \eta_5'''\frac{\partial^2}{\partial \eta_2'\partial \eta_3'} +$$
$$+\theta_3\eta_5\left(\frac{\partial}{\partial \eta_8'''} \hat{-} \frac{\partial}{\partial \eta_8}\right) + \eta_5\frac{\partial^2}{\partial \eta_2\partial \eta_3''} + \eta_3'\theta_5'\left(\frac{\partial}{\partial \eta_8''} + \frac{\partial}{\partial \eta_8'''}\right) \hat{-}\, \theta_5'\left(\frac{\partial^2}{\partial \theta_2'\partial \eta_3'''} + \frac{\partial^2}{\partial \eta_2'\partial \theta_3'''}\right) \hat{-}\, \eta_5'\frac{\partial^2}{\partial \eta_2'\partial \eta_3'''} \tag{40}$$

	$\underline{\mathcal{C}_2}$	$\eta_2\theta_3\theta_3''$	$\eta_2\theta_3\eta_3''$	$\eta_2\eta_3\theta_3''$	$\eta_2\eta_3\eta_3''$	$\theta_2'\eta_3'\theta_3'''$	$\theta_2'\eta_3'\eta_3'''$	$\eta_2'\eta_3'\theta_3'''$	$\eta_2'\eta_3'\eta_3'''$	η_8	η_8'	η_8''	η_8'''
$\underline{d_3}\downarrow$	$\underline{d_3\mathcal{C}_2}$	0	$\theta_3\eta_5$	$\hat{-}\theta_3''\eta_5''$	$\eta_3\eta_5\hat{-}\eta_3''\eta_5''$	0	$\hat{-}\eta_3'\theta_5'$	$\hat{-}\eta_3'\theta_5' + \theta_3'''\eta_5'''$	$\hat{-}\eta_3'\eta_5' + \eta_3'''\eta_5'''$	$\hat{-}\theta_3\eta_5\,\hat{-}\theta_3'''\eta_5'''$	$\theta_3''\eta_5'' + \theta_3'''\eta_5'''$	$\eta_3'\theta_5' + \theta_3''\eta_5''$	$\theta_3\eta_5 + \eta_3'\theta_5'$

$$\mathrm{Ker}(\underline{d_3}) = \Big\{\eta_2\theta_3\theta_3'',\ \ \theta_2'\eta_3'\theta_3''',\ \ \eta_8 + \eta_8' + \eta_2(\theta_3\eta_3'' + \eta_3\theta_3''),$$
$$\eta_8 + \eta_8''' + \eta_2'\eta_3'\theta_3''',\ \ \eta_2\eta_3\theta_3'' + \theta_2'\eta_3'\eta_3''' + \eta_8'',\ \ \eta_8 + \eta_8' - \eta_8'' + \eta_8'''\Big\},$$
$$\dim_q \mathrm{Ker}(\underline{d_3}) = 2q^{-1} + 4q,$$
$$\underline{H_2} = \mathrm{Ker}(\underline{d_3})/\mathrm{Im}(\underline{d_2}) = \Big\{\eta_2\eta_3\theta_3'' + \theta_2'\eta_3'\eta_3''' + \eta_8''\Big\}, \qquad \dim_q(\underline{H_2}) = q \tag{41}$$

and

$$\mathrm{Im}(\underline{d_3}) = \Big\{\theta_3\eta_5,\ \ \theta_5'\eta_3',\ \ \theta_3''\eta_5'',\ \ \theta_3'''\eta_5''',\ \ \eta_3\eta_5 - \eta_3''\eta_5'',\ \ \eta_3'\eta_5' - \eta_3'''\eta_5'''\Big\} \tag{42}$$

5.5. Differential $\underline{d_4}$ and the cohomologies $\underline{H_3}$ and $\underline{H_4}$

Finally, from

$$\underline{K_4} = \underline{p_8''''}\left(\ominus\frac{\partial^2}{\partial p_3\partial \underline{p_5}} + \frac{\partial^2}{\partial \underline{p_3'}\partial p_5'} \ominus \frac{\partial^2}{\partial p_3''\partial \underline{p_5''}} + \frac{\partial^2}{\partial p_3'''\partial \underline{p_5'''}}\right) \tag{43}$$

$$\underline{d_4} = \eta_8''''\left(\hat{-}\frac{\partial^2}{\partial \eta_3\partial \eta_5} + \frac{\partial^2}{\partial \eta_3'\partial \eta_5'} \hat{-} \frac{\partial^2}{\partial \eta_3''\partial \eta_5''} + \frac{\partial^2}{\partial \eta_3'''\partial \eta_5'''}\right) \tag{44}$$

Thus in the $4 \cdot 2 = 8$-dimensional space $C_4 = V_3 \otimes \underline{V_5} \oplus \underline{V_3} \otimes V_5 \oplus V_3 \otimes \underline{V_5} \oplus V_3 \otimes \underline{V_5}$ the 7-dimensional kernel

$$\ker(\underline{d_4}) = \Big\{\theta_3\eta_5,\ \eta_3'\theta_5',\ \theta_3''\eta_5'',\ \theta_3'''\eta_5''',\ \eta_3\eta_5 + \eta_3'\eta_5',\ \eta_3\eta_5 - \eta_3''\eta_5'',\ \eta_3'\eta_5' - \eta_3'''\eta_5'''\Big\} \tag{45}$$

Therefore

$$\underline{H_3} = \mathrm{Ker}(\underline{d_4})/\mathrm{Im}(\underline{d_3}) = \left\{\eta_3\eta_5 + \eta_3'\eta_5'\right\} \quad \text{and} \quad \dim_q(\underline{H_3}) = q^2 \tag{46}$$

The space $C_5 = \underline{V}$ is one-dimensional, the basis element is η_8'''' and it coincides with the $8-7=1$-dimensional image of $\underline{d_4}$. Therefore the coimage of $\underline{d_4}$ is empty, and the cohomology

$$\underline{H_4} = 0 \tag{47}$$

5.6. Reduced Jones superpolynomials for 3_1

Thus

$$\begin{aligned} \dim_q(\underline{H_0}) &= q^{-1}, \\ \dim_q(\underline{H_1}) &= 0, \\ \dim_q(\underline{H_2}) &= q, \\ \dim_q(\underline{H_3}) &= q^2, \\ \dim_q(\underline{H_4}) &= 0 \end{aligned} \tag{48}$$

and therefore

$$\underline{P_\Box^{3_1}} = q^{-1}\cdot q^4\sum_{i=0}^{4}(qT)^i\cdot\dim_q(\underline{H_i}) = \frac{q^4}{q}\left(q^{-1} + 0\cdot(qT) + q^1\cdot(qT)^2 + q^2\cdot(qT)^3\right) = \boxed{q^2 + q^6T^2 + q^8T^3} \tag{49}$$

what coincides with the answer (21), obtained from the 2-strand representation.

6. *A*-reduced differentials for the 4_1 knot

Now we repeat the same procedure in the case of the figure-eight knot.

6.1. Reduction of bosonic operator (30)

$$\underline{K_1} = (\underline{p_6'''} \ominus \underline{p_6'})\frac{\partial^2}{\partial \underline{p_3'}\partial p_3'''} \ominus p_5'\frac{\partial^2}{\partial p_2'\partial p_3'''} + \underline{p_5'''}\frac{\partial^2}{\partial p_2'\partial \underline{p_3'}}$$

$$\begin{aligned}\underline{K_2} = {} & \underline{p_2}p_4\frac{\partial}{\partial \underline{p_6'}} + \underline{p_2}p_4\frac{\partial}{\partial \underline{p_6'''}} + \underline{p_8}\left(\frac{\partial^2}{\partial p_2'\partial \underline{p_6'''}} \ominus \frac{\partial^2}{\partial p_3'''\partial \underline{p_5'''}}\right) + \underline{p_8'}\left(\frac{\partial^2}{\partial p_2'\partial \underline{p_6'}} + \frac{\partial^2}{\partial p_3'''\partial \underline{p_5'''}}\right) + \\ & + \underline{p_8''}\left(\frac{\partial^2}{\partial \underline{p_3'}\partial p_5'} \ominus \frac{\partial^2}{\partial p_2'\partial \underline{p_6'}}\right) + \underline{p_8'''}\left(\frac{\partial^2}{\partial p_2'\partial \underline{p_6'''}} + \frac{\partial^2}{\partial \underline{p_3'}\partial p_5'}\right) + \underline{p_8''''}\left(\frac{\partial^2}{\partial \underline{p_3'}\partial p_5'} + \frac{\partial^2}{\partial p_3'''\partial \underline{p_5'''}}\right)\end{aligned}$$

$$\begin{aligned}\underline{K_3} = {} & p_6\frac{\partial^2}{\partial p_2'\partial p_4} \ominus \underline{p_2}p_6\left(\frac{\partial}{\partial \underline{p_8}} + \frac{\partial}{\partial \underline{p_8'}}\right) + p_6''\frac{\partial^2}{\partial p_2'\partial p_4} + \underline{p_2}p_6''\left(\frac{\partial}{\partial \underline{p_8''}} \ominus \frac{\partial}{\partial \underline{p_8'''}}\right) + \\ & + p_3''\underline{p_5''}\left(\frac{\partial}{\partial \underline{p_8'}} + \frac{\partial}{\partial \underline{p_8''}} \ominus \frac{\partial}{\partial \underline{p_8''''}}\right) + p_3\underline{p_5}\left(\frac{\partial}{\partial \underline{p_8'''}} \ominus \frac{\partial}{\partial \underline{p_8}} \ominus \frac{\partial}{\partial \underline{p_8''''}}\right)\end{aligned}$$

$$\underline{K_4} = p_3p_3''\left(\frac{\partial}{\partial p_6''} \ominus \frac{\partial}{\partial p_6}\right) + \underline{p_2}p_3''\frac{\partial}{\partial \underline{p_5}} \ominus \underline{p_2}p_3\frac{\partial}{\partial \underline{p_5''}} \tag{50}$$

6.2. Differential $\underline{d_1}$ and the cohomology $\underline{H_0}$

As usual, we begin from the relevant bosonic operator

$$\underline{K_1} = (p_6''' \ominus \underline{p_6'})\frac{\partial^2}{\partial \underline{p_3'}\partial p_3'''} \ominus p_5'\frac{\partial^2}{\partial p_2'\partial p_3'''} + \underline{p_5'''}\frac{\partial}{\partial p_2'\partial \underline{p_3'}} \tag{51}$$

In this case the space $\underline{\mathcal{C}_0} = V_2' \otimes \underline{V_3'} \otimes V_3'''$ has dimension $\dim_q(\underline{\mathcal{C}_0}) = q^{-1} + 2q + q^3 = qD^2$. The differential

$$\underline{d_1} = -(\eta_6''' - \eta_6')\frac{\partial^2}{\partial \eta_3'\partial \eta_3'''} \hat{-} \theta_5'\left(\frac{\partial^2}{\partial \theta_2'\partial \eta_3'''} + \frac{\partial^2}{\partial \eta_2'\partial \theta_3'''}\right) \hat{-} \eta_5'\frac{\partial^2}{\partial \eta_2'\partial \eta_3'''} - \eta_5'''\frac{\partial^2}{\partial \eta_2'\partial \eta_3'} \tag{52}$$

converts the basis in this space as follows:

$\underline{d_1} \downarrow$ (53)

$\underline{\mathcal{C}_0}$	$\theta_2'\eta_3'\theta_3'''$	$\theta_2'\eta_3'\eta_3'''$	$\eta_2'\eta_3'\theta_3'''$	$\eta_2'\eta_3'\eta_3'''$
$\underline{d_1}\mathcal{C}_0$	0	$\theta_2'(\eta_6''' \hat{-} \eta_6') \hat{-} \eta_3'\theta_5'$	$\theta_3'''\eta_5''' \hat{-} \theta_3'\eta_5'$	$\eta_2'(\eta_6''' \hat{-} \eta_6') \hat{-} \eta_3'\eta_5' + \eta_3'''\eta_5'''$

Clearly,

$$\mathrm{Ker}(\underline{d_1}) = \{\theta_2'\eta_3'\theta_3'''\}, \qquad \dim_q(\underline{H_0}) = \mathrm{din}_q\mathrm{Ker}(\underline{d_1}) = q^{-1} \tag{54}$$

Note that the second item in $\underline{d_1}$ would annihilate also the linear combination $\theta_2'\eta_3'\eta_3''' - \eta_2'\eta_3'\theta_3'''$, but it is not annihilated by the first item in $\underline{d_1}$. The image

$$\mathrm{Im}(\underline{d_1}) = \left\{\theta_2'(\eta_6' \hat{-} \eta_6''') + \eta_3'\theta_5',\ \ \theta_3'\eta_5' \hat{-} \theta_3'''\eta_5''',\ \ \eta_2'(\eta_6' \hat{-} \eta_6''') + \eta_3'\eta_5' \hat{-} \eta_3'''\eta_5'''\right\}, \qquad \dim_q\mathrm{Im}(\underline{d_1}) = 2 + q^2 \tag{55}$$

6.3. Differential $\underline{d_2}$ and the cohomology $\underline{H_1}$

From

$$\underline{K_2} = p_2p_4\frac{\partial}{\partial \underline{p_6'}} + \underline{p_2p_4}\frac{\partial}{\partial p_6'''} + \underline{p_8}\left(\frac{\partial^2}{\partial p_2'\partial \underline{p_6'''}} \ominus \frac{\partial^2}{\partial p_3'''\partial \underline{p_5'''}}\right) + \underline{p_8'}\left(\frac{\partial^2}{\partial p_2'\partial \underline{p_6'}} + \frac{\partial^2}{\partial p_3'''\partial \underline{p_5'''}}\right) +$$
$$+\underline{p_8''}\left(\frac{\partial^2}{\partial \underline{p_3'}\partial p_5'} \ominus \frac{\partial^2}{\partial p_2'\partial \underline{p_6'}}\right) + \underline{p_8'''}\left(\frac{\partial^2}{\partial p_2'\partial \underline{p_6'''}} + \frac{\partial^2}{\partial \underline{p_3'}\partial p_5'}\right) + \underline{p_8''''}\left(\frac{\partial^2}{\partial \underline{p_3'}\partial p_5'} + \frac{\partial^2}{\partial p_3'''\partial \underline{p_5'''}}\right) \tag{56}$$

we get the differential:

$$\underline{d_2} = \eta_2\theta_4\left(\frac{\partial}{\partial \eta_6'} + \frac{\partial}{\partial \eta_6'''}\right) + \eta_8\left(\frac{\partial^2}{\partial \eta_2'\partial \eta_6'''} \hat{-} \frac{\partial^2}{\partial \eta_3'''\partial \eta_5'''}\right) + \eta_8'\left(\frac{\partial^2}{\partial \eta_2'\partial \eta_6'} + \frac{\partial^2}{\partial \eta_3'''\partial \eta_5'''}\right) +$$
$$+\eta_8''\left(\frac{\partial^2}{\partial \eta_3'\partial \eta_5'} \hat{-} \frac{\partial^2}{\partial \eta_2'\partial \eta_6'}\right) + \eta_8'''\left(\frac{\partial^2}{\partial \eta_2'\partial \eta_6'''} + \frac{\partial^2}{\partial \eta_3'\partial \eta_5'}\right) + \eta_8''''\left(\frac{\partial^2}{\partial \eta_3'\partial \eta_5'} + \frac{\partial^2}{\partial \eta_3'''\partial \eta_5'''}\right) \tag{57}$$

which acts on the space $\underline{\mathcal{C}_1}$ of dimension $\dim_q(\underline{\mathcal{C}_1}) = 4 + 4q^2 = 4qD$ as follows:

$\underline{d_2} \downarrow$

$\underline{\mathcal{C}_1}$	$\theta_2'\eta_6'$	$\eta_2'\eta_6'$	$\theta_2'\eta_6'''$	$\eta_2'\eta_6'''$	$\eta_3'\theta_5'$	$\eta_3'\eta_5'$	$\theta_3'''\eta_5'''$	$\eta_3'''\eta_5'''$
$\underline{d_2\mathcal{C}_1}$	$\eta_2\theta_2'\eta_4$	$\eta_2\eta_2'\theta_4 + \eta_8' \hat{-} \eta_8''$	$\eta_2\theta_2'\eta_4$	$\eta_2\eta_2'\theta_4 + \eta_8 + \eta_8'''$	0	$\eta_8'' + \eta_8''' + \eta_8''''$	0	$\hat{-}\eta_8 + \eta_8' + \eta_8''''$

so that

$$\mathrm{Ker}(\underline{d_2}) = \left\{\eta_3'\theta_5',\ \ \theta_3'''\eta_5''',\ \ \theta_2'(\eta_6' - \eta_6'''),\ \ \eta_2'(\eta_6' - \eta_6''') + \eta_3'\eta_5' - \eta_3'''\eta_5'''\right\},$$
$$\dim_q\mathrm{Ker}(\underline{d_2}) = 3 + q^2, \qquad \underline{H_1} = \mathrm{Ker}(\underline{d_2})/\mathrm{Im}(\underline{d_1}) = \left\{\theta_3'\eta_5'\right\}, \qquad \dim_q(\underline{H_1}) = 1 \tag{58}$$

and

$$\mathrm{Im}(\underline{d_2}) = \left\{\eta_2\theta_2'\theta_4,\ \ \eta_2\eta_2'\theta_4 + \eta_8 + \eta_8''',\ \ \eta_8 \hat{-} \eta_8' \hat{-} \eta_8'''',\ \ \eta_8'' + \eta_8''' + \eta_8''''\right\},$$

$$\dim_q \mathrm{Im}(\underline{d_2}) = q^{-1} + 3q, \qquad \dim_q \mathrm{Ker}(\underline{d_2}) + q \cdot \dim_q \mathrm{Im}(\underline{d_2}) = \dim_q(\mathcal{C}_1) = 4 + 4q^2 = 4qD \quad (59)$$

The combination $\eta_8 - \eta_8' \hat{+} \eta_8'' + \eta_8''' \in \mathrm{Im}(\underline{d_2})$ as a linear combination of the two last entries in (59).

6.4. *Differential $\underline{d_3}$ and the cohomology $\underline{H_2}$*

From

$$\underline{K_3} = p_6 \frac{\partial^2}{\partial p_2' \partial p_4} \ominus \underline{p_2} p_6 \left(\frac{\partial}{\partial \underline{p_8}} + \frac{\partial}{\partial \underline{p_8'}}\right) + p_6'' \frac{\partial^2}{\partial p_2' \partial p_4} + \underline{p_2} p_6'' \left(\frac{\partial}{\partial \underline{p_8''}} \ominus \frac{\partial}{\partial \underline{p_8'''}}\right) +$$
$$+ p_3'' \underline{p_5''} \left(\frac{\partial}{\partial \underline{p_8'}} + \frac{\partial}{\partial \underline{p_8''}} \ominus \frac{\partial}{\partial \underline{p_8''''}}\right) + p_3 \underline{p_5} \left(\frac{\partial}{\partial \underline{p_8'''}} \ominus \frac{\partial}{\partial \underline{p_8}} \ominus \frac{\partial}{\partial \underline{p_8''''}}\right) \quad (60)$$

we get:

$$\underline{d_3} = -(\theta_6 + \theta_6'') \left(\frac{\partial^2}{\partial \theta_2' \partial \eta_4} + \frac{\partial^2}{\partial \eta_2' \partial \theta_4}\right) - (\eta_6 + \eta_6'') \frac{\partial^2}{\partial \eta_2' \partial \eta_4} +$$
$$\hat{-} \eta_2\theta_6 \left(\frac{\partial}{\partial \eta_8} + \frac{\partial}{\partial \eta_8'}\right) + \eta_2\theta_6'' \left(\frac{\partial}{\partial \eta_8''} \hat{-} \frac{\partial}{\partial \eta_8'''}\right) + \eta_3\theta_5 \left(\frac{\partial}{\partial \eta_8'''} \hat{-} \frac{\partial}{\partial \eta_8} \hat{-} \frac{\partial}{\partial \eta_8''''}\right) + \eta_3''\theta_5'' \left(\frac{\partial}{\partial \eta_8''} + \frac{\partial}{\partial \eta_8'} \hat{-} \frac{\partial}{\partial \eta_8''''}\right) \quad (61)$$

acts on the space $\underline{\mathcal{C}_2}$ of dimension $\dim_q(\underline{\mathcal{C}_2}) = q^{-1} + (2+5)q + q^3 = q(D^2+5)$:

$\underline{d_3} \downarrow$				
$\underline{\mathcal{C}_2}$	$\eta_2\theta_2'\theta_4$	$\eta_2\theta_2'\eta_4$	$\eta_2\eta_2'\theta_4$	$\eta_2\eta_2'\eta_4$
$\underline{d_3\mathcal{C}_2}$	0	$\eta_2(\theta_6 + \theta_6'')$	$\eta_2(\theta_6 + \theta_6'')$	$\eta_2(\eta_6 + \eta_6'')$

η_8	η_8'	η_8''	η_8'''	η_8''''
$\hat{-}\eta_2\theta_6 \hat{-} \eta_3\theta_5$	$\hat{-}\eta_2\theta_6 + \eta_3''\theta_5''$	$\eta_2\theta_6'' + \eta_3''\theta_5''$	$\hat{-}\eta_2\theta_6'' + \eta_3\theta_5$	$\hat{-}\eta_3\theta_5 \hat{-} \eta_3''\theta_5''$

Therefore

$$\mathrm{Ker}(\underline{d_3}) = \left\{\eta_2\theta_2'\theta_4,\ \ \eta_2(\theta_2'\eta_4 - \eta_2'\theta_4),\ \ \eta_2\eta_2'\theta_4 + \eta_8 + \eta_8''',\ \ \eta_8 - \eta_8' - \eta_8'''',\ \ \eta_8'' + \eta_8''' + \eta_8''''\right\},$$

$$\dim_q \mathrm{Ker}(\underline{d_3}) = q^{-1} + 4q,$$
$$\underline{H_2} = \mathrm{Ker}(\underline{d_3})/\mathrm{Im}(\underline{d_2}) = \left\{\eta_2(\theta_2'\eta_4 - \eta_2'\theta_4)\right\}, \qquad \dim_q(\underline{H_2}) = q \quad (62)$$

and

$$\mathrm{Im}(\underline{d_3}) = \left\{\eta_2(\theta_6 + \theta_6''),\ \ \eta_3\theta_5 + \eta_3''\theta_5'',\ \ \eta_2\theta_6 + \eta_3\theta_5,\ \ \eta_2(\eta_6 + \eta_6'')\right\},$$

$$\dim_q \mathrm{Im}(\underline{d_3}) = 3 + q^2, \qquad \dim_q \mathrm{Ker}(\underline{d_3}) + q \cdot \dim_q \mathrm{Im}(\underline{d_3}) = \dim_q(\mathcal{C}_2) = q^{-1} + 7q + q^3 \quad (63)$$

6.5. Differential $\underline{d_4}$ and the cohomologies $\underline{H_3}$ and $\underline{H_4}$

Finally,

$$\underline{K_4} = p_3p_3''\left(\frac{\partial}{\partial p_6''}\ominus\frac{\partial}{\partial p_6}\right) + \underline{p_2}p_3''\frac{\partial}{\partial \underline{p_5}}\ominus \underline{p_2}p_3\frac{\partial}{\partial \underline{p_5''}} \tag{64}$$

The corresponding differential

$$\underline{d_4} = -\theta_3\theta_3''\left(\frac{\partial}{\partial\theta_6''}\,\hat{-}\,\frac{\partial}{\partial\theta_6}\right) - (\theta_3\eta_3''\,\hat{-}\,\eta_3\theta_3'')\left(\frac{\partial}{\partial\eta_6''}-\frac{\partial}{\partial\eta_6}\right) + \eta_2\theta_3''\frac{\partial}{\partial\eta_5}\,\hat{+}\,\eta_2\theta_3\frac{\partial}{\partial\eta_5''} \tag{65}$$

converts the $\underline{\mathcal{C}_3}$ of dimension $\dim_q(\underline{\mathcal{C}_3}) = 4+4q^2 = 4qD$:

$\underline{d_4}\downarrow$

$\underline{\mathcal{C}_3}$	$\eta_2\theta_6$	$\eta_2\eta_6$	$\eta_2\theta_6''$	$\eta_2\eta_6''$	$\theta_3\eta_5$	$\eta_3\eta_5$	$\theta_3''\eta_5''$	$\eta_3''\eta_5''$
$\underline{d_4\mathcal{C}_3}$	$\hat{-}\,\eta_2\theta_3\theta_3''$	$\hat{-}\,\eta_2(\theta_3\eta_3''+\eta_3\theta_3'')$	$\eta_2\theta_3\theta_3''$	$\eta_2(\theta_3\eta_3''+\eta_3\theta_3'')$	$\eta_2\theta_3\theta_3''$	$\eta_2\eta_3\theta_3''$	$\hat{-}\,\eta_2\theta_3\theta_3''$	$\hat{-}\,\eta_2\theta_3\eta_3''$

Thus

$$\mathrm{Ker}(\underline{d_4}) = \Big\{\eta_2(\theta_6+\theta_6''),\ \ \theta_3\eta_5+\theta_3''\eta_5'',\ \ \eta_2\theta_6+\theta_3\eta_5,\ \ \eta_2\eta_6+\eta_3\eta_5-\eta_3''\eta_5'',\ \ \eta_2(\eta_6+\eta_6'')\Big\},$$
$$\dim_q\mathrm{Ker}(\underline{d_4}) = 3+2q^2,$$
$$\underline{H_3} = \mathrm{Ker}(\underline{d_4})/\mathrm{Im}(\underline{d_3}) = \Big\{\eta_2\eta_6+\eta_3\eta_5-\eta_3''\eta_5''\Big\},\qquad \dim_q(\underline{H_3}) = q^2 \tag{66}$$

The space

$$\mathcal{C}_4 = \Big\{\eta_2\theta_3\theta_3'',\ \ \eta_2\theta_3\eta_3'',\ \ \eta_2\eta_3\theta_3'',\ \ \eta_2\eta_3\eta_3''\Big\},\qquad \dim_q(\underline{\mathcal{C}_4}) = q^{-1}+2q+q^3 = qD^2 \tag{67}$$

$$\mathrm{Im}(\underline{d_4}) = \Big\{\eta_2\theta_3\theta_3'',\ \ \eta_2\theta_3\eta_3'',\ \ \eta_2\eta_3\theta_3''\Big\}$$
$$\dim_q\mathrm{Im}(\underline{d_4}) = q^{-1}+2q,$$
$$\mathrm{Coim}(\underline{d_4}) = \{\eta_2\eta_3\eta_3''\},\qquad \dim_q(\underline{H_4}) = \dim_q\mathrm{Coim}(\underline{d_4}) = q^3 \tag{68}$$

6.6. Reduced Jones superpolynomials for 4_1

Thus we have

$$\begin{aligned}\dim_q(\underline{H_0}) &= q^{-1},\\ \dim_q(\underline{H_1}) &= 1,\\ \dim_q(\underline{H_2}) &= q,\\ \dim_q(\underline{H_3}) &= q^2,\\ \dim_q(\underline{H_4}) &= q^3\end{aligned} \tag{69}$$

so that the reduced Jones superpolynomial is

$$\underline{P_\square^{4_1}} = q^{-1}\frac{q^2}{(q^2T)^2}\sum_{i=0}^{4}(qT)^i\cdot\dim_\mathrm{q}(\underline{H_i}) = \frac{1}{q}\frac{q^2}{(q^2T)^2}\Big(q^{-1}+(qT)+q(qT)^2+q^2(qT)^3+q^3(qT)^4\Big) =$$
$$= \frac{1}{q^4T^2}+\frac{1}{q^2T}+1+q^2T+q^4T^2 \tag{70}$$

what coincides with the value of the superpolynomial [22, 34]

$$\boxed{P_\square^{4_1}(a|q|T) = 1+T^2a^2+q^2T+\frac{1}{q^2T}+\frac{1}{T^2a^2}} \tag{71}$$

at $a=q^2$.

7. Conclusion

This second part of review series explains the difference between reduced and unreduced superpolynomials from the perspective of Khovanov-Rozansky categorification approach. We did not discuss the Chern-Simons-theory origin of this reduction, as well as its (in)dependence on the choice of the "marked" edge in the link diagram. Instead, since for small knots reduced polynomials are considerably simpler, we used this chance to describe cohomology calculus in much more detail than it was done in [1].

Acknowledgments

Our work is partly supported by the Ministry of Education and Science of the Russian Federation under the contract 8207, by the grants NSh-3349.2012.2, RFBR-10-01-00538, by the joint grants 90453-Ukr, 11-01-92612-Royal Society, 12-02-92108-Yaf-a and by CNPq 400635/2012-7, the Brazil National Counsel of Scientific and Technological Development.

References

[1] V.Dolotin and A.Morozov, arXiv:1208.4994 v1
[2] V.F.R.Jones, Invent.Math. 72 (1983) 1; Bull.AMS 12 (1985) 103; Ann.Math. 126 (1987) 335; E.Witten, Commun.Math.Phys. 121 (1989) 351
[3] R.Gopakumar and C.Vafa, Adv.Theor.Math.Phys. 3 (1999) 1415-1443, hep-th/9811131; H.Ooguri and C.Vafa, Nucl.Phys. B577 (2000) 419-438, hep-th/9912123; M.Aganagic, A.Klemm, M.Marino and C.Vafa, Commun.Math.Phys. 254 (2005) 425-478, hep-th/0305132; A.Iqbal, C.Kozcaz and C.Vafa, JHEP 0910:069, 2009, hep-th/0701156; M.Aganagic, M.C.N.Cheng, R.Dijkgraaf, D.Krefl and C.Vafa arXiv:1105.0630
[4] L.Kauffman, Topology 26 (1987) 395-407; Trans.Amer.Math.Soc. 311 (1989) n.2, 697-710; L.Kauffman and P.Vogel, J.Knot Theory Ramifications 1 (1992) 59-104
[5] M.Khovanov. Duke Math.J. 101 (2000) no.3, 359426, math/9908171
[6] D.Bar-Natan, Algebraic and Geometric Topology 2 (2002) 337-370, math/0201043
[7] M.Khovanov, Experimental Math. 12 (2003) no.3, 365374, math/0201306; J.Knot theory and its Ramifications 14 (2005) no.1, 111–130, math/0302060; Algebr. Geom. Topol. 4 (2004) 1045-1081, math/0304375; Int. Journal of Math. 18 (2007), no. 8, 869885, math/0510265; math/0605339; arXiv:1008.5084
[8] M.Khovanov and L.Rozansky, Fund. Math. 199 (2008), no. 1, 191, math/0401268; Geom.Topol. 12 (2008), no. 3, 13871425, math/0505056; math/0701333
[9] A.Shumakovitch, arXiv:math/0405474
[10] D.Bar-Natan, Geom. Topol. 9 (2005) 1443-1499, math/0410495; J.Knot Theory Ramifications 16 (2007), no. 3, 243255, math/0606318
[11] L.Ng, Duke Math. J. 141 (2008) 365406, math/0407071
[12] S. Gukov, A. Schwarz, and C. Vafa, Lett. Math. Phys. 74 (2005) 5374, hep-th/0412243
[13] J.Rasmussen, Duke Math.J. 136 (2007) 551-583, math.GY/0508510; math.GT/0607544
[14] S.Gukov and J.Walcher, hep-th/0512298
[15] E.Wagner, *Sur l'homologie de KhovanovRozansky des graphes et des entrelacs*
[16] M.Stosic, Algebr.Geom.Topol. 7 (2007) 261284, math/0511532; math/0605579; Topology Appl. 156 (2009) no. 3, 533541, math/0606656
[17] P.Turner, Algebr. Geom. Topol. 8 (2008), no. 2, 869884, math/0606369; math/0606464
[18] L.Rozansky, arXiv:1005.3266
[19] B.Webster, arXiv:1005.4559
[20] A.Oblomkov, J.Rasmussen, and V.Shende, arXiv:1201.2115
[21] E.Gorsky, A.Oblomkov and J.Rasmussen, arXiv:1206.2226
[22] N.M.Dunfield, S.Gukov and J.Rasmussen, Experimental Math. 15 (2006), 129159, math/0505662
[23] S.Gukov, A.Iqbal, C.Kozcaz and C.Vafa, Comm.Math.Phys. 298 (2010) 757-785, arXiv:0705.1368
[24] E. Gorsky, Contemp.Math. 566, (AMS, Providence, 2012), 213-232, arXiv:1003.0916 ; arXiv:1110.1674; E. Gorsky and M. Mazin, J.Combinatorial Th. A 120 (2013), 4963, arXiv:1105.1151; arXiv:1204.5448
[25] D.Galakhov, A.Mironov, A.Morozov, A.Smirnov, Teor.Mat.Fiz. 172 (2012) 73-99, arXiv:1104.2589
[26] M.Aganagic and Sh.Shakirov, arXiv:1105.5117; arXiv:1202.2489
[27] P.Dunin-Barkovsky, A.Mironov, A.Morozov, A.Sleptsov and A.Smirnov, arXiv:1106.4305 v3
[28] N.Carqueville and D.Murfet, arXiv:1108.1081

[29] I.Cherednik, arXiv:1111.6195
[30] Sh.Shakirov, arXiv:1111.7035
[31] S.Gukov and M.Stosic, arXiv:1112.0030
[32] A.Mironov, A.Morozov, Sh.Shakirov and A.Sleptsov, JHEP 1205 (2012) 070, arXiv:1201.3339
[33] K.Kawagoe, *unpublished*
[34] H.Itoyama, A.Mironov, A.Morozov and And.Morozov, JHEP 1207 (2012) 131, arXiv:1203.5978
[35] M.Aganagic and C.Vafa, arXiv:1204.4709
[36] H.Fuji, S.Gukov and P.Sulkowski, arXiv:1203.2182; arXiv:1205.1515
[37] E.Gorsky, A.Oblomkov, J.Rasmussen and V.Shende, arXiv:1207.4523
[38] A.Mironov and A.Morozov, arXiv:1208.2282
[39] Anton Morozov, arXiv:1208.3544
[40] S.Nawata, P.Ramadevi, Zodinmawia and X.Sun, arXiv:1209.1409
[41] H.Fuji, S.Gukov, M.Stosic and P.Sulkowski, arXiv:1209.1416

Some vortex solutions in the extended Skyrme-Faddeev model

L. A. Ferreira[1], **M. Hayasaka**[2], **J. Jäykkä**[3], **N. Sawado**[2] **and K. Toda**[4]

[1] Instituto de F sica de São Carlos; IFSC/USP,Universidade de São Paulo - USP, Caixa Postal 369, CEP 13560-970, São Carlos-SP, Brazil
[2] Department of Physics, Tokyo University of Science, Noda, Chiba 278-8510, Japan
[3] Nordita, Roslagstullsbacken 23, SE-10691 Stockholm, Sweden
[4] Department of Mathematical Physics, Toyama Prefectural University, Kurokawa 5180, Imizu, Toyama, 939-0398, Japan

E-mail: `sawado@ph.noda.tus.ac.jp (Corresponding author)`

Abstract. Analytical and numerical vortex solutions for the extended Skyrme-Faddeev model in a $(3+1)$ dimensional Minkowski space-time are investigated. The extension is obtained by adding to the Lagrangian a quartic term, which is the square of the kinetic term, and a potential which breaks the $SO(3)$ symmetry down to $SO(2)$. The construction of the solutions has been done in twofold: one makes use of an axially symmetric ansatz and solves the resulting ODE by an analytical and a numerical way. The analytical vortices are obtained for special form of the potentials, and the numerical ones are computed using the successive over relaxation method for wider choice of the potentials. Another is based on a simulational technique named the simulated annealing method which is available to treat the non-axisymmetric shape of solutions. The crucial thing for determining the structure of vortices is *the type* of the potential.

1. Introduction

The so-called Skyrme-Faddeev model was introduced in the seventies [1] as a generalization to $(3+1)$ dimensions of the $O(3)$ non-linear sigma model in $(2+1)$ dimensions. The Skyrme term, quartic in derivatives of the eld, balances the quadratic kinetic term and according to Derrick's theorem, allows the existence of stable solutions with non-trivial Hopf topological charges. Due to the highly non-linear character of the model and the lack of symmetries, the rst soliton solutions were only constructed in the late nineties using numerical methods [2, 3, 4, 5]. Since then the interest in the model has increased considerably and it has found applications in many areas of physics due mainly to the knotted character of the solutions [6]. One of the aspects of the model that has attracted considerable attention has been its connection with gauge theories. Faddeev and Niemi have conjectured that it might describe the low energy limit of the pure $SU(2)$ Yang-Mills theory [7]. They based their argument on a decomposition of the physical degrees of freedom of the $SU(2)$ connection, proposed in the eighties by Cho [8], and involving a triplet of scalar elds $\vec{n}$ taking values on the sphere S^2 ($\vec{n}^2 = 1$). Gies [9] has calculated the Wilsonian one loop e ective action for the pure $SU(2)$ Yang-Mills theory assuming Cho's decomposition, and found that the Skyrme-Faddeev action is indeed part of it, but additional quartic terms in the derivatives of the triplet $\vec{n}$ are unavoidable. In fact, the rst numerical Hopf solitons were rst constructed for the Skyrme-Faddeev model modi ed by a quartic term

[2] which is the square of the kinetic term. However, the soliton solutions in [2] were constructed for a sector of the theory where the signs of the coupling constants disagree with those indicated by Gies' calculations. Therefore, it is worth investigating the model with correct sign of the coupling constants.

In this paper we consider an extended Skyrme-Faddeev model (ESF) de ned by the Lagrangian

$$\mathcal{L} = M^2\,\partial_\mu \vec{n} \quad \partial^\mu \vec{n} \quad \frac{1}{e^2}(\partial_\mu \vec{n} \wedge \partial \ \vec{n})^2 + \frac{}{2}\,(\partial_\mu \vec{n} \quad \partial^\mu \vec{n})^2 \quad V(n_3) \tag{1}$$

where $\vec{n}$ is a triplet of real scalar elds taking values on the sphere S^2, n_3 its third component, M is a coupling constant with dimension of (length) ^{1}, e^2 and are dimensionless coupling constants, and the potential is a functional of the third component n_3 of the triplet $\vec{n}$. Note that the potential breaks the $O(3)$ symmetry of the original Skyrme-Faddeev down to $O(2)$, the group of rotations on the plane $n_1\,n_2$, and so eliminating two of the three Goldstone boson degrees of freedom. In this paper the main role of potential is to stabilize the vortex solutions.

The static energy density ($\mathcal{H}_{\rm static} = \quad \mathcal{L}$) associated to (1) is positive de nite if $V > 0$, $M^2 > 0$, $e^2 > 0$ and $\quad < 0$. That is the sector explored in [2] and where Hopf soliton solutions were rst constructed (for $V = 0$). In addition, that is also the sector explored in [10] but with additional terms involving second derivatives of the $\vec{n}$ eld, and where Hopf soliton were also constructed. The static energy density of (1) is also positive de nite for $V > 0$ if

$$M^2 > 0\,; \qquad e^2 < 0\,; \qquad \quad < 0\,; \qquad \quad e^2 \quad 1 \tag{2}$$

That is the sector that agrees with the signature of the terms in the one loop e ective action calculated in [9] and it is the sector that we will consider in this paper. Static Hopf solitons were constructed in [11] for the sector (2) (with $V = 0$) and their quantum excitations, including comparison with glueball spectrum, were considered in [12]. An interesting feature of the Hopf solitons constructed in [11] is that they shrink in size and then disappear as $\quad e^2 \to 1$, which is exactly the point where the exact vortex solution exist [13].

The aim of the present paper is to investigate if vortex solutions for the model (1) continue to exist when the condition $\quad e^2 = 1$ is relaxed, and so if they co-exist with the Hopf solitons of [11]. In order to stabilize the solution, we shall introduce the types of potential

$$V(n_3) = \frac{^2}{2}\,v_b^a, \qquad v_b^a \quad (1 + n_3)^a (1 \quad n_3)^b \tag{3}$$

where $a + b =$ *non-zero integer*, and is a real coupling constant. A special choice of the parameters a, b we have holomorphic solutions of the model while for the other case we still have numerical solutions.

In this paper, rst we discuss the integrable holomorphic solutions of the model and next we shall perform the numerical stu . The numerical simulations are done by twofold: one is by solving a di erential equation which is accomplished in terms of *the standard successive over relaxation*. Another is based on energy minimization scheme called *the simulated annealing*. Especially the latter analysis demonstrates the detailed behavior of the symmetry breaking of the solutions by the change of the structure of the potential.

2. The integrable sector of the model

The rst exact vortex solutions for the theory (1) were constructed in [13] for the case where the potential vanishes, and by exploring the integrability properties of a submodel of (1). In order to describe those exact vortex solutions it is better to perform the stereographic projection of the target space S^2 onto the plane parameterized by the complex scalar eld u and related to $\vec{n}$ by

$$\vec{n} = (u + u^*, \quad i(u \quad u^*), u^2 \quad 1)/(1 + u^2) \tag{4}$$

It was shown in [13] that the eld con gurations of the form

$$u \quad u(z, y), \quad u^* \quad u^*(z^*, y), \quad \text{for} \quad e^2 = 1, \qquad V = 0 \tag{5}$$

are exact solutions of (1), where $z = x^1 + i\varepsilon_1 x^2$ and $y = x^3 \quad \varepsilon_2 x^0$, with $\varepsilon_a = \pm 1$, $a = 1, 2$, and x^μ, $= 0, 1, 2, 3$, are the Cartesian coordinates of the Minkowski space-time. The simplest solution is of the form $u = z^n e^{i\,k\,y}$, with n integer, and it corresponds to a vortex parallel to the x^3-axis and with waves traveling along it with the speed of light.

In terms of the complex scalar eld u introduced in 4 the Lagrangian 1 becomes

$$\mathcal{L} = 4M^2 \frac{\partial_\mu u \partial^\mu u^*}{(1 + u^2)^2} + \frac{8}{e^2} \frac{(\partial_\mu u)^2 (\partial \ u^*)^2}{(1 + u^2)^4} + (\ e^2 \quad 1) \frac{(\partial_\mu u \ \partial^\mu u^*)^2}{(1 + u^2)^4} \qquad V(\mid u \mid^2) \tag{6}$$

where we have used the fact that n_3 is a functional of $\mid u \mid^2$ only, and so is the potential. The Euler-Lagrange equations following from (6), or (1), reads

$$(1 + |u|^2) \partial^\mu \mathcal{K}_\mu \quad 2u^* \mathcal{K}_\mu \partial^\mu u = \quad \frac{u}{4} (1 + |u|^2)^3 \, V' \tag{7}$$

where $V' = \frac{\partial V}{\partial u^2}$, and

$$\mathcal{K}_\mu \quad M^2 \partial_\mu u + \frac{4}{e^2} \frac{(\partial \ u \partial \ u) \partial_\mu u^* + (\ e^2 \quad 1)(\partial \ u \partial \ u^*) \partial_\mu u}{(1 + u^2)^2} \tag{8}$$

We point out that the theory (6) possesses an integrable sector de ned by the condition

$$(\partial_\mu u)^2 = 0 \tag{9}$$

Such condition was rst discovered in the context of the CP^1 model using the generalized zero curvature condition for integrable theories in any dimension [14], and then applied to many models with target space being the sphere S^2, or CP^1 (see [15] for a review). It leads to an in nite number of local conserved currents. Indeed, (9) together with the equations of motion (7) imply the conservation of the in nity of currents given by

$$J^G_\mu \quad \mathcal{K}_\mu \frac{G}{u} \quad \mathcal{K}^*_\mu \frac{G}{u^*} \tag{10}$$

where G is any functional of $\mid u \mid^2$ only. For the case where the potential vanishes, the set of conserved currents is considerably enlarged since G can be an arbitrary functional of u and u^*, but not of their derivatives. If in addition to the condition (9) one takes $V = 0$ and $e^2 = 1$, then the equations of motion reduce to $\partial^2 u = 0$. It is in that integrable sector that the solutions (5) lie, and were studied in [13].

It is interesting to note that (7) with a special choice of the potential

$$V(n_3) = \frac{2}{2} (1 + n_3)^{2 \quad 2/n} (1 \quad n_3)^{2+2/n} \tag{11}$$

have an analytical, holomorphic solution for each topological charge as

$$u(\ , \varphi, z, \ ) = \left(\frac{}{a}\right)^n e^{i[\epsilon n \varphi + k(z+\tau)]} \tag{12}$$

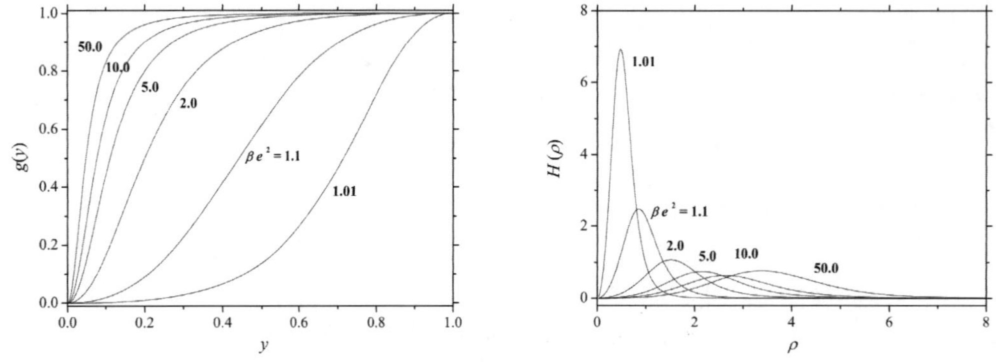

Figure 1. The $n = 2$ pro le $g(y)$ and the corresponding Hamiltonian density of the real space $H(\ )$ of $k^2 = 0.0$ for the constant $r_0^2\ ^2/M^2 = 1.0$.

where $\epsilon = \pm 1$ and a describes a scale of the solution. Here we used dimensionless polar coordinates $(\ , \varphi, z, \ )$ de ned by

$$x^0 = ct = r_0\ , \quad x^1 = r_0 \quad \cos\varphi, \quad x^2 = r_0 \quad \sin\varphi, \quad x^3 = r_0 z \tag{13}$$

and where we have introduced a length scale r_0 given by

$$r_0^2 = \frac{4}{M^2\ e^2}. \tag{14}$$

Substituting (12) into (7) the a can be determined such as

$$a = \mid n \mid \quad \frac{M^2(\ e^2 \quad 1)}{r_0^2\ ^2}^{1/4} = \mid n \mid \quad \frac{(\ e^2)(\ e^2 \quad 1)M^4}{4\ ^2}^{1/4} \tag{15}$$

Clearly, the special solution at $\ e^2 = 1$ is obtained if we take a proper limit of the vanishing potential, i.e. $e^2 \to 1$ and $\ ^2 \to 0$ with $(\ e^2 \quad 1)/\ ^2 =$ constant.

In the case with the potential, the current (10) is still conserved because

$$\begin{aligned} \partial^\mu J_\mu = & \frac{^2G}{u^2}\partial^\mu u\mathcal{K}_\mu + \frac{G}{u}\partial^\mu\mathcal{K}_\mu + \frac{^2G}{u^*\ u}\partial^\mu u^*\mathcal{K}_\mu \\ & \frac{^2G}{u^{*2}}\partial^\mu u^*\mathcal{K}^*_\mu \quad \frac{G}{u^*}\partial^\mu\mathcal{K}^*_\mu \quad \frac{^2G}{u\ u^*}\partial^\mu u^*\mathcal{K}^*_\mu = 0 \end{aligned} \tag{16}$$

where we have used the reduced integrable equation

$$\partial^\mu\mathcal{K}_\mu + \frac{2}{4}(1+|u|^2)^2\frac{\partial V(|u|^2)}{\partial u^*} = 0. \tag{17}$$

The Hamiltonian density associated to (6) is not positive de nite due to the quartic terms in time derivatives. We shall arrange the Legendre transform of each term in (6) to make explicit

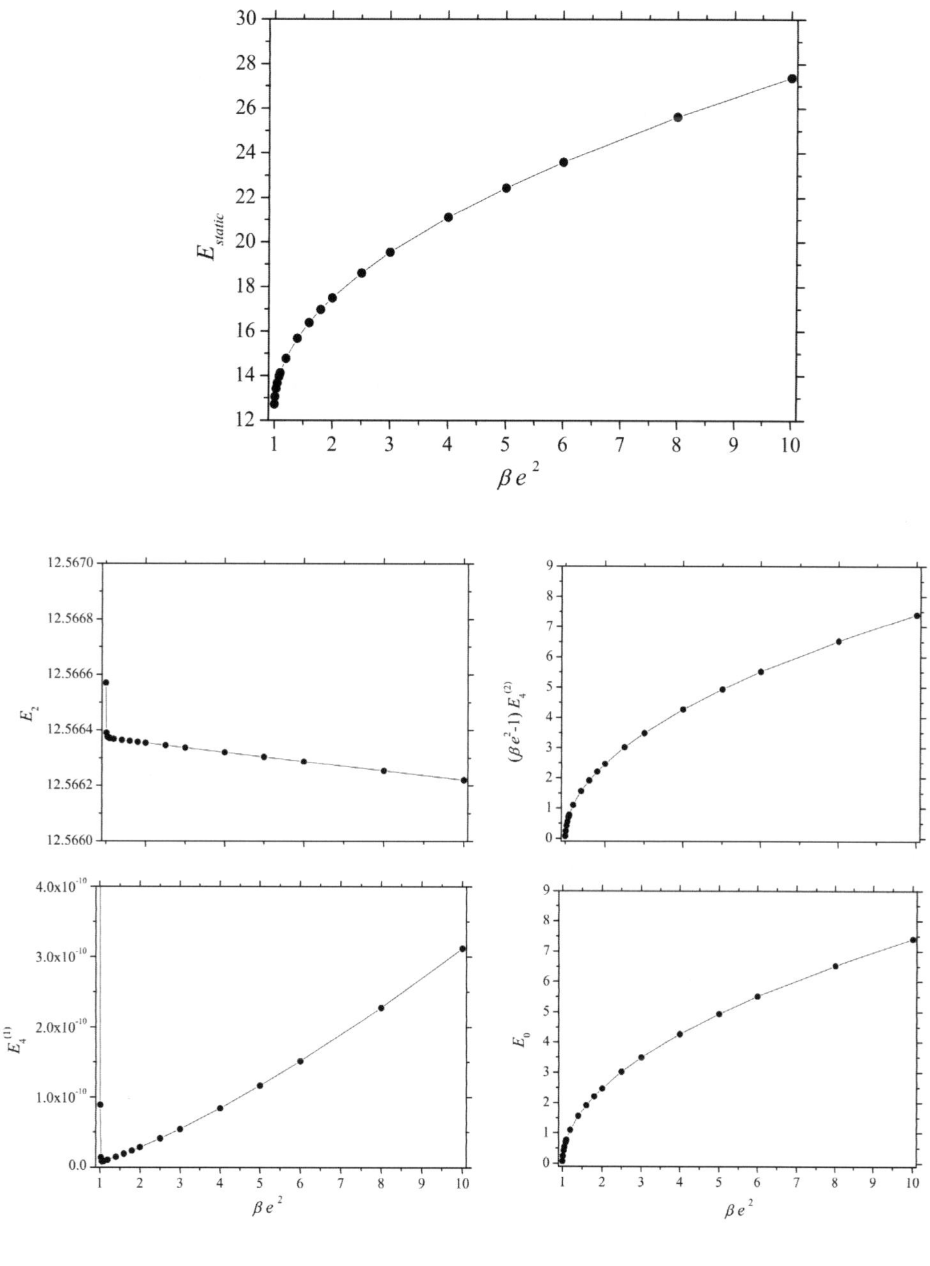

Figure 2. The static energy and its components corresponding to the solutions of Fig.1.

Journal of Physics: Conference Series **411** (2013) 012014 doi:10.1088/1742-6596/411/1/012014

such non positive contributions, and write the Hamiltonian density as (see [16] for details)

$$\mathcal{H} \;=\; 4M^2 \frac{\left[|\, \dot u \,|^2 + \vec\nabla u \;\; \vec\nabla u^* \right]}{(1+ u^2)^2} \quad \frac{24}{e^2} \frac{(\vec\nabla u)^2 (\vec\nabla u^*)^2}{(1+ u^2 2)^4} \quad \left(\frac{2}{3}\right)^2 \quad F^2$$

$$24 \frac{(\;\; e^2 \;\;\; 1)}{e^2} \frac{\left[|\, \dot u \,|^2 + \frac{1}{3} \vec\nabla u \;\; \vec\nabla u^* \right] \left[\vec\nabla u \;\; \vec\nabla u^* \;\;\; |\, \dot u \,|^2 \right]}{(1+ u^2)^4} + V(|\, u \,|^2) \qquad (18)$$

where $\dot u$ denotes the x^0-derivative of u, and $\vec\nabla u$ its spatial gradient, and where we have denoted

$$\frac{\dot u^2}{(\vec\nabla u)^2} \quad \frac{1}{3} + F e^{i\,\Phi} \qquad (19)$$

with $F > 0$ and $0 \quad \Phi \quad 2$, being functions of the space-time coordinates. The most of terms in (18) make positive contribution while the second term has some possibility to be negative. Note also, for static con gurations apparently it is positive de nite for the range of parameters given in (2).

3. The integrable and the non-integable sectors: numerical analysis by the SOR

Although in the previous section we used the polar coordinates, for the numerical study it is more convenient to use a new radial coordinate y ($0 \quad y \quad 1$), de ned by $\; = \sqrt{\frac{1 \;\; y}{y}}$. We introduce the solution ansatz of the form

$$u(\;\; , \varphi, z, \;\;) = \sqrt{\frac{1 \quad g(y)}{g(y)}} \, e^{i[\epsilon n \varphi + k(z+\tau)]} \qquad (20)$$

where the pro le function $g(y)$ is de ned at the period $0 \quad g \quad 1$. The equation can be written as

$$\frac{d}{dy} \; \frac{y(1 \quad y)}{g(1 \quad g)} g' R \; + \left(g \quad \frac{1}{2}\right) \frac{S}{y(1 \quad y)} \quad \Omega \quad\; \frac{y(1 \quad y)}{g(1 \quad g)} g' \quad ^2 \Big\}$$

$$= \frac{1}{y^2} \frac{r_0^2 \;\; ^2}{M^2} (1 \quad g)^{1 \;\; \frac{2}{n}} g^{1+\frac{2}{n}} \left\{ 4g \quad 2\left(1 + \frac{1}{n}\right) \right\} \qquad (21)$$

where the primes at this time indicate derivatives w.r.t.y and where

$$\Omega = n^2$$

$$S = 1 + \;\; e^2 g(1 \quad g) \frac{y}{1 \quad y} \quad \Omega + \quad \frac{y(1 \;\; y)}{g(1 \;\; g)} g' \quad ^2 \Big\}$$

$$R = 1 + g(1 \quad g) \frac{y}{1 \quad y} \quad (\;\; e^2 \quad 2)\Omega + \;\; e^2 \quad \frac{y(1 \;\; y)}{g(1 \;\; g)} g' \quad ^2 \Big\} \qquad (22)$$

The energy in the unit of $4M^2$ per unit length for the time-dependent vortex can be estimated in terms of following four parts of integrals of the dimensionless Hamiltonian $H := \mathcal{H}/4M^2$

$$E \;=\; 2 \int_0^\infty d \;\; H(\;\;) = E_2 + E_4^{(1)} + (\;\; e^2 \quad 1) E_4^{(2)} + \frac{r_0^2 \;\; ^2}{M^2} E_0 \qquad (23)$$

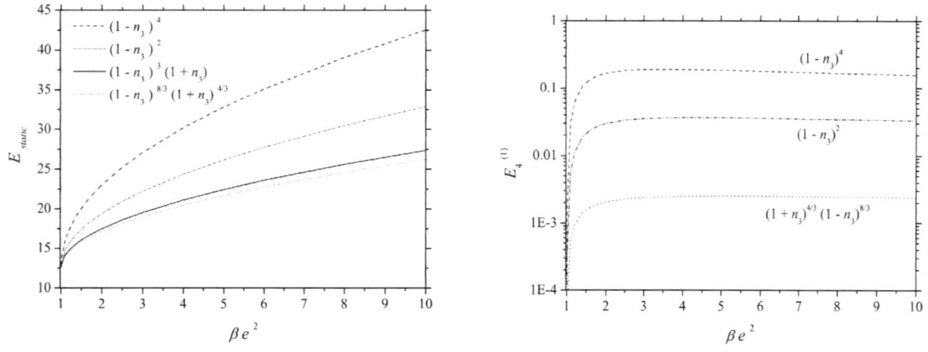

Figure 3. The $n = 2$ static energy and the component $E_4^{(1)}$ for several type of potentials v_j^i. $k^2 = 0$ and $r_0^2\ \ ^2/M^2 = 1.0$.

in which the components are de ned as

$$E_2 = \int_0^1 \frac{dy}{y(1\ \ y)} \quad 2k^2\frac{1\ \ y}{y} + n^2 + \quad \frac{y(1\ \ y)}{g(1\ \ g)}g' \quad ^2\Big\} g(1\ \ g) \tag{24}$$

$$E_4^{(1)} = \int_0^1 \frac{dy}{2(1\ \ y)^2} \quad 4k^2\frac{1\ \ y}{y} + n^2 \quad \frac{y(1\ \ y)}{g(1\ \ g)}g' \quad ^2\Big\} \\ n^2 \quad \frac{y(1\ \ y)}{g(1\ \ g)}g' \quad ^2\Big\}(g(1\ \ g))^2 \tag{25}$$

$$E_4^{(2)} = \int_0^1 \frac{dy}{2(1\ \ y)^2} \quad 4k^2\frac{1\ \ y}{y} + n^2 + \quad \frac{y(1\ \ y)}{g(1\ \ g)}g' \quad ^2\Big\} \\ n^2 + \quad \frac{y(1\ \ y)}{g(1\ \ g)}g' \quad ^2\Big\}(g(1\ \ g))^2 \tag{26}$$

$$E_0 = 2 \quad \int_0^1 \frac{dy}{y^2} g^{2+\frac{2}{n}}(1\ \ g)^{2\ \ \frac{2}{n}}. \tag{27}$$

It is easy to see that the k^2 term in (26) has positive contributions to the energy while in (25) the sign of the k^2 term depends on the spatial structure of the solution g. For the holomorphic solution (12), the energy (25) becomes zero and then the energy of the integrable sector keeps positive de nite for all values of k^2.

For $n = 2$, the explicit form the potential is

$$V_{n=2} = \frac{2}{2}(1 + n_3)(1\ \ n_3)^3. \tag{28}$$

The potential has zero at both the origin and the in nity thus it is so called a *new*-BS (Baby-Skyrmion) type. The equation (21),(22) are solved in terms of the standard successive over relaxation scheme. Fig.1 is the pro le function and the Hamiltonian density for $n = 2$. Fig.2 is the energy per unit length and its components for several values of $\ \ e^2$ and xed $\ \ ^2$. We con rmed that the value of the component $E_1^{(4)}$ is regarded as zero within the numerical uncertainty. This clearly indicates that the solution satis es the condition (9).

Although we have obtained the analytical solutions for a special form of the potential (3), we have many options for choice of the potential. We can obtain many numerical solutions for

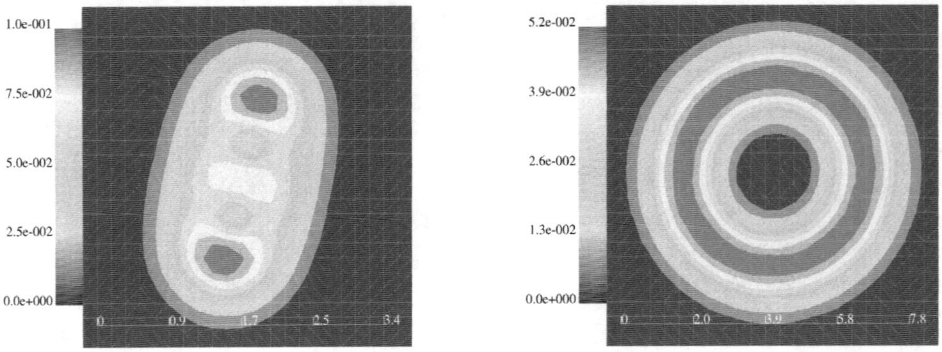

Figure 4. The energy density per unit length of the *old*-BS potential v_1^0 (*left*) and the *new*-BS potential v_1^1 (*right*).

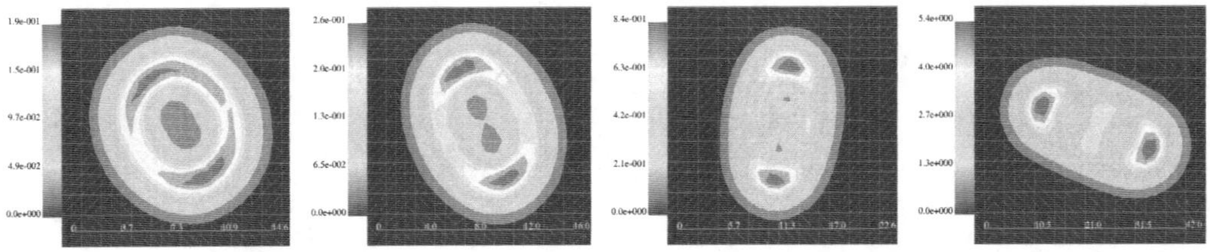

$e^2 = 1.01$ $e^2 = 1.1$ $e^2 = 2.0$ $e^2 = 20.0$

Figure 5. The energy density per unit length of the *old*-BS potential v_1^0 for several e^2.

the several types of the potentials. We show the result of $n = 2$ for the potentials $v_2^0, v_4^0, v_{8/3}^{4/3}$; of course these are not of the form of the analytical solution. Note that for such non-integrable solutions, the second term in (18) survives and has a possibility to be negative. Here we only consider the case for $k^2 = 0$. Fig.3 presents the energies and the component $E_4^{(1)}$ for these potentials. For $n = 2$, the *old*-BS potentials give higher total energy than the *new*-BS. This indicates that the same class of potentials gives the similar energy and then, for $n = 2$ the energy of the *new* type potential $v_{8/3}^{4/3}$ is closest to the integrable sector, which is also plotted in Fig.3 for reference.

4. Broken axisymmetric solutions: analysis by the SA

In the previous sections, we have assumed that the solution is invariant under the $O(2)$ internal symmetry $u \to e^{i\phi}u$ as well as the transformations of the Poincaré group given by rotation on the plane x^1x^2 and translations in the directions x^0 and x^3. However, there is a possibility of relaxing some of these symmetries. In [17], the authors generalized the *old*-BS type potential as $V = \frac{\mu^2}{2}(1 \quad n_3)^s$ $(0 < s \quad 4)$ and saw how the solution deforms from the rotational symmetry on the x^1x^2 plane by the change of or s.

For the deformed solution, the straightforward generalization of the ansatz (20) is

$$u = f(\ ,\varphi)e^{i\Theta(\rho,\varphi)}e^{ik(z+\tau)} \tag{29}$$

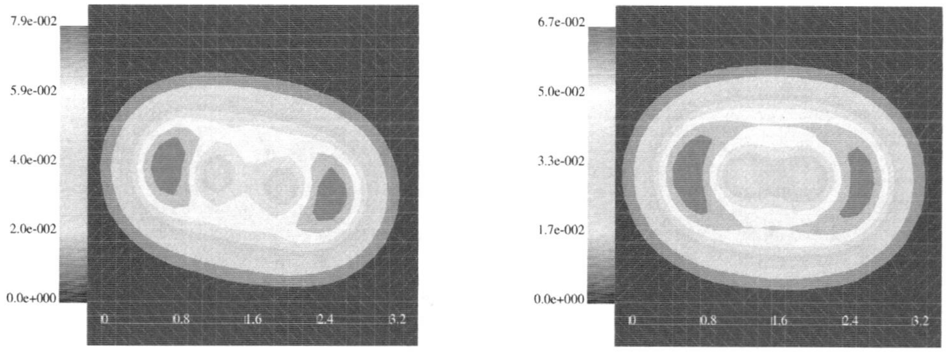

Figure 6. The energy density per unit length of $V_\epsilon = \frac{\mu^2}{2}(1 \quad n_3)(1+\epsilon n_3)$ for $\epsilon = 0.25$ (*left*) and $\epsilon = 0.5$ (*right*). The results for $\epsilon = 0, 0, 1.0$ are already presented in Fig.4.

or equivalently,

$$\vec{n} = (\sin F(\ ,\varphi)\cos[\ (\ ,\varphi) + k(z + \ )], \sin F(\ ,\varphi)\sin[\ (\ ,\varphi) + k(z+ \ )], \cos F(\ ,\varphi)) \quad (30)$$

where the $\vec{n}$ eld in terms of u is given in (4). Here $(\ ,\varphi) := n\varphi + \ _0(\ ,\varphi)$ and $_0$ is homotopic to the constant map. The method is a kind of the Monte-Carlo method in which one generates random numbers and properly change the value of the elds $\vec{n}$ by the numbers so as to drop the energy. However, a more sophisticated method may be applied to the problem. The simulated annealing method [18] is the application of the Metropolis algorithm which can successfully avoids the unwanted saddle points. The analysis has been done by minimizing the energy per unit length which is obtained by integrating the hamiltonian corresponding to (1) with the ansatz (30) into the $(\ ,\varphi)$ plane. We have con rmed that for the holomorphic solution (12) the hamiltonian is positive de nite. For the general solution, however, the energy positivity is supported only for a certain range of k. Here we present the results for $k = 0$. Also, we examine the case of $n = 3$ because the solution exhibits obvious symmetry breaking such as from axial to $\mathbb{Z}_2$-symmetry.

We study the following several cases:

(i) We start with two standard cases: the *old*-BS potential v_1^0 and the *new*-BS potential v_1^1. In Fig.4, we present the energy density per unit length. For the *old*-BS potential, the solution strongly deforms from the axial symmetry while for the *new*-BS the solution keeps the symmetry.

(ii) Next we shall see how the solution behaves for the change of the model parameters. In Fig.5, we plot the energy density per unit length of the *old*-BS potential for several values of the model parameter e^2. As is easily observed that for larger value of e^2, the deformation is enhanced.

(iii) From the results of (i),(ii), we get a new insight for the symmetry breaking. That is, most crucial thing for the deformation is that the potential is nite or not at the point antipodal to the vacuum. In order to investigate this criterion further, we introduce following two types of one-parameter family of potentials

$$V_\epsilon(n_3) := \frac{^2}{2}(1 \quad n_3)(1+\epsilon n_3), \quad 0 \quad \epsilon \quad 1 \quad (31)$$

$$V_\delta(n_3) := \frac{^2}{2}(1 \quad n_3)(1+n_3)^\delta, \quad 0 \qquad 1 \quad (32)$$

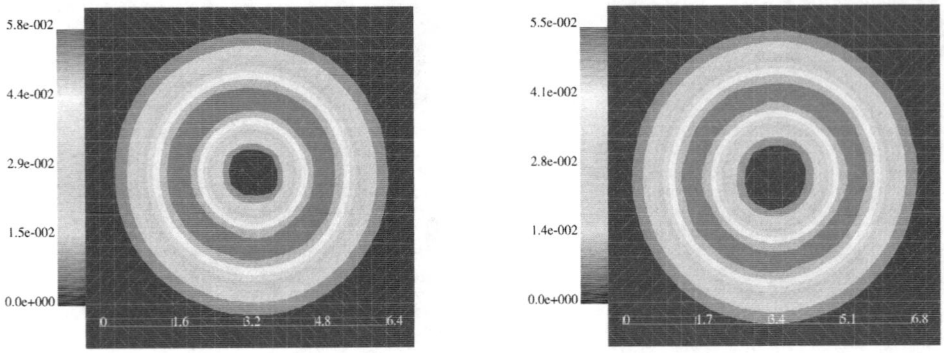

Figure 7. The energy density per unit length of $V_\delta = \frac{\mu^2}{2}(1 \quad n_3)(1+n_3)^\delta$ for $\quad = 0.25$ (*left*) and $\quad = 0.5$ (*right*). The results for $\quad = 0.0, 1.0$ are already presented in Fig.4.

Note that for $\epsilon, \quad = 0$ the potentials become the *old*-BS and $\epsilon, \quad = 1$ they are the *new*-BS. The results are stimulating. In the case of V_ϵ, the solutions always break the axial symmetry except only for $\epsilon = 1.0$ (Fig.6). On the other hand, in the V_δ, the solutions always keep the axial symmetry except only for $\quad = 0.0$ (Fig.7). Thus we con rmed the criterion for the mechanism of the symmetry breaking: if a potential has the vacuum value at the point antipodal to the true vacuum, the solution always exhibits the axial symmetry, and if there is no another zero except for the vacuum, the solution deforms.

We summarize our results: for $n = 1$, the potential for the integrable sector is the *old*-BS type and the solution naturally has the axial symmetry. For $n \quad 2$ the potential for the integrable sector is always the *new*-BS type, and in terms of the above simulational study, the solutions always should be axially symmetric. As a result, our prescription of the ansatz (20) is valid for all topological charges.

Acknowledgments Nobuyuki Sawado expresses his gratitude to the conference organizers of ISQS-20 for kind accommodation and hospitality. The authors would like to thank Wojtek Zakrzewski and Paweł Klimas for many useful discussions. We also acknowledge nancial support from FAPESP (Brazil). Juha Jäykkä appreciate the Flemish Science Foundation (FWO) for the support.

References

[1] L. D. Faddeev, "Quantization of solitons,", Princeton preprint IAS Print-75-QS70 (1975).
L. D. Faddeev, in *40 Years in Mathematical Physics* , (World Scienti c, 1995);
[2] J. Gladikowski and M. Hellmund, "Static solitons with non-zero Hopf number," Phys. Rev. D **56**, 5194 (1997) [arXiv:hep-th/9609035].
[3] L. D. Faddeev and A. J. Niemi, "Knots and particles," Nature **387**, 58 (1997) [arXiv:hep-th/9610193], "Toroidal con gurations as stable solitons," and arXiv:hep-th/9705176.
[4] R. A. Battye and P. M. Sutcli e, "Knots as stable soliton solutions in a three-dimensional classical eld theory," Phys. Rev. Lett. **81**, 4798 (1998) [arXiv:hep-th/9808129]; "Solitons, links and knots," Proc. Roy. Soc. Lond. **A455**, 4305-4331 (1999). [hep-th/9811077].
[5] J. Hietarinta and P. Salo, "Faddeev-Hopf knots: Dynamics of linked unknots," Phys. Lett. **B451**, 60-67 (1999). [hep-th/9811053]. "Ground state in the Faddeev-Skyrme model," Phys. Rev. D **62**, 081701(R) (2000).
[6] E. Babaev, L. D. Faddeev and A. J. Niemi, "Hidden symmetry and duality in a charged two-condensate Bose system," Phys. Rev. B **65**, 100512(R) (2002) [arXiv:cond-mat/0106152].
E. Babaev, Phys. Rev. Lett. **88**, 177002 (2002) [arXiv:cond-mat/0106360].

[7] L. D. Faddeev and A. J. Niemi, "Partially dual variables in SU(2) Yang-Mills theory," Phys. Rev. Lett. **82**, 1624 (1999) [arXiv:hep-th/9807069];

[8] Y. M. Cho, "A Restricted Gauge Theory," Phys. Rev. D **21**, 1080 (1980); and "Extended Gauge Theory And Its Mass Spectrum," Phys. Rev. D **23**, 2415 (1981).

[9] H. Gies, "Wilsonian e ective action for SU(2) Yang-Mills theory with Cho-Faddeev-Niemi-Shabanov decomposition," Phys. Rev. D **63**, 125023 (2001), hep-th/0102026

[10] N. Sawado, N. Shiiki and S. Tanaka, "Hopf soliton solutions from low energy e ective action of SU(2) Yang-Mills theory," Mod. Phys. Lett. A **21**, 1189 (2006) [arXiv:hep-ph/0511208].

[11] L. A. Ferreira, N. Sawado, K. Toda, "Static Hop ons in the extended Skyrme-Faddeev model," *Journal of High Energy Physics* JHEP **0911**, 124 (2009). [arXiv:0908.3672 [hep-th]].

[12] L. A. Ferreira, S. Kato, N. Sawado, K. Toda, "Quantum solitons in the extended Skyrme-Faddeev model," *Acta Polytechnica* **51**, issue 1, 47-49 (2011)

[13] L. A. Ferreira, "Exact vortex solutions in an extended Skyrme-Faddeev model," *Journal of High Energy Physics* JHEP **0905**, 001 (2009) [arXiv:0809.4303 [hep-th]].

[14] O. Alvarez, L. A. Ferreira and J. Sanchez Guillen, "A New approach to integrable theories in any dimension," Nucl. Phys. B **529**, 689 (1998) [arXiv:hep-th/9710147].

[15] O. Alvarez, L. A. Ferreira and J. Sanchez-Guillen, "Integrable theories and loop spaces: Fundamentals, applications and new developments," Int. J. Mod. Phys. A **24**, 1825 (2009) [arXiv:0901.1654 [hep-th]].

[16] L. A. Ferreira and A. C. Riserio do Bon m, JHEP **1003**, 119 (2010) [arXiv:0912.3404 [hep-th]].

[17] I. Hen and M. Karliner, "Rotational symmetry breaking in baby Skyrme models," Nonlinearity **21**, 399 (2008) [arXiv:0710.3939 [hep-th]].

[18] M. Hale, O. Schwindt and T. Weidig, "Simulated annealing for topological solitons," Phys. Rev. E **62**, 4333 (2000) [hep-th/0002058].

Qubits, superqubits and squbits

Andrzej M. Frydryszak

Institute of Theoretical Physics, University of Wroclaw,pl. Borna 9,50-205 Wroclaw, Poland

E-mail: `amfry@ift.uni.wroc.pl`

Abstract. We analyze recently proposed formalisms which use nilpotent variables to describe and/or generalize qubits and notion of entanglement. There are two types of them distinguished by the commutativity and or anti-commutativity of basics variables. While nilpotent commuting variables suit the new description of the qubits and entanglement, the anticommuting do not, but they can be used to describe generalized objects - superqubits. A squbit, in the present context, is a version of superqubit introduced to cure some of problematic properties of the superqubit.

1. Introduction

Recently there appeared new algebraic approaches to study qubit systems and their generalizations in the context of quantum entanglement. Depending on the basics assumption of the variables realizing the two-levelness of the quantum system, the final outcome is different. The idea to describe the two level system by specially chosen variables is around for some time (see account and references in [1]) The basics assumption is realized by the condition

$$1 = (\Gamma)^0, \qquad \Gamma, \qquad (\Gamma)^2 = 0, \tag{1}$$

which yelds the use of nilpotents with vanishing square (so called nilpotents of order one). Obviously one can consider higher order nilpotents to describe multi-level system e.g. qutrits [2]. Realization of nilpotency is very natural in the Grassmann algebra and variables of this type are used in formulation of supersymmetry since the late seventies of the Twentieth Century. Usually such variables are denoted by θ. Another possibility of realization of nilpotency, despite the graded commutative Grassmannian algebra, gives the commutative algebra with nilpotents: η, $\eta^2 = 0$. This option was used by physicists a bit later [3, 4, 2, 5, 6], firstly within the statistical physics to realize the Pauli exclusion principle for spinless objects like fluxes. Recenly there was developed classical [5] and quantum [2, 6] formalism based on such variables to describe multiqubit systems. This formalism in natural way allows to study multiqubit entanglement and structure of state spaces.

It turns out that anticommuting θ variables and the superspaces are not well suited to study the multiqubit systems and the questions of entanglement. I particular, natural realizations of supersymmetry concern rather different systems then qubits, moreover the conventional Berezinian/superdeterminant is not good for characterization of entanglement. It is worth noting that the first time the supersymmetrization of the qubit was proposed by Hruby [7], that was done within the supersymmetric field theoretical model where the qubit field has its anticommuting partner antiqubit. However there one uses rather field theoretical generalization of qubit to QFT-qubit. Here we are focused on nonrelativistic quantum mechanical objects.

Table 1. Comparison of the θ and η description of two qubit system.

Anticommutative	Commutative
$g(\theta_1, \theta_2) =$ $g_0 + \theta_1 g_1 + \theta_2 g_2 + \theta_1\theta_2 g_{12}$	$f(\eta_1, \eta_2) =$ $f_0 + \eta_1 f_1 + \eta_2 f_2 + \eta_1\eta_2 f_{12}$
$Berg = \begin{pmatrix} g_0 & g_1 \\ g_2 & g_{12} \end{pmatrix}$ $= (g_0 - g_1 g_{12}^{-1} g_2) g_{12}^{-1}$	$det f = \begin{pmatrix} f_0 & f_1 \\ f_2 & f_{12} \end{pmatrix}$ $= f_0 f_{12} - f_1 f_2$

Let us observe firstly that expansion coefficients g_1 and g_2 of the superfunction g have to be odd elements of a graded algebra and cannot be reduced to the complex or real numbers. For η-function one can take all coefficients f_i, f_{12} complex or real valued. This point is essential from the physical point of view. Namely, when treating $g(\theta_i)$ and $f(\eta_i)$ as a generalized wave functions and developing generalization of the Schrödinger realization for θ variables one gets Grassmann algebra valued scalar product [8, 9, 10] and graded algebra valued "probability amplitudes". Therefore some additional mapping is needed to get numerical information from such a Grassmann element. Usualy the so called *body mapping* [8] is used, other posibilities gives Banach-Grassmann algebra defined by Rogers [11, 12]. The η-functions do not yield such a complications. Such approaches were considered already in the context of supersymmetric quantum mechanics in super-Hilbert spaces cf. Refs. [13, 8, 9, 10]. Using superHilbert space approach one can consider e.g. coherent fermionic states which have properties analogous to the conventional coherent states, but are not physical in the usual sense. The supermatrix and matrix appearing in the Table 1 can be thought as a matrix of second order θ-superderivative and η-derivative. Again, in the case of the θ-function the properties of the Berezinian prevent us from getting information about factorization properties of the $g(\theta_1, \theta_2)$, while conventional determinant for the $f(\eta_1, \eta_2)$ does the job. In supersymmetric theories, but not only, the anticommuting variables are used to describe fermionic degrees of freedom. As was observed by Kitayev in the context of quanum information theory multi-fermion states are more nonlocal then bosonoc ones, therefore we can expect that objects like superqubit may produce more nonlocality.

One can expect that superspaces related to graded commutative algebra might be not right arena to describe qubits, but as it was shown in [14] specific supersymmetric generalization of qubit. Moreover for the superqubit systems the notion of entanglement has to be modified due to some algebraic artefacts. As it was noted above, from the physical point of view the use of anticommuting variables yields essential difficulties, that is why there was attempt to cure some of them, and modify the notion to the so called squbits [15] - prefix s comes here from abbreviated prefix super. Let us note that last word is already used in quite different context. The term squbit in nanotechnogy information devices denotes superconducting qubit - a device based on Josephson's junction.

In the following we shall sketch the description of qubits in terms of the η-variables, recall main properties of superqubits [14]. Then using results of Ref. [15] we will discuss why the modification of superqubit to the squbit might be mandatory.

Before going into details let us comment on chronology of the main concepts discussed in the present paper:

- 1976 - supersymmetric quantum mechanics introduced in 1976 by Nicolai [13]

- 1970 - origins of quantum computing, qubits
- 2006 - qubits described by commuting nilpotent variables [2, 1]
- 2009 - superqubits [14]
- 2010 - squbits [15]

2. Qubit - η-space description

The formalism of η-Hilbert spaces gives new tools to study multiqubit entanglement. Instead of considering the $\mathbb{C}^2$ Hilbert space and its tensor product structures one uses properties of functions of η-variables. Namely, if we take explicit form of one qubit and two qubit states in e.g. binary bases

$$\begin{aligned} \psi^{(1)}(x) &= \psi_0(x)|0\rangle + \psi(x)|1\rangle, & (2)\\ \psi^{(2)}(x) &= \psi_0(x)|00\rangle + \psi_1(x)|10\rangle + \psi_2(x)|01\rangle + \psi_{12}(x)|11\rangle, & (3) \end{aligned}$$

they can be written as a η-variable functions as follows

$$\psi(x,\eta) = \psi_0(x) + \eta\psi(x) \tag{4}$$

and similarly for two qubits

$$\psi^{(2)}(x,\eta_1,\eta_2) = \psi_0(x) + \eta_1\psi_1(x) + \eta_2\psi_2(x) + \eta_1\eta_2\psi_{12}(x) \tag{5}$$

In the case of pure states it provides strong tool - functional determinants and criteria for factorability of η-functions. This information is in direct correspondence to the known measures of entanglement.

The formalism can be briefly presented as follows. The η variables are nilpotent elements coming from the commutative algebra $\mathcal{N}$ generated by unit and the set of nilpotents of the first order. We can split given element to the numerical part and the rest $\mathcal{N} \ni \nu = b(\nu) + s(\nu)$ - the body and soul of an element. To describe qubit and many-qubit states one takes bimodule over $\mathcal{N}$. $\mathcal{H} \rightsquigarrow \mathcal{N}$-module with the $\mathcal{N}$-scalar product i.e. The generalized scalar product is defined as

$$< .,. >: \mathcal{H} \times \mathcal{H} \mapsto \mathcal{N} \tag{6}$$

such that for $F, G \in \mathcal{H}$

$$\begin{aligned} \nu^* < F,G > &= < \nu F, G > = < F, \nu^* G >, \quad \nu \in \mathcal{N} & (7)\\ < F,G > &= 0 \quad \forall G \in \mathcal{H} \Rightarrow F = 0 & (8)\\ b(< F,G >)^* &= b(< G,F >) & (9)\\ b(< F,F >) &\geq 0, \quad \forall F \in \mathcal{H} & (10) \end{aligned}$$

Particular realization of such $\mathcal{N}$-module is given by the function space $F[\vec{\eta}_n]$.

$$< F,\ G >_{\mathcal{N}} = \int F^*(\vec{\eta})G(\vec{\eta})e^{<\vec{\eta}^*,\vec{\eta}>}\, d\vec{\eta}^*\, d\vec{\eta}, = \int F^*(\vec{\eta})G(\vec{\eta})d\mu(\vec{\eta}^*,\ \vec{\eta}) \tag{11}$$

where

$$F^*(\vec{\eta}) = \sum_{k=0}^{n}\sum_{I_k} F^*_{I_k}\eta^{I_k *}, \tag{12}$$

and $I_k = (i_1, i_2, \ldots, \imath_k)$ is an ordered multi-index.

$$< F,\ G >_{\mathcal{N}} = \sum_{k=0}\sum_{I_k} F^*_{I_k}G_{I_k} \tag{13}$$

Note that when $F_{I_k} \in \mathbb{C}$ the generalized $\mathcal{N}$-scalar product takes complex values and such assumption is natural for qubit's states description.

2.1. Entanglement of two qubits
To illustrate how factorization criteria compare to entanglement let as take two qubit η-function (5). The w_{12} denotes the η-Wronskian with respect to η_1 and η_2 variables

$$w_{12} = detW_{12} = \begin{vmatrix} F & \frac{\partial F}{\partial \eta_1} \\ \frac{\partial F}{\partial \eta_2} & \frac{\partial^2 F}{\partial \eta_1 \partial \eta_2} \end{vmatrix} = \begin{vmatrix} F & \partial_1 F \\ \partial_2 F & \partial_{12} F \end{vmatrix} = F_0 F_{12} - F_1 F_2 \tag{14}$$

There is valid the following statement: for arbitrary function $F(\eta_1, \eta_2)$ vanishing of the Wronskian $w_{12}(F)$ is equivallent to factorization of the form $F(\eta_1, \eta_2) = G(\eta_1)\tilde{G}(\eta_2)$, for some G and $\tilde{G}$. Wronskian for the function of two η variables has numerical values. Considering the Werner state represented by the function $\psi_W(\eta_1, \eta_2) = \frac{1}{\sqrt{2}}(\eta_1 + \eta_2)$ we get that $w_{12}(\psi_W) = -\frac{1}{2}$. For the GHZ state $\psi_{GHZ}(\eta_1, \eta_2) = \frac{1}{\sqrt{2}}(1 + \eta_1 \eta_2)$ and $w_{12}(\psi_{GHZ}) = \frac{1}{2}$

The two-tangle expressed by the Wronskian takes the form

$$\tau(\psi) = 4|w_{12}(\psi)|, \tag{15}$$

e.g. $\tau(\psi_W) = 1$, $\tau(\psi_{GHZ}) = 1$. This approach works for multi-qubit states. For three qubits and more the Wronskians depend explicitly on η-variables, what yelds the array of conditions and selection of important invariants for multi-qubit states. What is intersting the constraints comming from these criteria are desirable, because they select a subfamily of important invariants from the very quickly growing with the number of qubits, the set of all invariants. Detailed account of this approach can be found in the Refs. [16, 6].

3. Superqubit - extension of qubit to $(2|1)$ superspace
The superspace formalism turns out to fit the description of the extension of qubit to the system which is symmetric with respect of the $Osp(1|2)$, the minimal supersymmetric extension of the $SL(2, \mathbb{C})$ which is the group SLOCC transformations playing the fundamental role in description of entanglement. This was the departure point in the definition of superqubit given in Ref. [14]. Let the $\mathcal{Q} = \mathcal{Q}_0 + \mathcal{Q}_1$ denotes the Banach-Grassmann algebra in the sense of Rogers, where $\mathcal{Q}_i$ is its even and odd part for $i = 0$ and $i = 1$ respectively. The numerical part of $q \in \mathcal{Q}$ is called body the rest is a soul of the element q: $q = b(q) + s(q)$. One considers a graded module $\mathbb{V} = \mathbb{V}_0 \oplus \mathbb{V}_1$ of dimension (2|1) over the algebra $\mathcal{Q}$ which extends the $\mathbb{C}^2$ Hilbert space of a qubit.

$$\mathbb{C}^2 \rightsquigarrow \mathbb{V}_{\mathcal{Q}}^{(2|1)} \tag{16}$$

with the $\mathcal{Q}$-scalar product i.e.

$$< .,. >: \mathbb{V} \times \mathbb{V} \mapsto \mathcal{Q} \tag{17}$$

such that for ψ,$\phi \in \mathbb{V}$

$$< \psi, \phi q > \;=\; < \psi, \phi > q \quad q \in \mathcal{Q} \tag{18}$$

$$< \psi, \phi >^{\#} \;=\; (-1)^{\psi(\phi+1)} < \phi, \psi >, \tag{19}$$

where $(q^{\#})^{\#} = (-)^{|q|} q$ for homogenous elements i.e. $q \in \mathcal{Q}_i$, $|q| = i$, $i = 0, 1$.

The one-superqubit system is described by supervector

$$\psi = |0\rangle\psi_0 + |1\rangle\psi_1 + |\bullet\rangle\psi_\bullet = |\psi\rangle = |k\rangle\psi_k + |\bullet\rangle\psi_\bullet \equiv \sum_X |X\rangle\psi_X \tag{20}$$

where $|i\rangle$, $i = 0, 1$ are even vectors, $\psi_i \in \mathcal{Q}_0$ and $|\bullet\rangle$ is an odd vector, $\psi_\bullet \in \mathcal{Q}_1$. In this way the two level system is extended to the three level system with one level being of different nature then the other two. Moreover, there is no balance between odd and even (bose and fermi) degrees of freedom as in usual supersymmetric models. The $\mathcal{Q}$-scalar square $\langle\psi,\psi\rangle = \delta^{ij}\psi_i^{\#}\psi_j - \psi_\bullet^{\#}\psi_\bullet$ gives value with the nontrivial soul and normalized superstate has the form

$$|\psi\rangle = (\delta^{i_1 i_2}\psi_{i_1}^{*}\psi_{i_2})^{-\frac{1}{2}}\left(|k\rangle(1+\frac{1}{2}(\delta^{i_1 i_2}\psi_{i_1}^{*}\psi_{i_2})^{-1}\psi_\bullet^{\#}\psi_\bullet)\psi_k + |\bullet\rangle\psi_\bullet\right) \tag{21}$$

3.1. Two superqubits

To see how questions concerning entanglement can be studied let us consider a two-superqubit states

$$|\psi\rangle = |jk\rangle\psi_{jk} + |j\bullet\rangle\psi_{j\bullet} + |\bullet k\rangle\psi_{\bullet k} + |\bullet\bullet\rangle\psi_{\bullet\bullet} \tag{22}$$

$dim\mathbb{V}^{(2)} = (5|4)$; $\psi_j k$, $\psi_{\bullet\bullet}$ are even and $\psi_{\bullet k}$, $\psi_{j\bullet}$ are odd. The super-SLOCC now is given by $OSp(2|1)\times OSp(2|1)$. To define the generalization of the entanglement for two-superqubits we can use the $(2|1)\times(2|1)$-supermatrix

$$\psi_{XY} = \begin{pmatrix} \psi_{jk} & \psi_{j\bullet} \\ \psi_{\bullet k} & \psi_{\bullet\bullet} \end{pmatrix}. \tag{23}$$

Natural generalization of determinant is the Berezinian, but it is not good candidate for super-generalization of 2-tangle. Its explicit form

$$Ber(\psi_{XY}) = det(\psi_{jk} - \psi_{j\bullet}\psi_{\bullet\bullet}^{-1}\psi_{\bullet k})\psi_{\bullet\bullet}^{-1} \tag{24}$$

shows that invertibility of $\psi_{\bullet\bullet}$ is too strong condition, restricting the possible two-superqubit states, since for example for the product state the $\psi_{\bullet\bullet}$ will be a product of two odd Grassmann numbers and therefore noninvertible. The solution given in Ref. [14] is based on the following observation done for conventional determinant

$$det(\psi_{jk}) = \frac{1}{2}tr((\psi\epsilon)^T\epsilon\psi) \tag{25}$$

then using the following modifications (SL(2) to Osp(2—1)) $\oplus$ (transpose $\to$ super-transpose) one can write

$$sdet(\psi_{XY}) = \frac{1}{2}str((\psi E)^{ST}E\psi), \tag{26}$$

what gives for two-superqubits explicitly

$$sdet(\psi_{XY}) = (\psi_{00}\psi_{11} - \psi_{01}\psi_{10} + \psi_{0\bullet}\psi_{1\bullet} + \psi_{\bullet 0}\psi_{\bullet 1}) - \frac{1}{2}\psi_{\bullet\bullet}^2 \tag{27}$$

This expression turned out to be good candidate for generalization of 2-tangle to the super 2-tangle

$$s\tau_{XY} = 4sdet\psi_{XY}(sdet\psi_{XY})^{\#} \tag{28}$$

It has basics property that $s\tau_{XY} = 0$ for product states. However its behaviour is different then we have used to for 2-tangle. Let us illustrate this using some examples:

Ex.1. The simple superstate [14]

$$|\phi\rangle = i|\bullet\bullet\rangle,$$

is maximally entangled in the sense of super 2-tangle: $s\tau_{XY} = 1$. This is an "algebraic artefact" because the separation of a superstate proportional to $|\bullet\bullet\rangle$ to product of one-superqubit states requires the nilpotent (a bodyless) coefficient in front of it being the product of two odd Grassmann numbers. Let us call such a state algebraically entangled.

Journal of Physics: Conference Series **411** (2013) 012015 doi:10.1088/1742-6596/411/1/012015

Ex.2. Maximally super-entangled superstate [14]

$$|\psi\rangle = \frac{1}{\sqrt{3}}(|00\rangle + |11\rangle + i|\bullet\bullet\rangle) \tag{29}$$

$$s\tau_{XY}(|\psi\rangle) = 4(\frac{1}{3} + \frac{1}{2}\frac{1}{3})^2 = 1,$$

Ex.3. similar superstate

$$|\psi\rangle = \frac{1}{\sqrt{3}}(|00\rangle + |11\rangle + |\bullet\bullet\rangle) \tag{30}$$

is not maximally entangled and $s\tau_{XY} = \frac{1}{9}$.

Ex.4. The following superstate

$$|\psi\rangle = \frac{1}{2}(|00\rangle + |11\rangle + \sqrt{2}|\bullet\bullet\rangle) = \frac{1}{\sqrt{2}}(\psi_{GHZ} + |\bullet\bullet\rangle) \tag{31}$$

has vanishing super 2-tangle, while it seems to be not separable (ψ_{GHZ} is the two qubit Greenberger Horne Zeilinger state).

Let us note that there exists generalization of the above construction to the three superqubits: $s\tau_{XYZ}$ given in the Ref. [14].

This approach is under development, recently there were proposed two-superqubit states that are "more nonlocal" then respective two-qubit states [17]. The analysis involves operation with Grassmann valued probabilities and mentioned before choice of the way of ascribing final numerical values to such an objects - that always was a problematic issue of SUSY-QM in superHilbert spaces.

4. Squbit - reshaped superqubit

Soon after the introduction of the superqubit there was proposed its modification [15], with aims to cure the following properties of superqubits:

- Grassmannian coordinates make difficult/impossible a probabilistic interpretation,
- presence of algebraically entangled states,

The proposed modification to ged rid of Grassmann valued probability amplitudes requires the introduction of auxiliary Grassmann variables θ_i in such a way that new superstate can be written in the form

$$|\psi\rangle = b|B\rangle + f_i\theta_i|F_i\rangle + b_{ij}\theta_i\theta_j|B_{ij}\rangle + f_{ijk}\theta_i\theta_j\theta_k|F_{ijk}\rangle + \cdots \tag{32}$$

b, f are complex; $|B\rangle$ are even; $|F\rangle$ are odd. Such θ variables "neutralize" odd states $|F\rangle$. Then

$$\begin{aligned}\langle\langle\psi|\psi\rangle\rangle &\equiv \int e^{\sum_i \bar{\theta}_i\theta_i}\langle\psi|\psi\rangle \; \Pi_i d\bar{\theta}_i d\theta_i = \\ & |b|^2 + \sum |f_i|^2 + \sum_{i<j}|b_{ij}|^2 + \sum_{i<j<k}|f_{ijk}|^2 + \cdots\end{aligned} \tag{33}$$

restores possibility of probabilistic interpretation. Let us note that it resembles earlier studied approach based on the commuting nilpotent η-variables.

Now, the superspace of superstates has the decomposition $\mathcal{H}_{BF} = \mathcal{H}_B \oplus \mathcal{H}_F$

- bosonic states: $|0\rangle, \theta_i\theta_j|0\rangle, \ldots$

- fermionic states: $\theta_i|0\rangle, \theta_i\theta_j\theta_k|0\rangle, ..,$

where $|0\rangle$ is the Fock vacuum. Moreover one complements set of θ_i by $\bar{\theta}^j$ in such a way that θ_i, $\bar{\theta}^j$ satisfy the Clifford algebra

$$\{\bar{\theta}^i, \theta_j\} = \delta^i_j, \quad \{\theta_i, \theta_j\} = 0, \quad \{\bar{\theta}^i, \bar{\theta}^j\} = 0 \tag{34}$$

and the new space $\mathcal{H}^\theta = \mathcal{H}^\theta_B \oplus \mathcal{H}^\theta_F$ is spanned on

$$\theta_i\theta_j \cdots |0\rangle \otimes \theta_k\theta_l \cdots |0\rangle = \theta_i\theta_j \cdots \theta_k\theta_l \cdots |0\rangle \tag{35}$$

To conlude the construction the extension of $\mathcal{H}_{BF}$ to $\mathcal{H}_{BF} \otimes \mathcal{H}^\theta$ is proposed

$$\mathcal{H} = \left(\mathcal{H}_B \otimes \mathcal{H}^\theta_B\right) \oplus \left(\mathcal{H}_F \otimes \mathcal{H}^\theta_F\right) \tag{36}$$

giving the bosonic and fermionic dimensions of this space balanced i.e.

- 2^{N-1} of $|B_I\rangle$
- 2^{N-1} of $|F_I\rangle$

Finally superstate defining the squbit can be written in the following form

$$|\psi\rangle = b|0\rangle \otimes |B\rangle + f_i\theta_i|0\rangle \otimes |F_i\rangle + b_{ij}\theta_i\theta_j|0\rangle \otimes |B_{[ij]}\rangle + f_{ijk}\theta_i\theta_j\theta_k|0\rangle \otimes |F_{[ijk]}\rangle + \ldots \tag{37}$$

What gives the simplest squbit:

$$|\psi\rangle = b|0\rangle \otimes |B\rangle + f\theta|0\rangle \otimes |F\rangle \tag{38}$$

Let us bring here the conclusion of the Authors of the Ref. [15]: *the auxiliary θ system allows the dynamical mixing between fermions and bosons of the $\mathcal{H}_{BF}$ system, with a well-defined positive norm, necessary for a probabilistic interpretation.*

As it is seen from this brief account, the squbit has ballanced fermionic and bosonic degrees of freedom and is more close to usual supersymmetric models. It also seems to departure too far from the original superqubit model.

Final Comments

The notion of qubit is deeply rooted in quantum mechanics and the importance of this simple system was discovered after realizing the existence of new resource available in nature - quantum entanglement. For decades the real value of quantum entanglement was neglected and occasionaly studied by people interested in quantum paradoxes and fundamental problems of the quantum mechanics. Presently it is intesively studied theoretically and its evidence is experimentally determined. Despite research based on conventional for quantum mechanics formalism of the Hilbert space, there are atempts to describe entanglement in different ways, using experiences from supermathematics and other algebraical structures. The η-Hilbert space formulation based on commuting nilpotent variables gives new functional tools to study many-qubit entanglement and to realize that many of interesting entangled states can be identified as elementary η-functions. On the other hand after four decades of supersymmetric theories there was proposed super-extension of the notion of qubit and super-entanglement of superstates with theoretically more nonlocality then in conventional many-qubit states. But physical meaning of this extension is still unclear.

References

[1] Frydryszak A M, Int. J. Mod. Phys. **A** 22 2513 (2007)
[2] Mandilara A, Akulin V M, Smilga A V, Viola L, Phys. Rev. A **74**, 022331 (2006)
[3] Ziegler K, Europhys. Lett. **9** 277 (1989)
[4] Palumbo F, Phys. Rev. **D** 50 2826 (1994)
[5] Frydryszak A M, Int. J. Mod. Phys. **A** 22 2513 (2007) [arXiv:hep-th/0609072]
[6] Frydryszak A M, Int. J. Mod. Phys. **A25** 951 (2010)
[7] Hruby J, *Supersymmetry and qubit field theory*, arXiv:quant-ph/0402188
[8] DeWitt B, *Supermanifolds* CUP, Cambridge, 1984. 2nd ed.
[9] Frydryszak A, Jakobczyk L, Lett. Math. Phys. **16** 101 (1988)
[10] Frydryszak A, Lett. Math. Phys. **26** 105 (1992)
[11] Rogers A, J. Math. Phys. **21** 1352 (1980)
[12] Rogers A, *Supermanifolds: Theory and Applications*, World Scientific, Singapore, 2007
[13] Nicolai H, J. Theor. Phys A: Math. Gen. **9** 1497 (1976)
[14] Borsten L, Dahanayake D, Duff M J, Rubens W, Phys.Rev. **D** 81 105023 (2010)
[15] Castellani L, Grassi P A, Sommovigo L, *Quantum computing with superqubits* arXiv:1001.3753
[16] Frydryszak A M, *Nilpotent quantum mechanics, qubits, and flavors of entanglement* arXiv:0810.3016
[17] Borsten L, Bradler K, Duff M J, *Tsirelson's bound and supersymmetric entangled states* arXiv:1206.6934

Noncommutative Instantons Revisited

Masashi Hamanaka

Department of Mathematics, Nagoya University, Furocho, Chikusa-ku, Nagoya, 464-8602, JAPAN

E-mail: `hamanaka@math.nagoya-u.ac.jp`

Toshio Nakatsu

Institute for Fundamental Sciences, Setsunan University, 17-8 Ikeda Nakamachi, Neyagawa, Osaka, 572-8508, JAPAN

E-mail: `nakatsu@mpg.setsunan.ac.jp`

Abstract. We discuss the Atiyah-Drinfeld-Hitchin-Manin (ADHM) construction of $U(N)$ instantons in noncommutative (NC) space and prove the one-to-one correspondence between moduli spaces of the noncommutative instantons and the ADHM data, together with an origin of the instanton number for $U(1)$. We also give a derivation of the ADHM construction from the viewpoint of the Nahm transformation of instantons on four-torus.

1. Introduction

Instantons are finite-action solutions of anti-self-dual (ASD) Yang-Mills equation and become exact solutions of classical Yang-Mills theories. They can reveal non-perturbative aspects of the quantum theories. Actually, the path-integrations, formulating the quantum theories, could reduce to finite-dimensional integrations over the instanton moduli spaces. The Atiyah-Drinfeld-Hitchin-Manin (ADHM) construction [2] is a powerful method to obtain the instantons. Furthermore, via the construction, the moduli space is mapped to the set of quadruple matrices. The aforementioned integration, being thereby an integration over the matrices, becomes tractable [10]. To evaluate the integration, the use of noncommutative instantons [34] is relevant so that a localization formula can be applied to the integration [33, 30]. In the procedures, various formulas and relations of the ADHM construction are required. Hence it is worthwhile to elucidate the noncommutative ADHM construction together with the group actions and to present all the ingredients in the construction explicitly.

In our paper [17], we discuss the ADHM construction of $U(N)$ instantons in noncommutative spaces and prove one-to-one correspondence between moduli spaces of the noncommutative instantons and the ADHM data. We also argue group actions (especially torus actions) on noncommutative instantons. The present article is a brief report of it with some co-supplement materials.

2. Noncommutative Field Theory

Noncommutative (NC) space is a space which coordinate ring is noncommutative. Let x^μ be the spacial coordinates. The noncommutativity is expressed by the following commutation relations.

$$[x^\mu, x^\nu] = i\theta^{\mu\nu}, \tag{1}$$

where $\theta^{\mu\nu}$ is a real antisymmetric tensor. When $\theta^{\mu\nu}$ vanishes identically, the coordinate ring is commutative and the underlying space reduces to a commutative one. The commutation relations (1), like the canonical commutation relations in quantum mechanics, lead to "space-space uncertainty relation." Singularities in commutative space could resolve in noncommutative space thereby. This is one of the prominent features of field theories on noncommutative space, NC field theories, for short, and yields various new physical objects such as $U(1)$ instantons.

In four-dimensional Euclidean space, the antisymmetric tensor takes the canonical form as

$$\theta^{\mu\nu} = \left(\begin{array}{cc|cc} 0 & -\theta_1 & 0 & 0 \\ \theta_1 & 0 & 0 & 0 \\ \hline 0 & 0 & 0 & -\theta_2 \\ 0 & 0 & \theta_2 & 0 \end{array}\right), \tag{2}$$

where the real numbers θ_1, θ_2 are called *noncommutative parameters*. The spacial coordinates x^μ has a Fock representation. To see this, introduce complex coordinates z_1, z_2 by $z_1 = x^2 + ix^1$ and $z_2 = x^4 + ix^3$. By using the complex coordinates, the noncommutativity is expressed as

$$[z_1, \bar{z}_1] = 2\theta_1, \ [z_2, \bar{z}_2] = 2\theta_2. \tag{3}$$

This shows that each pair of the complex coordinate and its complex conjugates gives annihilation and creation operators of a harmonic oscillator. In this article, we set $\theta_{j=1,2} \leq 0$. $a_j = (1/\sqrt{-2\theta_j})\bar{z}_j$ and $a_j^\dagger = (1/\sqrt{-2\theta_j})z_j$ are the annihilation and creation operators of harmonic oscillators.

Let $\mathcal{H} = \mathcal{H}_1 \otimes \mathcal{H}_2$ be a Fock space spanned by the occupation number basis $|n_1, n_2\rangle = |n_1\rangle \otimes |n_2\rangle$ $(n_1, n_2 = 0, 1, \dots)$. Field variables on the NC space can be represented as an infinite matrices of the form

$$\hat{f}(\hat{x}) = \sum_{m_1,m_2,n_1,n_2=0}^{\infty} f_{m_1,m_2,n_1,n_2}|m_1, m_2\rangle\langle n_1, n_2|. \tag{4}$$

In the above, the hat symbol is used to emphasize that these are operators. We remark that, as seen in subsection 4.4, to take the inverse of these operators is a delicate issue because the operator zero-modes, if they exist, need to be accounted.

For more details on noncommutative field theories, see e.g. [6, 11, 17, 19, 23, 39, 40].

3. Nahm Transformation and Origin of the ADHM Duality

The ADHM construction is a descendant of the twistor theory founded by R. Penrose. Twistor theoretical idea was first applied by R. Ward to instantons. It converts the self-duality of gauge fields on S^4 to the holomorphy of complex vector bundles on $\mathbb{C}P_3$ (see e.g. [44]), where the construction reduces to an algebro-geometric one, and finally Atiyah, Drinfeld, Hitchin and Manin obtained an algebraic method to generate *all* instanton solutions on S^4 [2]. (Instantons on S^4 are equivalent to those on $\mathbb{R}^4$ by the conformal invariance and Uhlenbeck's theorem.) The construction is based on one-to-one correspondence between the moduli space of instantons and that of the ADHM data. The correspondence is thereafter called the *ADHM duality*.

It was also applied to the case of the BPS monopoles by W. Nahm, which is called the *ADHMN construction* or the *Nahm construction.* Schenk [38] and Braam and van Baal [5], extracting the Fourier transform, known as the Nahm transformation, from the construction and applying to instantons on four-torus, obtained the duality between instantons on four-torus and its dual torus, which is also known as the *Nahm duality.* Nahm transformations of explicit ASD gauge fields are performed in e.g. [1, 16, 20, 43]. For surveys of the Nahm transformation, see e.g. [4, 9, 22].

The ADHM duality is also understood as a version of the Nahm duality. The ADHM duality in this line was first outlined in [7] and made rigorously in [25] with a generalization to ALE spaces. (A beautiful treatment can be also found in [9].) Noncommutative version of the ADHM duality is discussed in [17, 27, 32]. We start this section by giving a brief review on the Nahm transformation and then argue the ADHM/Nahm duality by taking certain limits.

3.1. Poincaré Line Bundle

The Poincaré line bundle plays a central role in the Nahm duality.

The Poincaré line bundle $\mathcal{P}$ is a $U(1)$ line bundle over the product space $T^4 \times \widehat{T}^4$, where T^4 is a four-torus and $\widehat{T}^4$ is the dual four-torus. The four torus is realized as $T^4 = \mathbb{R}^4/\Lambda$, where Λ is a maximal lattice in $\mathbb{R}^4$. The dual four-torus is $\widehat{T}^4 = \mathbb{R}^{4*}/2\pi\Lambda^*$, where $\mathbb{R}^{4*}$ is the dual vector space and Λ^* is the dual lattice. We use x^μ and ξ_μ as the coordinates of $\mathbb{R}^4$ and $\mathbb{R}^{4*}$ with the dual pairing $\xi \cdot x = \xi_\mu x^\mu$. The line bundle $\mathcal{P}$ is equipped with a $U(1)$ potential of the form

$$\omega(x,\xi) = i\xi_\mu dx^\mu. \tag{5}$$

Note that, taking another pair (x',ξ'), where $x' \in x + \Lambda$ and $\xi' \in \xi + 2\pi\Lambda^*$, amounts to $U(1)$ gauge transformation. Actually, it is apparent to see that, for $x' = x+\lambda$ ($\lambda \in \Lambda$) and $\xi' = \xi + 2\pi\mu$ ($\mu \in \Lambda^*$), two gauge potentials $\omega(x',\xi')$ and $\omega(x,\xi)$ are gauge equivalent.

$$\omega(x+\lambda, \xi + 2\pi\mu) = \omega(x,\xi) + g^{-1}(x,\xi)dg(x,\xi), \tag{6}$$

where $g(x,\xi) = e^{2\pi i\mu\cdot x}$ is single-valued on $T^4 \times \widehat{T}^4$.

The field strength $\Omega = d\omega$ becomes constant.

$$\Omega(x,\xi) = id\xi_\mu \wedge dx^\mu. \tag{7}$$

The following form of the gauge potential is also used conveniently.

$$\omega'(x,\xi) = -ix^\mu d\xi_\mu. \tag{8}$$

The above form can be obtained from (5) by the gauge transformation $\exp(-i\xi\cdot x)$ on $\mathbb{R}^4 \times \mathbb{R}^{4*}$.

$$\omega'(x,\xi) = \omega(x,\xi) + e^{i\xi\cdot x}de^{-i\xi\cdot x}. \tag{9}$$

Let π, $\hat{\pi}$ be natural projections from $T^4 \times \widehat{T}^4$ onto T^4 and $\widehat{T}^4$. We summarize on the Poincaré line bundle as follows.

$$\begin{array}{ccccc} & & \mathcal{P} & & \\ & & \downarrow & & \\ & & T^4 \times \widehat{T}^4 & & \\ & \overset{\pi}{\swarrow} & & \overset{\hat{\pi}}{\searrow} & \\ T^4 & & & & \widehat{T}^4 \end{array}$$

Journal of Physics: Conference Series **411** (2013) 012016 doi:10.1088/1742-6596/411/1/012016

3.2. Nahm Transformation

Let $E \to T^4$ be a $U(N)$ vector bundle with the 2nd Chern number $c_2(E) = k$ and A be a $U(N)$ gauge potential on E. The tensor product bundle $\pi^*E \otimes \mathcal{P}$ is the $U(N)$ vector bundle over $T^4 \times \widehat{T}^4$ equipped with the gauge potential $A \otimes 1_{[1]} + 1_{[N]} \otimes \omega$. Taking the slice along $T^4 \times \{\xi\}$, we regard $\pi^*E \otimes \mathcal{P}|_{T^4\times\{\xi\}} \to T^4 \times \{\xi\}$ as a $U(N)$ vector bundle over T^4. In particular, the gauge potential on $\pi^*E \otimes \mathcal{P}|_{T^4\times\{\xi\}}$ is given by $A_\xi = A \otimes 1_{[1]} + 1_{[N]} \otimes i\xi_\mu dx^\mu$. The field strength F_ξ of A_ξ equals to F of A.

We introduce the Dirac operator. Let $S^\pm \to T^4$ be a spinor bundle on T^4. The Dirac operator acting on the sections $\Gamma(T^4, S^\pm \otimes \pi^*E \otimes \mathcal{P}|_{T^4\times\{\xi\}})$ is given by

$$\begin{aligned} \mathcal{D}[A_\xi] &= e_\mu \otimes (\partial_\mu + A_\mu + i\xi_\mu), \\ \bar{\mathcal{D}}[A_\xi] &= \bar{e}_\mu \otimes (\partial_\mu + A_\mu + i\xi_\mu), \end{aligned} \tag{10}$$

where $e_\mu = (-i\sigma_i, 1)$, $\bar{e}_\mu = (i\sigma_i, 1)$. These are the *Weyl operators* rather than the Dirac operators, however, we use the word the "Dirac operator" for simplicity.

The dual vector bundle $\widehat{E} \to \widehat{T}^4$ can be constructed by using the Dirac zero modes. The fiber $\widehat{E}_\xi$ is identified with $\mathrm{Ker}\,\bar{\mathcal{D}}[A_\xi]$. Since the Atiyah-Singer family index theorem implies $\dim \mathrm{Ker}\,\bar{\mathcal{D}}[A_\xi] = k$, the rank of $\widehat{E}$ is k. Actually, $\widehat{E}$ is a $U(k)$ vector bundle equiped with the dual $U(k)$ gauge potential $\widehat{A}$. For the description of the dual gauge potential, we regard $\widehat{E}$ as a sub-bundle of $\widehat{H}$, where $\widehat{H} \to \widehat{T}^4$ is an infinite-dimensional trivial vector bundle with the fiber $\widehat{H}_\xi = L^2(T^4, S^+ \otimes \pi^*E \otimes \mathcal{P}|_{T^4\times\{\xi\}})$. (See Fig. 1.) By using the projection $P : \widehat{H} \to \widehat{E}$, diffrential $\widehat{d}$ on $\Gamma(\widehat{T}^4, \widehat{H})$ induces the covariant derivative $\widehat{D} = \widehat{d} + \widehat{A}$ as

$$\widehat{D} = P\widehat{d} : \Gamma(\widehat{T}^4, \widehat{E}) \to \Gamma(\widehat{T}^4, \Lambda^1 \otimes \widehat{E}). \tag{11}$$

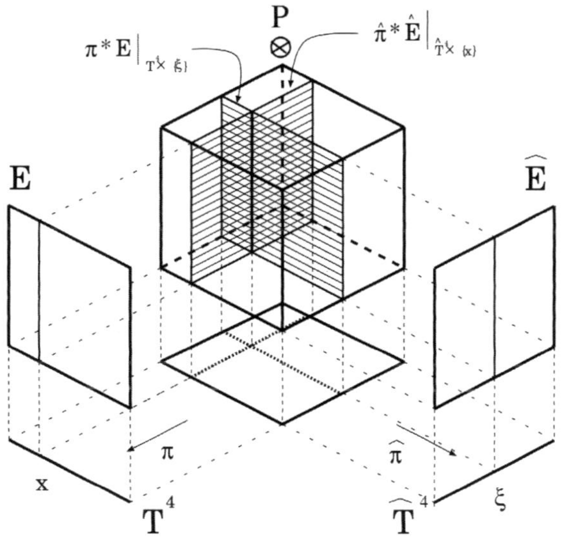

Figure 1. The stage of the Nahm transformation

The $U(k)$ gauge potential $\widehat{A}$ is expressed by using the Dirac zero modes as

$$\widehat{A}^{pq}_\mu(\xi) = \int_{T^4} d^4x\, \psi^{p\dagger}_\xi(x) \frac{\partial}{\partial \xi^\mu} \psi^q_\xi(x) \tag{12}$$

where ψ_ξ^p $(p = 1, 2, \ldots, k)$ are the normalized Dirac zero-modes.

$$\bar{\mathcal{D}}[A_\xi]\psi_\xi^p(x) = 0, \quad \int_{T^4} d^4x\ \psi_\xi^{p\dagger}(x)\psi_\xi^q(x) = \delta^{pq}. \tag{13}$$

The 2nd Chern number of $\widehat{E}$ turns out to be N [5, 38]. Thus we obtain the $U(k)$ vector bundle $\widehat{E} \to \widehat{T}^4$ with $c_2(\widehat{E}) = N$ and the $U(k)$ gauge potential $\widehat{A}$ on $\widehat{E}$. This gives the Nahm transformation $\mathcal{N} : (E, A) \mapsto (\widehat{E}, \widehat{A})$.

$$\begin{array}{ccccc} & & \pi^* E \otimes \mathcal{P} & & \\ & & \downarrow & & \\ E & & T^4 \times \widehat{T}^4 & & \widehat{E} \\ \downarrow & \overset{\pi}{\swarrow} & & \overset{\hat{\pi}}{\searrow} & \downarrow \\ T^4 & & & & \widehat{T}^4 \end{array}$$

The inverse transformation is carried out by following the same procedure. Let $\widehat{E} \to \widehat{T}^4$ be a $U(k)$ vector bundle with $c_2(\widehat{E}) = N$ and $\widehat{A}$ be a $U(k)$ gauge potential on $\widehat{E}$. We regard the slice of the tensor product bundle $\hat{\pi}^*\widehat{E} \otimes \mathcal{P}$ along $\{x\} \times \widehat{T}^4$ as a $U(k)$ vector bundle over $\widehat{T}^4$ equipped with the gauge potential $\widehat{A}_x = \widehat{A} \otimes 1_{[1]} - 1_{[k]} \otimes ix^\mu d\xi_\mu$. The field strength $\widehat{F}_x$ of $\widehat{A}_x$ equals to $\widehat{F}$ of $\widehat{A}$. The Dirac operator acting on the sections $\Gamma(\widehat{T}^4, \widehat{S}^\pm \otimes \pi^*\widehat{E} \otimes \mathcal{P}|_{\{x\}\times\widehat{T}^4})$ takes the form

$$\begin{aligned} \widehat{\mathcal{D}}[\widehat{A}_x] &= e_\mu \otimes (\widehat{\partial}^\mu + \widehat{A}^\mu - ix^\mu), \\ \widehat{\bar{\mathcal{D}}}[\widehat{A}_x] &= \bar{e}_\mu \otimes (\widehat{\partial}^\mu + \widehat{A}^\mu - ix^\mu). \end{aligned} \tag{14}$$

By using the zero modes of $\widehat{\bar{\mathcal{D}}}[\widehat{A}_x]$, consisting of N normalizable ones, we obtain a $U(N)$ vector bundle $E \to T^4$ with $c_2(E) = k$ and a $U(N)$ gauge potential A on E. This gives the inverse Nahm transformation $\widehat{\mathcal{N}} : (\widehat{E}, \widehat{A}) \mapsto (E, A)$.

$$\begin{array}{ccccc} & & \hat{\pi}^* \widehat{E} \otimes \mathcal{P} & & \\ & & \downarrow & & \\ E & & T^4 \times \widehat{T}^4 & & \widehat{E} \\ \downarrow & \overset{\pi}{\swarrow} & & \overset{\hat{\pi}}{\searrow} & \downarrow \\ T^4 & & & & \widehat{T}^4 \end{array}$$

It can be shown $\mathcal{N}\widehat{\mathcal{N}} = \widehat{\mathcal{N}}\mathcal{N} = id.$ [5, 38]. Thus, eventually, $\widehat{\mathcal{N}}$ turns out to be the inverse of $\mathcal{N}$. The Nahm duality is summarized in the following dictionary.

$G = U(N)$		$\widehat{G} = U(k)$
k-instanton on T^4 (coord. x^μ)	$\overset{1:1}{\longleftrightarrow}$	N-instanton on $\widehat{T}^4$ (coord. ξ^μ)
instanton : $A_{\mu[N]}$	massless Dirac eq. $\bar{\mathcal{D}}\psi = 0$ k solutions: $\psi(\xi, x)$ $\longrightarrow$	$\widehat{A}_{\mu[k]} = \int_{T^4} d^4x\ \psi^\dagger \frac{\partial}{\partial \xi^\mu}\psi$
$A_{\mu[N]} = \int_{\widehat{T}^4} d^4\xi\ v^\dagger \frac{\partial}{\partial x^\mu} v$	massless Dirac eq. $\widehat{\bar{\mathcal{D}}}v = 0$ N solutions: $v(x, \xi)$ $\longleftarrow$	instanton : $\widehat{A}_{\mu[k]}$

3.3. A Derivation of the ADHM construction from the Nahm transformation

We can understand the duality in the ADHM/Nahm construction in some limit of the Nahm duality.

- Taking all four radii of the four-torus infinity $\Rightarrow$ ADHM construction of instantons
  Then the radii of the dual torus become zero. Hence the dual torus shrinks into one point and derivatives become meaningless because derivatives measure difference between two points. As the result, all derivatives in the dual ASD equation and the dual massless Dirac equation drop out and the differential equations becomes matrix equations. This naively yields the ADHM duality: one-to-one correspondence between the moduli space of the ASD equations on $\mathbb{R}^4$ (=infinite-size torus) and the moduli space of the matrix equations. For more detailed discussion, see [43].
- Taking three radii infinity and the other radius zero $\Rightarrow$ Nahm construction of monopoles

4. ADHM Construction of (Noncommutative) Instantons

In this section, we discuss the ADHM construction of $U(N)$ k-instantons on commutative and noncommutative Euclidean $\mathbb{R}^4$, especially focusing on the case of $(N,k) = (2,1)$, that is, $U(2)$ 1-instantons. The construction reads as follows.

Firstly, we take the ADHM data for $U(N)$ k-instantons. They are quadruple complex matrices $\big(B_1, B_2, I, J\big)$ which satsifies the noncommutative ADHM equation

$$\mu_{\mathbb{R}} \equiv [B_1, B_1^\dagger] + [B_2, B_2^\dagger] + II^\dagger - J^\dagger J = \zeta, \quad \mu_{\mathbb{C}} \equiv [B_1, B_2] + IJ = 0, \tag{15}$$

where $B_{1,2}$ are $k \times k$ matrices, and I and J are $k \times N$ and $N \times k$ matrices. $\zeta \equiv -[z_1, \bar{z}_1] - [z_2, \bar{z}_2] = -2(\theta_1 + \theta_2)$ in the first equation of (15) is a non-negative constant originated in the noncommutativity (cf. section 2). When the antisymmetric tensor $\theta^{\mu\nu}$ is ASD, the constant vanishes and the noncommutative ADHM equation (15) coincides with the commutative one.

Secondly, with the ADHM data $\big(B_1, B_2, I, J\big)$, we associate "0-dimensional Dirac equation"

$$\nabla^\dagger V = 0, \quad \nabla(x) = \begin{pmatrix} I^\dagger & J \\ \bar{z}_2 - B_2^\dagger & -(z_1 - B_1) \\ \bar{z}_1 - B_1^\dagger & z_2 - B_2 \end{pmatrix}. \tag{16}$$

The solution V for equation (16) is a $(N+2k) \times N$ matrix. In the noncommutative case, we have to take care about the existence of operator zero-modes of V.

Finally, normalize V so that

$$V^\dagger V = 1_N. \tag{17}$$

Then, by using the above V, we get the corresponding instanton solution as (cf. (12))

$$A_\mu = V^\dagger \partial_\mu V. \tag{18}$$

Actually, due to the construction, it satisfies the (noncommutative) ASD Yang-Mills equation

$$F_{z_1\bar{z}_1} + F_{z_2\bar{z}_2} = 0, \; F_{z_1 z_2} = 0. \tag{19}$$

The instanton number is computed to be $-k$.

For reviews on noncommutative instantons, see [14, 15, 24, 26, 31, 37].

Journal of Physics: Conference Series **411** (2013) 012016 doi:10.1088/1742-6596/411/1/012016

4.1. Comments on Instanton Moduli Spaces

Geometry of the instanton moduli space depends on the constant ζ. [28, 29]. For the case of $(N,k) = (2,1)$, see Fig. 2.

- When $\zeta = 0$, the moduli space contains the small instanton singularities. This is also the case of the instantons on commutative $\mathbb{R}^4$.
- When $\zeta > 0$, the small instanton singularities disapper and replaced by a new class of smooth instantons, $U(1)$ instantons.

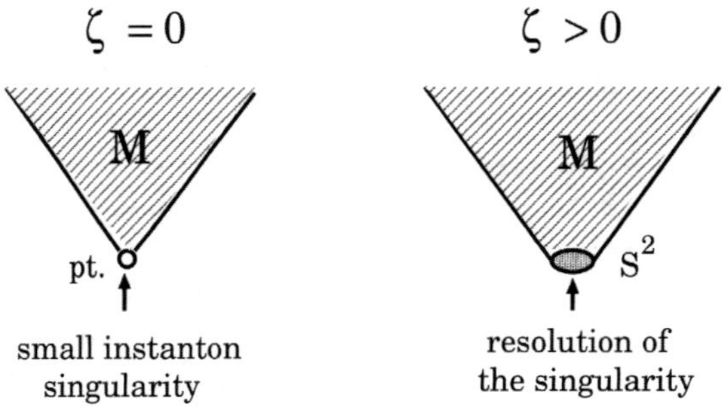

Figure 2. $(N,k) = (2,1)$ Instanton Moduli Space

4.2. ADHM Construction of Instantons

Here we present the ADHM construction of the Belavin-Polyakov-Schwartz-Tyupkin (BPST) instanton solution [3], which is the $SU(2)$ 1-instanton on commutative Euclidean $\mathbb{R}^4$.

In the case of $k=1$, the commutators in the ADHM equation (15) are dropped out and the equation becomes quite simple. $B_{1,2}$ are chosen as arbitrary complex numbers $\alpha_{1,2}$, and then I, J in the equation ($\zeta = 0$) are easily solved by using a real number ρ. Thus we have

$$B_1 = \alpha_1,\ B_2 = \alpha_2,\ I = (\rho, 0),\ J = \begin{pmatrix} 0 \\ \rho \end{pmatrix},\quad \alpha_{1,2} \in \mathbb{C},\ \rho \in \mathbb{R}. \tag{20}$$

The normalized solution of the 0-dimensional Dirac equation (16) takes the form

$$V = \frac{1}{\sqrt{\phi}} \begin{pmatrix} \bar{e}_\mu (x_\mu - b_\mu) \\ -\rho \quad 0 \\ 0 \quad -\rho \end{pmatrix},\quad \phi = |x-b|^2 + \rho^2, \tag{21}$$

where b_μ are the real and imaginary parts of $\alpha_{1,2}$ as $\alpha_1 = b_2 + ib_1$, $\alpha_2 = b_4 + ib_3$. The normalization factor ϕ is determined by (17).

Finally the instanton solution is constructed as follows

$$A_\mu = V^\dagger \partial_\mu V = \frac{i(x-b)^\nu \eta_{\mu\nu}}{(x-b)^2 + \rho^2}, \tag{22}$$

where $\eta_{\mu\nu} = \eta^i_{\mu\nu}\sigma_i$ denotes the contraction of 't Hooft's ASD eta symbol by the Pauli matrices. The field strength is computed to

$$F_{\mu\nu} = \frac{2i\rho^2}{(|z-\alpha|^2 + \rho^2)^2}\eta_{\mu\nu}. \tag{23}$$

This is exactly the BPST instanton solution [3]. Distribution of $\mathrm{Tr}F\tilde{F}$ is depicted at the upper right-hand side in Fig. 3. The moduli space is a five-dimensional space, where each point describes the positions b^μ and the size ρ of the instanton.

In the zero-size limit $\rho \to 0$, the field strength $F_{\mu\nu}$ concentrates at $x^\mu = b^\mu$ and becomes singular. This gives a singularity in the (complete) instanton moduli space, called the *small instanton singularity*. (See Fig. 3.) On noncommutative spaces, the small instanton singularity is resolved and a new class of instantons appears.

4.3. ADHM Construction of Noncommutative BPST Instantons

Noncommutative BPST solution is a noncommutative version of the BPST instanton and is obtained also by applying the ADHM procedure. A solution of the noncommutative ADHM equation is easily found:

$$B_{1,2} = 0, \quad I = (\sqrt{\rho^2 + \zeta}, 0), \quad J = \begin{pmatrix} 0 \\ \rho \end{pmatrix}. \tag{24}$$

By comparing with the previous solution, I is now deformed by the noncommutativity of the coordinates. This particularly implies that the size of instanton is larger than that of the commutative one. In fact, even in the $\rho \to 0$ limit, the configuration keeps to be smooth. This configuration is essentially the same as a $U(1)$ instanton solution which will be constructed more in detail in the next subsection.

BPST instantons on commutative and noncommutative spaces are summarized as follows.

BPST instanton		NC BPST instanton
$\mu_{\mathbb{R}} = 0,\ \mu_{\mathbb{C}} = 0$	ADHM equation	$\mu_{\mathbb{R}} = \zeta,\ \mu_{\mathbb{C}} = 0$
$B_{1,2} = \alpha_{1,2}$, $I = (\rho, 0), J^t = (0, \rho)$	ADHM data	$B_{1,2} = \alpha_{1,2}$, $I = (\sqrt{\rho^2 + \zeta}, 0), J^t = (0, \rho)$
$\mathbb{R}^4 \times$ orbifold $\mathbb{C}^2/\mathbb{Z}_2$ (singular)	moduli space	$\mathbb{R}^4 \times$ Eguchi-Hanson $\mathbb{C}^2/\mathbb{Z}_2$ (regular)
$F_{\mu\nu} \to$ (singular)	zero-size limit	$F_{\mu\nu} \to$ (regular)

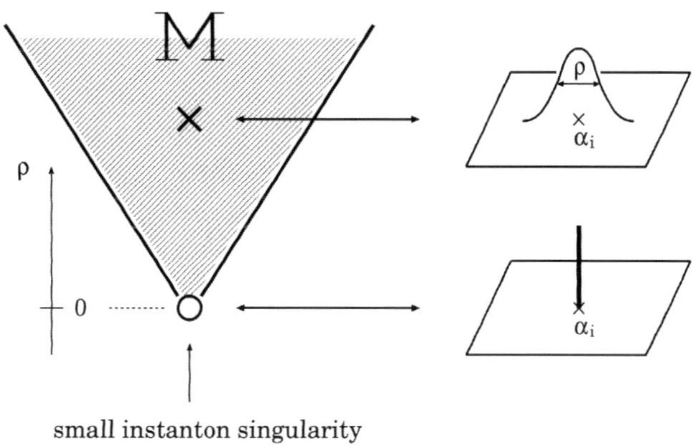

Figure 3. Instanton moduli space and the configuration of the BPST instanton

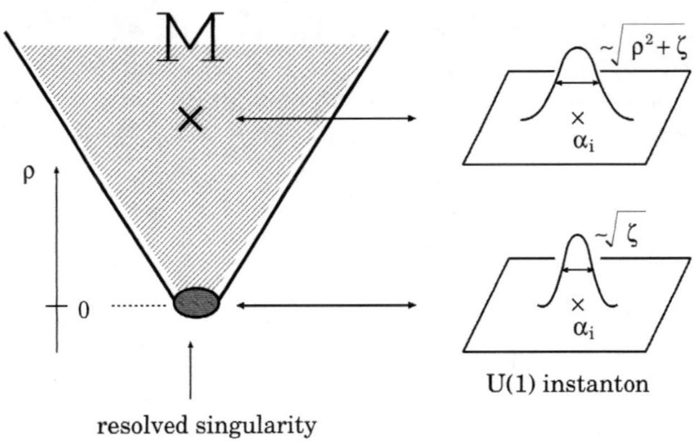

Figure 4. Instanton moduli space and the configutration of the NC BPST instanton

4.4. ADHM Construction of Noncommutative U(1) 1-instanton

We put $U(1)$ 1-instanton at the origin. Solution of the noncommutative ADHM equation (15) relevant to describe such $U(1)$ 1-instanton, is

$$B_{1,2} = 0, I = \sqrt{\zeta}, J = 0. \tag{25}$$

The 0-dimensional Dirac equation (16) reads

$$\hat{\nabla}^\dagger \hat{V} = \begin{pmatrix} \sqrt{\zeta} & \hat{z}_2 & \hat{z}_1 \\ 0 & -\hat{\bar{z}}_1 & \hat{\bar{z}}_2 \end{pmatrix} \hat{V} = 0. \tag{26}$$

Apart from the normalization, the following form of $\hat{V}$ gives a naive solution of (26).

$$\hat{V}_1 = \begin{pmatrix} \hat{z}_1\hat{\bar{z}}_1 + \hat{z}_2\hat{\bar{z}}_2 \\ -\sqrt{\zeta}\hat{\bar{z}}_2 \\ -\sqrt{\zeta}\hat{\bar{z}}_1 \end{pmatrix}. \quad \hat{\nabla}^\dagger \hat{V}_1 = 0. \tag{27}$$

$\hat{V}_1$ is an operator acting on the Fock space $\mathcal{H}$. As described in section 2, the coordinates $\hat{\bar{z}}_1$ and $\hat{\bar{z}}_2$ are the annihilation operators. Hence $\hat{V}_1$ has an operator zero-mode $|0,0\rangle$ and there exists no inverse of $\hat{V}_1^\dagger \hat{V}_1$ on $\mathcal{H}$. This means that $\hat{V}_1$ cannot be normalized on the whole of the Fock space in the operator sense.

K. Furuuchi [12] shows that $\hat{V}_1$ yields a smooth ASD instanton solution on the subspace $\mathcal{H}_1 \equiv \mathcal{H} - \mathbb{C}|0,0\rangle$. By using the shift operators, he also discussed that the solution can be converted to $\hat{V}$ which satisfies the normalization condition (17) on $\mathcal{H}$ [13]:

$$\hat{V} = \hat{V}_1 \hat{\beta}_1 \hat{U}_1^\dagger, \quad \hat{V}^\dagger \hat{V} = 1, \tag{28}$$

where

$$\hat{\beta}_1 = (1-\hat{P}_1)(\hat{V}_1^\dagger \hat{V}_1)^{-\frac{1}{2}}(1-\hat{P}_1) = \sum_{(n_1,n_2)\neq(0,0)} \frac{1}{\sqrt{(n_1+n_2)(n_1+n_2+\zeta)}} |n_1,n_2\rangle\langle n_1,n_2|, \tag{29}$$

$$\hat{U}_1 \hat{U}_1^\dagger = 1, \quad \hat{U}_1^\dagger \hat{U}_1 = 1 - |0,0\rangle\langle 0,0|. \tag{30}$$

The shift operator $\hat{U}_1$ converts operators on $\mathcal{H}_1$ into those on $\mathcal{H}$. The problem of operator zero-modes can be solved in this manner to give the correct operators on $\mathcal{H}$.

Finally, the smooth ASD instanton solution on $\mathcal{H}$ is obtained by further taking the ADHM procedure. In particular, the instanton number is computed to be -1. More discussion is seen in our paper [18].

4.5. Origin of Instanton Numbers

We apply Furuuchi's observation and several properties on noncommutative field theory to all the other ingredients of the ADHM construction and prove the existence of them in the operator sense. As a result, we obtain the following formula (a noncommutative version of the *Corrigan-Goddard-Osborn-Templeton or Osborn formula* [8, 35])

$$\int d^4x \ \mathrm{Tr}_N(F_{\mu\nu}F^{\mu\nu}) = -\int d^4x \partial^2\partial_\mu(\mathrm{Tr}_k f^{-1}\partial^\mu f), \ f := (\nabla^\dagger\nabla)^{-1}, \tag{31}$$

which is crucial to show that any ADHM data labeled by k yields k-instanton solutions even for the $U(1)$ case. This directly shows an origin of the instanton number in the context of the ADHM construction, which has been discussed by several authors for explicit instanton solutions e.g. [14, 21, 36, 41, 42].

Detailed proofs and other results are seen in our paper [17].

Acknowledgments

The authors thank the Yukawa Institute for Theoretical Physics at Kyoto University. Discussions during the YITP workshop on "Field Theory and String Theory" (YITP-W-12-05) were useful to complete this work. The authors are also grateful to A. Sako for fruitful discussion. MH is supported in part by the Toyoaki Scholarship Foundation and the Grant-in-Aid for Young Scientists (#23740182). TN is supported in part by the Grant-in-Aid for Scientific Research No. 24540223.

References

[1] Asakawa T, Carow-Watamura U, Teshima Y and Watamura S 2012, Boundary state analysis on the equivalence of T-duality and Nahm transformation in superstring theory *Prog. Theor. Phys.* **127** 665 (*Preprint* arXiv:1201.0125 [hep-th])

[2] Atiyah M F, Hitchin N J, Drinfeld V G and Manin Y I 1978, Construction of instantons *Phys. Lett.* A **65** 185

[3] Belavin A A, Polyakov A M, Schwartz A S and Tyupkin Y S 1975, Pseudoparticle solutions of the Yang-Mills equations *Phys. Lett.* B **59** 85

[4] Bartocci C, Bruzzo, U and Ruipérez D H 2009 *Fourier-Mukai and Nahm Transforms in Geometry and Mathematical Physics* (Birkhäuser)

[5] Braam P J and van Baal P 1989, Nahm's transformation for instantons *Commun. Math. Phys.* **122**, 267

[6] Chu C S 2005 Non-commutative geometry from strings *Preprint* hep-th/0502167

[7] Corrigan E and Goddard P 1984 Construction of instanton and monopole solutions and reciprocity *Annals Phys.* **154** 253

[8] Corrigan E, Goddard P, Osborn H and Templeton S 1979 Zeta function regularization and multi-instanton determinants *Nucl. Phys.* B **159** 469

[9] Donaldson S K and Kronheimer P B 1990 *The Geometry of Four-Manifolds* (Oxford UP)

[10] Dorey N, Hollowood T J, Khoze V V and Mattis M P 2002 The calculus of many instantons *Phys. Rept.* **371** 231 (*Preprint* hep-th/0206063)

[11] Douglas M R and Nekrasov N A 2002 Noncommutative field theory *Rev. Mod. Phys.* **73** 977 (*Preprint* hep-th/0106048)

[12] Furuuchi K 2000 Instantons on noncommutative $\mathbb{R}^4$ and projection operators *Prog. Theor. Phys.* **103** 1043 (*Preprint* hep-th/9912047)

[13] Furuuchi K 2001, Equivalence of projections as gauge equivalence on noncommutative space *Commun. Math. Phys.* **217** 579 (*Preprint* hep-th/0005199)

[14] Furuuchi K 2000 Topological charge of $U(1)$ instantons on noncommutative $\mathbb{R}^4$ *Preprint* hep-th/0010006

[15] Hamanaka M 2003 Noncommutative solitons and D-branes *Ph. D thesis* (University of Tokyo) *Preprint* hep-th/0303256

[16] Hamanaka M and Kajiura H 2003 Gauge fields on tori and T-duality *Phys. Lett.* B **551** 360 (*Preprint* hep-th/0208059)

[17] Hamanaka M and Nakatsu T 2012 ADHM Construction and Group Actions for Noncommutative Instantons *in preparation*

[18] Hamanaka M and Nakatsu T 2012 Exact Construction of Noncommutative Instantons *in preparation*

[19] Harvey J A 2001 Komaba lectures on noncommutative solitons and D-branes *Preprint* hep-th/0102076
[20] Hori K 1999 D-branes, T-duality, and index theory *Adv. Theor. Math. Phys.* **3** 281 (*Preprint* hep-th/9902102)
[21] Ishikawa T, Kuroki S I and Sako A 2002 Instanton number calculus on noncommutative $\mathbb{R}^4$ *J. High Energy Phys.* JHEP08(2002)028 (*Preprint* hep-th/0201196)
[22] Jardim M 2004 A survey on Nahm transform *J. Geom. Phys.* **52** 313 (*Preprint* math/0309305)
[23] Konechny A and Schwarz A 2002 Introduction to M(atrix) theory and noncommutative geometry *Phys. Rept.* **360** 353 (*Preprint* hep-th/0012145)
[24] Konechny A and Schwarz A 2002 Introduction to M(atrix) theory and noncommutative geometry, II *Phys. Rept.* **360** 353 (*Preprint* hep-th/0107251)
[25] Kronheimer P B and Nakajima H 1990 Yang-Mills instantons on ALE gravitational instantons *Math. Ann.* **288** 263
[26] Lechtenfeld O 2004 Noncommutative instantons and solitons *Fortsch. Phys.* **52** 596 (*Preprint* hep-th/0401158)
[27] Maeda Y and Sako A 2012 Noncommutative Deformation of Spinor Zero Mode and ADHM Construction *J. Math. Phys.* **53** 022303 (*Preprint* arXiv:0910.3441 [hep-th])
[28] Nakajima H 1994 Resolutions of moduli spaces of ideal instantons on $\mathbb{R}^4$ *Topology, Geometry and Field Theory* (Singapore, World Scientific) 129
[29] Nakajima H 1999 *Lectures on Hilbert Schemes of Points on Surfaces* (AMS)
[30] Nakajima H and Yoshioka K 2005 Instanton counting on blowup 1 *Invent. Math.* **162** 313 (*Preprint* math/0306198 [math-ag]).
[31] Nekrasov N A 2000 Trieste lectures on solitons in noncommutative gauge theories *Preprint* hep-th/0011095
[32] Nekrasov N A 2002 Lectures on open strings, and noncommutative gauge fields *Preprint* hep-th/0203109
[33] Nekrasov N A 2004 Seiberg-Witten prepotential from instanton counting," *Adv. Theor. Math. Phys.* **7** 831 (*Preprint* hep-th/0206161)
[34] Nekrasov N and Schwarz A 1998 Instantons on noncommutative $\mathbb{R}^4$ and (2,0) superconformal six dimensional theory *Commun. Math. Phys.* **198** 689 (*Preprint* hep-th/9802068)
[35] Osborn H 1979 Calculation of multi-instanton determinants *Nucl. Phys.* B **159** 497
[36] Sako A 2003 Instanton number of noncommutative $U(n)$ gauge theory *J. High Energy Phys.* JHEP04(2003)023 (*Preprint* hep-th/0209139)
[37] Schaposnik F A 2004 Noncommutative solitons and instantons," *Braz. J. Phys.* **34** 1349 (*Preprint* hep-th/0310202)
[38] Schenk H 1988 On a generalized Fourier transform of instantons over flat tori *Commun. Math. Phys.* **116** 177
[39] Seiberg N and Witten E 1999 String theory and noncommutative geometry *J. High Energy Phys.* JHEP09(1999)032 (*Preprint* hep-th/9908142)
[40] Szabo R J 2003 Quantum field theory on noncommutative spaces *Phys. Rept.* **378** 207 (*Preprint* hep-th/0109162)
[41] Tian Y 2004 Topological charge of ADHM instanton on $\mathbb{R}^2_{\rm NC} \times \mathbb{R}^2$ *Mod. Phys. Lett.* A **19** 1315 (*Preprint* hep-th/0404128)
[42] Tian Y, Zhu C J and Song X C 2003 Topological charge of noncommutative ADHM instanton *Mod. Phys. Lett.* A **18** 1691 (*Preprint* hep-th/0211225)
[43] van Baal P 1996 Instanton moduli for $T^3 \times \mathbb{R}$ *Nucl. Phys. Proc. Suppl.* **49** 238 (*Preprint* hep-th/9512223)
[44] Ward R S and Wells R O 1990 *Twistor Geometry and Field Theory* (Cambridge UP)

Journal of Physics: Conference Series **411** (2013) 012017 doi:10.1088/1742-6596/411/1/012017

Minimal Area Surfaces, Euclidean Wilson loops and Riemann Theta Functions

Riei Ishizeki, Martin Kruczenski and Sannah Ziama

Department of Physics, Purdue University, 525 Northwestern Avenue, W. Lafayette, IN 47907-2036, USA

E-mail: rishizek@purdue.edu, markru@purdue.edu, sziama@purdue.edu

Abstract.
The AdS/CFT correspondence relates Wilson loops in supersymmetric N=4 gauge theory to minimal area surfaces in anti-de Sitter space. In this paper we consider the case of Euclidean flat Wilson loops which are related to minimal area surfaces in Euclidean AdS3 space. Using known mathematical results for such minimal area surfaces we describe an infinite parameter family of analytic solutions for closed Wilson loops. Further, we also consider examples of surfaces with two boundaries dual to correlators of two Wilson loops.

1. Introduction

The computation of Wilson loops, which are dual to minimal area surfaces, plays a central role in understanding the AdS/CFT correspondence [1, 2]. Few explicit examples are known of minimal area surfaces in AdS space. For closed, Euclidean Wilson loops (with constant scalar) the most studied one is the circular Wilson loop [3] which is dual to a half-sphere. Another interesting one is the lens shaped discussed in [4]. For infinite Wilson loops, parallel lines [2] and the cusp are known [5]. Concerning multiple contours, the known one consists of two concentric circles [6]. For Minkowski signature, light-like lines play an important role starting with the light-like cusp [7] and culminating with a large recent activity [8] in relation to scattering amplitudes following [9, 10].

Here we show that, in the case of flat Euclidean Wilson loops which are dual to minimal area surfaces in Euclidean AdS_3, known results from the mathematical literature [11] allow us to find an infinite parameter family of solutions in terms of Riemann theta functions. The construction is based on [12] and follows previous results in [13, 14].

The presentation is organized as follows. We start by writing the equations of motion arriving at the cosh-Gordon equation. Next we solve the equations of motion in terms of theta functions and compute the regularized area. Finally we give examples of single curves and multiple curves together with the minimal area surfaces ending on them. More details can be found in the work [15].

2. Equations of motion

In this section we write the equations of motion for a minimal area surface in Euclidean AdS_3. We use embedding coordinates $X_{\mu=0\dots3}$ parameterizing a space $R^{3,1}$ and subjected to the constraint

$$X_0^2 - X_1^2 - X_2^2 - X_3^2 = 1 , \tag{1}$$

Journal of Physics: Conference Series **411** (2013) 012017 doi:10.1088/1742-6596/411/1/012017

Other useful coordinates are Poincare coordinates (X, Y, Z) given by:

$$X + iY = \frac{X_1 + iX_2}{X_0 - X_3}, \qquad Z = \frac{1}{X_0 - X_3}. \tag{2}$$

The boundary is an R^2 space and located at $Z = 0$. A string is parameterized by complex coordinates $z = \sigma + i\tau$, $\bar{z} = \sigma - i\tau$. The action in conformal gauge is given by

$$\begin{aligned} S &= \frac{1}{2}\int \left(\partial X_\mu \bar{\partial} X^\mu + \Lambda(X_\mu X^\mu - 1)\right)\, d\sigma\, d\tau & (3) \\ &= \frac{1}{2}\int \frac{1}{Z^2}\left(\partial_a X \partial^a X + \partial_a Y \partial^a Y + \partial_a Z \partial^a Z\right)\, d\sigma\, d\tau, & (4) \end{aligned}$$

where Λ is a Lagrange multiplier and the μ indices are raised and lowered with the $R^{3,1}$ metric. An Euclidean classical string is given by functions $X_\mu(z, \bar{z})$ obeying the equations of motion:

$$\partial\bar{\partial} X_\mu = \Lambda X_\mu \, , \tag{5}$$

where Λ, the Lagrange multiplier is given by

$$\Lambda = -\partial X_\mu \bar{\partial} X^\mu. \tag{6}$$

These equations should be supplemented by the Virasoro constraints which read

$$\partial X_\mu \partial X^\mu = 0 = \bar{\partial} X_\mu \bar{\partial} X^\mu. \tag{7}$$

Later on we will be interested in finding the solutions in Poincare coordinates (X, Y, Z) but for the moment it is convenient to study the problem in embedding coordinates X_μ. We can rewrite the equations using the matrix

$$\mathbb{X} = \begin{pmatrix} X_0 + X_3 & X_1 - iX_2 \\ X_1 + iX_2 & X_0 - X_3 \end{pmatrix} = X_0 + X_i \sigma^i \, , \tag{8}$$

where σ^i denote the Pauli matrices. Notice also that Poincare coordinates are simply given by

$$Z = \frac{1}{\mathbb{X}_{22}}, \qquad X + iY = \frac{\mathbb{X}_{21}}{\mathbb{X}_{22}}. \tag{9}$$

The matrix $\mathbb{X}$ satisfies

$$\mathbb{X}^\dagger = \mathbb{X}, \quad \det \mathbb{X} = 1, \quad \partial\bar{\partial}\mathbb{X} = \Lambda \mathbb{X}, \quad \det(\partial \mathbb{X}) = 0 = \det(\bar{\partial}\mathbb{X}) \, , \tag{10}$$

as follows from the definition of $\mathbb{X}$, the constraint (1), the equations of motion (5) and the Virasoro constraints (7). These are non-linear equations which in principle are difficult to solve. However, known procedures (see [15]) reduce the equations to the cosh-Gordon equation which has known solutions in terms or Riemann theta functions. Therefore the main equation that has to be solved is:

$$\frac{\partial^2}{\partial z \partial \bar{z}} \alpha(z, \bar{z}) = 2\cosh 2\alpha \, , \tag{11}$$

where $\alpha(z, \bar{z})$ is a real function defined on the world-sheet. In fact, α satisfies a more general equation that contains an arbitrary holomorphic function $f(z)$ that we have set to one. This can always be done locally but, doing it globally as we do here implies a restriction in the type of solution we consider. Once a solution for $\alpha(z, \bar{z})$ is found, the full solution $\mathbb{X}(z, \bar{z})$ can be reconstructed by a procedure discussed also in [15]. Instead of explaining this procedure in detail we give the solution in the next section. It is straight-forward to check that it satisfies the equations of motion.

3. Solutions

The solutions are written in terms of theta functions [11, 16]. We need two theta functions, one has zero characteristics and the other fixed characteristics that we call $\Delta_{1,2}$. The latter we denote as:

$$\hat{\theta}(\zeta) = \theta \begin{bmatrix} \Delta_1 \\ \Delta_2 \end{bmatrix} (\zeta) \tag{12}$$

In this paper, for definiteness, we use the convention of [16] as regards theta functions. The theta functions are associated to an auxiliary (genus g) hyperelliptic Riemann surface determined as a subspace of $\mathbb{C}^2$ by an equation

$$\mu^2 = \lambda \prod_{j=1}^{2g} (\lambda - \lambda_j) \tag{13}$$

where (μ, λ) parameterize $\mathbb{C}^2$. If $\hat{\theta}$ has an odd characteristic then

$$e^{2\alpha} = \frac{\theta^2(\zeta)}{\hat{\theta}^2(\zeta)} , \tag{14}$$

satisfies the cosh-Gordon equation provided we take

$$\zeta = 2\omega(p_1)\bar{z} + 2\omega(p_3)z. \tag{15}$$

where ω are the normalized holomorphic differentials and $p_{1,3}$ are two branch points of the Riemann surface such that

$$2\int_{p_1}^{p_3} \omega = \Delta_2 + \Pi\Delta_1 , \tag{16}$$

As discussed, associated to the solution of the cosh-Gordon equation one finds a solution for the minimal area surface which, in Poincare coordinates, is

$$Z = \left| \frac{\hat{\theta}(2\int_{p_1}^{p_4})}{\hat{\theta}(\int_{p_1}^{p_4})\theta(\int_{p_1}^{p_4})} \right| \frac{|\theta(0)\theta(\zeta)\hat{\theta}(\zeta)|\, |e^{\mu z+\nu\bar{z}}|^2}{|\hat{\theta}(\zeta - \int_{p_1}^{p_4})|^2 + |\theta(\zeta - \int_{p_1}^{p_4})|^2} , \tag{17}$$

$$X + iY = e^{2\bar{\mu}\bar{z}+2\bar{\nu}z} \frac{\theta(\zeta - \int_{p_1}^{p_4})\overline{\theta(\zeta + \int_{p_1}^{p_4})} - \hat{\theta}(\zeta - \int_{p_1}^{p_4})\overline{\hat{\theta}(\zeta + \int_{p_1}^{p_4})}}{|\hat{\theta}(\zeta - \int_{p_1}^{p_4})|^2 + |\theta(\zeta - \int_{p_1}^{p_4})|^2} , \tag{18}$$

Here p_4 denotes a point on the Riemann surface on which the solution depends. The notation $\int_{p_1}^{p_4}$ indicates

$$\int_{p_1}^{p_4} = \int_{p_1}^{p_4} \omega \tag{19}$$

where ω are the normalized holomorphic differentials. Finally ζ is given by eq.(15). This gives the shape of the surface. Now we need to compute the area.

4. Computation of the area

The expectation value of the Wilson loop is determined by the (regularized) area of the minimal surface we described. The area can be computed as

$$A = 4\int e^{2\alpha} d\sigma d\tau , \tag{20}$$

This integral actually diverges. To regulate the divergence we cut the surface at a value $Z = \epsilon > 0$. In the limit $\epsilon \to 0^+$ the area behaves as

$$A = \frac{L}{\epsilon} + A_f \ , \tag{21}$$

where L should be the length of the Wilson loop and A_f is the finite part which is identified with the expectation value of the Wilson loop through:

$$\langle W \rangle = e^{-\frac{\sqrt{\lambda}}{2\pi} A_f} \ . \tag{22}$$

This finite piece can be computed from the solution as

$$A_{\mathrm{f}} = -8 D_{p_1 p_3} \ln \theta(0) \oint (\sigma d\tau - \tau d\sigma) + 2 \oint \frac{D_{p_1 p_3} \hat{\theta}(\zeta)}{|D_{p_1} \hat{\theta}(\zeta)|} \, d\ell. \tag{23}$$

where $d\ell$ is the differential of arc in the world-sheet boundary.

This concludes our presentation of the analytical results obtained. Now we go into some examples to understand how these solutions look.

5. Example: single contour

We now illustrate the shape of the Wilson loops and their dual surfaces. Consider the function

$$\mu(\lambda) = i\sqrt{-i(\lambda+1-i)}\sqrt{-i(\lambda+1+i)}\sqrt{-i(\lambda-\frac{1+i}{2})}\sqrt{-i(\lambda-\frac{1-i}{2})}\sqrt{2-\lambda}\sqrt{\lambda}\sqrt{\lambda+\frac{1}{2}} \ , \tag{24}$$

where the square root is taken to have a cut on the negative real axis. The cuts in the complex plane are illustrated in fig.1. The period matrix is

$$\Pi = \begin{pmatrix} 0.5 + 0.64972i & 0.14972i & -0.5 \\ 0.14972i & -0.5 + 0.64972i & 0.5 \\ -0.5 & 0.5 & 0.639631i \end{pmatrix} . \tag{25}$$

To write the solution we choose the points $p_1 = 0$ and $p_3 = \infty$ which works well since a path between 0 and ∞ defines an odd characteristic that can be taken to be

$$\Delta_1 = \begin{bmatrix} 0 \\ 0 \\ 1 \end{bmatrix}, \quad \Delta_2 = \begin{bmatrix} 1 \\ 1 \\ 1 \end{bmatrix} . \tag{26}$$

We therefore have

$$\hat{\theta}(\zeta) = \theta \begin{bmatrix} 0 & 0 & 1 \\ 1 & 1 & 1 \end{bmatrix} (\zeta). \tag{27}$$

We can now write

$$\zeta = 2(\omega(\infty) z + \omega(0) \bar{z}) = \begin{bmatrix} -0.4903\sigma - 0.19069\tau \\ -0.4903\sigma + 0.19069\tau \\ 0.59321\tau \end{bmatrix} , \tag{28}$$

where we used $z = \sigma + i\tau$. The shape of the Wilson loop and its dual surface are illustrated in figures 2 and 3 for two different choices of the point p_4.

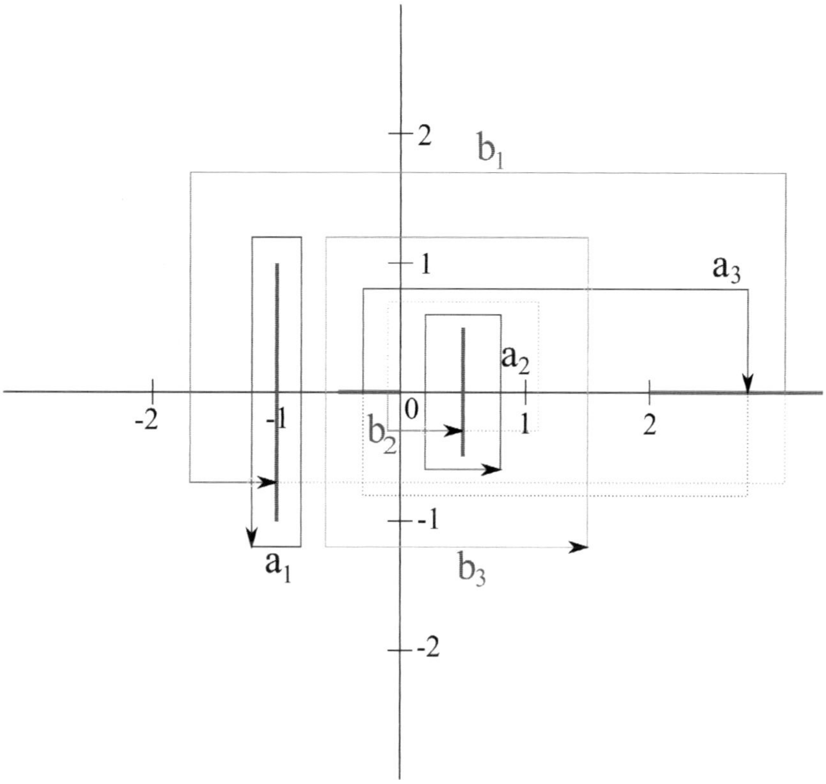

Figure 1: Genus three Riemann surface. A basis of fundamental cycles is depicted with solid paths being in the upper sheet and dotted lines in the lower sheet.

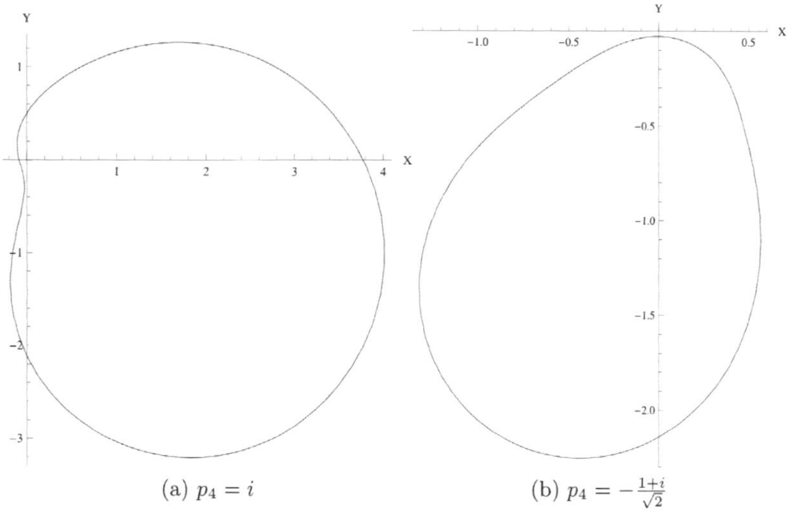

(a) $p_4 = i$ (b) $p_4 = -\frac{1+i}{\sqrt{2}}$

Figure 2: Shape of the Wilson loops for two different choices of the point p_4.

Journal of Physics: Conference Series **411** (2013) 012017 doi:10.1088/1742-6596/411/1/012017

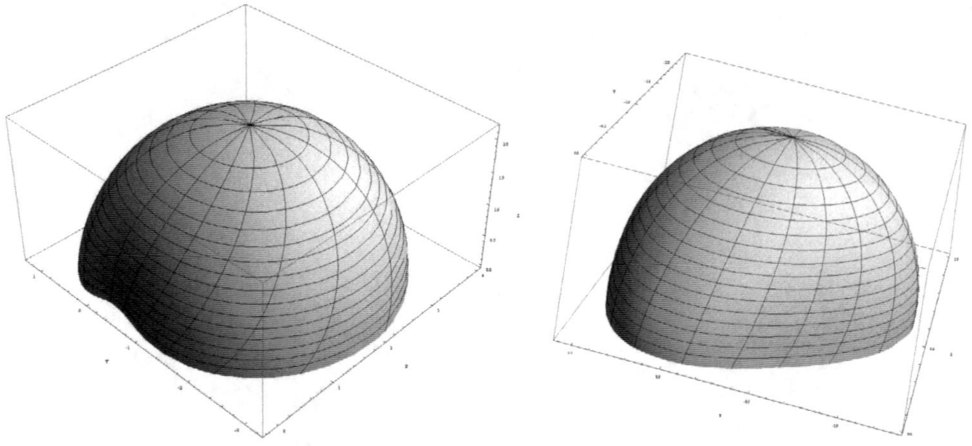

(a) $p_4 = i$ (b) $p_4 = -\frac{1+i}{\sqrt{2}}$

Figure 3: Minimal area surfaces ending on the contours illustrated in fig.2. We emphasize that the surfaces are known analytically.

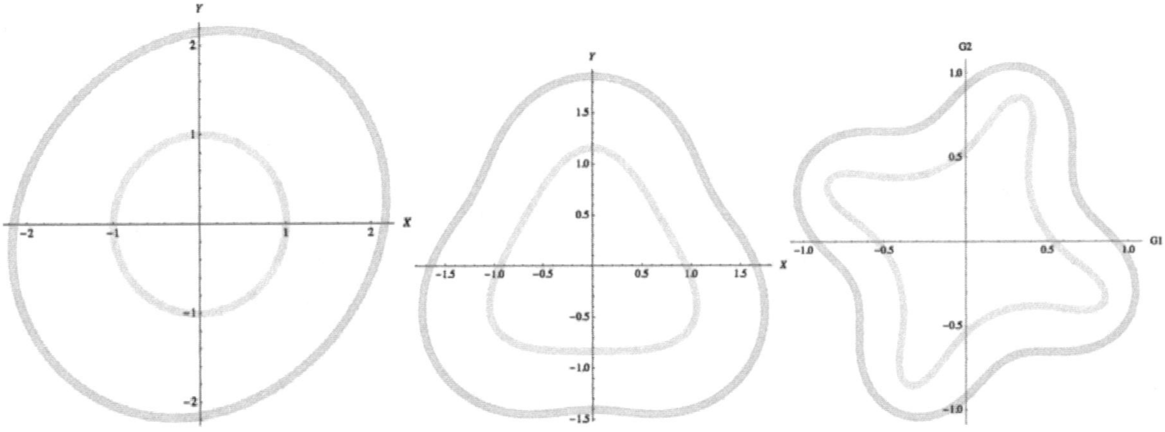

(a) (b) (c)

Figure 4: Shape of multiple contours. The minimal area surfaces connects the interior and exterior contours.

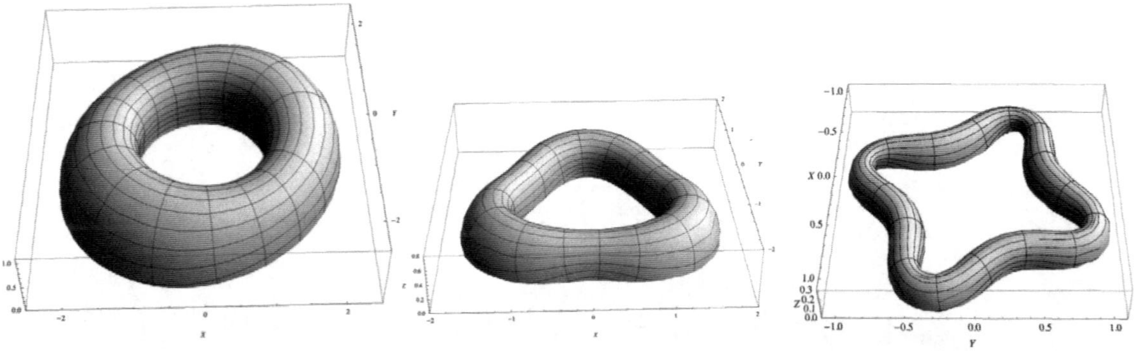

(a) (b) (c)

Figure 5: Shape of the minimal area surfaces ending on the contours in fig.4.

6. Example: Multiple contours

By changing the position of the cuts, one can find solutions ending on two contours. Since the construction is completely similar we just illustrate the results in figs. 4 and 5.

Acknowledgments

This work was supported in part by NSF through grants PHY-0805948, a CAREER Award PHY-0952630, a Graduate Fellowship (S.Z.) and an AGEP grant #0450373, by DOE through grant DE-FG02-91ER40681 and by the SLOAN Foundation. In addition, M.K. is grateful to Perimeter Institute and M.K. and S.Z. to the Simons Center for Geometry and Physics for hospitality while part of this work was being done.

7. References

[1] J. Maldacena, "The large N limit of superconformal field theories and supergravity," Adv. Theor. Math. Phys. **2**, 231 (1998) [Int. J. Theor. Phys. **38**, 1113 (1998)], hep-th/9711200,
S. S. Gubser, I. R. Klebanov and A. M. Polyakov, "Gauge theory correlators from non-critical string theory," Phys. Lett. B **428**, 105 (1998) [arXiv:hep-th/9802109],
E. Witten, "Anti-de Sitter space and holography," Adv. Theor. Math. Phys. **2**, 253 (1998) [arXiv:hep-th/9802150].

[2] J. M. Maldacena, "Wilson loops in large N field theories," Phys. Rev. Lett. **80**, 4859 (1998) [arXiv:hep-th/9803002],
S. J. Rey and J. T. Yee, "Macroscopic strings as heavy quarks in large N gauge theory and anti-de Sitter supergravity," Eur. Phys. J. C **22**, 379 (2001) [arXiv:hep-th/9803001].

[3] D. E. Berenstein, R. Corrado, W. Fischler and J. M. Maldacena, "The Operator product expansion for Wilson loops and surfaces in the large N limit," Phys. Rev. D **59**, 105023 (1999) [arXiv:hep-th/9809188],
D. J. Gross and H. Ooguri, "Aspects of large N gauge theory dynamics as seen by string theory," Phys. Rev. D **58**, 106002 (1998) [arXiv:hep-th/9805129],
J. K. Erickson, G. W. Semenoff and K. Zarembo, "Wilson loops in N = 4 supersymmetric Yang-Mills theory," Nucl. Phys. B **582**, 155 (2000) [arXiv:hep-th/0003055],
N. Drukker and D. J. Gross, "An exact prediction of N = 4 SUSYM theory for string theory," J. Math. Phys. **42**, 2896 (2001) [arXiv:hep-th/0010274],
V. Pestun, "Localization of gauge theory on a four-sphere and supersymmetric Wilson loops," arXiv:0712.2824 [hep-th].

[4] N. Drukker, S. Giombi, R. Ricci and D. Trancanelli, "Supersymmetric Wilson loops on S**3," JHEP **0805**, 017 (2008) [arXiv:0711.3226 [hep-th]].

[5] N. Drukker, D. J. Gross and H. Ooguri, "Wilson loops and minimal surfaces," Phys. Rev. D **60**, 125006 (1999) [arXiv:hep-th/9904191].

[6] N. Drukker and B. Fiol, "On the integrability of Wilson loops in AdS(5) x S**5: Some periodic ansatze," JHEP **0601**, 056 (2006) [arXiv:hep-th/0506058].

[7] M. Kruczenski, "A note on twist two operators in N = 4 SYM and Wilson loops in Minkowski signature," JHEP **0212**, 024 (2002) [arXiv:hep-th/0210115].

[8] See e.g.
J. Maldacena and A. Zhiboedov, "Form factors at strong coupling via a Y-system," JHEP **1011**, 104 (2010) [arXiv:1009.1139 [hep-th]],
L. F. Alday, B. Eden, G. P. Korchemsky, J. Maldacena and E. Sokatchev, "From correlation functions to Wilson loops," arXiv:1007.3243 [hep-th],
L. F. Alday, D. Gaiotto, J. Maldacena, A. Sever and P. Vieira, "An Operator Product Expansion for Polygonal null Wilson Loops," arXiv:1006.2788 [hep-th].

[9] L. F. Alday and J. M. Maldacena, "Gluon scattering amplitudes at strong coupling," JHEP **0706**, 064 (2007) [arXiv:0705.0303 [hep-th]].

[10] L. F. Alday and J. Maldacena, "Null polygonal Wilson loops and minimal surfaces in Anti-de-Sitter space," JHEP **0911**, 082 (2009) [arXiv:0904.0663 [hep-th]].

[11] M. Babich and A. Bobenko, "Willmore Tori with umbilic lines and minimal surfaces in hyperbolic space", Duke Mathematical Journal **72, No. 1**, 151 (1993).

[12] E. D. Belokolos, A. I. Bobenko,V. Z. Enol'skii, A. R. Its, V. B. Matveev, "Algebro-Geometric Approach to Nonlinear Integrable Equations," Springer-Verlag series in Non-linear Dynamics, Springer-Verlag Berlin Heidelberg NewYork (1994).

Journal of Physics: Conference Series **411** (2013) 012017 doi:10.1088/1742-6596/411/1/012017

[13] N. Dorey and B. Vicedo, "On the dynamics of finite-gap solutions in classical string theory," JHEP **0607**, 014 (2006) [arXiv:hep-th/0601194].

[14] K. Sakai and Y. Satoh, "Constant mean curvature surfaces in AdS_3," JHEP **1003**, 077 (2010) [arXiv:1001.1553 [hep-th]].

[15] R. Ishizeki, M. Kruczenski and S. Ziama, "Notes on Euclidean Wilson loops and Riemann Theta functions," Phys. Rev. D **85**, 106004 (2012) [arXiv:1104.3567 [hep-th]].

[16] H. M. Farkas and I. Kra, "Riemann Surfaces", Graduate Texts in Mathematics, Second Edition, Springer-Verlag, New, Berlin, Heidelberg (1991).

Journal of Physics: Conference Series **411** (2013) 012018 doi:10.1088/1742-6596/411/1/012018

Integrable chains with Jordan - Schwinger representations

Roland Kirschner

Institut für Theoretische Physik, Universität Leipzig, PF 100 920, D-04009 Leipzig, Germany

E-mail: roland.kirschner@itp.uni-leipzig.de

Abstract. The restiction to the class of Jordan - Schwinger representations of $s\ell(n+1)$ results in simple relations for the L matrices and in explicit expressions for the general Yang-Baxter operators as products of two parameter permutation operators. Limits are studied which are related to the finite dimensional representations and to degenerate Yangians. The analogy to the $s\ell(2)$ case leads to analogous forms of global spin chain operators.

1. Introduction

The algebra generated by $n+1$ Heisenberg pairs $x_s, \partial_s, s = 1, ..., n+1$ contains a $g\ell(n+1)$ subalgebra. This was noticed early in the context of angular momentum [1, 2, 3]. Representations of $g\ell(n+1)$ can be constructed by combining a representation of $g\ell(n)$ with such an algebra [4, 5, 6, 7, 8]. This iteration in rank can be conveniently formulated in terms of L matrices [9] and used in constructing Yang-Baxter operators [10].

In this contribution we focus on the $g\ell(n+1)$ algebra built merely from $n+1$ Heisenberg pairs and consider representations spanned by monomials in x_s. One encounters therein lowest weight representations, called Jordan-Schwinger representations, which are characterised by just one representation parameter, to be compared with $n+1$ parameters for generic representations. The complete iteration would take $\frac{1}{2}n(n+1)+1$ Heisenberg pairs to construct the generic generators, whereas $n+1$ are sufficient for the first elementary building block.

This means that the restriction to the Jordan-Schwinger class of representations allows the restriction to $n+1$ degrees of freedom (per site of a spin chain), and this restriction goes along with simplicity in structure, which we would like to emphasize here.

In physical applications of higher rank symmetries and related integrable structures the Jordan-Schwinger case plays an important role, because the composition of the symmetry generators by more elementary operators related to canonical variables is natural.

The studies of the generalised chiral Potts model provides an example where the features of factorisation and permutations have been observed in the construction of Yang-Baxter solutions [11, 12, 13, 14]. The context and the formulation differs much from our discussion and involves deformation at roots of unity, which is avoided here.

The example having attracted much interest recently is the observation of Yangian symmetry based on the superalgebra $s\ell(4|4)$ in the calculation of $\mathcal{N}=4$ super Yang-Mills amplitudes [15].

The generators of $g\ell(n+1)$ representations can be written in terms of the $n+1$ Heisenberg pairs in several ways. We select two versions, write the related L matrices and study relations for their products. This provides an easy way to operators obeying Yang-Baxter RLL relations.

Journal of Physics: Conference Series **411** (2013) 012018 doi:10.1088/1742-6596/411/1/012018

We obtain explicit expressions of the two R operators permuting the first and the second representation parameters in the product of two L matrices. Then the R operator intertwining two Jordan-Schwinger representations is obtained as the product of these two parameter permutation operators. We study the limit, where the representation parameter $2\ell_1$ of the first tensor factor approaches a non-negative integer value and a finite-dimensional invariant subspace emerges. We study also the asymptotics of large representation parameters, where degenerate Yangian algebra representations appear.

The various R operators as well as the L matrices provide building units of global operators of integrable spin chains. The relations between the building units imply relations of factorisation and commutativity between the global chain operators.

In this way we review some basic points of [16] and previous work by presenting the non-trivial extension of the $s\ell(2)$ case discussed there to the Jordan-Schwinger restricted higher rank case. In these papers the results on the R operators have been used to construct Baxter operators, prove Baxter relations and analyse the relations between different approaches to Baxter operators [17].

Baxter relations for the generic higher rank case have been considered recently in [18, 19]. They are not compatible in a simple way with the restriction to Jordan-Schwinger representations; the tensor product of such restricted representations does not decompose into restricted ones only.

2. Jordan-Schwinger representations of $g\ell(n+1)$

Choosing $n+1$ canonical pairs $x_s, \partial_s, s = 1, ..., n+1$ one can construct operators obeying the $g\ell(n+1)$ algebra relations. The straightforward form $E_{ij} = x_i \partial_j$ can be modified in several ways:

$$E_{ij}^J = \frac{x_i}{x_j}(N_j + \delta_j), \quad E_{ij}^{-J} = -\frac{x_j}{x_i}(N_j + \delta_j), \quad E_{ij}^{TJ} = -(N_i + \delta_i)\frac{x_j}{x_i}, \quad E_{ij}^{-TJ} = (N_i + \delta_i)\frac{x_i}{x_j}, \tag{1}$$

for $i, j = 1, ..., n+1$. $N_i = x_i \partial_i$ acts as infinitesimal dilatation operator on the coordinate operator x_i. In each case the algebra relations are

$$[E_{ij}^C, E_{jk}^C] = E_{ik}^C, \ i \neq k, \quad [E_{ij}^C, E_{ji}^C] = N_i + \delta_i - N_j - \delta_j \tag{2}$$

The L matrix with the matrix elements

$$L_{ij}^C = u\delta_{ij} + E_{ji}^C \tag{3}$$

obeys the fundamental RLL relation with Yang's $(n+1)^2 \times (n+1)^2$ R-matrix $\mathcal{R}_{12}(u) = u\mathrm{I} + \mathrm{P}_{12}$, where P_{12} denotes the the permutation matrix.

$$\mathcal{R}_{12}(u-v)L_1(u) \otimes L_2(v) = L_2(v) \otimes L_1(u)\mathcal{R}_{12}(u-v) \tag{4}$$

We shall use matrix notation. The canonical pairs are put into diagonal matrices, $X = diag(x_1, ..., x_{n+1}), \partial^x = diag(\partial_1, ..., \partial_{n+1})$. M is the matrix with all elements equal to 1. In the first relations to be discussed the latter matrix is relevant and also its relation involving diagonal matrices,

$$M \ XY \ M = M \ (X \cdot Y), \tag{5}$$

where $X \cdot Y = tr(XY)$. Y, ∂^y stand for the diagonal matrices involving another set of $n+1$ canonical pairs.

We restrict to a special case in the shifts δ and write the L matrices in matrix notation

$$L^J(u) = I(u-1) + \hat{\partial}M\hat{X}, \quad L^{TJ}(v) = Iv - \hat{Y}M\hat{\partial}^y. \tag{6}$$

Now we write down products of these matrices

$$L_x^J(u)L_y^{TJ}(v) = I(u-1)v + v\partial^x MX - uYM\partial^y - (X\cdot Y)\partial^x M\partial^y,$$

$$L_y^{TJ}(v)L_x^J(u) = I(u-1)v + v\partial^x MX - uYM\partial^y - YMX(\partial^y\cdot\partial^x),$$

and study the similarity transformation by $(X\cdot Y)^\lambda$ leading to the extension $\partial^x \to \partial^x + \lambda\frac{Y}{YX}, \partial^y \to \partial^y + \lambda\frac{X}{YX}$. We notice that the result can be written in terms of L matrix products again as

$$(X\cdot Y)^{-\lambda}L_x^J(u)L_y^{TJ}(v)(X\cdot Y)^\lambda = L_x^J(u+\lambda)L_y^{TJ}(v-\lambda) - \lambda(1-\lambda+v-u)(I - \frac{YMX}{Y\cdot X}), \quad (7)$$

$$(X\cdot Y)^{-\lambda}L_y^{TJ}(v)L_x^J(u)(X\cdot Y)^\lambda = L_y^{JS}(v)L_x^J(u) - \frac{YMX}{Y\cdot X}\lambda(\lambda-1+u-v+N^y+N^x), \quad (8)$$

$N^x = X\cdot\partial^x, N^y = \partial^y\cdot Y$. Corresponding relations for the product of L matrices of one and the same representation can be obtained by noticing that (6) implies

$$L^{TJ}(u;x,\partial_x) = L^J(u-1;\partial_x,-x)$$

and performing the canonical transformation interchanging the canonical pairs

$$K_0(x_s,\partial_s)K_0 = (\partial_s,-x_s).$$

The interesting particular case appears at $\lambda = u-v-1$ where the second term of one relation vanishes. With a simple parameter substitution we obtain the Yang-Baxter relation

$$(X\cdot\partial_y)^{u-v}L_x^J(u)L_y^J(v) = L_x^J(v)L_y^J(u)(X\cdot\partial_y)^{u-v}, \quad (9)$$

$$(\partial_x\cdot Y)^{u-v}L_x^{TJ}(u)L_y^{TJ}(v) = L_x^{TJ}(v)L_y^{TJ}(u)(\partial_x\cdot Y)^{u-v}.$$

The vanishing of the remainder in the other relation can happen at $\lambda \neq 0$ only if the representation space is restricted to eigenfunctions of $N^x + N^y$.

The simple explicit expressions of the L matrices allows to write an explicit expression for the ordinary transfer matrix for a corresponding spin chain. The definition of the general transfer matrix as a trace of a product of e.g. with $R_{i0}(u) = (\partial_i X_0)^u$ can be given only after specifying the representation space labelled by 0.

As suggested by the notation one may consider representations on functions of $x_i, i = 1,...,n+1$ spanned by $\prod x_i^{\alpha_i+m_i}$ with m_i running over all integers and $\alpha_1,...,\alpha_{n+1}$ fixed. Such a representation decomposes into ones where the $U(1)$ representation generated by $N^x = \sum_1^{n+1} x_s\partial_s$ is irreducible. In the particular case where for a distinguished index value i we set $\alpha_i = 2\ell$, $\sum_s m_s = 0$ and $\alpha_s = 0, s\neq i$ it can be reduced to the subspace spanned by the monomials with non-negative integer $m_s, s\neq i$, which has the lowest weight vector $x_i^{2\ell}$. Regarding the general labelling of $g\ell(n+1)$ representations of lowest weight by the eigenvalues of E_{rr}^J in their action on the lowest weight vector the considered Jordan-Schwinger case has the feature of all these labels vanishing but one. The involved representation of $s\ell(n+1)$ is irreducible for a generic value of 2ℓ but contains a $2\ell+1$ dimensional invariant subspace for non-negative integer values of 2ℓ, just as in the rank $n=1$ case. This known fact can be checked easily. The separation of the $s\ell(n+1)$ subalgebra representation will be done explicitly in the next section by transformation to $x'_s, \partial'_s, s\neq i$, where the new coordinates are defined by the ratios $x'_s = \frac{x_s}{x_i}$.

3. Restriction to $s\ell(n+1)$

As the first step towards the reduction to $s\ell(n+1)$ we rewrite the L matrices and their relations in coordinate ratios and separate the dilatation operators $N^x = \sum_1^{n+1} x_s \partial_s^x, \tilde{N}^y = \sum_1^{n+1} \partial_s^y y_s$. This takes distinguishing one direction in the $n+1$ dimensional space, e.g. the one related to the index i or j respectively:

$$x'_s = \frac{x_s}{x_i}, \quad N_s^x = N_s^{x'} = x'_s \partial'_s, s \neq i, \quad N_i = N^x - \hat{m}_x, \quad \hat{m}_x \sum_{s \neq i} N'_s.$$

$$y'_s = \frac{y_s}{y_j}, \quad \tilde{N}_s^y = \tilde{N}_s^{y'} = \partial_d^y y'_s, s \neq j, \quad \tilde{N}_j^y = \tilde{N}^y - \hat{m}_y, \quad \hat{m}_y = \sum_{s \neq j} \tilde{N}'_s$$

The distinguished component of the derivative has to be transformed with care:

$$\partial_i = x_i^{-1} N_i = x_i^{-1}(N^x - \hat{m}_x) = \mathcal{N}^x \, x_i^{-1}, \quad \mathcal{N}^x = N^x + 1 - \hat{m}_x$$

$$\partial_j = \tilde{N}_j y_j^{-1} = y_j^{-1} \mathcal{N}^y, \quad \mathcal{N}^y = \tilde{N}^y - 1 - \hat{m}_y$$

We introduce diagonal matrices $X', N', \partial'^x, \tilde{\partial}^x$ with the corresponding components $x'_s, N'_s, \partial'^x_s, \tilde{\partial}^x_s, s \neq i$ on the diagonal and with 1 as the ith diagonal element. Correpondingly, the diagonal matrices $Y', \tilde{N}', \partial'^y, \tilde{\partial}^y$ are introduced. The definitions have been put in such a way to have

$$X = X' \cdot x_i, \; \partial^x = \tilde{\partial}^x \cdot \mathcal{N}^x x_i^{-1}, \quad Y = Y' \cdot x_j, \; \partial^y = y_i^{-1} \mathcal{N}^y \tilde{\partial}^y.$$

With these notations we write factorised forms of the L matrices

$$L_x^J(u) = L_x^{J/i}(u-1, u+N^x) = X'^{-1} m_i^{-1} K_i(u-1, u+N^x; N') m_i X' = \tag{10}$$

$$\tilde{\partial}^x m_i^t K_i^t(u-1, u+N^x; N') m_i^{t-1} \tilde{\partial}^{-1\,x},$$

$$L_y^{TJ}(v) = L_y^{TJ/j}(v, v+1-\tilde{N}^y) = Y' m_j^t K_j^t(v, v+1-\tilde{N}^y; -\tilde{N}') m_j^{t-1} Y'^{-1} = \tag{11}$$

$$\tilde{\partial}^{y\,-1} m_j^{-1} K_j(v, v+1-\tilde{N}^y; -\tilde{N}^y) m_j \tilde{\partial}^y,$$

$$K_i(u_1, u_2; N) = (I + I_i(u_2 - 1))\,(I + N c_i)\,(I u_1 - I_i(u_1 - 1)), \quad m_i = I + r_i.$$

Auxiliary matrices I_i, c_i, r_i have been introduced here. c_i, r_i have most elements vanishing besides of the elements on the column i or row i respectively. These elements are equal 1, but the corresponding diagonal element is zero too. Further we denote $I_i = \hat{e}_{ii}$. These standard matrices obey the algebraic relations

$$c_i^T = r_i, \quad c_i^2 = 0, \quad I_i c_i = 0, \quad r_i I_i = 0, \quad c_i I_i = c_i, \quad I_i r_i = r_i, \tag{12}$$

$$r_i c_i = n I_i, \quad r_i \hat{Z} c_i = I_i \sum_{s \neq i} Z_s, \quad c_i r_i = M - I_i - r_i - c_i = M_i.$$

Here Z stands for any diagonal matrix. The dependence on the $U(1)$ or dilatation generators N^x or N^y has been separated; it enters the coordinate factorized form only by the diagonal matrix factor $I + I_i(u_2 - 1)$ or $I + I_j(v_2 - 1)$ via $u_2 = u + N^x$ or $v_2 = v + 1 - \tilde{N}^y$. These generators are substituted by corresponding numbers $N^x \to 2\ell_x$ or $\tilde{N}^y \to -2\ell_y$ upon restriction to irreducible $U(1)$ representations. In physical terms, only the n degrees of freedom expressed by the ratio coordinates and their conjugate momenta are left active now. The ∂ factorised form plays an auxiliary role in establishing the relation between the J and TJ forms of L matrices. Notice that in the latter form the dilatation generator enters also via the definition of $\tilde{\partial}$.

In this way we obtain the similarity transformations of the L matrix products with $(X' \cdot Y')^{\lambda}$ by argument shift on the l.h.s. from the relations (7), (8),

$$(X' \cdot Y')^{-\lambda} L_x^{J/i}(u-1, u+N^x+\lambda) L_y^{TJ/j}(v, v+1-\tilde{N}^y-\lambda)(X' \cdot Y')^{\lambda} = \tag{13}$$

$$L_x^{J/i}(u-1+\lambda, u+N^x+\lambda) L_y^{TJ/j}(v-\lambda, v+1-\tilde{N}^y-\lambda) - \lambda(1-\lambda+v-u)(I - \frac{Y'MX'}{Y' \cdot X'}),$$

$$(X' \cdot Y')^{-\lambda} L_y^{TJ/j}(v, v+1-\tilde{N}^y-\lambda) L_x^{J/i}(u-1, u+N^x+\lambda)(X' \cdot Y')^{\lambda} = \tag{14}$$

$$L_y^{TJ/j}(v, v+1-\tilde{N}^y) L_x^{J/i}(u-1, u+N^x) - \frac{Y'MX'}{Y' \cdot X'} \lambda(\lambda-1+u-v+\tilde{N}^y+N^x).$$

Notice that the exponent λ in the similarity operator may depend on $N^x, \tilde{N}^y$ since the latter commute with the ratio coordinates. Therefore the case of the vanishing of the extra term can be considered for both relations now and this results in parameter permutation relations.

4. Yang-Baxter relations

The similarity of the factorised representations of the Lax matrices (10), (11) of different types, in particular the coordinate TJ and the derivative J forms lead to

$$L^{TJ/j}(v, v+1-\tilde{N}; Y', \tilde{N}') = L^{J/j}(v, v+1-\tilde{N}, \tilde{\partial} \to Y', N' \to -\tilde{N}'). \tag{15}$$

The representation parameters coincide and the indicated substitutitons have to be done. As pointed out above the derivative factorised expressions depend on the dilatation $N = \sum_1^{n+1} N_s$ not only via the representation parameter argument but also by the factor $\mathcal{N}$ involved in the definition of $\tilde{\partial}$. In the case J

$$\mathcal{N} = N + 1 - \sum N'_s = \hat{u}_2 - u_1 - \sum N'_s.$$

Thus $\mathcal{N}$ is connected to the representation parameter arguments of the L matrix, not in universal way. In the substitution of the L^{TJ} matrix (15) we have

$$\mathcal{N} = \hat{v}_2 - v_1 - \sum N'_s = 1 - \tilde{N}^y - \sum N'_s$$

One should be careful to read $\tilde{N}$ according to the TJ convention and N'_s in the last term according to the J convention.

The indicated substitution can be expressed as

$$(\tilde{\partial}, N') = (\Gamma(\mathcal{N}))^{-1}(\partial', \ \ N')\Gamma(\mathcal{N}),$$

and

$$(Y, -\tilde{N}') = K'_0(\partial', N')K'_0.$$

The latter denotes the elementary canonical transformation of the reduced n degrees of freedom. Thus,

$$L^{TJ/j}(v, v+1-\tilde{N}; Y', \tilde{N}') = K'_0(\Gamma(\mathcal{N}))\tilde{\partial} m_j^t K_j^t(v, v+1-\tilde{N}^y; N') m_j^{t-1} \tilde{\partial}^{-1}(\Gamma(\mathcal{N}))^{-1} K'_0 =$$

$$K'_0(\Gamma(\mathcal{N})) L^{J/j}(v, v+1-\tilde{N}^y; Y', N') \ (\Gamma(\mathcal{N}))^{-1} K'_0$$

This relation can be used to derive from the permutation relations (13), (14) involving L matrices of different form (J and TJ) the ones with only J . Substituting L^{TJ} the operator K'_0 can be carried to the other operators leading to the substitution $Y' \to \partial'_y$ in the similarity operator and the remainder. It is the important consequence of the above discussion that the substituted similarity transformations $\Gamma(\mathcal{N})$ carry different arguments $\mathcal{N}$ on both sides.

$$\tilde{R}(\lambda) L_x^{J/i}(u-1, u+N^x+\lambda) L_y^{J/j}(v, v+1-\tilde{N}^y-\lambda) =$$

$$L_x^{J/i}(u-1+\lambda, u+N^x+\lambda) L_y^{J/j}(v-\lambda, v+1-\tilde{N}^y-\lambda)\tilde{R}(\lambda)$$

$$-\lambda(1-\lambda+v-u)(I - \frac{\partial' M X'}{\partial' \cdot X'})\tilde{R}(\lambda),$$

$$\tilde{R}(\lambda) L_y^{J/j}(v, v+1-\tilde{N}^y-\lambda) L_x^{J/i}(u-1, u+N^x+\lambda) =$$

$$L_y^{J/j}(v, v+1-\tilde{N}^y) L_x^{J/i}(u-1, u+N^x)\tilde{R}(\lambda)$$

$$-\frac{\partial'_y M X'}{\partial'_y \cdot X'}\lambda(\lambda-1+u-v+\tilde{N}^y+N^x)\tilde{R}(\lambda),$$

with

$$\tilde{R}(\lambda) = \Gamma(1-\tilde{N}^y - \sum N'_s)(X' \cdot \partial')^{-\lambda}\Gamma^{-1}(1-\tilde{N}^y-\lambda-\hat{m}_y).$$

We abbreviate $\hat{m}_y = \sum_{s \neq j} \tilde{N}_s^{y\prime}$. In the first relation we choose $\lambda = v - u + 1$ to remove the remainder and denote the second representation arguments on both sides by u_2 and v_2 and put finally $u_1 = u - 1, v_1 = v$. In this way the arguments of the Γ functions become $1 - \tilde{N}^y = v_2 - u_1$, $1 - \tilde{N}^y - v + u - 1 = v_2 - v_1$. We obtain the standard first parameter permutation relation (compare [20])

$$\mathrm{R}^1_{xy}(u_1|v_1,v_2) L^J_x(u_1,u_2) L^J_y(v_1,v_2) = L^J_x(v_1,u_2) L^J_y(u_1,v_2)\mathrm{R}^1_{xy}(u_1|v_1,v_2),$$

where

$$\mathrm{R}^1_{xy}(u_1|v_1,v_2) = \Gamma^{-1}(v_2-u_1-\hat{m}_y)(X' \cdot \partial')^{u_1-v_1}\Gamma(v_2-v_1-\hat{m}_y). \tag{16}$$

In the second relation we choose $\lambda = v - u + 1 - \tilde{N}^y - N^x$ to remove the remainder. We denote now $u_2 = u + N^x, v_2 = v + 1 - \tilde{N}^y, u_1 = u - 1, v_1 = v$. Then the arguments of the Γ functions are $1 - \tilde{N}^y = v_2 - v_1, 1 - \tilde{N}^y - v + u - 1 = u_2 - v_2$. However the first arguments in the L matrices appear reversed compared to the standard parameter permutation (compare [20]). Renaming v_1, u_1 by u_1, v_1 leads to the standard form of the second parameter permutation relation

$$\mathrm{R}^2_{yx}(u_1,u_2|v_2) L^J_y(u_1,u_2) L^J_x(v_1,v_2) = L^J_y(u_1,v_2) L^J_x(v_1,u_2)\mathrm{R}^2_{yx}(u_1,u_2|v_2),$$

where

$$\mathrm{R}^2_{yx}(u_1,u_2|v_2) = \Gamma^{-1}(v_2-u_1-\hat{m}_y)(X' \cdot \partial')^{u_2-v_2}\Gamma(u_2-u_1-\hat{m}_y). \tag{17}$$

The general Yang-Baxter operator $\mathrm{R}_{12}(u-v;\ell_1,\ell_2) = \mathrm{R}_{12}(u_1,u_2|v_1,v_2)$ acting on the tensor product of $s\ell(n+1)$ representation modules of the considered JS type with weights ℓ_1, ℓ_2 $\mathcal{U}_{1,\ell_1} \otimes \mathcal{U}_{2,\ell_2} \to \mathcal{U}_{1,\ell_2} \otimes \mathcal{U}_{2,\ell_1}$ and obeying the RLL relation with Jordan-Schwinger type L matrices

$$\mathrm{R}_{12}(u_1,u_2|v_1,v_2) L_1^{J/i}(u_1,u_2) L_2^{J/j}(v_1,v_2) = L_1^{J/i}(v_1,v_2) L_2^{J/j}(u_1,u_2)\mathrm{R}_{12}(u_1,u_2|v_1,v_2),$$

$u_1 = u - 1, u_2 = u + 2\ell_1$, factorises into the constructed parameter permutation operators,

$$\mathrm{R}_{12}(u_1,u_2|v_1,v_2) = \mathrm{R}^1_{12}(u_1|v_1,u_2)\mathrm{R}^2_{12}(u_1,u_2|v_2) = \mathrm{R}^2_{12}(v_1,u_2|v_2)\mathrm{R}^1_{12}(u_1|v_1,v_2). \tag{18}$$

This factorisation is a specific feature of the considered Jordan-Schwinger type representations. The general Yang-Baxter operator acting on generic $s\ell(n+1)$ representations decomposes into

$n+1$ factors corresponding to the $n+1$ representation labels of the L matrix of generic representation (cf. [18]).

In the case of coinciding distinguished directions $i=j$ the obtained expressions for the factor operators (16), (17) can be rewritten as

$$\mathrm{R}^1_{12}(u_1|v_1,v_2) = U_2^{-1}(v_2-u_1)\, D_{21}^{(u_1-v_1)}\, U_2(v_2-v_1), \tag{19}$$

$$\mathrm{R}^2_{12}(u_1,u_2|v_2) = U_1^{-1}(v_2-u_1)\, D_{12}^{(u_2-v_2)}\, U_1(u_2-u_1),$$

where

$$D_{12} = \prod_{s\neq i} x'_{1s} - \sum_{r\neq i} x'_{2r} \prod_{s\neq i,r} x'_{1s},$$

$$U_1(\alpha) = (\prod_{s\neq i} x'_{1s})^{-\alpha} \prod_{s\neq i} \Gamma(N'_{1s}+1)\, \Gamma^{-1}(1+\sum_{s\neq i} N'_{1s}-\alpha) = (\prod_{s\neq i} x'_{1s})^{-\alpha} V_1(\alpha).$$

In the case of rank $n=1$ the expressions coincide with the ones in [20, 16].

The complete analogy of the factorisation and Yang Baxter relations between R_{12}, R^1_{12}, R^2_{12}, L in the Jordan-Schwinger $s\ell(n+1)$ case to the case of $s\ell(2)$ implies in both cases the same relations of factorisation and commutativity between the the corresponding operators describing a spin chain: the ordinary transfer matrix, the general transfer matrix,

$$\mathrm{T}_s(u) = tr_0(P_{10}R_{10}(u)...P_{N0}R_{N0}(u))$$

and two Baxter operators

$$Q_1(u) = tr_0(P_{10}R^1_{10}(u_1|v_1,0)...P_{N0}R^1_{N0}(u_1|v_1,0)),$$

$$Q_2(u) = tr_0(P_{10}R^2_{10}(u_1,u_2|0)...P_{N0}R^2_{N0}(u_1,u_2|0)).$$

The trace in the representation space labelled by 0 with the weight denoted by spin s is well defined for generic values of $2s$. In particular we have

$$\mathrm{PT}_s(u) = \mathrm{Q}_2(u-s)\mathrm{Q}_1(u+s+1) = \mathrm{Q}_1(u+s+1)\mathrm{Q}_2(u-s).$$

P denotes the cyclic shift in the chain.

Starting with the second parameter permutation relation involving R^2_{12} and its explicit expression (17) one can derive

$$(I+r_iX_1)\mathrm{R}^2_{12}(u_1,u_2|0)L^J_1(u_1,u_2)(I-r_iX_2) = \{...\}c_i + \tag{20}$$

$$P_{X12}\mathrm{R}^2_{12}(u_1+1,u_2+1)|0) + u_1u_2I_i\mathrm{R}^2_{12}(u_1-1,u_2-1|0) +$$

$$\{u_1(I-I_i-P_{X12}) + P_{X12}B\}\,\mathrm{R}^2_{12}(u_1,u_2|0).$$

One-dimensional matrix projectors $I_i, P_{X12}, P_{X12}B$ are involved,

$$P_{X12} = \frac{X_1^{-1}M_iX_2}{(X_1^{-1}\cdot X_2)}, \quad B = (X_1^{-1}\cdot X_2)[X_1^{-1}M_iX_1N_1 - N_1X_2^{-1}M_iX_2](X_1^{-1}\cdot X_2)^{-1}.$$

In the lowest rank case $n=1$ this becomes the intermediate step towards the Baxter relation for the product of Q_2 with the ordinary transfer matrix [16]. Indeed, in this case $B=0$ and $P_{X12} = I - I_i$. The unspecified term proportional to c_i is irrelevant for this purpose.

It is not clear whether this relation can be exploited for deriving relations for global chain operators in the higher rank case in the framework of Jordan-Schwinger type representations. This could be of interest because the known treatments of Baxter relations expand over generic representations.

Journal of Physics: Conference Series **411** (2013) 012018 doi:10.1088/1742-6596/411/1/012018

5. Integer limits

We study the Yang-Baxter operator in the case $u_2 - u_1 = 2\ell_1 + 1 = m + 1 + \varepsilon$ where m is a non-negative integer. We shall see how the appearence of a $m+1$ dimensional invariant subspace in the first tensor factor is reflected in the limiting behaviour of the Yang-Baxter operator and shall obtain its restriction to the finite-dimensional invariant subspace for the first tensor factor in the limit.

In the first factorised representation (18) we have the factors $U_2^{-1}(2\ell_1 + 1), U_1(2\ell_1 + 1)$. In the limit two contributions appear naturally.

$$U_1(2\ell_1 + 1) = \prod x_{1s}'^{-m-1} \Gamma(1 + N'_{1s})$$

$$\left\{ \Pi(m \geq \hat{m}_1) \Gamma(1 + m + \varepsilon - \hat{m}_1)(-1)^{m+1-\hat{m}_1} \ \varepsilon + \Pi(\hat{m}_1 > m) \frac{1}{\Gamma(\hat{m}_1 - m - \varepsilon)} \right\},$$

$$U_2^{-1}(2\ell_1 + 1) = \left\{ \Pi(m \geq \hat{m}_2) \frac{(-1)^{m+1-\hat{m}_2}}{\Gamma(1 + m + \varepsilon - \hat{m}_2) \ \varepsilon} + \Pi(\hat{m}_2 > m) \Gamma(\hat{m}_2 - m - \varepsilon) \right\}$$

$$\prod \frac{1}{\Gamma(1 + N'_{2s})} x_{1s}'^{m+1}.$$

Here we abbreviate $\hat{m}_1 = \sum_{s \neq i} N'_{1s}$ and $\Pi(...)$ stand for projectors with the projection rule indicated in the argument.

We obtain the Yang-Baxter operator $\mathbf{R}_{12}$ restricted to the finite dimensional subspace in the first tensor factor at $2\ell_1 = m$ as

$$\mathrm{P}_{12} \mathbf{R}_{12}(u - v | \frac{n}{2}, \ell_2) = \lim_{\varepsilon \to 0} \mathrm{R}_{12}(u - v | \frac{1}{2}(m + \varepsilon), \ell_2) \Pi_1^m = \tag{21}$$

$$\Pi_2^m \frac{(-1)^{\hat{m}_2}}{\Gamma(1 + m - \hat{m}_2) \prod \Gamma(1 + N_{2s})} (1 - \sum \frac{x_{\prime 1r}}{x_{\prime 2r}})^{u_1 - v_1} V_2(u_2 - v_1)$$

$$V_1^{-1}(v_2 u_1)(1 - \sum \frac{x_{2r}}{x_{1r}})^{u_2 - v_2} \prod \Gamma(1 + N_{1s}) \ \Pi_1^m \Gamma(1 + m - \hat{m}_1)(-1)^{\hat{m}_1}.$$

Here Π_1^m denotes the projector on the subspace spanned by $\prod_{s \neq i} x_{1s}'^{m_s}$ with $\sum_{s \neq i} m_s \leq m$.

In the case $m = 1$ we expect that $\mathbf{R}_{12}$ reproduces the L matrix in the symmetry basis,

$$\mathbf{R}_{12}(u - v | \frac{1}{2}, \ell_2) = C \cdot L_2(u - v + \frac{1}{2} | \ell_2), \quad C = -\frac{\Gamma(u - v - \frac{1}{2} - \ell_2)}{\Gamma(\frac{1}{2} - u + v - \ell_2)} \tag{22}$$

We calculate easily the matrix element of $\mathbf{R}_{12}$ with the lowest weight state 1 and identify the coefficient C,

$$\mathbf{R}_{12} \cdot 1 = \frac{\Gamma(u_2 - v_2 - 1}{\Gamma(v_1 - u_1 - 1)} = (u_1 - v_1 + 1) \ C.$$

(22) can be checked by proceeding this calculation with the action on $1 \cdot x_{2s}^p$ and $x_{1r} \cdot x_{2s}^p$. In the calculation we notice that besides of the explicit projectors the factors $\prod \Gamma^{-1}(1 + N_s)$ restrict the powers of the monomials; negative powers are removed. The generalisation of the L matrix to the case of other values of m can be obtained in the same way.

We have seen that the restriction to the finite-dimensional subspace at integer values of 2ℓ goes in analogy to the rank $n = 1$ case. The analogy ends if regarding the complement to the invariant finite-dimensional subspace. Unlike the rank 1 case it is not simply related to just one Jordan-Schwinger type representation with $\ell = \frac{1}{2}m$ replaced by another value. This implies that the arguments for deriving Baxter relations cannot be copied from the rank 1 case in a straightforward way.

6. Degeneracy limits

Consider the Lax matrix (10) and identify $L^{J/i\pm}$.

$$L^{J/i}(u_1,u_2) = L^{J/i-)}(u_2)(Iu_1 - I_i(u_1-1))(I+r_iX') = (I-r_iX')(I+I_i(u_2-1))\, L^{J/i+}(u_1), \quad (23)$$

$$L^{J/i-}(u_2) = \lim_{u_1\to\infty} L^{(i)}(u_1,u_2)(Iu_1 - I_i(u_1-1))^{-1},$$

$$L^{J/i+}(u_1) = \lim_{u_2\to\infty} (I + I_i(u_2-1))^{-1} L^{(i)}(u_1,u_2).$$

By the limit relations it is clear that $L^{J/i\pm}$ obey the fundamental RLL relation (4) and thus their matrix elements generate algebra representations of the Yangian different from the ordinary one generated by the elements of L^J.

$L^{J/i-}$ coincide with the elementary degenerate L matrices used by Bazhanov et al. [19] as "partonic" building blocks. Their Baxter operator construction starts from the decomposition of a generic $s\ell(n+1)$ representation L matrix in terms of a product of $n+1$ L^{i-} matrices. The generic representation L matrix can be also obtained from the considered Jordan-Schwinger L matrices by iteration as discussed in [9, 10]. The relation (23) of L^J to L^{J-} shows how the connection of our approach to the one by Bazhanov et al. [17, 19] extends from the rank $n=1$ case discussed in [16] to higher rank.

R operators intertwining the degenerate $L^{\pm}$ with L and among each other can be obtained similar to the $s\ell(2)$ case. Consider the first parameter permutation RLL relation (16) in the case of coinciding distinguished directions in the asymptotics $v_1 \to \infty$.

$$\mathrm{R}^1_{12}(u_1|v_1,v_2)L^J_1(u_1,u_2)L^J_2(v_1,v_2) = L^J_1(v_1,u_2)L^J_2(u_1,v_2)\mathrm{R}^1_{12}(u_1|v_1,v_2),$$

$$L^{J/i}(v_1,v_2) = L^{J/i-}(v_2)(v_1 I - I_i(v_1-1))(1+\mathcal{O}(v_1^{-1})).$$

The symmetry of the L matrices implies

$$(v_1 I - I_i(v_1-1))L_2(u_1,v_2)(v_1 I - I_i(v_1-1))^{-1} = v_1^{-\hat{m}_2} L_2(u_1,v_2) v_1^{\hat{m}_2}, \ \hat{m}_2 = \sum_{s\neq i} N'_{2s}.$$

In this way we obtain

$$\mathrm{r}^+_{12}(u_1-v_2)L^{J/i}_1(u_1,u_2)L^{J/i-}_2(v_2) = L^{J/i-}_1(u_2)L^J_2(u_1,v_2)\mathrm{r}^+_{12}(u_1-v_2),$$

where

$$v_1^{\hat{m}_2}\mathrm{R}^1_{12}(u_1|v_1,v_2) \to \mathrm{r}^+_{12}(u_1-v_2).$$

With the explicit expression (16) we obtain

$$\mathrm{r}^+_{12}(u) = \Gamma(1+\hat{m}_2+u)\exp(\sum_{r\neq i} x_{1r}\partial_{2r}). \quad (24)$$

The result appears quite similar to the rank $n=1$ case and this analogy extends to the other degenracy limit Yang-Baxter operators and the relations among them.

$$\mathrm{R}^+_{12}(u_1,u_2|v_2)L^{J/i}_1(u_1,u_2)L^{J/i-}_2(v_2) = L^{J/i-}_1(v_2)L^J_2(u_1,u_2)\mathrm{R}^+_{12}(u_1,u_2|v_2),$$

where

$$v_1^{\hat{m}_2}\mathrm{R}_{12}(u_1,u_2|v_1,v_2) \to \mathrm{R}^+_{12}(u_1,u_2|v_2) = \mathrm{r}^+_{12}(u_1-u_2)\mathrm{R}^2_{12}(u_1,u_2|v_2).$$

$$\mathrm{R}^-_{12}(u_1,u_2|v_1)L^{J/i}_1(u_1,u_2)L^{J/i+}_2(v_1) = L^{J/i-}_1(v_1)L^J_2(u_1,u_2)\mathrm{R}^-_{12}(u_1,u_2|v_1),$$

where

$$\mathrm{R}_{12}(u_1, u_2|v_1, v_2)v_2^{-\hat{m}_1} \rightarrow \mathrm{R}^-_{12}(u_1, u_2|v_1) = \mathrm{R}^1_{12}(u_1|v_1, u_2)\mathrm{r}^-_{12}(u_1 - u_2),$$

$$\mathrm{r}^-_{12}(u) = \exp(-\sum_{r \neq i} x_{2r}\partial_{1r})\frac{(-1)^{\hat{m}_1}}{\Gamma(1 + \hat{m}_1 + u)}.$$

Operators describing a spin chain can be built from these Yang-Baxter operators as well. The experience of the rank $n = 1$ case tells us that here the trace definition should include a regularisation,

$$\mathrm{Q}_\pm(u) = tr_0[q^{\hat{m}_0}\mathrm{P}_{10}\mathrm{R}^\pm_{10}(u_1, u_2|0)...\mathrm{P}_{N0}\mathrm{R}^\pm_{N0}(u_1, u_2|0)].$$

The commutativity and factorisation properties hold, in particular

$$\frac{1}{1 - q^n}\mathrm{T}_s(u|q) = \mathrm{Q}_+(u - s - 1)\mathrm{Q}_-(u + s) = \mathrm{Q}_-(u + s)\mathrm{Q}_+(u - s - 1)$$

where $\mathrm{T}_s(u|q)$ is the q regularised version of the general transfer operator.

Acknowledgments

The author is grateful to D. Chicherin, S. Derkachov, D. Karakhanyan and P. Valinevich for collaboration on related problems.

References

[1] Jordan P 1935 *Zeitschr. f. Physik* **94** 331
[2] Schwinger J 1952 *notes reprinted in: Quantum Mechanics of Angular Momentum* Biedenharn L C and van Dam H (eds.) (London: Academic Press 1965)
[3] Holstein T and Primakoff H 1940 *Phys. Rev.* **58** 1098
[4] Gelfand I M and Naimark M A 1950 *Unitary representations of the classical groups, Trudy Math. Inst. Steklov, Vol.* **36**, (Moscow-Leningrad); (German translation: 1957 Berlin:Akademie Verlag)
[5] Borel A and Weil A 1954 *Sem. Bourbaki, Representations lineaires et espaces homogenes Kählerians des groupes de Lie compactes*
[6] Hecht K T 1987 *The vector coherent state method and its application to problems of higher symmetry. Lecture Notes in Physics* **290** (Berlin:Springer)
[7] Biedenharn L C and Lohe M A 1992 *Commun. Math. Phys.* **146** 483; 1995 *Quantum group symmetry and q-tensor algebras* (Singapore:World Scientific)
[8] Dobrev V K, Truini P and Biedenharn L C 1994 *J. Math. Phys.* **35** 6058 [arXiv:q-alg/9502001]
[9] Derkachov S, Karakhanyan D, Kirschner R and Valinevich P 2008 *Lett. Math. Phys.* **85** 221 [arXiv:0805.4724 [hep-th]]
[10] Karakhanyan D and Kirschner R 2010 *SIGMA* **6** 029 [arXiv:0910.5144 [hep-th]]
[11] Kashaev R M, Mangazeev V V, and Nakanishi T 1991 *Nucl. Phys. B* **362** 563
[12] Bazhanov V V, Kashaev R M, Mangazeev V V and Stroganov Yu G 1991 *Commun. Math. Phys* **138** 393
[13] Date E, Jimbo M, Miki K and Miwa T 1991 *Commun. Math. Phys.* **137** 133
[14] Tarasov V 1993 *Commun. Math. Phys.* **158** 459 [arXiv:hep-th/9211105]
[15] Drummond J M, Henn J, Korchemsky G P and Sokatchev E 2010 *Nucl. Phys. B* **828** 317 [arXiv:0807.1095 [hep-th]];
Drummond J M, Henn J and Plefka J 2009 *JHEP* **0905** 046 [arXiv:0902.2987 [hep-th]];
Drummond J M and Ferro L 2010 *JHEP* **1007** 027 [arXiv:1001.3348 [hep-th]]
[16] Chicherin D, Derkachov S, Karakhanyan D and Kirschner R 2012 *Nucl. Phys. B* **854** 393 and 433 [arXiv:1106.4991 [hep-th]] and [arXiv:1107.0643 [hep-th]]
[17] Bazhanov V V, Lukowski T, Meneghelli C and Staudacher M 2010 *J. Stat. Mech.* **1011** P11002 [arXiv:1005.3261 [hep-th]]
[18] Derkachov S E and Manashov A N 2006 *SIGMA* **2** 084 [arXiv:nlin/0612003]; 2011 *Lett. Math. Phys.* **97** 185 [arXiv:1008.4734 [nlin.SI]]
[19] Bazhanov V V, Frassek R, Lukowski T, Meneghelli C and Staudacher M 2011 *Nucl. Phys. B* **850** 148 [arXiv:1010.3699 [math-ph]]
[20] Derkachov S, Karakhanyan D and Kirschner R 2007 *Nucl. Phys. B* **785** 263 [arXiv:hep-th/0703076]

XXth International Conference on Integrable Systems and Quantum Symmetries (ISQS-20) IOP Publishing
Journal of Physics: Conference Series **411** (2013) 012019 doi:10.1088/1742-6596/411/1/012019

$\mathcal{N}$=2 supersymmetric $\mathbb{S}^2 \to \mathbb{CP}^3 \to \mathbb{S}^4$ fibration viewed as superparticle mechanics

N. Kozyrev[1], **S. Krivonos**[1] **and O. Lechtenfeld**[2]

[1] Bogoliubov Laboratory of Theoretical Physics, JINR, 141980 Dubna, Russia
[2] Institute of Theoretical Physics and Riemann Center for Geometry and Physics, Leibniz Universität Hannover, Appelstrasse 2, D-30167 Hannover, Germany

Abstract. We discuss a Hamiltonian reduction procedure that relates the mechanics of an $\mathcal{N}$=2 particle on $\mathbb{CP}^3$ with the motion of such a superparticle on $\mathbb{S}^4$ in the presence of an instanton background. The key ingredients of the bosonic fibration $\mathbb{S}^2 \to \mathbb{CP}^3 \to \mathbb{S}^4$ are recalled from the viewpoint of particle mechanics on $\mathbb{CP}^3$. We describe an $\mathcal{N}$=2 supersymmetric extension which allows for a Hamiltonian reduction. The $\mathbb{S}^2$ degrees of freedom are encoded in the supercharges via SU(2) currents. Finally, we present the Hamiltonian of our system and its superfield Lagrangian.

1. Introduction

The mechanics of a particle moving on a manifold E is of a special nature if E allows for a fibration, $F \to E \to B$, i.e. if E is fibered over a base manifold B with fibers F. If we represent the coordinates of the total manifold E in terms of base and fiber coordinates, then the kinetic energy splits into a base part and a fiber part. If the fibration is nontrivial, e.g., if the total space is not a product of base and fiber, then particular interaction terms will appear. These terms are proportional to the particle velocities and, therefore, describe an interaction with some (possibly non-Abelian) magnetic field.

A prototypical fibration is $\mathbb{S}^2 \to \mathbb{CP}^3 \to \mathbb{S}^4$, with $\mathbb{CP}^3$ as total space, $\mathbb{S}^4$ as base and $\mathbb{S}^2$ as fiber. The problem we address in this letter is the construction of a supersymmetric extension of this fibration, viewed as a Hamiltonian reduction in $\mathcal{N}$=2 supersymmetric mechanics. Immediate questions are:

- Why do we limit ourselves to $\mathcal{N}$=2 supersymmetry?
- Does the construction require new features?

The first question is easy to answer: We are not aware of any suitable supersymmetric mechanics with $\mathcal{N}$>2 describing particle motion on $\mathbb{S}^{4k+3}$ and $\mathbb{HP}^k$. The main reason is the absence of a complex structure on these manifolds, which is crucial for $\mathcal{N}$=4 supersymmetry.[1] Keeping in mind a possible extension of our approach to the fibrations $\mathbb{S}^{4k+3} \to \mathbb{CP}^{2k+1} \to \mathbb{HP}^k$ we limit ourselves to $\mathcal{N}$=2 supersymmetry.

The answer to the second question comes from the paper [1] where the particular fibration $\mathbb{S}^2 \to \mathbb{CP}^3 \to \mathbb{HP}^1$ has been considered in the bosonic case. The crucial ingredient implicitly used in this paper is the representation of $\mathbb{CP}^3$ as the coset SO(5)/U(2) instead of the commonly used SU(4)/U(3) one. This parametrization asked for a special splitting of the standard complex $\mathbb{CP}^3$ coordinates $\{z_\alpha, \bar z^\alpha \mid \alpha = 1,2,3\}$, transforming under the fundamental representation of SU(3), into the sets $\{w_a, \bar w^a \mid a = 1,2\}$ and $\{u, \bar u\}$, which form an SU(2) doublet and singlet, respectively, in the coset SO(5)/U(2).

Let us recall the construction of [1] in more detail. Starting from the Lagrangian of a free bosonic particle on $\mathbb{CP}^3$,

$$\mathcal{L}_{CP3} = g^\alpha{}_\beta \dot z^\beta \dot{\bar z}_\alpha, \qquad g^\alpha{}_\beta = \frac{\delta^\alpha{}_\beta}{1+z^\gamma \bar z_\gamma} - \frac{z^\alpha \bar z_\beta}{\left(1+z^\gamma \bar z_\gamma\right)^2}, \tag{1.1}$$

and splitting the three complex coordinates z^α on $\mathbb{CP}^3$ into coordinates on $\mathbb{S}^4$ and $\mathbb{S}^2$ via [2]

$$\begin{aligned} z^1 &= \bar w^1 - \bar u\, w_2, & z^2 &= -\bar w^2 - \bar u\, w_1, & z^3 &= \bar u, \\ \bar z_1 &= w_1 - u\, \bar w^2, & \bar z_2 &= -w_2 - u\, \bar w^1, & \bar z_3 &= u, \end{aligned} \tag{1.2}$$

one may rewrite the Lagrangian (1.1) in the form

$$\mathcal{L}_{CP3} = \frac{\dot w_a \dot{\bar w}^a}{\left(1+w\cdot\bar w\right)^2} + \frac{\left(\dot u - \mathcal{A}\right)\left(\dot{\bar u} - \bar{\mathcal{A}}\right)}{\left(1+u\bar u\right)^2}, \tag{1.3}$$

where

$$\mathcal{A} = \frac{w_a \dot w^a - u\left(w_a \dot{\bar w}^a - \dot w_a \bar w^a\right) + u^2 \bar w_a \dot{\bar w}^a}{1+w\cdot\bar w}. \tag{1.4}$$

The corresponding Hamiltonian,

$$\mathcal{H}_{CP3} = \left(1+w\cdot\bar w\right)^2 P^a \bar P_a + \left(T\bar T - U^2\right) \tag{1.5}$$

[1] The cases of $\mathbb{S}^2$, $\mathbb{S}^3$ and $\mathbb{S}^4 \sim \mathbb{HP}^1$ are rather exceptional, and the corresponding $\mathcal{N}$=4 supersymmetric mechanics have been constructed in [2, 3, 4, 5, 6].

[2] $\mathbb{S}^2$ is a complex manifold, thus any of z^α coordinates defines a chart on it. We choose z_3 for this purpose.

with

$$P^a = p^a + \frac{U\,\bar{w}^a - T\,w^a}{1 + w\cdot\bar{w}}, \qquad \bar{P}_a = \bar{p}_a - \frac{U\,w_a - \overline{T}\,\bar{w}_a}{1 + w\cdot\bar{w}} \tag{1.6}$$

and

$$T = p_u + \bar{u}^2\bar{p}_u, \qquad \overline{T} = \bar{p}_u + u^2 p_u, \qquad U = up_u - \bar{u}\bar{p}_u, \tag{1.7}$$

displays an interesting feature: the coordinates $\{u, \bar{u}\}$ and corresponding momenta $\{p_u, \bar{p}_u\}$ enter the Hamiltonian only through the SU(2) currents $T, \overline{T}$ and U. The latter satisfy

$$\{U,\, T\} = T, \qquad \{U,\, \overline{T}\} = -\overline{T}, \qquad \{T, \overline{T}\} = -2U \tag{1.8}$$

with respect to the standard Poisson brackets

$$\{w_a, p^b\} = \delta^b_a, \qquad \{\bar{w}^b, \bar{p}_a\} = \delta^b_a, \qquad \{u, p_u\} = 1, \qquad \{\bar{u}, \bar{p}_u\} = 1. \tag{1.9}$$

This SU(2) is not a symmetry of the system, but the corresponding Casimir operator

$$\mathcal{C}_{su(2)} = T\overline{T} - U^2 \tag{1.10}$$

Poisson-commutes with the $\mathbb{CP}^3$ Hamiltonian (1.5) and, therefore, can be fixed to a constant m. The currents T, $\overline{T}$ and U do not Poisson-commute with (1.5), but their brackets are closed, and they fully encode the $\mathbb{S}^2$ degrees of freedom in the Hamiltonian. Hence, we may consistently perform the reduction $\{T,\, \overline{T},\, U\} \to 0$ and arrive at a purely $\mathbb{S}^4$ Hamiltonian.

In what follows, we will construct an $\mathcal{N}$=2 supersymmetric extension of CP^3 mechanics, in which the $\mathbb{S}^2$ coordinates and momenta enter the supercharges (and hence the Hamiltonian) only through the SU(2) currents (1.7). For demonstration we start from the simplest example of $\mathcal{N}$=2 supersymmetric mechanics on $\mathbb{S}^2$.

2. $\mathcal{N}$=2 supersymmetric mechanics on $\mathbb{S}^2$

2.1. The standard approach

In the standard description of $\mathcal{N}$=2 mechanics on the sphere $\mathbb{S}^2$, which is based on chiral superfields, we need, besides bosonic coordinates $\{u, \bar{u}\}$ and the corresponding momenta $\{p_u, \bar{p}_u\}$, a pair of fermionic coordinates $\{\xi, \bar{\xi}\}$ subject to the standard brackets

$$\{\xi, \bar{\xi}\} = \mathrm{i}. \tag{2.1}$$

It is straightforward to check that the supercharges

$$Q = (1 + u\bar{u})\, p_u\bar{\xi}, \qquad \overline{Q} = (1 + u\bar{u})\, \bar{p}_u\xi, \tag{2.2}$$

form an $\mathcal{N}$=2 super Poincare algebra

$$\{Q, Q\} = \{\overline{Q}, \overline{Q}\} = 0, \qquad \{Q, \overline{Q}\} = \mathrm{i}H \tag{2.3}$$

with the Hamiltonian

$$H = (1 + u\bar{u})^2\Big(p_u - \mathrm{i}\frac{\bar{u}\xi\bar{\xi}}{1 + u\bar{u}}\Big)\Big(\bar{p}_u + \mathrm{i}\frac{u\xi\bar{\xi}}{1 + u\bar{u}}\Big). \tag{2.4}$$

One may check that the supercharges (2.2) and the Hamiltonian (2.4) commute with the SU(2) generators

$$\tilde{T} = p_u + \bar{u}^2\bar{p}_u + \mathrm{i}\bar{u}\xi\bar{\xi}, \qquad \tilde{\overline{T}} = \bar{p}_u + u^2 p_u - \mathrm{i}u\xi\bar{\xi}, \qquad \tilde{U} = up_u - \bar{u}\bar{p}_u - \mathrm{i}\xi\bar{\xi} \tag{2.5}$$

and that, moreover,

$$H = \tilde{\mathcal{C}}_{su(2)} = \tilde{T}\tilde{\overline{T}} - \tilde{U}^2. \tag{2.6}$$

Clearly the supercharges (2.2) can not be expressed through the SU(2) currents (2.2) alone. Thus, the standard formulation of $\mathcal{N}$=2 supersymmetric mechanics on $\mathbb{S}^2$ is not suitable for our purposes.

2.2. *Extended version*

The proper and seemingly unique way towards a suitable $\mathcal{N}$=2 supersymmetric mechanics on $\mathbb{S}^2$ extends the number of fermionic variables. We introduce a second pair of fermionic coordinates $\{\eta, \bar{\eta}\}$ obeying the canonical brackets

$$\{\eta, \bar{\eta}\} = \mathrm{i}. \tag{2.7}$$

It is easily checked that the (extended) supercharges

$$\mathcal{Q} = T\bar{\xi} + U\eta - \mathrm{i}\eta\xi\bar{\xi}, \qquad \overline{\mathcal{Q}} = \overline{T}\xi - U\bar{\eta} + \mathrm{i}\bar{\eta}\xi\bar{\xi} \tag{2.8}$$

constructed from the bosonic currents (1.7) also generate an $\mathcal{N}$=2 super Poincare algebra (2.3), but with the purely bosonic Hamiltonian

$$\tfrac{1}{\mathrm{i}}\{\mathcal{Q}, \overline{\mathcal{Q}}\} = \mathcal{H} = \mathcal{C}_{su(2)} = T\overline{T} - U^2, \tag{2.9}$$

which is just the bosonic part of H. This is the version of supersymmetric mechanics we are looking for.

The Hamiltonian (2.9) resembles the one of $\mathcal{N}$=4 supersymmetric mechanics on $\mathbb{S}^2$ constructed from a non-linear chiral supermultiplet [2, 7]. This similarity is not occasional: one easily finds two additional supercharges

$$\mathcal{S} = T\bar{\eta} - U\xi + \mathrm{i}\xi\eta\bar{\eta}, \qquad \overline{\mathcal{S}} = \overline{T}\eta + U\bar{\xi} - \mathrm{i}\bar{\xi}\eta\bar{\eta} \tag{2.10}$$

which commute with the supercharges $\mathcal{Q}$ and $\overline{\mathcal{Q}}$ in (2.8) and form an $\mathcal{N}$=2 super Poincare algebra with the same Hamiltonian,

$$\tfrac{1}{\mathrm{i}}\{\mathcal{S}, \overline{\mathcal{S}}\} = \mathcal{H} = \mathcal{C}_{su(2)} = T\overline{T} - U^2. \tag{2.11}$$

Thus, we have $\mathcal{N}$=4 supersymmetric mechanics on $\mathbb{S}^2$ exactly the same as in [2, 7]. We stress that from the $\mathcal{N}$=2 supersymmetric point of view one needs a general complex $\mathcal{N}$=2 bosonic superfield (to have four fermions) instead of chiral one more commonly used.

3. A novel $\mathcal{N}$=2 supersymmetric mechanics on $\mathbb{CP}^3$

To construct an $\mathcal{N}$=2 supersymmetric extension of $\mathbb{CP}^3$ mechanics with the Hamiltonian (1.5), one should introduce, besides the fermions $\{\xi, \bar{\xi}, \eta, \bar{\eta}\}$ of the previous section, a doublet of fermionic coordinates $\{\psi_a, \bar{\psi}^a\}$ with canonical brackets

$$\left\{\psi_a, \bar{\psi}^b\right\} = \mathrm{i}\delta_a^b, \qquad (\psi_a)^\dagger = \bar{\psi}^a. \tag{3.1}$$

These fermionic coordinates accompany the bosonic coordinates $\{w_a, \bar{w}^a\}$ to render our system $\mathcal{N}$=2 supersymmetric.

We are ready to present the supercharges,

$$Q = (1 + w\cdot\bar{w})\, P^a\psi_a - \mathrm{i}\,\psi_a\bar{w}^a\,\psi_b\bar{\psi}^b + \alpha\left(T\bar{\xi} + U\eta - \mathrm{i}\,\eta\,\xi\bar{\xi}\right) + \mathrm{i}\psi_a w^a\eta\xi - \tfrac{\mathrm{i}}{\alpha}\psi_a\psi^a\xi - \mathrm{i}\psi_a\bar{w}^a\xi\bar{\xi},$$
$$\overline{Q} = (1 + w\cdot\bar{w})\, \overline{P}_a\bar{\psi}^a + \mathrm{i}\, w_a\bar{\psi}^a\,\psi_b\bar{\psi}^b + \alpha\left(\overline{T}\xi - U\bar{\eta} + \mathrm{i}\,\bar{\eta}\,\xi\bar{\xi}\right) - \mathrm{i}\bar{w}_a\bar{\psi}^a\bar{\eta}\bar{\xi} - \tfrac{\mathrm{i}}{\alpha}\bar{\psi}_a\bar{\psi}^a\bar{\xi} + \mathrm{i}w_a\bar{\psi}^a\xi\bar{\xi} \tag{3.2}$$

where the momenta $\{P^a, \overline{P}_a\}$ were defined in (1.6), and α is a real parameter. Its sign is irrelevant, due to the involution $\{\eta, \bar{\eta}, \xi, \bar{\xi}\} \to \{-\eta, -\bar{\eta}, -\xi, -\bar{\xi}\}$ which maps $Q(\alpha)$ to $Q(-\alpha)$.

The supercharges (3.2) commute as needed,

$$\{Q, Q\} = \{\overline{Q}, \overline{Q}\} = 0, \qquad \{Q, \overline{Q}\} = \mathrm{i}H, \tag{3.3}$$

with the Hamiltonian

$$H = (1+w\cdot\bar{w})^2 \left(P^a - \mathrm{i}A^a\right)\left(\overline{P}_a + \mathrm{i}\overline{A}_a\right) + \alpha^2\left(T + \tfrac{\mathrm{i}}{\alpha}B\right)\left(\overline{T} - \tfrac{\mathrm{i}}{\alpha}\overline{B}\right) - \alpha^2\left(U + \tfrac{\mathrm{i}}{2\alpha^2}B_u\right)^2 + H_{4f}. \tag{3.4}$$

Here,

$$\begin{aligned} A^a &= \frac{1}{1+w\cdot\bar{w}}\left(\tfrac{2}{\alpha}\psi^a\xi + w^a\xi\eta + \bar{w}^a(\xi\bar{\xi} + \psi_b\bar{\psi}^b) - 2\bar{w}^b\psi_b\bar{\psi}^a\right), \\ \overline{A}_a &= \frac{1}{1+w\cdot\bar{w}}\left(\tfrac{2}{\alpha}\bar{\psi}_a\bar{\xi} + \bar{w}_a\bar{\xi}\bar{\eta} + w_a(\xi\bar{\xi} + \psi_b\bar{\psi}^b) + 2w_b\bar{\psi}^b\psi_a\right), \\ B &= \bar{w}^a\bar{\psi}_a\eta, \qquad \overline{B} = w_a\psi^a\bar{\eta}, \qquad B_u = 2\psi_a\bar{\psi}^a - \alpha w_a\psi^a\xi - \alpha\bar{w}^a\bar{\psi}_a\bar{\xi}, \end{aligned} \tag{3.5}$$

and the four-fermion term takes the form

$$\begin{aligned} H_{4f} = {} & \left(\alpha w_a\psi^a\xi + \alpha\bar{w}^a\bar{\psi}_a\bar{\xi} + 2w_a\bar{w}^a\psi_b\bar{\psi}^b - 2\bar{w}^a\psi_a w_b\bar{\psi}^b\right)\eta\bar{\eta} + \\ & \left(\tfrac{3}{2}\bar{w}^a\psi_a w_b\bar{\psi}^b - \tfrac{3}{2}w_a\bar{w}^a\psi_b\bar{\psi}^b - 2\psi_a\bar{\psi}^a\right)\xi\bar{\xi} + \\ & \left(-\xi\bar{\eta} - \tfrac{3}{2\alpha}w_b\bar{\psi}^b\xi + \tfrac{1}{\alpha}\bar{w}^b\bar{\psi}_b\bar{\eta}\right)\psi_a\psi^a + \left(\bar{\xi}\eta + \tfrac{3}{2\alpha}\bar{w}^b\psi_b\bar{\xi} + \tfrac{1}{\alpha}w_b\psi^b\eta\right)\bar{\psi}^a\bar{\psi}_a + \\ & + \tfrac{1}{\alpha^2}(1 - 2\alpha^2 + \alpha^2 w\cdot\bar{w})\,\psi_a\bar{\psi}^a\,\psi_b\bar{\psi}^b. \end{aligned} \tag{3.6}$$

The bosonic part of the Hamiltonian H (3.4) reads

$$H_{bos} = (1+w\cdot\bar{w})^2\,P^a\overline{P}_a + \alpha^2\left(T\overline{T} - U^2\right). \tag{3.7}$$

For α=1 this Hamiltonian coincides with the Casimir operator of the SU(4) group and, therefore, with the Hamiltonian of bosonic $\mathbb{CP}^3$ mechanics. Another value, $\alpha=\sqrt{2}$, corresponds to the Casimir operator of SO(5).

Clearly, the reduction

$$\{T, \overline{T}, U, \eta, \bar{\eta}, \xi, \bar{\xi}\} \to 0 \tag{3.8}$$

proceeds nicely and yields the reduced supercharges

$$Q_{red} = (1+w\cdot\bar{w})\,p_w^a\psi_a - \mathrm{i}\,\psi_a\bar{w}^a\,\psi_b\bar{\psi}^b, \qquad \overline{Q}_{red} = (1+w\cdot\bar{w})\,\bar{p}_{wa}\bar{\psi}^a + \mathrm{i}\,w_a\bar{\psi}^a\,\psi_b\bar{\psi}^b. \tag{3.9}$$

They properly commute to the reduced Hamiltonian

$$\tfrac{1}{\mathrm{i}}\left\{Q_{red}, \overline{Q}_{red}\right\} = H_{red} = (1+w\cdot\bar{w})^2\left(p_w^a - \mathrm{i}\hat{A}^a\right)\left(\bar{p}_{wa} + \mathrm{i}\hat{\overline{A}}_a\right) + (-2 + w\cdot\bar{w})\,\psi_a\bar{\psi}^a\,\psi_b\bar{\psi}^b \tag{3.10}$$

with

$$\hat{A}^a = \frac{1}{1+w\cdot\bar{w}}\left(\bar{w}^a\psi_b\bar{\psi}^b - 2\bar{w}^b\psi_b\bar{\psi}^a\right), \qquad \hat{\overline{A}}_a = \frac{1}{1+w\cdot\bar{w}}\left(w_a\psi_b\bar{\psi}^b + 2w_b\bar{\psi}^b\psi_a\right). \tag{3.11}$$

We have produced a $\mathcal{N}$=2 supersymmetric extension of mechanics on $\mathbb{S}^4$.

4. Superfield description

4.1. Component Lagrangian

Before constructing a superfield action for the system presented, it is instructive to find the component Lagrangian from our Hamiltonian (3.4) at α=1. This can be done by performing a

Legendre transformation over the bosonic variables and by adding fermionic kinetic terms which produce the Dirac brackets (2.1), (2.7) and (3.1). The resulting Lagrangian reads

$$L = \frac{\dot w_a \dot{\bar w}^a}{(1+w\cdot\bar w)^2} + \frac{\mathrm{i}}{2}(\dot\psi_a\bar\psi^a - \psi_a\dot{\bar\psi}^a) + \frac{\mathrm{i}}{2}(\dot\xi\bar\xi - \xi\dot{\bar\xi}) + \frac{\mathrm{i}}{2}(\dot\eta\bar\eta - \eta\dot{\bar\eta}) + \\ + \mathrm{i}A^a\dot w_a - \mathrm{i}\bar A_a\dot{\bar w}^a + \frac{(\dot u+\Lambda)(\dot{\bar u}+\bar\Lambda)}{(1+u\bar u)^2} - B\bar B - \frac{B_u^2}{4} - H_{4f}, \tag{4.12}$$

where

$$\Lambda = \frac{(w^a - u\bar w^a)\,\dot w_a + (uw_a - u^2\bar w_a)\,\dot{\bar w}^a}{1+w\cdot\bar w} - \mathrm{i}\left(Bu^2 - \bar B - uB_u\right), \\ \bar\Lambda = \frac{(\bar u^2 w^a + \bar u\bar w^a)\,\dot w_a - (\bar u w_a + \bar w_a)\,\dot{\bar w}^a}{1+w\cdot\bar w} - \mathrm{i}\left(B - \bar B\bar u^2 + \bar u B_u\right). \tag{4.13}$$

Since the key ingredient for the superfield description of the Lagrangian (4.12) is a superfield Lagrangian for the extended version of $\mathcal{N}$=2 mechanics on $\mathbb{S}^2$ discussed in Subsection 2.2, we begin with this case.

4.2. Superfield action for $\mathbb{S}^2$

As we already stressed, chiral superfields are not sufficient for our case. Thus we start from two general $\mathcal{N}$=2 superfields $\{u,\bar u\}$ with an ansatz for the action,

$$S_{S2} = -\int \mathrm{d}t\mathrm{d}\theta\mathrm{d}\bar\theta\left(F_1 Du\bar D\bar u + F_2 D\bar u\bar D u\right). \tag{4.14}$$

The superfields $\{u,\bar u\}$ are unconstrained and, therefore, contain the following components,

$$u = u|,\quad \bar u = \bar u|,\quad \hat\xi = i\bar D\bar u|,\quad \hat{\bar\xi} = iDu|,\quad \hat\eta = iD\bar u|,\quad \hat{\bar\eta} = i\bar Du|,\quad A = \tfrac12\left[D,\bar D\right]u|,\quad \bar A = \tfrac12\left[D,\bar D\right]\bar u|, \tag{4.15}$$

where, as usual, | denotes the $\theta,\bar\theta \to 0$ limit.

Our task is to fix the arbitrary functions F_1 and F_2 entering (4.14) in such a way as to get the proper component Lagrangian [3]

$$\mathcal{L}_{S2} = \frac{\dot u\dot{\bar u}}{(1+u\bar u)^2} + \frac{\mathrm{i}}{2}(\dot\xi\bar\xi - \xi\dot{\bar\xi}) + \frac{\mathrm{i}}{2}(\dot\eta\bar\eta - \eta\dot{\bar\eta}) \tag{4.16}$$

after integration over θ and $\bar\theta$ in (4.14) and elimination of the auxiliary components A and $\bar A$ (4.15). Finally, one has to determine the relation between $\{\hat\xi,\hat{\bar\xi},\hat\eta,\hat{\bar\eta}\}$ and $\{\xi,\bar\xi,\eta,\bar\eta\}$. Comparing the transformations properties of our superfields $\{u,\bar u\}$ to those which follow from our explicit construction of supercharges in (2.8), one finds that

$$\hat\xi = \sqrt2(\xi + \bar u\bar\eta),\qquad \hat{\bar\xi} = \sqrt2(\bar\xi + u\eta),\qquad \hat\eta = -\sqrt2(\bar u\eta - \bar u^2\bar\xi),\qquad \hat{\bar\eta} = \sqrt2(u\bar\eta - u^2\xi). \tag{4.17}$$

After integration over θ and $\bar\theta$ in (4.14) and eliminating the auxiliary fields by their equations of motion, we arrive at the Lagrangian

$$\mathcal{L}_2 = \left[F_1 + F_2 - \frac{(F_1-F_2)^2}{F_1+F_2}\right]\dot u\dot{\bar u} + \mathrm{i}F_1\big(\dot{\hat\xi}\hat{\bar\xi} - \hat\xi\dot{\hat{\bar\xi}}\big) + \mathrm{i}F_2\big(\dot{\hat\eta}\hat{\bar\eta} - \hat\eta\dot{\hat{\bar\eta}}\big) - \mathrm{i}(\dot u\bar u - u\dot{\bar u})\big(F_1'\hat\xi\hat{\bar\xi} + F_2'\hat\eta\hat{\bar\eta}\big) + \\ + (F_1' + u\bar uF_1'' + F_2' + u\bar uF_2'')\hat\xi\hat{\bar\xi}\hat\eta\hat{\bar\eta} + \mathrm{i}\left[F_1' - F_2' - \frac{F_1-F_2}{F_1+F_2}(F_1'+F_2')\right]\left(u\dot u\hat\eta\hat\xi - \bar u\dot{\bar u}\hat{\bar\xi}\hat{\bar\eta}\right) \tag{4.18}$$

[3] This Lagrangian follows from (4.12) in the limit $\{w_a,\psi_a\} \to 0$.

After passing to the proper fermions via (4.17), one may easily check that the Lagrangian (4.18) coincides with (4.16), if we will fix our functions F_1 and F_2 as

$$4\,F_1 = \frac{1}{1+u\bar{u}}, \qquad 4\,F_2 = \frac{1}{u\bar{u}\,(1+u\bar{u})}. \tag{4.19}$$

This proves that the superfield action (4.14) indeed yields the component Lagrangian (4.16) upon the choice (4.19) for F_1 and F_2.

How unique is this result? One may wonder about adding further terms to the superfield action (4.14), such as

$$F_3\,u^2 D\bar{u}\bar{D}\bar{u}, \qquad F_4\,\bar{u}^2 Du\bar{D}u, \qquad \mathrm{i}F_5\,(\dot{u}\bar{u} - \dot{\bar{u}}u). \tag{4.20}$$

A detailed analysis shows that these terms do not play any role: to reproduce (4.16), the functions F_3 and F_4 have to vanish while the F_5 term can be absorbed into (4.18) by a suitable modification of F_1 and F_2. Hence, the superfield action (4.14) is completely fixed.

4.3. Complete superfield action

Having at hands the superfield Lagrangian for our $\mathcal{N}$=2 mechanics on $\mathbb{S}^2$, it is straightforward to construct the full superfield Lagrangian for the complete system on $\mathbb{CP}^3$,

$$\mathcal{L} = -\frac{1}{4}\left[\frac{Dw_a\bar{D}\bar{w}^a}{(1+w\cdot\bar{w})^2} + \frac{(Du+M^aDw_a)(\bar{D}\bar{u}+\bar{M}_a\bar{D}\bar{w}^a)}{1+u\bar{u}} + \frac{(D\bar{u}+N^aDw_a)(\bar{D}u+\bar{N}_a\bar{D}\bar{w}^a)}{u\bar{u}(1+u\bar{u})}\right] \tag{4.21}$$

with

$$M^a = \frac{w^a - u\bar{w}^a}{1+w\cdot\bar{w}}, \qquad N^a = \frac{\bar{u}^2 w^a + \bar{u}\bar{w}^a}{1+w\cdot\bar{w}}. \tag{4.22}$$

Coincidence with the full component Lagrangian (4.12) is achieved upon the following identification of the physical components in our superfields,

$$\hat{\xi} = \mathrm{i}(Du + M^aDw_a)|, \qquad \hat{\eta} = \mathrm{i}(D\bar{u} + N^aDw_a)|, \qquad \psi_a = \frac{\mathrm{i}Dw_a}{\sqrt{2}(1+w\cdot\bar{w})}|, \tag{4.23}$$

and subsequent redefinitions (4.17).

5. Conclusion

In this Letter we have proposed a Hamiltonian reduction procedure that relates the mechanics of an $\mathcal{N}$=2 particle on $\mathbb{CP}^3$ with the motion of such a superparticle on $\mathbb{S}^4$ in the presence of an instanton background. The key ingredient for the existence of the $\mathcal{N}$=2 extension of the bosonic reduction was a novel action for the $\mathcal{N}$=2 supersymmetric particle moving on $\mathbb{S}^2$. It turned out that the latter is nothing but $\mathcal{N}$=4 supersymmetric mechanics with a non-linear chiral supermultiplet, written in terms of $\mathcal{N}$=2 superfields. The supercharges contain the bosonic variables and their momenta only via SU(2) currents. Therefore, the Hamiltonian reduction procedure paralleled the purely bosonic one. The full system, for which we have constructed the Hamiltonian and supercharges, is a new variant of $\mathcal{N}$=2 mechanics for a superparticle on $\mathbb{CP}^3$. It contains twice as many fermions as the standard version based on $\mathcal{N}$=2 chiral superfields. We also provided a superspace description for our system.

Our construction may be extended to reductions over $\mathbb{S}^3$ as well as to $\mathbb{S}^{4k+3} \to \mathbb{CP}^{2k+1} \to \mathbb{HP}^k$ fibrations. We will consider these extensions elsewhere.

Acknowledgements

We are indebted to Armen Nersessian for valuable discussions. This work was partly supported by Volkswagen Foundation grant I/84 496, by RFBR grants 11-02-01335-a, 11-02-90445-Ukr, 12-02-00517-a.

[1] S. Bellucci, P.-Y. Casteill, A. Nersessian, *Four-dimensional Hall mechanics as a particle on* CP^3, Phys. Lett. B574 (2003) 121, `arXiv:hep-th/0306277`.

[2] E. Ivanov, S. Krivonos, O. Lechtenfeld, *N=4, d=1 supermultiplets from nonlinear realizations of* $D(2,1;\alpha)$, Class. Quant. Grav. 21 (2004) 1031, `arXiv:hep-th/0310299`.

[3] M. de Crombrugghe, V. Rittenberg, *Supersymmetric quantum mechanics*, Ann. Phys. 151 (1983) 99; E.A. Ivanov, A.V. Smilga, *Supersymmetric gauge quantum mechanics: superfield description*, Phys. Lett. B257 (1991) 79; V.P. Berezovoj, A.I. Pashnev, *Three-dimensional N=4 extended supersymmetrical quantum mechanics*, Class. Quant. Grav. 8 (1991) 2141; A. Maloney, M. Spradlin, A. Strominger, *Superconformal multi-black hole moduli spaces in four dimensions*, JHEP 0204 (2002) 003, `arXiv:hep-th/9911001`.

[4] S. Bellucci, S. Krivonos, V. Ohanyan, *N=4 supersymmetric MICZ-Kepler systems on* S^3, Phys. Rev. D76 (2007) 105023, `arXiv:0706.1469[hep-th]`.

[5] M. Konyushikhin, A.V. Smilga, *Self-duality and supersymmetry*, Phys. Lett. B689 (2010) 95, `arXiv:0910.5162[hep-th]`; E.A. Ivanov, A.V. Smilga, *Dirac operator on complex manifolds and supersymmetric quantum mechanics*, `arXiv:1012.2069[hep-th]`; A.V. Smilga, *Dolbeault complex on* $S^4\backslash\{.\}$ *and* $S^6\backslash\{.\}$ *through supersymmetric glasses*, SIGMA 7 (2011), 105, `arXiv:1105.3935[hep-th]`; E.A. Ivanov, M.A. Konyushikhin, A.V. Smilga, *SQM with Non-Abelian Self-Dual Fields: Harmonic Superspace Description*, JHEP 1005 (2010) 033, `arXiv:0912.3289[hep-th]`.

[6] S. Krivonos, O. Lechtenfeld, A. Sutulin, *N=4 supersymmetry and the BPST instanton*, Phys. Rev. D81 (2010) 085021, `arXiv:1001.2659[hep-th]`.

[7] S. Bellucci, A. Beylin, S. Krivonos, A. Nersessian, E. Orazi, *N=4 supersymmetric mechanics with nonlinear chiral supermultiplet*, Phys. Lett. B616 (2005) 228, `arXiv:hep-th/0503244`.

Journal of Physics: Conference Series **411** (2013) 012020 doi:10.1088/1742-6596/411/1/012020

XXZ-type Bethe ansatz equations and quasi-polynomials

Jian Rong Li[⋆] **and Vitaly Tarasov**[*]

[⋆] Department of Mathematics, Lanzhou University, Lanzhou 730000, P. R. China

[*] Department of Mathematical Sciences, Indiana University – Purdue University Indianapolis, 402 North Blackford St, Indianapolis, IN 46202-3216, USA

[*] St. Petersburg Branch of Steklov Mathematical Institute, Fontanka 27, St. Petersburg, 191023, Russia

Abstract. We study solutions of the Bethe ansatz equation for the XXZ-type integrable model associated with the Lie algebra $\mathfrak{sl}_N$. We give a correspondence between solutions of the Bethe ansatz equations and collections of quasi-polynomials. This extends the results of E. Mukhin and A. Varchenko for the XXX-type model and the trigonometric Gaudin model.

1. Introduction

In this paper we study solutions of the Bethe ansatz equation for the XXZ-type integrable models associated with the Lie algebra $\mathfrak{sl}_N$, see (1). These equations arise in the Bethe ansatz method of computing eigenvalues and eigenvectors of commuting Hamiltonians of integrable models. The method gives the eigenvalues and eigenvectors by evaluating explicit rational functions on solutions of the Bethe ansatz equations, see for instance [KBI] and references therein.

Solutions of the Bethe ansatz equations for the Gaudin model, both rational and trigonometric, as well for the XXX-type model have been studied by E. Mukhin and A. Varchenko in [MV1] – [MV3]. They established a correspondence between solutions of the Bethe ansatz equations and spaces of polynomials, quasi-polynomials or quasi-exponentials with certain properties. In this paper we extend the results of [MV2], [MV3] to the case of the XXZ-type Bethe ansatz equations.

Our method to construct a collection of quasi-polynomials corresponding to a solution of the Bethe ansatz equations is not completely analogous to that of [MV2], [MV3]. As a result, we managed to weaken technical assumptions on solutions of the Bethe ansatz equations compared with those imposed in [MV3]. Another advantage is that our method works withut restrictions for the root of unity case.

The plan of the paper is as follows. In Section 2, we describe the XXZ-type Bethe ansatz equations and refine the problem. In particular, we define regular and admissible solutions of the Bethe ansatz equations. In Section 3, we introduce regular collections of quasi-polynomials and show that each such a collection gives a regular solution of the Bethe ansatz equations. We also formulate there the main result of the paper, Theorem 3.4, which says that every admissible regular solution of the Bethe ansatz equations comes from a regular collection of quasi-polynomials. We prove Theorem 3.4 in Section 4. In Section 5, for a collection of quasi-polynomials $\mathcal{U}$ we consider the monic difference operator $D_{\mathcal{U}}$ whose kernel is generated by $\mathcal{U}$. We show that the operator $D_{\mathcal{U}}$ has rational coefficients if and only if there is a regular collection of quasi-polynomials $\widetilde{\mathcal{U}}$ such that $D_{\widetilde{\mathcal{U}}} = D_{\mathcal{U}}$. Also, for each solution $\boldsymbol{t}$ of the Bethe ansatz equations we define a difference operator $D^{\boldsymbol{t}}$ and show that the collection of quasi-polynomials $\mathcal{U}$ associated with $\boldsymbol{t}$ gives a basis of the kernel of $D^{\boldsymbol{t}}$, that is, $D^{\boldsymbol{t}} = D_{\mathcal{U}}$. The Appendix contains necessary technical identities.

The authors thank E. Mukhin for helpful discussions.

2. Bethe ansatz equations and quasi-polynomials

2.1. Bethe ansatz equations

Throughout the paper we fix a complex number q such that $q \neq 0, \pm 1$. More precisely, we fix the value of $\log q$ and for any complex number α set $q^{\alpha} = \exp(\alpha \log q)$.

Fix an integer $N \geqslant 2$. Given a collection $\boldsymbol{l} = (l_1, \dots, l_{N-1})$ of nonnegative integers, consider the space $\mathbb{C}^{\boldsymbol{l}} = \mathbb{C}^{l_1} \oplus \ldots \oplus \mathbb{C}^{l_{N-1}}$. We label coordinates on this space by two indices, the superscript enumerating the summands and the subscript enumerating coordinates within a summand. That is, we write

$$\boldsymbol{t} = (t_1^{(1)}, \dots, t_{l_1}^{(1)}, \ \dots \ , t_1^{(N-1)}, \dots, t_{l_{N-1}}^{(N-1)}) \in \mathbb{C}^{\boldsymbol{l}}.$$

The product of the symmetric groups $S_{\boldsymbol{l}} = S_{l_1} \times \ldots \times S_{l_{N-1}}$ acts on $\mathbb{C}^{\boldsymbol{l}}$ by permuting coordinates with the same superscript. The $S_{\boldsymbol{l}}$-orbit of a single point will be called an *elementary* $S_{\boldsymbol{l}}$*-orbit*.

Let $T_1(x), \dots, T_{N-1}(x)$ be monic polynomials in one variable and $\lambda_1, \dots, \lambda_N$ be complex numbers. Consider a system of equations on the variables $t_b^{(a)}$, $a = 1, \dots, N-1$, $b = 1, \dots, l_a$,

$$T_i(t_j^{(i)} q^2) \prod_{r=1}^{l_{i-1}} (t_j^{(i)} q - t_r^{(i-1)} q^{-1}) \prod_{s=1}^{l_{i+1}} (t_j^{(i)} - t_s^{(i+1)}) \prod_{\substack{k=1 \\ k \neq j}}^{l_i} (t_j^{(i)} q^{-1} - t_k^{(i)} q) = \tag{1}$$

$$= q^{2(\lambda_{i+1} - \lambda_i + l_i) - l_{i-1} - l_{i+1}} \, T_i(t_j^{(i)}) \prod_{r=1}^{l_{i-1}} (t_j^{(i)} - t_r^{(i-1)}) \prod_{s=1}^{l_{i+1}} (t_j^{(i)} q^{-1} - t_s^{(i+1)} q) \prod_{\substack{k=1 \\ k \neq j}}^{l_i} (t_j^{(i)} q - t_k^{(i)} q^{-1}),$$

$i = 1, \dots, N-1$, $j = 1, \dots, l_i$. Here and later we assume that $l_0 = l_N = 0$, unless otherwise stated. Equations (1) are called the *XXZ*-type Bethe ansatz equations associated with the data $\boldsymbol{l} = (l_1, \dots, l_{N-1})$, $\boldsymbol{T} = (T_1, \dots, T_{N-1})$, $\boldsymbol{\lambda} = (\lambda_1, \dots, \lambda_N)$.

The group $S_{\boldsymbol{l}}$ acts on solutions of equations (1) — for every solution $\boldsymbol{t}$ of (1), all points of the $S_{\boldsymbol{l}}$-orbit of $\boldsymbol{t}$ are solutions of (1) as well.

Equations (1) arise in the Bethe ansatz method for the quantum integrable model defined on the irreducible finite-dimensional representation of the quantum affine algebra $U_q(\widehat{\mathfrak{sl}_N})$. The polynomials $T_1(x), T_2(xq^{-2}), \dots, T_{N-1}(xq^{-2(N-1)})$ are the Drinfeld polynomials of the representation and the parameters $q^{2\lambda_1}, \dots, q^{2\lambda_N}$ describe the algebra of commuting Hamiltonians of the model. Notice that in the literature equations (1) are usually written in the form

$$\frac{T_i(t_j^{(i)} q^2)}{T_i(t_j^{(i)})} \prod_{r=1}^{l_{i-1}} \frac{t_j^{(i)} q - t_r^{(i-1)} q^{-1}}{t_j^{(i)} - t_r^{(i-1)}} \prod_{s=1}^{l_{i+1}} \frac{t_j^{(i)} - t_s^{(i+1)}}{t_j^{(i)} q^{-1} - t_s^{(i+1)} q} \prod_{\substack{k=1 \\ k \neq j}}^{l_i} \frac{t_j^{(i)} q^{-1} - t_k^{(i)} q}{t_j^{(i)} q - t_k^{(i)} q^{-1}} = \tag{2}$$

$$= q^{2(\lambda_{i+1} - \lambda_i + l_i) - l_{i-1} - l_{i+1}},$$

$i = 1, \dots, N-1$, $j = 1, \dots, l_i$.

2.2. Regular solutions of Bethe ansatz equations

For $i = 1, \ldots, N-1$, define the polynomials

$$p_i(x;\boldsymbol{t}) = \prod_{j=1}^{l_i} (x - t_j^{(i)}) \tag{3}$$

and

$$P_i(x;\boldsymbol{t}) = q^{2\lambda_{i+1}} p_i(xq^2;\boldsymbol{t})\, p_{i-1}(x;\boldsymbol{t})\, p_{i+1}(xq^{-2};\boldsymbol{t})\, T_i(x) + \\ + q^{2\lambda_i} p_i(xq^{-2};\boldsymbol{t})\, p_{i-1}(xq^2;\boldsymbol{t})\, p_{i+1}(x;\boldsymbol{t})\, T_i(xq^2)\,. \tag{4}$$

Then equations (1) take the form

$$P_i(t_j^{(i)};\boldsymbol{t}) = 0\,, \qquad i = 1, \ldots, N-1\,, \quad j = 1, \ldots, l_i\,. \tag{5}$$

A solution $\boldsymbol{t}$ of equations (1) is called *regular* if for any $i = 1, \ldots, N-1$, and any subset $I \subset \{1, \ldots, l_i\}$, the following condition holds: suppose that all the coordinates $t_j^{(i)}$, $j \in I$ are equal with the common value $t_I^{(i)}$; then

$$\frac{d^k}{dx^k} P_i(x;\boldsymbol{t})\big|_{x=t_I^{(i)}} = 0\,, \qquad k = 0, \ldots, |I|-1\,, \tag{6}$$

where and $|I|$ is the cardinality of I. Clearly, if for each $i = 1, \ldots, N-1$, all the coordinates $t_j^{(i)}$, $j = 1, \ldots, l_i$, are distinct, the solution $\boldsymbol{t}$ is regular.

If $\boldsymbol{t}$ is a regular solution of equations (1), then all points of the $S_{\boldsymbol{l}}$-orbit of $\boldsymbol{t}$ are regular solutions of equations (1) as well.

The next proposition is valid by definition of a regular solution of equations (1).

Proposition 2.1. *A point* $\boldsymbol{t}$ *is a regular solution of equations* (1) *if and only if for any* $i = 1, \ldots, N-1$, *the polynomial* $p_i(x;\boldsymbol{t})$ *divides the polynomial* $P_i(x;\boldsymbol{t})$.

3. Quasi-polynomials

3.1. Quasi-polynomials

A quasi-polynomial $f(x)$ of type α is an expression of the form $f(x) = x^\alpha p(x, \log x)$, where $\alpha \in \mathbb{C}$ and $p(x,s)$ is a polynomial. Call the quasi-polynomial $f(x)$ *log-free* if $p(x,s)$ does not depend on s. Say that a polynomial $r(x)$ divides $f(x)$ if $r(x)$ divides $p(x,s)$. The product of quasi-polynomials of types α and β is supposed to be of type $\alpha+\beta$. Clearly, the product of two nonzero quasi-polynomials is not zero.

We will use only algebraic properties of quasi-polynomials. The key relation is: if $f(x) = x^\alpha p(x, \log x)$, then

$$f(xq^\beta) = q^{\alpha\beta} x^\alpha p(xq^\beta, \log x + \beta \log q)\,.$$

Given functions $g_1(x), \ldots, g_k(x)$, their discrete Wronskian $W_k[g_1, \ldots, g_k]$ is defined by the rule

$$W_k[g_1, \ldots, g_k](x) = \det\big(g_i(xq^{-2(j-1)})\big)_{i,j=1}^k. \tag{7}$$

Let $\boldsymbol{\lambda} = (\lambda_1, \ldots, \lambda_N)$ be a collection of complex numbers. By definition, a *collection of quasi-polynomials of type* $\boldsymbol{\lambda}$ is a sequence of quasi-polynomials $\mathcal{U} = (u_1, \ldots, u_N)$ such that for every

$i = 1,\dots,N$, the quasi-polynomial u_i has type λ_i, and $W_N[u_1,\dots,u_N] \neq 0$. We call the collection $\mathcal{U}$ *semiregular* if $W_N[u_1,\dots,u_N]$ is log-free.

For a semiregular collection of quasi-polynomials $\mathcal{U} = (u_1,\dots,u_N)$ of type $\boldsymbol{\lambda}$, a sequence of monic polynomials $\mathcal{T} = (T_1,\dots,T_N)$ is called a *preframe* of $\mathcal{U}$ if for any $k = 1,\dots,N-1$ and any subset $\{i_1,\dots,i_k\} \subset \{1,\dots,N\}$, the product $\prod_{i=1}^{k}\prod_{j=0}^{k-i} T_{N-i+1}(xq^{-2j})$ divides the quasi-polynomial $W_k[u_{i_1},\dots,u_{i_k}](x)$ of type $\lambda_{i_1}+\dots+\lambda_{i_k}$, and

$$W_N[u_1,\dots,u_N](x) = \text{const}\, x^{\lambda_1+\dots+\lambda_N} \prod_{i=1}^{N}\prod_{j=0}^{N-i} T_{N-i+1}(xq^{-2j}). \tag{8}$$

Denote

$$Q_k^{\mathcal{T}}(x) = \prod_{i=1}^{k}\prod_{j=0}^{k-i} T_{N-i+1}(xq^{-2j}), \qquad k = 1,\dots,N. \tag{9}$$

Say that a preframe $\mathcal{T}$ is stronger than a preframe $\widetilde{\mathcal{T}} = (\widetilde{T}_1,\dots,\widetilde{T}_N)$ if $Q_k^{\widetilde{\mathcal{T}}}(x)$ divides $Q_k^{\mathcal{T}}(x)$ for any $k = 1,\dots,N-1$.

Until the end of this section we fix $\boldsymbol{\lambda} = (\lambda_1,\dots,\lambda_N)$ and assume that every collection of quasi-polynomials is of type $\boldsymbol{\lambda}$.

Lemma 3.1. *Let $\mathcal{U} = (u_1,\dots,u_N)$ be a semiregular collection of quasi-polynomials. There is a preframe $\mathcal{T} = (T_1,\dots,T_N)$ of $\mathcal{U}$ such that for every $k = 1,\dots,N$, the polynomial $Q_k^{\mathcal{T}}(x)$, see (9), is the greatest common divisor of the quasi-polynomials $W_k[u_{i_1},\dots,u_{i_k}](x)$, where $\{i_1,\dots,i_k\}$ runs over all k-element subsets of $\{1,\dots,N\}$ and $W_k[u_{i_1},\dots,u_{i_k}](x)$ has type $\lambda_{i_1}+\dots+\lambda_{i_k}$.*

Proof. The proof is similar to that of Lemma 4.9 in [MV2]. □

The preframe $\mathcal{T}$ defined in Lemma 3.1 is clearly the strongest preframe of $\mathcal{U}$. It is called the *frame* of $\mathcal{U}$.

For a semiregular collection of quasi-polynomials $\mathcal{U} = (u_1,\dots,u_N)$ and a preframe $\mathcal{T} = (T_1,\dots,T_N)$ of $\mathcal{U}$, define quasi-polynomials $y_0,\dots,y_{N-1}$ by the rule $u_N = y_{N-1}Q_1^{\mathcal{T}}$,

$$W_{N-i}[u_{i+1},\dots,u_N](x) = y_i(x)\,Q_{N-i}^{\mathcal{T}}(x), \qquad i = 0,\dots,N-2, \tag{10}$$

where $Q_1^{\mathcal{T}},\dots,Q_N^{\mathcal{T}}$ are given by (9). Notice that $y_0(x)$ is proportional to $x^{\lambda_1+\dots+\lambda_N}$. Call the semiregular collection $\mathcal{U}$ *regular* if the quasi-polynomials $y_1,\dots,y_{N-1}$ are log-free.

Assume that the collection $\mathcal{U}$ is regular. For $i = 1,\dots,N-1$, let l_i be the degree of the polynomial $x^{-\lambda_{i+1}-\dots-\lambda_N}y_i(x)$, and $t_1^{(i)},\dots,t_{l_i}^{(i)}$ be its roots. That is,

$$y_i(x) = c_i\,x^{\lambda_{i+1}+\dots+\lambda_N}\prod_{j=1}^{l_i}(x-t_j^{(i)}), \qquad i = 1,\dots,N-1, \tag{11}$$

for some nonzero complex numbers $c_1,\dots,c_{N-1}$. Set $\boldsymbol{l} = \boldsymbol{l}_{\mathcal{U},\mathcal{T}} = (l_1,\dots,l_{N-1})$, and denote by $X_{\mathcal{U},\mathcal{T}}$ the $S_{\boldsymbol{l}}$-orbit of the point $(t_j^{(i)})_{i=1,\dots,N-1,\,j=1,\dots,l_i}$ in $\mathbb{C}^{\boldsymbol{l}}$.

Journal of Physics: Conference Series **411** (2013) 012020 doi:10.1088/1742-6596/411/1/012020

Theorem 3.2. *Let $\mathcal{U}$ be a regular collection of quasi-polynomials of type $\boldsymbol{\lambda}$, $\mathcal{T} = (T_1, \dots, T_N)$ be a preframe of $\mathcal{U}$, and $\boldsymbol{l} = \boldsymbol{l}_{\mathcal{U},\mathcal{T}} = (l_1, \dots, l_{N-1})$. Then $X_{\mathcal{U},\mathcal{T}}$ is an elementary $S_{\boldsymbol{l}}$-orbit of regular solutions of equations (1) associated with the data $l_1, \dots, l_{N-1}$, $T_1, \dots, T_{N-1}$, $\lambda_1, \dots, \lambda_N$.*

Proof. Define the quasi-polynomials $\tilde{y}_1, \dots, \tilde{y}_{N-1}$ by the rule

$$W_{N-i}[\,u_i, u_{i+2}, \dots, u_N](x) \;=\; \tilde{y}_i(x)\, Q^{\mathcal{T}}_{N-i}(x)\,. \tag{12}$$

By Lemmas Appendix A.1, Appendix A.2, and formula (9), we have

$$W_2[\,\tilde{y}_i, y_i](x) \;=\; y_{i-1}(x)\, y_{i+1}(xq^{-2})\, T_i(x)\,, \tag{13}$$

that is,

$$\tilde{y}_i(x)\, y_i(xq^{-2}) - \tilde{y}_i(xq^{-2})\, y_i(x) \;=\; y_{i-1}(x)\, y_{i+1}(xq^{-2})\, T_i(x)\,.$$

Therefore

$$\begin{aligned} & y_i(x)\,\big(\tilde{y}_i(xq^2)\, y_i(xq^{-2}) - \tilde{y}_i(xq^{-2})\, y_i(xq^2)\big) \;= \\ & = \; y_i(xq^2)\, y_{i-1}(x)\, y_{i+1}(xq^{-2})\, T_i(x) + y_i(xq^{-2})\, y_{i-1}(xq^2)\, y_{i+1}(x)\, T_i(xq^2)\,. \end{aligned} \tag{14}$$

If the point $\boldsymbol{t} = (t^{(i)}_j)$ is defined by (10), (11), the polynomials $p_i(x;\boldsymbol{t})$, $P_i(x;\boldsymbol{t})$ are given by (3), (4), and the polynomial $\tilde{p}_i(x)$ equals $x^{-\lambda_i-\lambda_{i+2}-\dots-\lambda_N}\tilde{y}(x)$, then relation (14) is equivalent to

$$P_i(x;\boldsymbol{t}) \;=\; p_i(x)\,\big(q^{2(\lambda_i-\lambda_{i+1})}\tilde{p}_i(xq^2)\, p_i(xq^{-2}) - q^{2(\lambda_{i+1}-\lambda_i)}\tilde{p}_i(xq^{-2})\, p_i(xq^2)\big)\,.$$

By Proposition 2.1, this proves Theorem 3.2. □

Remark. By Lemma Appendix A.1, a collection of quasi-polynomials $\mathcal{U} = (u_1, \dots, u_N)$ has a preframe $\mathcal{T} = (T_1, \dots, T_N)$ if and only if $\widetilde{\mathcal{U}} = (u_1/T_N, \dots, u_N/T_N)$ is a collection of quasi-polynomials and has a preframe $\widetilde{\mathcal{T}} = (T_1, \dots, T_{N-1}, 1)$. Clearly, $X_{\mathcal{U},\mathcal{T}} = X_{\widetilde{\mathcal{U}},\widetilde{\mathcal{T}}}$, and for both $\mathcal{U}$ and $\widetilde{\mathcal{U}}$ equations (1) are the same since they involve only the polynomials $T_1, \dots, T_{N-1}$. Thus discussing relations between collections of quasi-polynomials and solutions of Bethe ansatz equations (1) we can restrict ourselves without loss of generality to preframes of the form $(T_1, \dots, T_{N-1}, 1)$.

Say that a point $\boldsymbol{t} = (t^{(i)}_j)$ is *generic* with respect to the polynomials $\boldsymbol{T} = (T_1, \dots, T_{N-1})$ if $t^{(i)}_j \neq t^{(i+1)}_k$ for all $i = 1, \dots, N-2$, $j = 1, \dots, l_i$, $k = 1, \dots, l_{i+1}$, and $T_i(t^{(i)}_j) \neq 0$ for all $i = 1, \dots, N-1$, $j = 1, \dots, l_i$. Clearly, all points of the $S_{\boldsymbol{l}}$-orbit of a generic point $\boldsymbol{t}$ are generic.

Lemma 3.3. *Let $\mathcal{U}$ be a regular collection of quasi-polynomials of type $\boldsymbol{\lambda}$, $\mathcal{T} = (T_1, \dots, T_N)$ be a preframe of $\mathcal{U}$, and $\boldsymbol{l} = \boldsymbol{l}_{\mathcal{U},\mathcal{T}} = (l_1, \dots, l_{N-1})$. Assume that $X_{\mathcal{U},\mathcal{T}}$ is the $S_{\boldsymbol{l}}$-orbit of a generic point with respect to the polynomials $T_1, \dots, T_{N-1}$. Then $\mathcal{T}$ is the frame of $\mathcal{U}$.*

Proof. Let $p_i(x) = x^{-\lambda_{i+1}-\dots-\lambda_N}\, y_i(x)$. Suppose $\widetilde{\mathcal{T}} = (\widetilde{T}_1, \dots, \widetilde{T}_N)$ is a preframe of $\mathcal{U}$ strictly stronger than $\mathcal{T}$. Let i be the largest number such that $\widetilde{T}_i \neq T_i$. Notice that $i > 1$. There is a number a such that $\widetilde{T}_i(a) = 0$ and $T_i(a) \neq 0$. This implies that $p_{i-1}(a) = 0$ and $p_{i-2}(a)\, T_{i-1}(a) = 0$, see (9), (10). However, if $X_{\mathcal{U},\mathcal{T}}$ is the $S_{\boldsymbol{l}}$-orbit of a generic point, then for every $j = 1, , \dots, N-1$, the polynomials $p_{j-1} T_j$ and p_j are coprime. The claim follows. □

3.2. Main result

A point $\boldsymbol{t} = (t_j^{(i)})_{i=1,\dots,N-1,\, j=1,\dots,l_i}$ is called *admissible* if

$$t_j^{(i)} \neq t_k^{(i)} q^2, \qquad i = 1, \dots, N-1, \quad j, k = 1, \dots, l_i \,. \tag{15}$$

In particular, $t_j^{(i)} \neq 0$ for all $i = 1, \dots, N-1$, $j = 1, \dots, l_i$. Clearly, all points of the $S_{\boldsymbol{l}}$-orbit of an admissible point $\boldsymbol{t}$ are admissible.

Fix collections of complex numbers $\boldsymbol{\lambda} = (\lambda_1, \dots, \lambda_N)$, nonnegative integers $\boldsymbol{l} = (l_1, \dots, l_{N-1})$ and monic polynomials $\boldsymbol{T} = (T_1, \dots, T_{N-1})$. The following theorem is the main result of the paper.

Theorem 3.4. *Let* $\boldsymbol{t}$ *be an admissible regular solution of equations* (1) *associated with the data* $\boldsymbol{l}, \boldsymbol{T}, \boldsymbol{\lambda}$. *Then there is a regular collection of quasi-polynomials* $\mathcal{U}$ *of type* $\boldsymbol{\lambda}$ *such that* $\mathcal{T} = (T_1, \dots, T_{N-1}, 1)$ *is a preframe of* $\mathcal{U}$ *and the* $S_{\boldsymbol{l}}$*-orbit of* $\boldsymbol{t}$ *equals* $X_{\mathcal{U},\mathcal{T}}$.

Theorem 3.4 will be proved in Section 4. The proof is going in three steps. First we prove the theorem for $N = 2$. The obtained statement is employed then at the second step to construct the required collection of quasi-polynomials $\mathcal{U}$ for general N. The final step is to show that $\mathcal{T} = (T_1, \dots, T_{N-1}, 1)$ is a preframe of the constructed collection $\mathcal{U}$.

Corollary 3.5. *Let* $\boldsymbol{t}$ *be a generic admissible regular solution of equations* (1) *associated with the data* $\boldsymbol{l}, \boldsymbol{T}, \boldsymbol{\lambda}$. *Then there is a regular collection of quasi-polynomials* $\mathcal{U}$ *of type* $\boldsymbol{\lambda}$ *such that* $\mathcal{T} = (T_1, \dots, T_{N-1}, 1)$ *is the frame of* $\mathcal{U}$ *and the* $S_{\boldsymbol{l}}$*-orbit of* $\boldsymbol{t}$ *equals* $X_{\mathcal{U},\mathcal{T}}$.

Proof. The statement follows from Theorem 3.4 and Lemma 3.3. □

4. Proof of Theorem 3.4

4.1. Proof of Theorem 3.4 for $N = 2$

Let $y(x) = x^\alpha p(x)$ be a log-free quasi-polynomial of type α. We call $y(x)$ admissible if the polynomials $p(x)$ and $p(xq^{-2})$ are coprime. In particular, this implies that $p(0) \neq 0$. Thus for an admissible quasi-polynomial $y(x)$, the number α and the polynomial $p(x)$ are determined uniquely.

Let $y(x) = x^\alpha p(x)$ be an admissible quasi-polynomial of type α. Since $p(x)$ and $p(xq^{-2})$ are coprime, it is known that there are unique polynomials $r(x)$ and $s(x)$ of degree at most $\deg p$ such that $r(x)p(x) + s(x)p(xq^{-2}) = 1$. Define the quasi-polynomials $\mathcal{A}[y]$ and $\mathcal{B}[y]$ of type $-\alpha$ by the rule

$$\mathcal{A}[y](x) \,=\, x^{-\alpha} r(x)\,, \qquad \mathcal{B}[y](x) \,=\, x^{-\alpha} s(x)\,, \tag{16}$$

so that

$$y(x)\,\mathcal{A}[y](x) + y(xq^{-2})\,\mathcal{B}[y](x) \,=\, 1\,. \tag{17}$$

For a polynomial $P(s)$ and a number c, let $\mathcal{I}[P,c]$ be the unique polynomial such that

$$\mathcal{I}[P,c](s) - c\,\mathcal{I}[P,c](s - 2\log q) \,=\, P(s)\,,$$

$\deg \mathcal{I}[P,c] = \deg P$ if $c \neq 1$, and $\deg \mathcal{I}[P,1] = 1 + \deg P$, $\mathcal{I}[P,1](0) = 0$. For example,

$$\mathcal{I}[1,1](s) \,=\, \frac{s}{2\log q}\,, \qquad \mathcal{I}[1,c](s) \,=\, \frac{1}{1-c}\,, \quad c \neq 1\,.$$

For a quasi-polynomial $f(x) = x^\alpha \sum_i x^i P_i(\log x)$ of type α, define the quasi-polynomial $\mathcal{I}[f]$ of type α by the rule $\mathcal{I}[f](x) = x^\alpha \sum_i x^i \mathcal{I}[P_i, q^{-2\alpha}](\log x)$, so that

$$\mathcal{I}[f](x) - \mathcal{I}[f](xq^{-2}) = f(x).$$

For example,

$$\mathcal{I}[1] = \frac{\log x}{\log q^2}, \qquad \mathcal{I}[x^\alpha] = \frac{x^\alpha}{1-q^{-2\alpha}}, \quad q^{-2\alpha} \neq 1.$$

For a quasi-polynomial $f(x) = x^\alpha p(x, \log x)$ of type α and a log-free quasi-polynomial $g(x) = x^\beta r(x)$ of type β, define the quasi-polynomial $\langle f(x)/g(x)\rangle_+ = x^{\alpha-\beta} h(x, \log x)$ of type $\alpha - \beta$ by requiring that $h(x,s)$ is the polynomial part of the ratio $p(x,s)/r(x)$, that is, $\deg_x\bigl(p(x,s) - r(x)\,v(x,s)\bigr) < \deg r(x)$. If $f(x) = g(x)\,[f(x)/g(x)]_+$, we say that $g(x)$ divides $f(x)$.

For an admissible quasi-polynomial $y(x) = x^\alpha p(x)$ of type α and a quasi-polynomial $V(x)$ of type β, define the quasi-polynomial $\mathcal{F}[y,V]$ of type $\beta - \alpha$ as follows. Let $a = \mathcal{A}[y]$ and $b = \mathcal{B}[y]$, see (16). Consider the quasi-polynomial

$$v(x) = \left\langle \frac{a(x)\,V(x) + b(xq^{-2})\,V(xq^{-2})}{y(xq^{-2})} \right\rangle_+$$

of type $\beta - 2\alpha$. Set $\mathcal{J}[y,V] = \mathcal{I}[v]$ and

$$\mathcal{F}[y,V](x) = V(x)\,\mathcal{B}[y](x) + y(x)\,\mathcal{J}[y,V](x). \tag{18}$$

Proposition 4.1. *Let $y(x)$ be an admissible quasi-polynomial of type α, $V(x)$ be a quasi-polynomial of type β. Let $Y = \mathcal{F}[y,V]$. Assume that $y(x)$ divides the quasi-polynomial $y(xq^2)\,V(x) + y(xq^{-2})\,V(xq^2)$ of type $\alpha+\beta$. Then $W_2[Y,y] = V$.*

Proof. Let $a = \mathcal{A}[y]$, $b = \mathcal{B}[y]$, $f = \mathcal{J}[y,V]$. By (17),

$$a(x)\,y(x) + b(x)\,y(xq^{-2}) = 1 \tag{19}$$

and

$$\begin{aligned} & a(xq^2)\,b(x)\,\bigl(y(xq^2)\,V(x) + y(xq^{-2})\,V(xq^2)\bigr) = \\ & = a(xq^2)\,V(xq^2) + b(x)\,V(x) - \bigl(a(x)\,a(xq^2)\,V(xq^2) + b(x)\,b(xq^2)\,V(x)\,y(x) \end{aligned}$$

so $y(x)$ divides the quasi-polynomial $a(xq^2)\,V(xq^2) + b(x)\,V(x)$ of type $\beta - \alpha$. Hence

$$y(xq^{-2})\,\bigl(f(x) - f(xq^{-2})\bigr) = a(x)\,V(x) + b(xq^{-2})\,V(xq^{-2}),$$

and $W_2[Y,y] = V$ by an easy simplification using (19). □

Proof of Theorem 3.4 for $N = 2$. Let $\boldsymbol{t} = (t_j^{(i)})$ be an admissible regular solution of equations (1) and $p_1(x;\boldsymbol{t})$ be the polynomial given by (3). Define an admissible quasi-polynomial $y(x) = x^{\lambda_2} p_1(x;\boldsymbol{t})$ of type $\alpha = \lambda_2$ and a quasi-polynomial $V(x) = x^{\lambda_1+\lambda_2}\,T_1(x)$ of type $\beta = \lambda_1 + \lambda_2$.

Let $u_2 = y$ and $u_1 = \mathcal{F}[y,V]$. Proposition 2.1 shows that $y(x)$ divides the quasi-polynomial $y(xq^2)\,V(x) + y(xq^{-2})\,V(xq^2)$ of type $\alpha+\beta$. Hence by Proposition 4.1,

$$W_2[u_1,u_2] = x^{\lambda_1+\lambda_2}\,T_1(x),$$

that is, $\mathcal{U} = (u_1,u_2)$ is the required collection of quasi-polynomials of type $\boldsymbol{\lambda} = (\lambda_1,\lambda_2)$. □

4.2. Construction of a collection of quasi-polynomials

Let $\boldsymbol{t} = (t_j^{(i)})$ be an admissible regular solution of equations (1) associated with the data $\boldsymbol{l}, \boldsymbol{T}, \boldsymbol{\lambda}$. Let

$$p_i(x) = \prod_{j=1}^{l_i} (x - t_j^{(i)}), \qquad i = 1, \dots, N-1, \tag{20}$$

cf. (3), and $p_0(x) = p_N(x) = 1$. Then for every $i = 0, \dots, N$,

$$y_i(x) = x^{\lambda_{i+1}+\ldots+\lambda_N} p_i(x)$$

is an admissible quasi-polynomial of type $\lambda_{i+1} + \ldots + \lambda_N$. Set $a_i = \mathcal{A}[y_i]$ and $b_i = \mathcal{B}[y_i]$, see (16), so that

$$a_i(x)\, y_i(x) + b_i(x)\, y_i(xq^{-2}) = 1. \tag{21}$$

The next lemma is equivalent to Proposition 2.1.

Lemma 4.2. *For any* $i = 1, \dots, N-1$, *the quasi-polynomial* $y_i(x)$ *divides the quasi-polynomial*

$$A_i(x) = y_i(xq^2)\, y_{i-1}(x)\, y_{i+1}(xq^{-2}) T_i(x) + y_i(xq^{-2})\, y_{i-1}(xq^2)\, y_{i+1}(x)\, T_i(xq^2) \tag{22}$$

of type $\lambda_i + 2\lambda_{i+1} + 3\,(\lambda_{i+2} + \ldots + \lambda_N)$.

Set $Q_1(x) = 1$ and

$$Q_k(x) = \prod_{i=2}^{k} \prod_{j=0}^{k-i} T_{N-i+1}(xq^{-2j}), \qquad k = 2, \dots, N, \tag{23}$$

cf. (9). We will construct a collection of quasi-polynomials $\mathcal{U} = (u_1, \dots, u_N)$ of type $\boldsymbol{\lambda}$ such that $u_N = y_{N-1}$ and

$$W_{N-i}[u_{i+1}, \dots, u_N](x) = y_i(x)\, Q_{N-i}(x), \qquad i = 0, \dots, N-2, \tag{24}$$

cf. (10). We employ the recursive procedure described below. Given the quasi-polynomials u_{i+1}, $\dots, u_N$, we obtain the quasi-polynomial u_i by formula (31) and verify relation (24) in Proposition 4.5.

For the first step of the process, set $u_{N-1} = \mathcal{F}[y_{N-1}, y_{N-2} T_{N-1}]$. Then Lemma 4.2 for $i = N-1$ and Proposition 4.1 yield

$$W_2[u_{N-1}, u_N] = W_2[u_{N-1}, y_{N-1}] = y_{N-2} T_{N-1} = y_{N-2}\, Q_2, \tag{25}$$

which is relation (24) for $i = N-2$.

Lemma 4.3. *The quasi-polynomial* $y_{N-2}(x)$ *divides the quasi-polynomial*

$$B(x) = y_{N-2}(xq^2)\, y_{N-3}(x)\, u_{N-1}(xq^{-2})\, T_{N-2}(x) + y_{N-2}(xq^{-2})\, y_{N-3}(xq^2)\, u_{N-1}(x)\, T_{N-2}(xq^2)$$

of type $\lambda_{N-2} + 3\lambda_{N-1} + 2\lambda_N$.

Proof. By Lemma 4.2, $y_{N-2}(x)$ divides the quasi-polynomial $A_{N-2}(x)$, see (22), of type $\lambda_{N-2} + 2\lambda_{N-1} + 3\lambda_N$. Relations (21) and (25) yield

$$B(x) = A_{N-2}(x)\big(a_{N-1}(x)\,u_{N-1}(x) + b_{N-1}(x)\,u_{N-1}(xq^{-2})\big) + \\ + y_{N-2}(x)\,T_{N-1}(x)\big(b_{N-1}(x) - a_{N-1}(x)\big),$$

which proves Lemma 4.3. □

Assume that the quasi-polynomials $u_{i+1},\dots,u_N$ are constructed already, and the following properties hold.

A. For any $j = i,\dots,N$,

$$W_{N-j}[u_{j+1},\dots,u_N](x) = y_j(x)\,Q_{N-j}(x). \tag{26}$$

B. For any $j = i+1,\dots,N$, there is a quasi-polynomial $w_{i+1,j}(x)$ of type $\lambda_{i+1}+\dots+\lambda_N - \lambda_j$ such that

$$W_{N-i-1}[u_{i+1},\dots,u_{j-1},u_{j+1},\dots,u_N](x) = w_{i+1,j}(x)\,Q_{N-i-1}(x). \tag{27}$$

In particular, $w_{i+1,i+1}(x) = y_{i+1}(x)$, see (26) for $j = i+1$.

C. For any $j = i+1,\dots,N$, the quasi-polynomial $y_i(x)$ divides the quasi-polynomial

$$B_{ij}(x) = y_i(xq^2)\,y_{i-1}(x)\,w_{i+1,j}(xq^{-2})\,T_i(x) + y_i(xq^{-2})\,y_{i-1}(xq^2)\,w_{i+1,j}(x)\,T_i(xq^2)$$

of type $\lambda_i - \lambda_j + 3(\lambda_{i+1}+\dots+\lambda_N)$. In particular, $B_{i+1,i+1}(x) = A_{i+1}(x)$, see (22).

For $i = N-2$, property A coincide with formula (25), property B is straightforward: $w_{N-1,N-1} = u_N$, $w_{N-1,N} = u_{N-1}$, and property C follows from Lemma 4.2 for $i = N-2$, and Lemma 4.3.

Define the quasi-polynomials $w_{ij}(x)$, $j = i,\dots,N$, by the rule

$$w_{ij} = \mathcal{F}[y_i,\,y_{i-1}\,\check{w}_{i+1,j}\,T_i], \tag{28}$$

see (18), where $\check{w}_{i+1,j}(x) = \check{w}_{i+1,j}(xq^{-2})$. Property C and Proposition 4.1 yield

$$W_2[w_{ij},y_i](x) = y_{i-1}(x)\,w_{i+1,j}(xq^{-2})\,T_i(x). \tag{29}$$

Lemma 4.4. *The quasi-polynomial* $y_i(x)$ *divides the quasi-polynomial* $\sum_{j=i+1}^{N}(-1)^j\,w_{ij}(x)\,u_j(x)$ *of type* $\lambda_i+\dots+\lambda_N$.

Proof. Let $b_i = \mathcal{B}[y_i]$, see (16). By (28) and (18),

$$\sum_{j=i+1}^{N}(-1)^j\,w_{ij}(x)\,u_j(x) = y_i(x)\sum_{j=i+1}^{N}(-1)^j\,\mathcal{J}[y_i,\,y_{i-1}\,\check{w}_{i+1,j}\,T_i](x) + \\ + b_i(x)\sum_{j=i+1}^{N}(-1)^j\,y_{i-1}(x)\,T_i(x)\,w_{i+1,j}(xq^{-2})\,u_j(x).$$

Then formula (27), Lemma Appendix A.3 and formula (26) for $j = i$ give

$$\sum_{j=i+1}^{N} (-1)^{j-i-1} w_{i+1,j}(xq^{-2})\, u_j(x) = \frac{W_{N-i}[u_{i+1}, \dots, u_N](x)}{Q_{N-i-1}(xq^{-2})} = y_i(x) \prod_{k=i+1}^{N-1} T_k(x), \qquad (30)$$

which proves the lemma. □

Set $c_{ij}(x) = w_{ij}(x)/y_i(x)$, $j = i, \dots, N$, and

$$u_i(x) = \sum_{j=i+1}^{N} (-1)^{j-i-1} c_{ij}(x)\, u_j(x). \qquad (31)$$

By Lemma 4.4, $u_i(x)$ is a quasi-polynomial of type λ_i.

Proposition 4.5. $W_{N-i+1}[u_i, \dots, u_N](x) = y_{i-1}(x)\, Q_{N-i+1}(x)$.

Proof. By (29) and (27),

$$c_{ij}(x) - c_{ij}(xq^{-2}) = \frac{y_{i-1}(x)\, w_{i+1,j}(xq^{-2})\, T_i(x)}{y_i(x)\, y_i(xq^{-2})}$$

$$= \frac{y_{i-1}(x)\, T_i(x)\, W_{N-i-1}[u_{i+1}, \dots, u_{j-1}, u_{j+1}, \dots, u_N](xq^{-2})}{y_i(x)\, y_i(xq^{-2})\, Q_{N-i-1}(xq^{-2})}.$$

So Lemma Appendix A.3 yields

$$\sum_{j=i+1}^{N} (-1)^j \left(c_{ij}(x) - c_{ij}(xq^{-2})\right) u_j(xq^{-2l}) = 0, \qquad l = 1, \dots, N-i-1,$$

and

$$\sum_{j=i+1}^{N} (-1)^{N-j} \left(c_{ij}(x) - c_{ij}(xq^{-2})\right) u_j(xq^{2(i-N)}) =$$

$$= \frac{y_{i-1}(x)\, T_i(x)\, W_{N-i}[u_{i+1}, \dots, u_N](xq^{-2})}{y_i(x)\, y_i(xq^{-2})\, Q_{N-i-1}(xq^{-2})} = \frac{y_{i-1}(x)\, Q_{N-i+1}(x)}{W_{N-i}[u_{i+1}, \dots, u_N](x)},$$

where the last equality also uses formula (26) for $j = i$ and formula (23). These relations together with (31) give

$$u_i(xq^{-2l}) = \sum_{j=i+1}^{N} (-1)^{j-i-1} c_{ij}(x)\, u_j(xq^{-2l}), \qquad l = 1, \dots, N-i-1, \qquad (32)$$

and

$$u_i(xq^{2(i-N)}) = \frac{(-1)^{N-i} y_{i-1}(x)\, Q_{N-i+1}(x)}{W_{N-i}[u_{i+1}, \dots, u_N](x)} + \sum_{j=i+1}^{N} (-1)^{j-i-1} c_{ij}(x)\, u_j(xq^{2(i-N)}). \qquad (33)$$

Using equalities (32), (33) in the definition of $W_{N-i+1}[u_i, \dots, u_N]$, see (7) completes the proof of the proposition. □

Journal of Physics: Conference Series **411** (2013) 012020 doi:10.1088/1742-6596/411/1/012020

Proposition 4.6. *For any* $j = i, \ldots, N$,

$$W_{N-i}[u_i, \ldots, u_{j-1}, u_{j+1}, \ldots, u_N](x) = w_{ij}(x)\, Q_{N-i}(x),$$

and the quasi-polynomial $y_{i-1}(x)$ *divides the quasi-polynomial*

$$B_{i-1,j}(x) = y_{i-1}(xq^2)\, y_{i-2}(x)\, w_{ij}(xq^{-2})\, T_{i-1}(x) + y_{i-1}(xq^{-2})\, y_{i-2}(xq^2)\, w_{ij}(x)\, T_{i-1}(xq^2)$$

of type $\lambda_{i-1} - \lambda_j + 3(\lambda_i + \ldots + \lambda_N)$.

Proof. Using (32) in the definition of $W_{N-i}[u_i, \ldots, u_{j-1}, u_{j+1}, \ldots, u_N](x)$, we get

$$W_{N-i}[u_i, \ldots, u_{j-1}, u_{j+1}, \ldots, u_N](x) = c_{ij}(x)\, W_{N-i}[u_{i+1}, \ldots, u_N](x) = w_{ij}(x)\, Q_{N-i}(x).$$

By Lemma 4.2, $y_{i-1}(x)$ divides the quasi-polynomial $A_{i-1}(x)$, see (22), of type $\lambda_{i-1} + 2\lambda_i + 3(\lambda_{i+1} + \ldots + \lambda_N)$. Relations (21) and (29) yield

$$B_{i-1,j}(x) = A_{i-1}(x)\big(a_i(x)\, w_{ij}(x) + b_i(x)\, w_{ij}(xq^{-2})\big) + \\ + y_{i-1}(x)\, T_i(x)\, w_{i+1,j}(xq^{-2})\big(b_i(x) - a_i(x)\big),$$

which proves the second part of the proposition. □

Propositions 4.5 and 4.6 shows that properties A–C with i replaced by $i-1$ are valid. So we can construct recursively all quasi-polynomials $u_1, \ldots, u_N$ satisfying relations (24).

4.3. Proof of Theorem 3.4

In this section we will show that the sequence of monic polynomials $\mathcal{T} = (T_1, \ldots, T_{N-1}, 1)$ is a preframe of the collection of quasi-polynomials $\mathcal{U} = (u_1, \ldots, u_N)$ constructed in Section 4.2.

It is shown in Section 4.2 that the quasi-polynomials $u_1, \ldots, u_N$ satisfy relations (24), see Proposition 4.5. Moreover, for any $1 \leqslant i \leqslant j \leqslant N$, there is a quasi-polynomial $w_{ij}(x)$ of type $\lambda_i + \ldots + \lambda_N - \lambda_j$ such that

$$W_{N-i}[u_i, \ldots, u_{j-1}, u_{j+1}, \ldots, u_N](x) = w_{ij}(x)\, Q_{N-i}(x), \tag{34}$$

see Proposition 4.6, and

$$W_2[w_{ij}, y_i](x) = y_{i-1}(x)\, w_{i+1,j}(xq^{-2})\, T_i(x), \tag{35}$$

cf. (29).

Lemma 4.7. *For any* $1 \leqslant i \leqslant j < k \leqslant N$, *there is a quasi-polynomial* $\widetilde{w}_{i;j,k}(x)$ *of type* $\lambda_i + \ldots + \lambda_N - \lambda_j - \lambda_k$ *such that*

$$W_2[w_{ij}, w_{ik}](x) = y_{i-1}(x)\, \widetilde{w}_{i;j,k}(x)\, T_i(x).$$

Proof. For $j = i$, we have $w_{ii}(x) = y_i(x)$ and $\widetilde{w}_{i;i,k}(x) = -w_{i+1,k}(xq^{-2})$ by (35), so the statement holds. For $j > i$, by Lemma Appendix A.3 and formula (21) we have

$$\begin{aligned} W_2[w_{ij}, w_{ik}](x) \;=\; & \big(a_i(x)\, w_{ik}(x) + b_i(x)\, w_{ik}(xq^{-2})\big)\, W_2[w_{ij}, y_i](x) - \\ & - \big(a_i(x)\, w_{ij}(x) + b_i(x)\, w_{ij}(xq^{-2})\big)\, W_2[w_{ik}, y_i](x)\,, \end{aligned}$$

which proves Lemma 4.7. □

Proposition 4.8. *For any* $1 \leqslant i \leqslant j_1 < \ldots < j_k \leqslant N$, *there is a quasi-polynomial* $\widetilde{w}_{i;j_1,\ldots,j_k}(x)$ *of type* $\lambda_i + \ldots + \lambda_N - \lambda_{j_1} - \ldots - \lambda_{j_k}$ *such that*

$$W_k[w_{i,j_1}, \ldots, w_{i,j_k}](x) \;=\; \widetilde{w}_{i;j_1,\ldots,j_k}(x) \prod_{l=0}^{k-2} \Big(y_{i-1}(xq^{-2l}) \prod_{m=i}^{i+l} T_m(xq^{-2l}) \Big)\,.$$

Proof. We prove the statement by induction with respect to k. The case $k = 2$ is the base of induction, see Lemma 4.7.

Set $f_{ij} = W[y_i, w_{ij}]$. Assume that $j_1 = i$. By Lemma Appendix A.2 and formula (35),

$$\begin{aligned} W_k[y_i, w_{i,j_2}, \ldots, w_{i,j_k}](x) \;&=\; \frac{W_{k-1}[f_{i,j_2}, \ldots, f_{i,j_k}](x)}{y_i(xq^{-2}) \ldots y_i(xq^{-2k})} \;= \\ &= (-1)^{k-1} W_{k-1}[w_{i+1,j_2}, \ldots, w_{i+1,j_k}](xq^{-2}) \prod_{l=0}^{k-2} \frac{y_{i-1}(xq^{-2l})\, T_i(xq^{-2l})}{y_i(xq^{-2(l+1)})} \\ &= (-1)^{k-1}\, \widetilde{w}_{i+1;j_2,\ldots,j_k}(x) \prod_{l=0}^{k-2} \Big(y_{i-1}(xq^{-2l}) \prod_{m=i}^{i+l} T_m(xq^{-2l}) \Big)\,, \end{aligned}$$

where for the last equality we use the induction assumption. For $j_1 > i$, by Lemma Appendix A.3 and formula (21) we have

$$\begin{aligned} & W_k[w_{i,j_1}, \ldots, w_{i,j_k}](x) \;= \\ & = \sum_{l=1}^{k} (-1)^{l-1} \big(a_i(x)\, w_{i,j_l}(x) + b_i(x)\, w_{i,j_l}(xq^{-2})\big)\, W_k[y_i, w_{i,j_1}, \ldots, w_{i,j_{l-1}}, w_{i,j_{l+1}}, \ldots, w_{i,j_k}](x)\,, \end{aligned}$$

which proves the proposition. □

To complete the proof of Theorem 3.4 we should show that for any $k = 1, \ldots, N-1$, and any k-element subset $\{i_1, \ldots, i_k\} \subset \{1, \ldots, N\}$, the polynomial $Q_k(x)$ divides the quasi-polynomial $W_k[u_{i_1}, \ldots, u_{i_k}](x)$ of type $\lambda_{i_1} + \ldots + \lambda_{i_k}$.

Let $\{j_1, \ldots, j_{N-k}\}$ be the complement of $\{i_1, \ldots, i_k\}$ in $\{1, \ldots, N\}$. Then by formula (34), Lemmas Appendix A.1, Appendix A.4, and formula (24),

$$\begin{aligned} W_k[u_{i_1}, \ldots, u_{i_k}](x) \;=\; & \mathrm{const} \cdot x^{(k+1-N)(\lambda_1 + \ldots + \lambda_N)} \;\times \\ & \times\; W_{N-k}[w_{1,j_1}, \ldots, w_{1,j_{N-k}}](xq^{2(N-k-1)})\, Q_{N-1}(x) \prod_{l=1}^{N-k-1} \frac{Q_{N-1}(xq^{2l})}{Q_N(xq^{2l})}\,. \end{aligned}$$

Then by Proposition 4.8,

$$W_k[u_{i_1},\dots,u_{i_k}](x) \;=\; \mathrm{const}\cdot \widetilde{w}_{1;j_1,\dots,j_{N-k}}(xq^{2(N-k-1)})\,Q_k(x)\,,$$

where the quasi-polynomial $\widetilde{w}_{1;j_1,\dots,j_{N-k}}$ has type $\lambda_{i_1}+\dots+\lambda_{i_k}$. Theorem 3.4 is proved.

5. Difference operators

5.1. Difference operator of a collection of quasi-polynomials

Recall that $q\neq 0,\pm 1$. Denote by τ the multiplicative shift operator that acts on functions of x by the rule

$$(\tau f)(x) \;=\; f(xq^{-2})\,.$$

A function $f(x)$ is called a quasi-constant if $\tau f = f$.

An operator $D = a_0(x) + a_1(x)\,\tau + \dots + a_N(x)\,\tau^N$, where $a_0,\dots,a_N$ are functions and a_N is not identically zero, is called a *difference operator of order* N. The functions $a_0,\dots,a_N$ are the *coefficients* of D. If $a_N=1$, the operator D is called *monic*.

Recall that for any collection of quasi-polynomials $\mathcal{U}=(u_1,\dots,u_N)$ by definition in Section 3.1, $W_N[u_1,\dots,u_N]\neq 0$.

Lemma 5.1. *For any collection of quasi-polynomials* $\mathcal{U}=(u_1,\dots,u_N)$ *there exists a unique monic difference operator* $D_{\mathcal{U}}$ *such that* $D_{\mathcal{U}}\,u_i=0$ *for all* $i=1,\dots,N$.

Proof. Define the operator $D_{\mathcal{U}}$ by the rule

$$D_{\mathcal{U}}\,f \;=\; \frac{W_{N+1}[u_1,\dots,u_N,f]}{W_N[u_1,\dots,u_N]}\,. \tag{36}$$

Then clearly $D_{\mathcal{U}}\,u_i \;=\; 0$ for all $i \;=\; 1,\dots,N$. On the other hand, write $D_{\mathcal{U}} \;=\; a_0+\dots+a_{N-1}\,\tau^{N-1}+\tau^N$. Then equalities $D_{\mathcal{U}}\,u_i=0$, $i=1,\dots,N$, amount to a system of linear equations on $a_0,\dots,a_{N-1}$:

$$a_0\,u_i + a_1\,\tau u_i + \dots + a_{N-1}\,\tau^{N-1}u_i \;=\; -\,\tau^N u_i\,, \qquad i=1,\dots,N. \tag{37}$$

Since $W_N[u_1,\dots,u_N]\neq 0$, the matrix $(\tau^{j-1}u_i)_{i,j=1}^N$ is invertible and solution of system (37) has is unique. □

The operator $D_{\mathcal{U}}$ is called the *fundamental difference operator* of the collection $\mathcal{U}$. Notice that by (36),

$$a_0(x) \;=\; (-1)^N\,\frac{W_N[u_1,\dots,u_N](xq^{-2})}{W_N[u_1,\dots,u_N](x)}\,. \tag{38}$$

Lemma 5.2. *Let* $\mathcal{U} = (u_1,\dots,u_N)$ *be a collection of quasi-polynomials. Any solution* f *of the difference equation* $D_{\mathcal{U}}\,f = 0$ *is a linear combination of* $u_1,\dots,u_N$ *with quasi-constant coefficients.*

Proof. Let

$$c_i(x) \;=\; -\,\frac{W_N[u_1,\dots,u_{i-1},f,u_{i+1},\dots,u_N](x)}{W_N[u_1,\dots,u_N](x)}\,, \qquad i=1,\dots,N. \tag{39}$$

By Lemma Appendix A.4, $f = c_1 u_1 + \ldots + c_N u_N$. If $c_i \neq 0$, consider a collection $\widetilde{\mathfrak{U}}_i = (u_1, \ldots, u_{i-1}, f, u_{i+1}, \ldots, u_N)$. Lemma 5.1 implies that $D_{\widetilde{\mathfrak{U}}_i} = D_{\mathfrak{U}}$, and by formula (38),

$$\frac{W_N[u_1, \ldots, u_{i-1}, f, u_{i+1}, \ldots, u_N](xq^{-2})}{W_N[u_1, \ldots, u_{i-1}, f, u_{i+1}, \ldots, u_N](x)} = \frac{W_N[u_1, \ldots, u_N](xq^{-2})}{W_N[u_1, \ldots, u_N](x)}.$$

Thus $c_i(xq^{-2}) = c_i(x)$ for all $i = 1, \ldots, N$. □

Lemma 5.3. *For a collection of quasi-polynomials* $\mathfrak{U} = (u_1, \ldots, u_N)$, *set* $v_N = u_N$ *and*

$$v_i = \frac{W_{N-i+1}[u_i, \ldots, u_N]}{W_{N-i}[u_{i+1}, \ldots, u_N]}, \qquad i = 1, \ldots, N-1. \tag{40}$$

Then

$$D_{\mathfrak{U}} = \Big(\tau - \frac{\tau v_1}{v_1}\Big) \ldots \Big(\tau - \frac{\tau v_N}{v_N}\Big). \tag{41}$$

Proof. By applying repeatedly Lemma Appendix A.2, we get

$$\Big(\tau - \frac{\tau v_i}{v_i}\Big) \ldots \Big(\tau - \frac{\tau v_N}{v_N}\Big) f = \frac{W_{N-i+2}[u_i, \ldots, u_N, f]}{W_{N-i+1}[u_i, \ldots, u_N]}, \qquad i = 1, \ldots, N.$$

Comparing this formula with formula (36) completes the proof. □

Corollary 5.4. *Let* $\mathfrak{U}$ *be a regular collection of quasi-polynomials. Then the operator* $D_{\mathfrak{U}}$ *has rational coefficients.*

Proof. For a regular collection of quasi-polynomials $\mathfrak{U}$, the expressions $\tau v_i / v_i$ in formula (41) are rational functions, which proves the claim. □

Proposition 5.5. *Let* $\mathfrak{U} = (u_1, \ldots, u_N)$ *be a collection of quasi-polynomials such that the operator* $D_{\mathfrak{U}}$ *has rational coefficients. Then the collection* $\mathfrak{U}$ *is semiregular, that is, the quasi-polynomial* $W_N[u_1, \ldots, u_N]$ *is* log*-free.*

We will prove Proposition 5.5, as well as Propositions 5.6 – 5.9 below, in Section 5.3.

Proposition 5.6. *Let* $\mathfrak{U}$ *be a collection of quasi-polynomials such that the operator* $D_{\mathfrak{U}}$ *has rational coefficients. Then there is a regular collection of quasi-polynomials* $\widetilde{\mathfrak{U}}$ *such that* $D_{\widetilde{\mathfrak{U}}} = D_{\mathfrak{U}}$.

Say that $\boldsymbol{\lambda} = (\lambda_1, \ldots, \lambda_N)$ is *dominance-free* if $q^{2(\lambda_i - \lambda_j)} \neq q^{2s}$ for all $1 \leqslant i < j \leqslant N$ and $s \in \mathbb{Z}_{\geqslant 1}$.

Proposition 5.7. *Let* $\mathfrak{U}$ *be a collection of quasi-polynomials of type* $\boldsymbol{\lambda}$ *such that the operator* $D_{\mathfrak{U}}$ *has rational coefficients. Assume that* $\boldsymbol{\lambda}$ *is dominance-free, or* q *is a root of unity. Then there is a regular collection of quasi-polynomials* $\widetilde{\mathfrak{U}}$ *of type* $\boldsymbol{\lambda}$ *such that* $D_{\widetilde{\mathfrak{U}}} = D_{\mathfrak{U}}$.

Example. Let $\mathfrak{U} = (u_1, u_2)$ be the collection of quasi-polynomials of type $\boldsymbol{\lambda} = (1, 0)$:

$$u_1(x) = x, \qquad u_2(x) = \frac{1}{1 - q^{-2}} + \frac{q^2 x \log x}{2 \log q}.$$

Then

$$W_2[u_1, u_2](x) = x - x^2, \qquad D_{\mathfrak{U}} = \tau^2 - \Big(q^{-2} + \frac{1 - xq^{-2}}{1 - x}\Big)\tau + q^{-2}\frac{1 - xq^{-2}}{1 - x},$$

and $\widetilde{\mathcal{U}} = (u_2, u_1)$ is a regular collection of quasi-polynomials such that $D_{\widetilde{\mathcal{U}}} = D_{\mathcal{U}}$. However if q is not a root of unity, $D_{\mathcal{U}}$ is not the fundamental difference operator of any regular collection of quasi-polynomials of type $\boldsymbol{\lambda} = (1,0)$. If $q^{2\ell} = 1$ for a positive integer ℓ, set $u_2'(x) = x^\ell u_2(x)$. Then $\widetilde{\mathcal{U}}' = (u_2', u_1)$ is a regular collection of quasi-polynomials of type $\boldsymbol{\lambda} = (1,0)$ such that $D_{\widetilde{\mathcal{U}}'} = D_{\mathcal{U}}$.

Say that $\boldsymbol{\lambda} = (\lambda_1, \dots, \lambda_N)$ is *generic* if $q^{2(\lambda_i - \lambda_j)} \neq q^{2s}$ for all $1 \leqslant i < j \leqslant N$ and $s \in \mathbb{Z}$.

Proposition 5.8. *Let $\mathcal{U} = (u_1, \dots, u_N)$ be a collection of quasi-polynomials of type $\boldsymbol{\lambda}$. such that the operator $D_{\mathcal{U}}$ has rational coefficients. Assume that $\boldsymbol{\lambda}$ is generic. Then the quasi-polynomials $u_1, \dots, u_N$ are* log*-free. In particular, the collection $\mathcal{U}$ is regular.*

Proposition 5.9. *Let $\mathcal{U} = (u_1, \dots, u_N)$ and $\widetilde{\mathcal{U}} = (\tilde{u}_1, \dots, \tilde{u}_N)$ be collections of quasi-polynomials of type $\boldsymbol{\lambda}$ such that $D_{\mathcal{U}} = D_{\widetilde{\mathcal{U}}}$ and the operator $D_{\mathcal{U}}$ has rational coefficients. Assume that $\boldsymbol{\lambda}$ is generic. Then there are quasi-constants $c_1, \dots, c_N$ such that $\tilde{u}_i = c_i u_i$, $i = 1, \dots, N$. If q is not a root of unity, then the quasi-constants $c_1, \dots, c_N$ are constants.*

5.2. Difference operator of a solution of Bethe ansatz equations

Fix collections of complex numbers $\boldsymbol{\lambda} = (\lambda_1, \dots, \lambda_N)$, nonnegative integers $\boldsymbol{l} = (l_1, \dots, l_{N-1})$ and monic polynomials $\boldsymbol{T} = (T_1, \dots, T_{N-1})$.

Given $\boldsymbol{t} \in \mathbb{C}^{\boldsymbol{l}}$, define the polynomials $p_1(x), \dots, p_{N-1}(x)$ by the rule

$$p_i(x) = \prod_{j=1}^{l_i} (x - t_j^{(i)}), \qquad i = 1, \dots, N-1, \tag{42}$$

cf. (20), and set $p_0(x) = p_N(x) = 1$. Let

$$R_i(x) = \frac{p_{i-1}(x)}{p_i(x)} \prod_{j=i}^{N-1} T_j(xq^{2(i-j)}), \qquad i = 1, \dots, N, \tag{43}$$

Define the *fundamental difference operator* $D^{\boldsymbol{t}}$ of the point $\boldsymbol{t}$ by the rule

$$D^{\boldsymbol{t}} = \Big(\tau - q^{-2\lambda_1} \frac{R_1(xq^{-2})}{R_1(x)}\Big) \dots \Big(\tau - q^{-2\lambda_N} \frac{R_N(xq^{-2})}{R_N(x)}\Big). \tag{44}$$

Theorem 5.10. *Let $\boldsymbol{t} \in \mathbb{C}^{\boldsymbol{l}}$ be a solution of equations* (1) *associated with the data $\boldsymbol{l}, \boldsymbol{T}, \boldsymbol{\lambda}$. Let $\mathcal{U}$ be a collection of quasi-polynomials of type $\boldsymbol{\lambda}$ such that $\mathcal{T} = (T_1, \dots, T_{N-1}, 1)$ is a preframe of $\mathcal{U}$ and the $S_{\boldsymbol{l}}$-orbit of $\boldsymbol{t}$ equals $X_{\mathcal{U},\mathcal{T}}$. Then $D^{\boldsymbol{t}} = D_{\mathcal{U}}$.*

Proof. Let $\mathcal{U} = (u_1, \dots, u_N)$. Since the $S_{\boldsymbol{l}}$-orbit of $\boldsymbol{t}$ equals $X_{\mathcal{U},\mathcal{T}}$, formulae (9), (10) imply

$$x^{\lambda_i} R_i(x) = \frac{W_{N-i+1}[u_i, \dots, u_N](x)}{W_{N-i}[u_{i+1}, \dots, u_N](x)},$$

Thus $D^{\boldsymbol{t}} = D_{\mathcal{U}}$, see formulae (40), (41), (44). □

Recall that a point $\boldsymbol{t} \in \mathbb{C}^{\boldsymbol{l}}$ is admissible if it satisfies conditions (15).

Journal of Physics: Conference Series **411** (2013) 012020 doi:10.1088/1742-6596/411/1/012020

Corollary 5.11. *Let $\boldsymbol{t} \in \mathbb{C}^{\boldsymbol{l}}$ be an admissible solution of equations* (1) *associated with the data $\boldsymbol{l}, \boldsymbol{T}, \boldsymbol{\lambda}$. Then there exists a collection of quasi-polynomials $\mathcal{U} = (u_1, \dots, u_N)$ of type $\boldsymbol{\lambda}$ such that $D^{\boldsymbol{t}} = D_{\mathcal{U}}$ and*

$$W_N[u_1, \dots, u_N](x) = x^{\lambda_1 + \dots + \lambda_N} \prod_{i=1}^{N} \prod_{j=0}^{N-i} T_{N-i+1}(xq^{-2j}). \tag{45}$$

Proof. By Theorem 3.4, the collection of quasi-polynomials $\mathcal{U} = (u_1, \dots, u_N)$ constructed in Section 4.2 has a preframe $\mathcal{T} = (T_1, \dots, T_{N-1}, 1)$ and the $S_{\boldsymbol{l}}$-orbit of $\boldsymbol{t}$ equals $X_{\mathcal{U},\mathcal{T}}$. Hence, formula (45) holds, cf. (8), and $D^{\boldsymbol{t}} = D_{\mathcal{U}}$ by Theorem 5.10. □

Theorem 5.12. *Let $\boldsymbol{t} \in \mathbb{C}^{\boldsymbol{l}}$ be an admissible solution of equations* (1) *associated with the data $\boldsymbol{l}, \boldsymbol{T}, \boldsymbol{\lambda}$. Let $\mathcal{U}$ be a collection of quasi-polynomials of type $\boldsymbol{\lambda}$ such that $D^{\boldsymbol{t}} = D_{\mathcal{U}}$. Assume that q is not a root of unity and $\boldsymbol{\lambda}$ is generic. Then $\mathcal{T} = (T_1, \dots, T_{N-1}, 1)$ is a preframe of $\mathcal{U}$ and the $S_{\boldsymbol{l}}$-orbit of $\boldsymbol{t}$ equals $X_{\mathcal{U},\mathcal{T}}$.*

Proof. Let $\mathcal{U} = (u_1, \dots, u_N)$, and $\widetilde{\mathcal{U}} = (\tilde{u}_1, \dots, \tilde{u}_N)$ be the collection of quasi-polynomials of type $\boldsymbol{\lambda}$ constructed in the proof of Theorem 3.4, see Section 4.2. By Theorem 5.10 $D_{\widetilde{\mathcal{U}}} = D^{\boldsymbol{t}} = D_{\mathcal{U}}$. Then by Lemma 5.9, for any $i = 1, \dots, N$, the quasi-polynomials u_i and $\tilde{u}_i$ are proportional, which proves the theorem. □

Theorem 5.12 shows that for q being not a root of unity, generic $\boldsymbol{\lambda}$, and an admissible solution $\boldsymbol{t}$ of the Bethe ansatz equations (1) solving the difference equation $D^{\boldsymbol{t}} f = 0$ allows one to produce a collection of quasi-polynomials $\mathcal{U}$ of type $\boldsymbol{\lambda}$ such that the $S_{\boldsymbol{l}}$-orbit of $\boldsymbol{t}$ equals $X_{\mathcal{U},\mathcal{T}}$. This is an alternative way to the construction of such a collection of quasi-polynomials given in Section 4.2. Moreover, Theorem 5.12 and Lemma 5.9 imply that such a collection of quasi-polynomials is unique up to rescaling of individual quasi-polynomials.

5.3. Proofs of Propositions 5.5 – 5.9

Recall that $q \neq 0, \pm 1$. Most of the technicalities in this section are related to the fact that q can be a root of unity.

Lemma 5.13. *Let $c(x)$ be a quasi-constant of the form $c(x) = x^{\alpha} r(x, \log x)$, where $\alpha \in \mathbb{C}$ and $r(x, y)$ is a rational function in x and a polynomial in y. Then $r(x, y)$ does not depend on y.*

Proof. Let $r(x, y) = r_0(x) + r_1(x)\, y + \dots + r_k(x)\, y^k$ and $k \geqslant 1$. The equality $c(x) = c(xq^{-2})$ imply that

$$r_k(x) = q^{-2\alpha} r_k(xq^{-2}), \qquad r_{k-1}(x) - q^{-2\alpha} r_{k-1}(xq^{-2}) = -2k\, r_k(x) \log q. \tag{46}$$

Denote by d_i the order of $r_i(x)$ at $x = 0$, that is, $\lim_{x \to 0}(x^{-d_i} r_i(x)) \neq 0, \infty$. Then the first equality in (46) gives $q^{2(\alpha + d_k)} = 1$, which makes impossible matching the orders of the left and right sides in the second equality. □

For a quasi-polynomial $f(x) = x^{\alpha}(P_0(x) + P_1(x) \log x + \dots + P_k(x)(\log x)^k)$, where $P_0, \dots,$ P_k are polynomials, $P_k \neq 0$, denote by $\langle f \rangle(x) = x^{\alpha} P_k(x)$ the top part of $f(x)$.

Lemma 5.14. *Let D be a difference operator with rational coefficients. If a quasi-polynomial $f(x)$ satisfies the equation $Df = 0$, then $D\langle f \rangle = 0$ too.*

Proof. Since $\tau(x^{\alpha} x^i (\log x)^j) = q^{-2(\alpha+i)} x^{\alpha} x^i r(\log x)$ for some monic polynomial $r(s)$ of degree j, the claim follows. □

Proof of Proposition 5.5. Let $c(x) = W_N[u_1, \ldots, u_N](x)/\langle W_N[u_1, \ldots, u_N]\rangle(x)$. By Lemma 5.14, $c(x)$ is a quasi-constant, and $c(x)$ satisfies the assumption of Lemma 5.13. Since the quasi-polynomial $\langle W_N[u_1, \ldots, u_N]\rangle$ is log-free, Proposition 5.5 follows. □

Proof of Propositions 5.6, 5.7. We will prove the Propositions by induction on the number of quasi-polynomials in the collection $\mathcal{U} = (u_1, \ldots, u_N)$.

If the quasi-polynomial u_N is not log-free, let $f = \langle u_N \rangle$. By Lemma 5.14, $D_{\mathcal{U}} f = 0$. If $W_N[u_1, \ldots, u_{N-1}, f] \neq 0$, set $\widehat{\mathcal{U}} = (u_1, \ldots, u_{N-1}, f)$. Then the collection $\widehat{\mathcal{U}}$ has type $\boldsymbol{\lambda}$, $D_{\widehat{\mathcal{U}}} = D_{\mathcal{U}}$, and the quasi-polynomial f is log-free.

If $W_N[u_1, \ldots, u_{N-1}, f] = 0$, then $W_N[u_1, \ldots, u_{i-1}, f, u_{i+1}, \ldots, u_N] \neq 0$ for some $i = 1, \ldots, N-1$, see Lemma 39. Then for the collection $\widehat{\mathcal{U}} = (u_1, \ldots, u_{i-1}, u_N, u_{i+1}, \ldots, u_{N-1}, f)$, $D_{\widehat{\mathcal{U}}} = D_{\mathcal{U}}$, and the quasi-polynomial f is log-free.

The quasi-constants $c_i(x)$ given by (39) have the form $c_i(x) = x^{\lambda_N - \lambda_i} r_i(x)$ for some rational functions $r_i(x)$. Let d_i be the order of $r_i(x)$ at $x = 0$. Since $c_i(x)$ is a quasi-constant, $q^{2(\lambda_N - \lambda_i + d_i)} = 1$. For a dominance-free $\boldsymbol{\lambda}$, we have $d_i \leqslant 0$, and we set $\ell = 0$. If q is a root of unity, we find an integer $\ell \geqslant d_i$ such that $q^{2\ell} = 1$. Then $x^{\lambda_i - \lambda_N + \ell - d_i}$ is a quasi-constant, and $\hat{u}_i(x) = x^{\lambda_i - \lambda_N} x^{\ell - d_i} u_N(x)$ is a quasi-polynomials of type λ_i. Thus $\widehat{\mathcal{U}}' = (u_1, \ldots, u_{i-1}, \hat{u}_i, u_{i+1}, \ldots, u_{N-1}, f)$ is a collection of quasi-polynomials of type $\boldsymbol{\lambda}$, $D_{\widehat{\mathcal{U}}'} = D_{\mathcal{U}}$, and the quasi-polynomial f is log-free.

Assume now that the quasi-polynomial u_N is log-free, so $r(x) = u_N(xq^{-2})/u_N(x)$ is a rational function. Let the functions $v_1, \ldots, v_N$ be given by (40). Set

$$D = \left(\tau - \frac{\tau v_1}{v_1}\right) \ldots \left(\tau - \frac{\tau v_{N-1}}{v_{N-1}}\right) = \tau^{N-1} + b_{N-2}\,\tau^{N-2} + \ldots + b_0\,.$$

By Lemma 5.3, the operator $D_{\mathcal{U}}$ factors,

$$D \cdot \big(\tau - r(x)\big) = D_{\mathcal{U}} = \tau^N + a_{N-1}\,\tau^{N-1} + \ldots + a_0\,,$$

so the coefficients $b_0, , \ldots, b_{N-2}$ are determined by the equations

$$a_i(x) = b_{i-1}(x) - b_i(x)\, r(xq^{-2i})\,, \qquad i = 0, \ldots, N-1\,,$$

where $b_{-1}(x) = 0$, and $b_{N-1}(x) = 1$. Therefore, $b_0(x), , \ldots, b_{N-1}(x)$ are rational functions.

Let $u'_i = W_2[u_i, u_N]$, and $\mathcal{U}' = (u'_1, \ldots, u'_{N-1})$. By Lemmas 5.3 and 5.1, the operator $D_{\mathcal{U}'}$ acts as follows

$$D_{\mathcal{U}'} f(x) = \frac{1}{u_N(xq^{2(1-N)})}\, D\big(f(x)u_N(x)\big)\,,$$

and hence, has rational coefficients. By the induction assumption, there is a regular collection of quasi-polynomials $\mathcal{U}'' = (u''_1, \ldots, u''_{N-1})$ such that $D_{\mathcal{U}''} = D_{\mathcal{U}'}$. Set

$$c_{ij}(x) = -\frac{W_{N-1}[u'_1, \ldots, u'_{j-1}, u''_i, u'_{j+1}, \ldots, u'_{N-1}](x)}{W_{N-1}[u'_1, \ldots, u'_{N-1}](x)}\,, \qquad i, j = 1, \ldots, N-1\,. \quad (47)$$

Similarly to the proofs of Proposition 5.5 and Lemma 5.2, one can show that

$$u''_i(x) = c_{i,1}(x)\, u'_1(x) + \ldots + c_{i,N-1}(x)\, u'_{N-1}(x)\,, \qquad i = 1, \ldots, N-1\,, \quad (48)$$

Journal of Physics: Conference Series **411** (2013) 012020 doi:10.1088/1742-6596/411/1/012020

the Wronskians in (47) are log-free quasi-polynomials, and $c_{ij}(x)$ are quasi-constants of the form $c_{ij}(x) = x^{\alpha_{ij}} r_{ij}(x)$ for some rational functions $r_{ij}(x)$. Let d_{ij} be the order of $r_{ij}(x)$ at $x = 0$. Since $c_{ij}(x)$ is a quasi-constant, $q^{2(\alpha_{ij}+d_{ij})} = 1$, and the function $x^{-d_{ij}} r_{ij}(x)$ is also a quasi-constant. Denote by $P(x)$ the least common denominator of the functions $x^{-d_{ij}} r_{ij}(x)$, $i, j = 1, \dots, N-1$. The polynomial $P(x)$ is a quasi-constant as well.

Define the collection of quasi-polynomials $\widetilde{\mathfrak{U}} = (\tilde{u}_1, \dots, \tilde{u}_N)$ by the rule: $\tilde{u}_N = u_N$ and

$$\tilde{u}_i(x) = P(x)\big(c_{i,1}(x)\, u_1(x) + \ldots + c_{i,N-1}(x)\, u_{N-1}(x)\big)\,, \qquad i = 1, \dots, N-1\,.$$

Then $W_2[\tilde{u}_i, \tilde{u}_N](x) = P(x)\, u_i''(x)$, $i = 1, \dots, N-1$, see (48). By Lemmas Appendix A.1, Appendix A.2,

$$W_{N-i+1}[\tilde{u}_i, \dots, \tilde{u}_N](x) \prod_{j=1}^{N-i-1} \tilde{u}_N(xq^{-2j}) = W_{N-i}[u_i'', \dots, u_{N-1}''](x) \prod_{k=0}^{N-i} P(xq^{-2k})\,,$$

$i = 1, \dots, N-2$, which implies that the quasi-polynomials $W_{N-i+1}[\tilde{u}_i, \dots, \tilde{u}_N]$ are log-free, and $\widetilde{\mathfrak{U}}$ is a regular collection of quasi-polynomials. Since $D_{\widetilde{\mathfrak{U}}} = D_{\mathfrak{U}}$ by Lemma 5.1, Proposition 5.6 is proved. □

Proof of Proposition 5.9. By Proposition 5.5, the quasi-polynomial $W_N[u_1, \dots, u_N]$ is log-free. Since $D_{\mathfrak{U}}\, \tilde{u}_i = 0$, the quasi-polynomials $W_N[u_1, \dots, u_{j-1}, \tilde{u}_i, u_{j+1}, \dots, u_N]$ are log-free as well. Set

$$c_{ij}(x) = -\frac{W_N[u_1, \dots, u_{j-1}, \tilde{u}_i, u_{j+1}, \dots, u_N](x)}{W_N[u_1, \dots, u_N](x)}\,, \qquad i, j = 1, \dots, N\,.$$

Similarly to the proof of Lemma 5.2, $c_{ij}(x)$ are quasi-constants of the form $c_{ij}(x) = x^{\lambda_i - \lambda_j} r_{ij}(x)$ for some rational functions $r_{ij}(x)$. Since $\boldsymbol{\lambda}$ is generic, that is, $q^{2(\lambda_i - \lambda_j)} \neq q^{2s}$ for $i \neq j$ and any $s \in \mathbb{Z}$, we get that $c_{ij} = 0$ for $i \neq j$, and $\tilde{u}_i(x) = r_{ii}(x)\, u_i(x)$, $i = 1, \dots, N$, where $r_{ii}(x)$ are rational functions and quasi-constants.

If q is not a root of unity, then the only rational functions that are quasi-constants are constant functions. □

Proof of Proposition 5.8. Let $\tilde{u}_i = \langle u_i \rangle$, $i = 1, \dots, N$. Then $\widetilde{\mathfrak{U}} = (\tilde{u}_1, \dots, \tilde{u}_N)$ is a collection of quasi-polynomials of type $\boldsymbol{\lambda}$. By Lemmas 5.14 and 5.1, $D_{\widetilde{\mathfrak{U}}} = D_{\mathfrak{U}}$. By the proof of Proposition 5.9 there are rational functions $r_{ii}(x)$ such that $\tilde{u}_i(x) = r_{ii}(x)\, u_i(x)$, $i = 1, \dots, N$. Since the quasi-polynomials $\tilde{u}_1, \dots, \tilde{u}_N$ are log-free, the quasi-polynomials $u_1, \dots, u_N$ are log-free as well. □

Appendix A. The Wronskian identities

Lemma Appendix A.1. *Given functions $f_1(x), \dots, f_k(x)$ and $g(x)$, we have*

$$W_k[gf_1, \dots, gf_k](x) = W_k[f_1, \dots, f_k](x) \prod_{i=0}^{k-1} g(xq^{-2i})\,.$$

The proof is straightforward.

Lemma Appendix A.2. *Given functions* $f_1(x),\dots,f_j(x)$, $g_1(x),\dots,g_k(x)$, *let*

$$h_i(x) = W_{j+1}[g_i, f_1,\dots,f_j](x)\,, \qquad i=1,\dots,k\,.$$

Then

$$W_k[h_1,\dots,h_k](x) = W_{j+k}[g_1,\dots,g_k,f_1,\dots,f_j](x)\prod_{l=1}^{k-1} W_j[f_1,\dots,f_j](xq^{-2l})\,.$$

The proof is similar to that of Lemma 9.4 in [MV2].

Lemma Appendix A.3. *Given functions* $f_1(x),\dots,f_k(x)$, *we have*

$$\sum_{i=1}^{k}(-1)^i\, W_{k-1}[f_1,\dots,f_{i-1},f_{i+1},\dots,f_k](x)\, f_i(xq^{-2l}) = 0\,, \qquad l=0,\dots,k-2\,,$$

$$\sum_{i=1}^{k}(-1)^{k-i}\, W_{k-1}[f_1,\dots,f_{i-1},f_{i+1},\dots,f_k](x)\, f_i(xq^{2-2k}) = W_k[f_1,\dots,f_k](x)\,.$$

Proof. Consider a $k\times k$ matrix M with entries $M_{ij} = f_i(xq^{-2(j-1)})$ for $j<k$ and $M_{il} = f_i(xq^{-2l})$. If $l=0,\dots,k-2$, two rows of M are the same, hence $\det M = 0$. If $l=k-1$, then $\det M = W_k[f_1,\dots,f_k]$. Expanding the determinant in the last row yields the claim. □

Lemma Appendix A.4. *Given functions* $f_1(x),\dots,f_k(x)$, *let*

$$g_i(x) = W_{k-1}[f_1,\dots,f_{i-1},f_{i+1},\dots,f_k](x)\,, \qquad i=1,\dots,k\,.$$

Then

$$W_j[g_j,\dots,g_1](x) = W_{k-j}[f_{j+1},\dots,f_k](xq^{2-2j})\prod_{l=0}^{k-2} W_s[f_1,\dots,f_s](xq^{-2l})\,.$$

The proof is similar to that of Lemma 9.5 in [MV2].

References

[KBI] V. E. Korepin, N. M. Bogoliubov and A. G. Izergin, *Quantum inverse scattering method and correlation functions*, Cambridge University Press, 1993.

[KS] P. P.Kulish and E. K. Sklyanin, *Quantum spectral transform method. Recent developments* Lect. Notes in Phys. **151** (1982) 61–119.

[MV1] E. Mukhin, A. Varchenko, *Critical points of master functions and flag varieties*, Commun. Contemp. Math. **6** (2004), no. 1, 111–163.

[MV2] E. Mukhin, A. Varchenko, *Solutions to the XXX type Bethe ansatz equations and flag varieties*, Cent. Eur. J. Math. **1** (2003), no. 2, 238–271.

[MV3] E. Mukhin, A. Varchenko, *Quasi-polynomials and the Bethe ansatz*, Geom. Topol. Monogr. **13** (2008), 385–420.

XXth International Conference on Integrable Systems and Quantum Symmetries (ISQS-20) IOP Publishing
Journal of Physics: Conference Series **411** (2013) 012021 doi:10.1088/1742-6596/411/1/012021

Bilinear equations, Bell polynomials and linear superposition principle

Wen-Xiu Ma

Department of Mathematics and Statistics, University of South Florida, Tampa, FL 33620-5700, USA

E-mail: mawx@cas.usf.edu

Abstract. A class of bilinear differential operators is introduced through assigning appropriate signs and used to create bilinear differential equations which generalize Hirota bilinear equations. The resulting bilinear differential equations are characterized by a special kind of Bell polynomials and the linear superposition principle is applied to the construction of their linear subspaces of solutions. Illustrative examples are made by an algorithm using weights of dependent variables.

1. Introduction

Nonlinear differential equations play a significant role in exploring physical phenomena in depth. Hirota presented a direct method to solve a kind of specific bilinear differential equations [1]; and soliton solutions are, despite their diversity, a universal phenomenon that Hirota bilinear equations describe [2].

It is known that under $u = 2(\ln f)_{xx}$, the KdV equation

$$u_t + 6uu_x + u_{xxx} = 0 \tag{1.1}$$

can be transformed into

$$(D_x D_t + D_x^4) f \cdot f = 0, \tag{1.2}$$

which reads

$$f_{xt} f - f_x f_t + f_{xxxx} f - 4 f_{xxx} f_x + 3 f_{xx}^2 = 0. \tag{1.3}$$

Through this bilinear form, general Wronskian solutions, including solitons and complexitons, are presented for the KdV equation [3, 4]. The Hirota D-operators [5] are defined to be

$$D_x^n f \cdot g = (\partial_x - \partial_{x'})^n f(x) g(x')|_{x'=x} = \partial_{x'}^n f(x+x') g(x-x')|_{x'=0}. \tag{1.4}$$

For example, we have

$$\begin{cases} D_x f \cdot g = f_x g - f g_x, \\ D_x^2 f \cdot g = f_{xx} g - 2 f_x g_x + f g_{xx}, \\ D_x^3 f \cdot g = f_{xxx} g - 3 f_{xx} g_x + 3 f_x g_{xx} - f g_{xxx}. \end{cases} \tag{1.5}$$

It is very interesting that most integrable equations possess the Hirota bilinear form. Soliton solutions, particularly three-soliton solutions and Wronskian, Grammian and Pfaffian solutions, to Hirota bilinear equations can be generated by the Hirota perturbation and Pfaffian techniques [4, 6, 7, 8].

However, Hirota bilinear equations are special and there are many other bilinear differential equations which are not written in the Hirota bilinear form. This report will introduce a kind of generalized bilinear differential operators and their corresponding bilinear equations, which still possess nice mathematical properties. More importantly, we will talk about links of the presented bilinear equations with multivariate Bell exponential polynomials and their linear subspaces of solutions by the linear superposition principle.

2. Bilinear differential operators and bilinear equations

2.1. Bilinear D_p-operators

Let p be a given natural number. We introduce bilinear differential operators as follows:

$$(D_{p,x}^n f\cdot g)(x) = (\partial_x+\alpha\partial_{x'})^n f(x)g(x')\big|_{x'=x} = \sum_{i=0}^{n}\binom{n}{i}\alpha^i(\partial_x^{n-i}f)(x)(\partial_x^i g)(x),\ n\geq 1, \tag{2.1}$$

where the powers of α are determined by

$$\alpha^i = (-1)^{r(i)},\ \text{where } i = r(i) \bmod p \text{ with } 0\leq r(i)<p,\ i\geq 0. \tag{2.2}$$

Obviously, the case of $p=1$ gives the normal derivatives, and the cases of $p=2k,\ k\in\mathbb{N}$, reduce to Hirota bilinear operators.

We can observe that the powers α^i read

$$p=3:\ +,\ -,\ +,\ +,\ -,\ +,\ \cdots \text{ for } i=0,1,2,\cdots; \tag{2.3}$$

$$p=5:\ +,\ -,\ +,\ -,\ +,\ +,\ -,\ +,\ -,\ +,\ \cdots \text{ for } i=0,1,2,\cdots; \tag{2.4}$$

$$p=7:\ +,\ -,\ +,\ -,\ +,\ -,\ +,\ +,\ -,\ +,\ -,\ +,\ -,\ +,\ \cdots \text{ for } i=0,1,2,\cdots; \tag{2.5}$$

and thus

$$D_{3,x}^3 f\cdot g = f_{3x}g-3f_{2x}g_x+3f_xg_{2x}+fg_{3x},$$
$$D_{5,x}^5 f\cdot g = f_{5x}g-5f_{4x}g_x+10f_{3x}g_{2x}-10f_{2x}g_{3x}+5f_xg_{4x}+fg_{5x},$$
$$D_{7,x}^7 f\cdot g = f_{7x}g-7f_{6x}g_x+21f_{5x}g_{2x}-35f_{4x}g_{3x}+35f_{3x}g_{4x}-21f_{2x}g_{5x}+7f_xg_{6x}+fg_{7x},$$

which are different from the Hirota bilinear differential expressions [1].

A common feature that the D_p-operators share is the Taylor expansion

$$f(x+\delta)g(x+\alpha\delta) = \sum_{i=0}^{\infty}\frac{1}{i!}(D_{p,x}^i f\cdot g)\delta^i, \tag{2.6}$$

if we define

$$g(x+\alpha\delta) = \sum_{i=0}^{\infty}\frac{(\partial_x^i g)(x)}{i!}\alpha^i\delta^i. \tag{2.7}$$

The case of bilinear operators with more than one dependent variables can be similarly defined as follows:

$$\begin{aligned}&(D_{p,x_1}^{n_1}\cdots D_{p,x_l}^{n_l} f\cdot g)(x_1,\cdots,x_l)\\ &= (\partial_{x_1}+\alpha\partial_{x_1'})^{n_1}\cdots(\partial_{x_l}+\alpha\partial_{x_l'})^{n_l} f(x_1,\cdots,x_l)g(x_1',\cdots,x_l')|_{x_i'=x_i},\ n_1,\cdots,n_l\geq 1.\end{aligned} \tag{2.8}$$

2.2. Bilinear equations

A bilinear differential equation associated with a multivariate polynomial $\mathcal{F} = \mathcal{F}(x_1, \cdots, x_l)$ is defined by

$$\mathcal{F}(D_{p,x_1}, \cdots, D_{p,x_l}) f \cdot f = 0, \tag{2.9}$$

which reduces a Hirota bilinear equation if $p = 2k,\ k \in \mathbb{N}$. When $p = 5$, we particularly have the generalized bilinear KdV equation

$$(D_{5,x}D_{5,t} + D_{5,x}^4) f \cdot f = 2\, f_{xt} f - 2\, f_x f_t + 2 f_{4x} f - 8 f_{3x} f_x + 6\, f_{2x}^2 = 0, \tag{2.10}$$

the generalized bilinear Boussinesq equation

$$(D_{5,t}^2 + D_{5,x}^4) f \cdot f = 2\, f_{2t} f - 2\, f_t^2 + 2 f_{4x} f - 8 f_{3x} f_x + 6\, f_{2x}^2 = 0, \tag{2.11}$$

and the generalized bilinear KP equation

$$(D_{5,t}D_{5,x} + D_{5,x}^4 + D_{5,y}^2) f \cdot f = 2\, f_{xt} f - 2\, f_x f_t + 2 f_{4x} f - 8 f_{3x} f_x + 6\, f_{2x}^2 + 2\, f_{2y} f - 2\, f_y^2 = 0. \tag{2.12}$$

Such generalized bilinear equations have two common characteristics:

- Bilinear: The nearest neighbors to linear equations.
- Using the D_p-operators: Nice mathematical operators.

Two basic questions in the mathematical theory of bilinear equations are:

- How can one characterize bilinear equations defined by (2.9)?
- What kind of exact solutions are there to bilinear equations defined by (2.9)?

In this report, we would like to discuss those two questions, and provide solutions to both questions through the Bell exponential polynomials and the linear superposition principle, respectively.

3. Relations with Bell exponential polynomials

3.1. Bell polynomials

To begin with, let y be a C^∞ function of x and introduce

$$y_r = \partial_x^r y,\ r \geq 1. \tag{3.1}$$

The Bell polynomials are defined by

$$Y_{nx}(y) = Y_n(y_1, \cdots, y_n) = e^{-y}\partial_x^n e^y,\ n \geq 1, \tag{3.2}$$

in combinatorial mathematics [9]. A direct computation tells

$$\begin{cases} Y_1 = y_1,\ Y_2 = y_1^2 + y_2,\ Y_3 = y_1^3 + 3y_1y_2 + y_3, \\ Y_4 = y_1^4 + 6y_1^2y_2 + 4y_1y_3 + 3y_2^2 + y_4, \\ Y_5 = y_1^5 + 10y_1^3y_2 + 10y_1^2y_3 + 15y_1y_2^2 + 5y_1y_4 + 10y_2y_3 + y_5. \end{cases} \tag{3.3}$$

A special case of the Faà di Bruno formula (see, e.g., [10]) presents the Bell polynomials precisely:

$$Y_{nx}(y) = \sum \frac{n!}{m_1! \cdots m_n!(1!)^{m_1} \cdots (n!)^{m_n}} y_1^{m_1} \cdots y_n^{m_n}, \tag{3.4}$$

where the sum is over all n-tuples of nonnegative integers $(m_1, \cdots, m_n)$ satisfying the constraint $m_1 + 2m_2 + \cdots + nm_n = n$. The Bell polynomials can also be computed from

$$\exp(\sum_{r=1}^{\infty} \frac{y_r}{r!} t^r) = 1 + \sum_{n=1}^{\infty} \frac{Y_n(y_1, \cdots, y_n)}{n!} t^n. \tag{3.5}$$

The general formula (3.4) immediately tells the homogeneous property

$$Y_n(\alpha y_1, \alpha^2 y_2, \cdots, \alpha^n y_n) = \alpha^n Y_n(y_1, \cdots, y_n), \tag{3.6}$$

whose left-hand side is evaluated through first substituting all $\alpha y_1, \alpha^2 y_2, \cdots, \alpha^n y_n$ into Y_n and then collecting powers of α and computing them by the rule (2.2). On the other hand, the general Leibniz rule

$$(fg)^{-1}\partial_x^n (fg) = \sum_{i=0}^{n} \binom{n}{i} (f^{-1}\partial_x^{n-i} f)(g^{-1}\partial_x^i g) \tag{3.7}$$

shows the addition formula for the Bell polynomials:

$$Y_{nx}(y + y') = \sum_{i=0}^{n} \binom{n}{i} Y_{(n-i)x}(y) Y_{ix}(y'). \tag{3.8}$$

Those two properties will be used to link bilinear equations to a special kind of Bell polynomials.

3.2. Binary Bell polynomials

We first explore a relation of the Bell polynomials to the D_p-operators. For the sake of computational convenience, we assume that

$$f = e^{\xi(x)}, \ g = e^{\eta(x)}. \tag{3.9}$$

Then using (3.6) and (3.8), we have

$$\begin{aligned} (fg)^{-1} D_{p,x}^n f \cdot g &= \sum_{i=0}^{n} \alpha^i \binom{n}{i} (f^{-1}\partial_x^{n-i} f)(g^{-1}\partial_x^i g) \\ &= \sum_{i=0}^{n} \alpha^i \binom{n}{i} Y_{(n-i)x}(\xi) Y_{ix}(\eta) \\ &= Y_n(y_1, \cdots, y_n)|_{y_r = \xi_{rx} + \alpha^r \eta_{rx}}, \end{aligned} \tag{3.10}$$

where $\xi_{rx} = \partial_x^r \xi$ and $\eta_{rx} = \partial_x^r \eta, \ r \geq 1$.

Similarly to the case of the Hirota D-operators [11], we introduce binary Bell polynomials

$$\mathcal{Y}_{p;nx}(v, w) = Y_n(y_1, \cdots, y_n)|_{y_r = \frac{1}{2}(w_{rx} + v_{rx}) + \frac{1}{2}\alpha^r (w_{rx} - v_{rx})}, \ n \geq 1, \tag{3.11}$$

where $v_{rx} = \partial_x^r v$ and $w_{rx} = \partial_x^r w, \ r \geq 1$. For example, we have

$$\begin{cases} \mathcal{Y}_{3;x}(v,w) = v_x, \ \mathcal{Y}_{3;2x}(v,w) = v_x^2 + w_{2x}, \ \mathcal{Y}_{3;3x}(v,w) = v_x^3 + 3v_x w_{2x} + w_{3x}, \\ \mathcal{Y}_{3;4x}(v,w) = v_x^4 + 6v_x^2 w_{2x} + 3w_{2x}^2 + 4v_x w_{3x} + v_{4x}, \\ \mathcal{Y}_{3;5x}(v,w) = {v_x}^5 + 10v_x^3 w_{2x} + 15 v_x w_{2x}^2 + 10 v_x^2 w_{3x} + 10 w_{2x} w_{3x} + 5 v_x v_{4x} + w_{5x}. \end{cases} \tag{3.12}$$

This way, upon setting that

$$w = \xi + \eta, \ v = \xi - \eta, \tag{3.13}$$

from (3.10), we have a combinatorial formula for the D_p-operators:

$$(fg)^{-1}D_{p,x}^n f\cdot g = \mathcal{Y}_{p;nx}(v=\ln\frac{f}{g}, w=\ln fg). \tag{3.14}$$

To characterize bilinear equations, we further introduce $\mathcal{P}$-polynomials:

$$\mathcal{P}_{p;nx}(q) = \mathcal{Y}_{p;nx}(0,q), \tag{3.15}$$

the first few of which in the case of $p=3$ read

$$\begin{cases} \mathcal{P}_{3;x}(q)=0,\ \mathcal{P}_{3;2x}(q)=q_{2x},\ \mathcal{P}_{3;3x}(q)=q_{3x},\ \mathcal{P}_{3;4x}(q)=3q_{2x}^2, \\ \mathcal{P}_{3;5x}=q_{5x}+10q_{2x}q_{3x},\ P_{3;6x}=q_{6x}+15q_{2x}^2+10q_{3x}^2. \end{cases} \tag{3.16}$$

In terms of

$$q=w-v=2\ln g,\ v=\ln\frac{f}{g}, \tag{3.17}$$

the combinatorial formula (3.14) becomes

$$(fg)^{-1}D_{p,x}^n f\cdot g = \mathcal{Y}_{p;nx}(v, v+q). \tag{3.18}$$

Letting $f=g$, this tells a relation between bilinear expressions and the $\mathcal{P}$-polynomials:

$$f^{-2}D_{p,x}^n f\cdot f = \mathcal{P}_{p;nx}(q=2\ln f). \tag{3.19}$$

Therefore, a bilinear equation

$$\mathcal{F}(D_{p,x})f\cdot f=0 \ \text{ with } \ \mathcal{F}(x)=\sum_{i=0}^{n} c_i x^i \tag{3.20}$$

is equivalent to an equation on a linear combination of $\mathcal{P}$-polynomials in $q=2\ln f$:

$$\sum_{i=0}^{n} c_i\mathcal{P}_{p;ix}(q=2\ln f)=0. \tag{3.21}$$

This is a characterization for our generalized bilinear equations in one dimensional case.

3.3. Multivariate binary Bell polynomials

For a C^∞ function $y=y(x_1,\cdots,x_l)$, define the variables [12]:

$$y_{r_1,\cdots,r_l}=y_{r_1x_1,\cdots,r_lx_l}=\partial_{x_1}^{r_1}\cdots\partial_{x_l}^{r_l}y(x_1,\cdots,x_l),\ r_1,\cdots,r_l\ge 0,\ \sum_{i=1}^{l}r_i\ge 1, \tag{3.22}$$

and the multivariate Bell polynomials

$$Y_{n_1x_1,\cdots,n_lx_l}(y)=Y_{n_1,\cdots,n_l}(y_{r_1,\cdots,r_l})=e^{-y}\partial_{x_1}^{n_1}\cdots\partial_{x_l}^{n_l}e^y,\ n_1,\cdots,n_l\ge 0,\ \sum_{i=1}^{l}n_i\ge 1, \tag{3.23}$$

which can be computed through

$$\exp(\sum_{\substack{r_1+\cdots+r_l\ge 1\\ r_1,\cdots,r_l\ge 0}}\frac{y_{r_1,\cdots,r_l}}{r_1!\cdots r_l!}t_1^{r_1}\cdots t_l^{r_l})=1+\sum_{\substack{n_1+\cdots+n_l\ge 1\\ n_1,\cdots,n_l\ge 0}}\frac{Y_{n_1,\cdots,n_l}}{n_1!\cdots n_l!}t_1^{n_1}\cdots t_l^{n_l}. \tag{3.24}$$

Three examples in differential polynomial form are

$$\begin{cases} Y_{x,t} = y_{xt} + y_x y_t, \ Y_{2x,t} = y_{2x,t} + y_{2x} y_t + 2y_{xt} y_x + y_x^2 y_t, \\ Y_{3x,t} = y_{3x,t} + y_{3x} y_t + 3y_{2x,t} y_x + 3y_{2x} y_{xt} + 3y_{2x} y_x y_t + 3y_x^2 y_{xt} + y_x^3 y_t. \end{cases} \tag{3.25}$$

Based on (3.24), we can show the homogeneous property:

$$Y_{n_1,\cdots,n_l}(\alpha^{r_1+\cdots r_l} y_{r_1,\cdots,r_l}) = \alpha^{n_1+\cdots+n_l} Y_{n_1,\cdots,n_l}(y_{r_1,\cdots,r_l}), \tag{3.26}$$

and the general Lebnitz rule

$$(fg)^{-1}\partial_{x_1}^{n_1}\cdots\partial_{x_l}^{n_l} fg = \sum_{i_1=0}^{n_1}\cdots\sum_{i_l=0}^{n_l}\prod_{j=1}^{l}\binom{n_j}{i_j}(f^{-1}\partial_{x_1}^{n_1-i_1}\cdots\partial_{x_l}^{n_l-i_l} f)(g^{-1}\partial_{x_1}^{i_1}\cdots\partial_{x_l}^{i_l} g) \tag{3.27}$$

implies the addition formula for the multivariate Bell polynomials:

$$Y_{n_1x_1,\cdots,n_lx_l}(y+y') = \sum_{i_1=0}^{n_1}\cdots\sum_{i_l=0}^{n_l}\prod_{j=1}^{l}\binom{n_j}{i_j} Y_{(n_1-i_1)x_1,\cdots,(n_l-i_l)x_l}(y) Y_{i_1x_1,\cdots,i_lx_l}(y'). \tag{3.28}$$

Similarly for the sake of computational convenience, we assume that

$$f = e^{\xi(x_1,\cdots,x_l)}, \ g = e^{\eta(x_1,\cdots,x_l)}. \tag{3.29}$$

Then by (3.26) and (3.28), we can compute that

$$\begin{aligned} &(fg)^{-1} D_{p,x_1}^{n_1}\cdots D_{p,x_l}^{n_l} f\cdot g \\ &= \sum_{i_1=0}^{n_1}\cdots\sum_{i_l=0}^{n_l}\prod_{j=1}^{l}\alpha^{i_j}\binom{n_j}{i_j}(e^{-\xi}\partial_{x_1}^{n_1-i_1}\cdots\partial_{x_l}^{n_l-i_l} e^{\xi})(e^{-\eta}\partial_{x_1}^{i_1}\cdots\partial_{x_l}^{i_l} e^{\eta}) \\ &= \sum_{i_1=0}^{n_1}\cdots\sum_{i_l=0}^{n_l}\prod_{j=1}^{l}\alpha^{i_j}\binom{n_j}{i_j} Y_{(n_1-i_1)x_1,\cdots,(n_l-i_l)x_l}(\xi) Y_{i_1x_1,\cdots,i_lx_l}(\eta) \\ &= \sum_{i_1=0}^{n_1}\cdots\sum_{i_l=0}^{n_l}\prod_{j=1}^{l}\binom{n_j}{i_j} \ Y_{(n_1-i_1)x_1,\cdots,(n_l-i_l)x_l}(\xi_{r_1,\cdots,r_l}) Y_{i_1x_1,\cdots,i_lx_l}(\alpha^{r_1+\cdots+r_l}\eta_{r_1,\cdots,r_l}) \\ &= Y_{n_1,\cdots,n_l}(y_{r_1,\cdots,r_l} = \xi_{r_1,\cdots,r_l} + \alpha^{r_1+\cdots+r_l}\eta_{r_1,\cdots,r_l}). \end{aligned} \tag{3.30}$$

Let us now introduce binary multivariate Bell polynomials in differential polynomial form:

$$\mathcal{Y}_{p;n_1x_1,\cdots,n_lx_l}(v,w) = Y_{n_1,\cdots,n_l}\left(y_{r_1,\cdots,r_l} = \xi_{r_1x_1,\cdots,r_lx_l} + \alpha^{r_1+\cdots+r_l}\eta_{r_1x_1,\cdots,r_lx_l}\right), \tag{3.31}$$

where

$$w = \xi + \eta, \ v = \xi - \eta. \tag{3.32}$$

Then from (3.30), a combinatorial formula follows

$$(fg)^{-1} D_{p,x_1}^{n_1}\cdots D_{p,x_l}^{n_l} f\cdot g = \mathcal{Y}_{p;n_1x_1,\cdots,n_lx_l}(v = \ln\frac{f}{g}, w = \ln fg). \tag{3.33}$$

Further setting the following multivariate $\mathcal{P}$-polynomials:

$$\mathcal{P}_{p;n_1x_1,\cdots,n_lx_l}(q) = \mathcal{Y}_{p;n_1x_1,\cdots,n_lx_l}(v = 0, w = q). \tag{3.34}$$

For example, we have

$$\mathcal{P}_{3;x,t}(q) = 0,\ \mathcal{P}_{3;2x,t} = \frac{1}{4}q_x^2 q_t,\ \mathcal{P}_{3;3x,t} = \frac{3}{4}q_x^2 q_{xt} + \frac{3}{8}q_x^3 q_t + \frac{3}{4}q_x q_{2x} q_t. \tag{3.35}$$

It now follows that

$$f^{-2} D_{p,x_1}^{n_1} \cdots D_{p,x_l}^{n_l} f \cdot f = \mathcal{P}_{p;n_1x_1,\cdots,n_lx_l}(q = 2\ln f). \tag{3.36}$$

Thus, a bilinear equation

$$\mathcal{F}(D_{p,x_1}, \cdots D_{p,x_l}) f \cdot f = 0 \ \text{ with } \ \mathcal{F}(x_1, \cdots, x_l) = \sum_{i_1,\cdots,i_l=0}^{n} c_{i_1,\cdots,i_l} x_1^{i_1} \cdots x_l^{i_l} \tag{3.37}$$

is equivalent to an equation on a linear combination of multivariate $\mathcal{P}$-polynomials in $q = 2\ln f$:

$$\sum_{i_1,\cdots,i_l=0}^{n} c_{i_1,\cdots,i_l} \mathcal{P}_{p;i_1x_1,\cdots,i_lx_l}(q = 2\ln f) = 0, \tag{3.38}$$

where the coefficients $c_{i_1,\cdots,i_l}$'s are constants. This is a characterization for generalized bilinear equations defined through the D_p-operators.

4. Linear superposition principle

4.1. Subspaces of solutions

Let $\mathcal{F}(x_1, \cdots, x_l)$ be a multivariate polynomial. Consider a bilinear equation

$$\mathcal{F}(D_{p,x_1}, \cdots, D_{p,x_l}) f \cdot f = 0. \tag{4.1}$$

Define a set of N wave variables

$$\theta_i = k_{1,i}x_1 + \cdots + k_{l,i}x_l,\ 1 \le i \le N, \tag{4.2}$$

where the $k_{j,i}$'s are constants, and form a linear combination of N exponential waves

$$f = \sum_{i=1}^{N} \varepsilon_i\, e^{\theta_i} = \sum_{i=1}^{N} \varepsilon_i\, e^{k_{1,i}x_1 + \cdots + k_{l,i}x_l}, \tag{4.3}$$

where all ε_i's are arbitrary constants.

Note that we have the bilinear identities:

$$\mathcal{F}(D_{p,x_1}, \cdots, D_{p,x_l})\, e^{\theta_i} \cdot e^{\theta_j} = \mathcal{F}(k_{1,i} + \alpha k_{1,j}, \cdots, k_{l,i} + \alpha k_{l,j})\, e^{\theta_i + \theta_j},\ 1 \le i, j \le N, \tag{4.4}$$

where the powers of α obey the rule (2.2). Therefore, we can have the following criterion for obtaining the linear subspaces of solutions defined by (4.3) (see also [13, 14]).

Theorem: *Let $N \ge 1$. An arbitrary linear combination of N exponential waves defined by (4.3) solves the generalized bilinear equation (4.1) iff the constants $k_{j,i}$'s satisfy*

$$\mathcal{F}(k_{1,i} + \alpha k_{1,j}, \cdots, k_{l,i} + \alpha k_{l,j}) + \mathcal{F}(k_{1,j} + \alpha k_{1,i}, \cdots, k_{l,j} + \alpha k_{l,i}) = 0,\ 1 \le i \le j \le N. \tag{4.5}$$

Let $\mathcal{F}$ be a multivariate polynomial defined as in (3.37). Obviously, the formula (3.33) yields

$$D_{p,x_1}^{i_1} \cdots D_{p,x_l}^{i_l} e^{\theta_i} \cdot e^{\theta_j} = \mathcal{Y}_{p;i_1x_1,\cdots,i_lx_l}(\theta_i - \theta_j, \theta_i + \theta_j)\, e^{\theta_i+\theta_j}, \ 1 \le i, j \le N. \tag{4.6}$$

Thus, we obtain an equivalent theorem on the linear subspaces of exponential N-wave solutions.

Theorem': *Let $\mathcal{F}$ be defined by (3.37) and $N \ge 1$. An arbitrary linear combination of N exponential waves defined by (4.3) presents a solution to the generalized bilinear equation (4.1) iff the wave variables θ_i's satisfy*

$$\sum_{i_1,\cdots,i_l=0}^{n} c_{i_1,\cdots,i_l}[\mathcal{Y}_{p;i_1x_1,\cdots,i_lx_l}(\theta_i - \theta_j, \theta_i + \theta_j) + \mathcal{Y}_{p;i_1x_1,\cdots,i_lx_l}(\theta_j - \theta_i, \theta_j + \theta_i)] = 0, \ 1 \le i \le j \le N. \tag{4.7}$$

This theorem has an advantage that the wave variables θ_i's can be nonlinear functions of dependent variables, but the first theorem only works for linear wave variables θ_i's.

Given a multivariate polynomial $\mathcal{F}$, below is one way of solving the system (4.5) or (4.7) for $k_{j,i}$ and $c_{i_1,\cdots,i_l}$, in order to obtain bilinear equations and their linear subspaces of solutions (see, e.g., [13, 15]). We adopt a kind of parameterization for wave numbers and frequencies and list the sequential solution procedure as follows:

- Introduce weights for the independent variables:

$$(w(x_1), \cdots, w(x_l)) = (w_1, \cdots, w_l), \tag{4.8}$$

  where the weights w_i's can be both positive and negative.
- Form a homogeneous polynomial $\mathcal{F}(x_1, \cdots, x_l)$, defined by (3.37), in some weight.
- Parameterize $k_{1,i}, \cdots, k_{l,i}$ using a parameter k_i:

$$k_{j,i} = b_j k_i^{w_j}, \ 1 \le j \le l, \tag{4.9}$$

  and then determine the proportional constants b_j's and the coefficients $c_{i_1,\cdots,i_l}$'s by solving the system (4.5) or (4.7).

4.2. Illustrative examples

To present illustrative examples, we consider the $(3+1)$-dimensional case with

$$(w(x), w(y), w(z), w(t)) = (w_x, w_y, w_z, w_t), \tag{4.10}$$

and

$$\theta_i = k_i x + l_i y + m_i z - \omega_i t, \ l_i = b_1 k_i^{w_y}, \ m_i = b_2 k_i^{w_z}, \ \omega_i = -b_3 k_i^{w_t}, \ 1 \le i \le N. \tag{4.11}$$

Then, upon forming a homogeneous multivariate polynomial in some weight

$$\mathcal{F} = \sum_{i_1,i_2,i_3,i_4=1}^{n} c_{i_1,i_2,i_3,i_4} x^{i_1} y^{i_2} z^{i_3} t^{i_4}, \tag{4.12}$$

we solve the system (4.5) or (4.7) for the proportional constants b_1, b_2, b_3 and the coefficients c_{i_1,i_2,i_3,i_4}'s so that we can determine the corresponding bilinear equations and their associated linear subspaces of solutions consisting of linear combinations of exponential waves. We are going to present two concrete illustrative examples by applying this general idea below.

Journal of Physics: Conference Series **411** (2013) 012021 doi:10.1088/1742-6596/411/1/012021

Example 1 - Example with positive weights

Let us set the weights of independent variables:

$$(w(x), w(y), w(z), w(t)) = (1, 2, 4, 5), \tag{4.13}$$

and consider a polynomial being homogeneous in weight 6:

$$\mathcal{F} = c_1 x^6 + c_2 x^4 y + c_3 x^2 z + c_4 x t + c_5 y z. \tag{4.14}$$

Following the parameterization of wave numbers and frequencies in (4.11), we set the wave variables

$$\theta_i = k_i x + b_1 k_i^2 y + b_2 k_i^4 z + b_3 k_i^5 t,\ 1 \le i \le N, \tag{4.15}$$

where k_i, $1 \le i \le N$, are arbitrary constants but the proportional constants b_1, b_2 and b_3 are to be determined by (4.5).

Now, a direct computation shows that the corresponding bilinear equation reads

$$\begin{aligned}\mathcal{F}(D_{3,x}, D_{3,y}, D_{3,z}, D_{3,t}) f \cdot f &= 2c_1 f_{6x} f + 20 c_1 f_{3x}^2 + 2c_2 f_{4x,y} f - 2c_2 f_{4x} f_y - 8c_2 f_{3x,y} f_x \\ &+ 8c_2 f_{3x} f_{xy} + 12 c_2 f_{2x,y} f_{2x} + 2c_3 f_{2x,z} f + 2c_4 f_{xt} f - 2c_4 f_x f_t + 2c_5 f_{yz} f - 2c_5 f_y f_z = 0. \end{aligned} \tag{4.16}$$

The corresponding linear subspace of N-wave solutions is given by

$$f = \sum_{i=1}^N \varepsilon_i\, e^{\theta_i} = \sum_{i=1}^N \varepsilon_i\, e^{k_i x + b_1 k_i^2 y + b_2 k_i^3 z + b_3 k_i^4 t}, \tag{4.17}$$

where ε_i, $1 \le i \le N$, are arbitrary constants but the proportional constants b_1, b_2 and b_3 are defined by

$$b_1 = -\frac{25 c_3}{8 c_5},\ b_2 = \frac{5 c_2}{c_5},\ b_3 = \frac{10 c_1}{c_4}, \tag{4.18}$$

when the coefficients of the polynomial $\mathcal{F}$ satisfy

$$4 c_1 c_5 = 5 c_2 c_3. \tag{4.19}$$

Example 2 - Example with positive and negative weights

Let us set the weights of independent variables:

$$(w(x), w(y), w(z), w(t)) = (1, -1, -2, 3), \tag{4.20}$$

and consider a polynomial being homogeneous in weight 3:

$$\mathcal{F} = c_1 x^3 + c_2 x^5 z + c_3 x y t. \tag{4.21}$$

Following the parameterization of wave numbers and frequencies in (4.11), we set the wave variables

$$\theta_i = k_i x + b_1 k_i^{-1} y + b_2 k_i^{-2} z + b_3 k_i^3 t,\ 1 \le i \le N, \tag{4.22}$$

where k_i, $1 \le i \le N$, are arbitrary constants but the proportional constants b_1, b_2 and b_3 are to be determined by (4.5).

Similarly, a similar direct computation shows that the corresponding bilinear equation reads

$$\mathcal{F}(D_{3,x}, D_{3,y}, D_{3,z}, D_{3,t}) f \cdot f = 2c_1 f_{3x} f + 2c_2 f_{5x,z} f + 20 c_2 f_{3x} f_{2x,z} + 2c_3 f_{xyt} f = 0, \tag{4.23}$$

and it possesses the linear subspace of N-wave solutions determined by

$$f = \sum_{i=1}^N \varepsilon_i\, e^{\theta_i} = \sum_{i=1}^N \varepsilon_i\, e^{k_i x + b_1 k_i^{-1} y + b_2 k_i^{-2} z + b_3 k_i^3 t}, \tag{4.24}$$

where the ε_i's and k_i's are arbitrary but b_1, b_2 and b_3 need to satisfy

$$c_1 + 11 c_2 b_2 + c_3 b_1 b_3 = 0. \tag{4.25}$$

5. Conclusion and further questions

We created a kind of generalized bilinear differential operators $D_{p,x}$, discussed their links with the Bell polynomials, and applied the linear superposition principle to the corresponding bilinear equations. Two illustrative examples in the case of $p = 3$ were made to shed light on the general framework.

There are, however, many other questions which are worth further investigation. Below is just a few of them.

Question 1 - Mixed bilinear equations

We can mix D_p-operators with different natural numbers p to formulate a more general bilinear equation, for example,

$$\sum_{p_1,\cdots,p_l=1}^{m} \sum_{i_1,\cdots,i_l=0}^{n} c_{i_1,\cdots,i_l}^{p_1,\cdots,p_l} D_{p_1,x_1}^{i_1} \cdots D_{p_l,x_l}^{i_l} f \cdot f = 0,$$

where the coefficients $c_{i_1,\cdots,i_l}^{p_1,\cdots,p_l}$'s are constants. This kind of mixed combinations will bring diversity in establishing links with binary Bell polynomials and formulations for linear subspaces of solutions.

Question 2 - Geometries related to multivariate polynomials

What kind of geometries of a multivariate polynomial $\mathcal{F}$ does the equation

$$\mathcal{F}(k_1 + \alpha k_1', \cdots, k_l + \alpha k_l') + \mathcal{F}(k_1' + \alpha k_1, \cdots, k_l' + \alpha k_l) = 0$$

define? It determines an affine geometry of $\mathcal{F}$ when $p = 2k$.

Question 3 - Parameterizations achieved by multiple parameters

Parameterize $k_{1,i}, \cdots, k_{l,i}$ using multiple parameters, for example, two parameters k_i and l_i:

$$k_{j,i} = \sum_{r=0}^{w_i} b_{j,r} k_i^r l_i^{w_i - r}, \ 1 \le j \le l.$$

What kind of spaces can exist for the proportional constants $b_{j,r}$ which will solve the system

$$\mathcal{F}(k_{1,i} + \alpha k_{1,j}, \cdots, k_{l,i} + \alpha k_{l,j}) + \mathcal{F}(k_{1,j} + \alpha k_{1,i}, \cdots, k_{l,j} + \alpha k_{l,i}) = 0, \ 1 \le i \le j \le N?$$

Question 4 - Bilinear Bäcklund transformations

In the case of the Hirota D-operators, binary Bell polynomials are used to build bilinear Bäcklund transformations for soliton equations [16]. Is there any theory in the general case of D_p-operators? This case should be a little bit more difficult than the Hirota case, noticing a fact that the sign function $(-1)^{r(i)}$ in the definition (2.2) does not satisfy

$$(-1)^{r(i+j)} = (-1)^{r(i)+r(j)}, \ i, j \ge 0,$$

when $p > 1$ is odd, while it is true when p is one or even. It is obvious that the above property holds when p is one or even; but it does not hold because we have

$$(-1)^{r(p)} \ne (-1)^{r(p-1)+r(1)}, \text{ due to } (-1)^{r(p)} = 1, \ (-1)^{r(p-1)} = 1, \ (-1)^{r(1)} = -1,$$

when $p > 1$ is odd. The property is also crucial in deriving Lax pairs from bilinear Bäcklund transformations (see, e.g., [11, 12]).

Question 5 - Criterion for multivariate polynomials with one zero

While we used multivariate polynomials, which have one zero and only one zero, to determine bilinear equations with given linear subspaces of solutions, we came up with an interesting question [15]: How can one determine if a multivariate polynomial $\mathcal{F}(x_1, \cdots, x_l)$ with real coefficients has one and only one zero in $\mathbb{R}^l$? Two examples of such multivariate polynomials are as follows:

$$x^2 + y^2 - 2y + 1, \text{ zero } (x, y) = (0, 1);$$

$$5x^2 + 4xy + y^2 - 2x - 2y + 2, \text{ zero } (x, y) = (-1, 3).$$

This problem is more general than Hilbert's 17th problem, since all such multivariate polynomials satisfy all the conditions in Hilbert's 17th problem. It is hoped that there would be a definitive answer to the question.

Acknowledgments
The work was supported in part by the State Administration of Foreign Experts Affairs of China, the National Natural Science Foundation of China (Nos. 61072147 and 11071159), Chunhui Plan of the Ministry of Education of China, the Natural Science Foundation of Shanghai (No. 09ZR1410800) and the Shanghai Leading Academic Discipline Project (No. J50101).

References
[1] Hirota R 2004 *The Direct Method in Soliton Theory* (Cambridge: Cambridge University Press)
[2] Hietarinta J 2005 *Phys. AUC* **15** (part 1) 31
[3] Ma W X 2002 *Phys. Lett.* A **301** 35
[4] Ma W X and You Y 2005 *Trans. Amer. Math. Soc.* **357** 1753
[5] Hirota R 1974 *Prog. Theor. Phys.* **52** 1498
[6] Hirota R 1989 *J. Phys. Soc. Jpn.* **58** 2285
[7] Hu X B and Wang H Y 2006 *Inverse Problems* **22** 1903
[8] Ma W X, Abdeljabbar A and Asaad M G 2011 *Appl. Math. Comput.* **217** 10016
[9] Bell E T 1934 *Ann. Math.* **35** 258
[10] Craik A D D 2005 *Amer. Math. Monthly* **112** 217
[11] Lambert F, Springael J and Willox R 1994 *J. Phys. A: Math. Gen.* **27** 5325
[12] Gilson C, Lambert F, Nimmo J and Willox R 1996 *Proc. R. Soc. Lond.* A **452** 223
[13] Ma W X and Fan E G 2011 *Comput. Math. Appl.* **61** 950
[14] Ma W X 2012 *World Appl. Sci. J.* to appear
[15] Ma W X, Zhang Y, Tang Y N and Tu J Y 2012 *Appl. Math. Comput.* **218** 7174
[16] Lambert F and Springael J 1997 *J. Phys. Soc. Jpn.* **66** 2211

Casimir Energies for Spheres in n-dimensional Minkowski Space and Generalized Bernoulli Polynomials

Patrick Moylan

Department of Physics, The Pennsylvania State University, Abington College, Abington, PA, 19001, USA

E-mail: pjm11@psu.edu

Abstract. We consider a mathematical version of conformal quantum field theory in n-dimensions which give a unified approach to quantum field theory on Minkowski space and the Einstein universe. This unifying treatment enables us to relate massless fields in Minkowski space with massless fields in the Einstein universe into which such fields uniquely and conformally extend. Apart from certain global differences, fields in the Einstein universe approximate fields in Minkowski space as the radius of the Einstein universe tends to infinity and we utilize this fact to describe what seems to be a precise method for determining Casimir energies on spheres of arbitrary radii in n-dimensional Minkowski space.

Specifically, our exact results for Casimir energies of massless scalar fields in the n dimensional Einstein universe involving generalized Bernoulli polynomials together with results on conformal covariance properties of massless fields and a scaling property of the Casimir energy for n even should enable us to obtain exact results for Casimir energies for massless scalar fields with Dirichlet boundary conditions on spheres in n dimensional Minkowski space, at least for n even. We illustrate our calculational method in detail for the $n = 2$ case, which is completely representative of the arbitrary n case (n even).

1. Conformal geometry in two dimensions and massless representations of $SO(2,2)$

For presentational clarity we treat the $n = 2$ case in complete detail, since it exhibits all of the essential aspects of our approach for the higher dimensional cases. Let $\mathbb{R}P^3$ denote the real projective three space and consider the projective quadric $\mathbf{M}(2)$ in $\mathbb{R}P^3$ defined by the the equation $Q(u,u) = u_{-1}^2 + u_0^2 - u_1^2 - u_2^2 = 0$ where an equivalence class of points of the form $u = (u_{-1}, u_0, u_1, u_2) \in \mathbb{R}^4$ denotes a point in $\mathbb{R}P^3$. Noting that u and $-u$ denote the same point of Q, it follows that $\mathbf{M}(2)$ is homeomorphic to $(S^1 \times S^1)/\mathbb{Z}_2$. Clearly the standard action of $SO(2,2)$ on $\mathbb{R}^4$ induces action on $\mathbf{M}(2)$. $\mathbf{M}(2)$ is stable under this action, and it is well-known that this action is transitive on $\mathbf{M}(2)$.

Next consider the double cover of $\mathbf{M}(2)$ which is $S^1 \times S^1$. We denote this space-time by $\bar{\mathbf{M}}$ or, more specifically, by $\bar{\mathbf{M}}(2)$. A point u in $\bar{\mathbf{M}}(2)$ is clearly determined by $u_{-1}^2 + u_0^2 = u_1^2 + u_2^2 = 1$ and we may introduce spherical coordinates on $\bar{\mathbf{M}}(2)$ as follows:

$$u_{-1} = cost\ , \quad u_0 = sint\ , \quad u_1 = sin\rho\ , \ u_2 = cos\rho \tag{1.1}$$

with the ranges of the angular parameters being $0 < t < 2\pi$, $0 < \rho < 2\pi$. An $S^1 \times S^1$ invariant metric on $\bar{\mathbf{M}}(2)$ is

$$d\sigma^2 = dt^2 - d\rho^2 \tag{1.2}$$

where dt and $d\rho$ denote arc length in radians on S^1. The standard *curved measure* on $\bar{\mathbf{M}}(2)$ is given by the form

$$d^2u = dt \wedge d\rho \ . \tag{1.3}$$

Now define two dimensional Minkowski space, $\mathbf{M}_0(2)$, to be $\mathbb{R}^2$ as a vector space, but whose metrical structure is that of a two dimensional Lorentzian manifold with infinitesmal arc length ds in an inertial coordinate system determined by

$$ds^2 = dx_0^2 - dx_1^2 \ . \tag{1.4}$$

The above defined action of $SO(2,2)$ on $\mathbf{M}(2)$ is that of (global) conformal transformations on $\bar{\mathbf{M}}(2)$, and there is a conformal embedding π^{-1} from $\mathbf{M}_0(2)$ into $\bar{\mathbf{M}}(2)$, given by $\pi^{-1}(x) = u$, with [1]

$$u_{-1} = p\,F,\ u_0 = p\,x_0\ ,\quad u_1 = p\,x_1\ ,\ u_2 = p\,D \tag{1.5}$$

where

$$p = \{F^2 + x_0^2\}^{-\frac{1}{2}}\ ,\quad F = 1 - \frac{x_M^2}{4}\ ,\ D = 1 + \frac{x_M^2}{4} \tag{1.6}$$

with $x_M^2 = x_0^2 - x_1^2$. The embedding map π^{-1} is conformal, since

$$d\sigma^2 = p^2\, ds^2 \ . \tag{1.7}$$

The standard volume form on $\mathbf{M}_0(2)$ is given by

$$d^2x = dx_0 \wedge dx_1 \ . \tag{1.8}$$

Using π^{-1} we find

$$d^2x = (p)^{-2}\, d^2u \ . \tag{1.9}$$

Note that equations (1.5) and (1.1) give

$$x_0 = 2\frac{sint}{(cost + cos\rho)},\ x_1 = \frac{2sin\rho}{(cost + cos\rho)} \ . \tag{1.10}$$

The Einstein energy, H, is just the generator of rotations on the first S^1 factor, i.e.

$$H = \frac{\partial}{\partial\,\tau} \tag{1.11}$$

on $\bar{\mathbf{M}}(2)$, and thus the action of the one parameter group of conformal transformations generated by H, sends a wave function $\phi(\tau,\rho)$ into $\phi(\tau - s, \rho)$, i.e., under Einstein temporal evolution

$$\tau \longrightarrow \tau - s \tag{1.12}$$

with $s \in I\!R$. The action of the one parameter group of time translations in Minkowski space is generated by

$$P_0 = \frac{\partial}{\partial x_0} \ . \tag{1.13}$$

P_0 is the Minkowski energy.

Let $C_K^0(\overline{\mathbf{M}})$ be the space of all K finite elements of $C^\infty(\overline{\mathbf{M}})$ for which $\phi(u) = \phi(-u)$ $(u \in \overline{\mathbf{M}})$. $(K \stackrel{\sim}{=} SO(2) \times SO(2).)$ Now for $u \in \overline{\mathbf{M}}$ we have $u = (\tau, \rho)$ with τ and $\rho \in S^1$. A basis for $C_K^0(\overline{\mathbf{M}})$ is given by the $K-$finite functions

$$\phi_{n,m}(\tau,\ \rho)\ =\ e^{i(n\tau + m\rho)} \tag{1.14}$$

with $n,\ m\ =\ 0,\ \pm 1,\ \pm 2,\ \pm 3, . \ , \ . \ , \ .$, and with $(-1)^{(n+m)} = 1$ which comes from the above condition that $\phi \in C_K^0(\overline{\mathbf{M}})$ must be even under the antipodal map, $u \to -u$, on $\overline{\mathbf{M}}$.

The subspace of all ϕ in $C_K^0(\overline{\mathbf{M}})$ for which

$$\Big(\Delta_{S^1} - \Delta'_{S^1}\Big)\phi(u) := \Big(\frac{\partial^2}{\partial\ \tau^2}\ -\ \frac{\partial^2}{\partial\ \rho^2}\Big)\phi(u)\ =\ 0 \tag{1.15}$$

is an $so(2,2)$ module. Denote this subspace by H_K and the representation of $so(2,2)$ on it by $d\pi_0$. H_K is the space of massless scalar fields on $\bar{\mathbf{M}}$ and we have established the following result:

Proposition:
H_K decomposes into the following K-finite subspaces:

$$H_K\ =\ (H_K^+\ \oplus\ H_K^-)\ \oplus\ H_0\ \oplus\ (H_K^{+'}\ \oplus\ H_K^{-'}) \tag{1.16}$$

where H_K^+ and H_K^- are the spaces spanned by the basis elements (1.14) for which $n\ =\ m$ $(m > 0)$ and $n\ =\ -\ m$ $(m < 0)$, respectively, and $H_K^{+'}$ and $H_K^{-'}$ are the spaces spanned by the basis elements (1.14) for which $-n\ =\ m$ $(m > 0)$ and $-n\ =\ -\ m$ $(m < 0)$, respectively, and H_0 is the one dimensional space spanned by the constant function, which is given by (1.14) with $n = m = 0$. The irreducible quotients of the $so(2,2)$ invariant subspaces H_K^+ and H_K^- ($H_K^{+'}$ and $H_K^{-'}$) by subspace H_0 are infinitesmally unitarizable, irreducible lowest (highest) weight representations of $so(2,2)$, respectively. The restrictions of these four quotients to $so(1,2)$ are holomorphic or anti-holomorphic discrete series series representations, and their completion with respect to the $so(2,1)$ discrete series norms are representations spaces for irreducible, positive and negative energy representations of $SO_0(2,2)$. The $H_K^{\pm}/H_0$ and $H_K^{\pm'}/H_0$ are spaces for positive and negative energy representations of $SO_0(2,2)$, respectively. The positive energy representations describes massless particles and the negative energy representations describe the corresponding antiparticles on $\overline{\mathbf{M}}$.

In order to transfer fields on $\overline{\mathbf{M}}$ (i.e. functions in $C_K^0(\overline{\mathbf{M}})$) to fields "living" on $\mathbf{M}_0(2)$ we need to introduce the "flat parallelization." For this we make use of the Bruhat decomposition of $SO(2,2)$ along lines similar to the treatment given in [2], [3] for the four dimensional case. For scalar fields of weight d, i.e. representations of $SO(2,2)$ induced from one dimensional representations of the scale extended Poincaré group of $\mathbf{M}_0$ which act trivially on the translations, it leads to the following [1]:

$$\phi_0(x_0(\tau,\rho), x_1(t,\rho))\ =\ p^d\ \phi(\tau,\rho) \tag{1.17}$$

where $\phi_0(x_0, x_1)$ is the function on $\mathbf{M}_0(2)$ defined by this equation, $\phi(\tau,\rho) \in C_K^0(\overline{\mathbf{M}})$ and $p = \frac{1}{2}(u_{-1} + u_2)$ (cf. Eq. (1.6)). For the massless fields considered in the Proposition the conformal weight, d, is zero.

2. Casimir energy of a massless scalar field on $\mathbf{M}_0(2)$ vanishing on the endpoints of an interval and its relationship to the corresponding problem on $\overline{\mathbf{M}}(2)$

The parallelization map (Eq. (1.17) with $d = 0$) establishes a precise relationship between fields on $\overline{\mathbf{M}}(2)$ and fields on $\mathbf{M}_0(2)$ and it enables us to relate the quantum field theories on the respective manifolds. The Minkowski energy operator for a massless scalar field on $\mathbf{M}_0(2)$ is

$$H_0 \;=\; \frac{1}{2}\int_{I\!R}[(\partial\phi_0/\partial x_0)^2 \;+\; (\partial\phi_0/\partial x_1)^2]\;dx_1 \;\;. \tag{2.1}$$

The energy operator (Einstein energy operator) for a massless scalar field on $\overline{\mathbf{M}}(2)$ is given by

$$H \;=\; \frac{1}{2}\int_{S^1}[(\partial\phi/\partial\tau)^2 \;+\; (\partial\phi/\partial\rho)^2]\;ds \tag{2.2}$$

For the Casimir energy on $\overline{\mathbf{M}}(2)$ we consider a real massless scalar field $\phi(t,s)$ on $\overline{\mathbf{M}}(2)$ and consider the interval which is an arc of the circle, S^1, of radius R satisfying $0 \le s \le L$, S^1 denoting the spatial part of the Einstein universe. (In what was done above we worked with R=1 for computational facility. We continue this practice except for final results or when it is important to note the dependence on the radius R of $\overline{\mathbf{M}}(2)$, in which case we reinstate the explicit R dependence.) We shall impose Dirichlet boundary conditions on the region $0 \le s \le L$:

$$\phi(t,0) \;=\; \phi(t,L) \;=\; 0 \;\;. \tag{2.3}$$

It suffices to take L to be a half circle with arc length $\pi R \;=\; C/2$, where C denotes the circumference of S^1. The condition $\phi(t,0) \;=\; 0$ forces $\phi(t,s)$ to be an odd function of s. Thus by taking suitable linear combinations of the K-finite functions (1.14) we obtain the following positive and negative energy solutions of the wave equation (1.15):

$$\phi_n^{\pm}(t,s) \;=\; \left(\frac{c}{2\pi R\omega_n}\right)^{1/2} e^{\pm i\omega_n t}\sin k_n s \;, \tag{2.4}$$

$$\omega_n \;=\; ck_n \;\;,\;\; k_n \;=\; \frac{2\pi n}{C} \;\;. \tag{2.5}$$

Notice that the boundary conditions (2.3) impose no restrictions on the allowed values of n for the functions $\phi_n^{\pm}(t,s)$ of Eq. (2.4) i.e. the boundary conditions are automatically satisfied due to the periodicity of the sine function on S^1. The $\phi_n^{\pm}(t,s)$ are orthonormal with respect to the following $SO(2,2)$ invariant sesquilinear form:

$$(\phi_1,\,\phi_2) \;=\; \frac{i}{c}\int_{S^1} ds\,(\phi_1^*\,\partial_t\,\phi_2 \;-\; \phi_2^*\,\partial_t\,\phi_1) \;\;. \tag{2.6}$$

The standard quantization of the field ϕ is performed by means of the expansion

$$\phi(t,s) \;=\; \sum_{n=1}^{\infty}\left\{a_n\phi_n^-(t,s) \;+\; a_n^{\dagger}\phi_n^+(t,s)\right\} \;, \tag{2.7}$$

where now $\phi(t,s)$ is an operator valued distribution which for fixed $(t,\;s) \in \overline{\mathbf{M}}(2)$ acts on the Fock space as a densely defined self-adjoint operator. The quantities a_n and $a_n^{\dagger}$ satisfy the commutation relations

$$[a_n, a_m^{\dagger}] = \delta_{n,m} \tag{2.8}$$

and

$$[a_n, a_m] = [a_n^\dagger, a_m^\dagger] = 0 \quad . \tag{2.9}$$

Substitution of $\phi(t,s)$ into Eq. (2.2) gives after some standard calculations

$$H \;=\; \sum_{n=1}^{\infty} \hbar\omega_n \left[a_n^\dagger a_n \;+\; \frac{1}{2}\right] \;\cdot\; I \tag{2.10}$$

where I is the identity operator on Fock space. The vacuum state is defined by the equation

$$a_n|0>= 0 \quad . \tag{2.11}$$

Thus the total vacuum or zero point energy is thus

$$E =< 0|H|0 >= \frac{1}{2}\hbar \sum_{n=1}^{\infty} \omega_n \; . \tag{2.12}$$

In order to determine the Casimir energy for the massless, scalar field we regularize Eq. (2.12) by inserting an exponential damping function $e^{-n\alpha/R}$ inside the summation, and obtain for the finite part in the limit $\alpha \to 0$ the result:

$$E(R) = -\frac{1}{12}\frac{\hbar c}{R}. \tag{2.13}$$

Now we analyze the problem in the Minkowski space parallelization. According to Eqs. (1.5) an arc of the circle of length $L = \pi R$ corresponds to a line segment of length $4R$ in the spatial part of $\mathbf{M}_0(2)$, so we want to relate our results to the standard result for the Casimir energy of a massless scalar field with Dirichlet boundary conditions on the endpoints of an interval of length $L = 4R$ in $I\!R^1$ (the spatial part of $\mathbf{M}_0(2)$), which is:

$$E_0(L) = -\frac{\pi}{24}\frac{\hbar c}{L}. \tag{2.14}$$

(Note the scaling property of Eqn. (2.14): $E_0(L)$ is an homogeneous function of L of degree -1. This scaling property seems to hold true in all even dimensions i.e. for n even [4], [5], [6].)

Using Eqs. (1.9) and (1.10) and (1.17) with $d = 0$ we rewrite the Einstein energy operator, H, in Eq. (2.2) as an operator on $\mathbf{M}_0(2)$:

$$H = \frac{1}{2}\int_{I\!R} p^{-1}[(\partial\phi_0/\partial x_0)^2 \;+\; (\partial\phi_0/\partial x_1)^2]\; dx_1 \quad . \tag{2.2$'$}$$

Due to temporal invariance of Eq. (2.6) (cf. [2] (p. 123)) we shall work with τ close to zero, and so $\tau \overset{\sim}{=} \sin\tau = \frac{x_0}{1+x_1^2/4}$. Using this we find

$$\frac{d}{dx_0}e^{\pm i\omega_n \arcsin\left(\frac{x_0}{1+x_1^2/4}\right)} = \frac{\pm i\omega_n}{(1+x_1^2/4)}e^{\pm i\omega_n \arcsin\left(\frac{x_0}{1+x_1^2/4}\right)} \;+\; O(x_0^2) =$$

$$= \pm i\omega_n p(x_0 = 0, x_1)e^{\pm i\omega_n\rho} \;+\; O(x_0^2). \tag{2.15}$$

For the term in Eq. (2.2$'$) involving the spatial derivative, we have $\sin\rho = \frac{x_1}{1+x_1^2/4}$ on the surface $\tau = 0$ and we compute

$$\frac{d}{dx_1}\sin\left(k_n\rho\right) = -\left(\frac{k_n}{1+x_1^2/4}\right)\cos\left(k_n \arcsin\left(\frac{x_1}{1+x_1^2/4}\right)\right) =$$

$$= -k_n\, p(x_0 = 0, x_1)\ \cos(k_n \rho)\ . \tag{2.16}$$

Substitution of Eqs. (2.15) and (2.16) into Eq. (2.2′) and then computation of the vacuum expectation value of Eq. (2.2′) gives:

$$E(R) = \frac{1}{2}\int_{I\!R} dx_1 p^{-1} \frac{\hbar c}{2\pi R}\sum_{n=1}^{\infty} n = -\frac{1}{12}\frac{\hbar c}{R}\ , \tag{2.17}$$

which agrees with Eq. (2.13). In obtaining the last term we have regularized the infinite sum over n by the same method as before (exponential damping function) and extracted the relevant finite part.

Next we compute the vacuum expectation value of the Minkowski energy operator H_0 in the Minkowski space parallelization. We obtain

$$< 0|H_0|0 > \ = \frac{1}{2}\int_{I\!R} < 0|[(\partial\phi_0/\partial x_0)^2\ +\ (\partial\phi_0/\partial x_1)^2]\ dx_1\ \ |0 > \ = \frac{1}{2} < 0|H|0 >, \tag{2.18}$$

where the last step follows from use of Eqs. (2.15) and (2.16) and then the fact that $\int_{I\!R} p^2 dx_1 = \frac{1}{2}\int_{I\!R} p dx_1$. Thus the Casimir energy corresponding to H_0 equals $\frac{1}{2}E(R)$.

We now compute the regularized vacuum expectation value of the Minkowski energy operator as a power series in inverse powers of R to lowest order in $1/R$, and show that the resulting Casimir energy agrees with Eq. (2.14) for an interval of length $L = 4R$. For large R, we have:

$$\sin(k_n s) = k_n x_1 + O(1/R^2) \tag{2.19}$$

and

$$\tau = \frac{x_0}{R} + O(1/R^2)\ . \tag{2.20}$$

By use of these equations in Eq. (2.4) we obtain to order $O(1/R^2)$:

$$\phi_n^{\pm}(t,s)\ =\ \left(\frac{c}{2\pi R\omega_n}\right)^{1/2}\ e^{\pm i\omega_n x_0}\sin(k_n x_1)\ . \tag{2.21}$$

With these results it is easy to see that the problem of computation of the Casimir energy for the operator H_0 in the flat parallelization is exactly the same as the computation of the Casimir energy on an interval of length $4R$ in Minkowski space, and our result for the Casimir energy agrees, for $L = 4R$, with Eq. (2.14) up to higher order terms in the curvature, provided we use the same Fock space defined by Eqs. (2.8) and (2.9) and the same definition of the vacuum given in Eq. (2.11).

3. Further developments and conclusions

Almost everything described above can be generalized to arbitrary n. $SO(2,2)$ becomes $SO(2,n)$ and the generalization of the harmonic component of $C_K^0(\overline{\mathbf{M}})$, i.e. the analog of H_K, becomes the "minimal representation" of $SO(2,n)$ at least for n large enough [7]. Sine functions are replaced by Gegenbauer polynomials and we know their asymptotic expansions [8].

We now give a brief outline of how the above described method should work for higher dimensions with an aim to determine the Casimir energy for a massless scalar field in n dimensional Minkowski space (n even) with Dirichlet boundary conditions on a sphere of arbitrary radius. We start with our exact analytical expression for the Casimir energy of a massless scalar field on the n dimensional Einstein universe ($n > 3$)with Dirichlet boundary conditions on a sphere of maximal radius, namely [9]:

$$E(R,n) = -\frac{\hbar c}{2n!R} B_n^{\{n-1\}}(\frac{n-2}{2})$$

where the generalized Bernoulli polynomials $B_n^{\{\sigma\}}(x)$ are explicitly given by [10]: $B_n^{\{\sigma\}}(x) =$

$$\sum_{k=0}^{n} \binom{n}{k} \binom{\sigma+k-1}{k} \frac{k!}{(2k)!} \sum_{j=0}^{k} (-1)^j \binom{k}{j} j^{2k} (x+j)^{n-k} \times \quad {}_2F_1[k-n, k-\sigma; 2k+1; j/(x+j)]$$

with ${}_2F_1[k-n, k-\sigma; 2k+1; j/(x+j)]$ being the Gaussian hypergeometric function (cf. [8]). Using the appropriate asymptotic expansions of Gegenbauer polynomials, which occur in the problem (cf. ref. [11]), we are able to obtain an expansion of the vacuum expectation value of the Minkowski energy operator in inverse powers of R, just as in the $n = 2$ case. We should be able to relate the relevant finite part of the regularization of the term of lowest order in $1/R$ to the Casimir energy for a massless scalar field in n dimensional Minkowski space with Dirichlet boundary conditions on a sphere of radius $2R$ just as we did in the $n = 2$ case. Finally, to obtain the value of the Casimir energy for a sphere of arbitrary radius, we recall the scaling property of the Casimir energy, which was mentioned above, and seems to hold true at least for even n [4], [5], [6]. Standard results for Casimir energies on spheres obtained by asymptotic expansions and numerical evaluation are tabulated in ref. [4] (cf. also [12]) up to $n = 6$, and we should be able to compare our results with theirs.

The difficulties with n odd seems to be related to the lack of a rigorous prescription for the removal of divergences in the conventional quantum field theory on Minkowski space. A detailed treatment of the $n = 3$ case is given in [13]. There it is argued that a change in the sign of the Casimir force occurs in the neighborhood of a critical radius R_c. When $R > R_c$, the Casimir force is attractive while for $R < R_c$ the force becomes repulsive. Similar pathologies seems to hold generally for odd n [4]. Clearly the scaling property of the Casimir energy is lost in this case, so it seems that all we can get out of our method for the odd n cases are approximations of the Casimir energies for spheres of cosmic proportions in the spatial part of Minkowski space.

References

[1] Orsted B, Segal I E 1989 *Jour. Funct. Anal.* **83** 150-184.
[2] Paneitz S M, Segal I E 1982 *Jour. Funct. Anal.* **47** 78-142.
[3] Moylan P 1995 *Jour. Math. Phys.* **36** 6 2826-2879.
[4] Bordag M, Mohideen U, Mostepanenko V M 2001 *Phys. Rep.* **353** 1.
[5] Miltão M S R, Farias F A 2010 *Advances in High Energy Physics* 120964.
[6] Nesterenko V V, Pirozhenko I G 1997 arXiv:hep-th/9707253v1.
[7] Kobayashi T, Mano G 2007 *Proc. Japan Acad.* **83** Ser. A 27-31.
[8] Gradshteyn I C, Rhyshik I M 2000 *Tables of Integrals, Sums, Series and Derivatives*, (Academic Press, New York).
[9] Moylan P to appear in *Proceedings of the 9th International Workshop "Lie Theory and Its Application in Physics"* (LT-9, 2011) (Springer-Verlag).
[10] Sirvastava H M, Todorov P G 1988 *Jour. Math. Anal. and Appl.* **130** 509–513.
[11] Moylan P 2008 *Jour. Phys. Conf. Ser.* **128** 012010.
[12] Valuyan M A, Gousheh S S 2009 arXiv:0911.3578v1 [hep-th].
[13] Sen S 1981 *J. Math. Phys.* **22** 2968.

XXth International Conference on Integrable Systems and Quantum Symmetries (ISQS-20) IOP Publishing
Journal of Physics: Conference Series **411** (2013) 012023 doi:10.1088/1742-6596/411/1/012023

Construction of cyclic calorons

Daichi Muranaka[1], Atsushi Nakamula[2] and Nobuyuki Sawado[3]

[1]Department of Mathematics, Nagoya University, Nagoya, Aichi 464-8602, Japan
[2]Department of Physics, Kitasato University, Sagamihara, Kanagawa 252-0373, Japan
[3]Department of Physics, Tokyo University of Science, Noda, Chiba 278-8510, Japan

E-mail: [1]`d12003k@math.nagoya-u.ac.jp`,[2]`nakamula@sci.kitasato-u.ac.jp` (Corresponding author), [3]`sawado@ph.noda.tus.ac.jp`

Abstract. Analytic Nahm data of calorons, *i.e.*, Yang-Mills instantons on $\mathbb{R}^3 \times S^1$, with spatial C_N-symmetries are considered. The Nahm equations for the bulk data are reduced to the periodic Toda equation by applying the C_N-symmetric ansatz for monopoles by Sutcliffe. It is found that the bulk data are solved by elliptic theta functions which enjoy the hermiticity and the reality conditions. The defining relation to the boundary data are given and the C_3-symmetric Nahm data are found as an example.

1. Introduction

The Atiyah-Drinfeld-Hitchin-Manin (ADHM) [1] and the Nahm [2] constructions are powerful tool for getting the information of the moduli space to the anti-selfdual (ASD) Yang-Mills solitons, such as the instantons and the Bogomoln'yi-Prasad-Sommerfield (BPS) monopoles [3]. The principal ingredients of the ADHM/Nahm construction are the ADHM/Nahm data, respectively. One can obtain the gauge fields through the so called Nahm transform once the ADHM/Nahm data are determined. Although the ADHM/Nahm data are significant objects in the analysis on the ASD Yang-Mills solitons, the construction of the exact forms to the ADHM/Nahm data is unexplored field.

Among the ASD Yang-Mills solitons, calorons, or periodic instantons, are quite interesting object [4]. They are ASD instantons on $\mathbb{R}^3 \times S^1$, so that if we take the circumference of S^1 be infinitely large, the calorons are reduced to instantons. On the other hand, it is naturally expected to be the BPS monopoles, if we take the ratio of the size of the calorons to the circumference be infinity. Hence, calorons are expected to give the connection between instantons and BPS monopoles. There are articles demonstrating this interpolation in analytic [5, 6, 7, 8, 9].

In this paper, we give an outline for the construction of the Nahm data of calorons with instanton charge N associated to spatial C_N-symmetries around an axis. It is known that the Nahm data of calorons are composed of the bulk data and the boundary data, respectively. We will apply the C_N-symmetric ansatz for monopole Nahm data given by Sutcliffe [10, 11] as the bulk Nahm data of calorons, and find that the defining equations for the bulk data are the well-known periodic Toda equations. As we will see, it is not appropriate to fix the boundary data in the basis on which the Toda equations are solved. Hence we will make a unitary transformation into another basis, in which the reality conditions are manifest. We will give a conjecture on the unitary transformations between these basis. We restrict ourselves to consider the calorons of the $SU(2)$ gauge theory, and of trivial holonomy cases.

This paper is organized as follows. In section 2, we find the bulk Nahm data by solving the reduced Nahm equations, and confirm its hermiticity. In section 3, we consider the reality conditions for the bulk data. We give a conjecture on the manifest form of the unitary transformation from the C_N-symmetric basis, on which the Toda equations are solved, into the reality basis. In section 4, we consider the fixing of the boundary data. As an illustration, we give the C_3-symmetric boundary data explicitly. Section 5 is devoted to the conclusion.

2. C_N-symmetric ansatz for the bulk data

The Nahm data of calorons are compounded by two parts, $T_j(s)$ and W, where $j = 1, 2, 3$. The former is three $N \times N$ matrix valued regular functions periodic in s, refered to as the bulk data; we take the fundamental period $[-\mu, \mu]$ here. Throughout this work, we take the gauge in which $T_0(s)$ is gauged away. The latter is an N-dimensional row vector of quaternion entries, refered to as the boundary data. In the Nahm construction of calorons [2], the Nahm data $\{T_j(s), W\}$ are defined by the following four set of conditions,

$$T'_j(s) = \frac{i}{2}\epsilon_{jkl}[T_j(s), T_k(s)], \tag{1}$$

$$T_j^\dagger(s) = T_j(s), \tag{2}$$

$$T_j^t(s) = T_j(-s), \tag{3}$$

$$T_j(-\mu) - T_j(\mu) = \frac{1}{2}\mathrm{tr}_2\, \sigma_j W^\dagger W, \tag{4}$$

where i, j and k run through 1 to 3, ϵ_{ijk} is totally anti-symmetric tensor, and the derivative is with respect to the variable s. The first conditions (1) are differential equaitons for the bulk data, known as Nahm equations. The second (2) and the third (3) are the hermiticity conditions and the reality conditions for the bulk data, respectively. The reality conditions are necessary due to the group isomorphism $SU(2) \simeq Sp(1)$. Those conditions (1)–(3) are commonly required to the definition of the Nahm data of the $SU(2)$ BPS monopoles [12]. The fourth conditions (4) are the relations between the bulk data and the boundary data, called the matching conditions, where the trace is taken for the quaternions. In the Nahm construction for the BPS monopoles, the matching conditions (4) are taken place by the boundary conditions on the bulk data. The $SU(2)$ caloron gauge fields can be obtained by the Nahm data $\{T_j(s), W\}$ through the Nahm transform.

Let us now consider the symmetry of caloron Nahm data under the action of $SO(3)$, a rotation in the configuration space. Let us denote R an element of the subgroup of $SO(3)$, *i.e.*, for a spatial rotation of a position vector we have $x_j \mapsto x'_j = R_{jk}x_k$, where R_{jk} is an image of R in the 3-dimentional orthogonal representation of $SO(3)$. We also denote R_2 the image of R in the 2-dimensional irreducible representation of $SU(2)$, which gives rise to a spatial rotation of quaternions, $x \mapsto x' = R_2 x R_2^{-1}$, where $x = x_\mu e_\mu$ and the quaternion basis are defined by $e_\mu = (1, -i\sigma_1, -i\sigma_2, -i\sigma_3)$.

A caloron Nahm data is said to be symmetric under the action of R, if the Nahm data $\{T_j, W\}$ enjoys

$$R_N T_j R_N^{-1} = R_{jk} T_k, \tag{5}$$

and

$$R_N \otimes R_2\, W^\dagger = W^\dagger \hat{q}, \tag{6}$$

where R_N denotes the image of R in $SL(n, \mathbb{C})$, and $\hat{q}$ is a unit quaternion, such that $\hat{q}^\dagger \hat{q} = 1$. For the case of C_N-symmetris, it is given by

$$R_N = \omega^l\, \mathrm{diag.}[\omega^{N-1}, \omega^{N-2}, \ldots, \omega, 1], \tag{7}$$

where ω is an N-th root of unity and $l = 0, 1, \ldots, N-1$.

So far, we have given a very short review on the caloron Nahm data with a spatial symmetry. Now we determine the bulk Nahm data satisfying the C_N-symmetric condition (5). For this purpose, we will apply the ansatz for the monopole Nahm data with C_N-symmetries given by Sutcliffe over a decade ago [10, 11], whose uniqueness is proved in [13]. The form of the $N \times N$ bulk data is given in terms of the functions $f_j(s)$ and $p_j(s)$ $(j = 0, 1, 2 \ldots, N-1)$ as

$$T_1 = \frac{1}{2} \begin{bmatrix} & f_1 & & & f_0 \\ f_1 & & f_2 & & \\ & f_2 & & \ddots & \\ & & \ddots & & f_{N-1} \\ f_0 & & & f_{N-1} & \end{bmatrix}, \tag{8}$$

$$T_2 = \frac{i}{2} \begin{bmatrix} & -f_1 & & & f_0 \\ f_1 & & -f_2 & & \\ & f_2 & & \ddots & \\ & & \ddots & & -f_{N-1} \\ -f_0 & & & f_{N-1} & \end{bmatrix}, \tag{9}$$

$$T_3 = \frac{1}{2}\, \mathrm{diag.}\, [p_1, p_2, \cdots, p_{N-1}, p_0], \tag{10}$$

where we have omitted the argument of the functions. One can find that the hermiticity conditions (2) are enjoyed if $f_j, p_j \in \mathbb{R}$. Substituting the ansatz (8)–(10) into the Nahm equations (1), we obtain the differential equations for $f_j(s)$'s and $p_j(s)$'s

$$f_j' = \frac{1}{2} f_j \left(p_{j+1} - p_j\right) \tag{11}$$

$$p_j' = f_{j-1}^2 - f_j^2, \tag{12}$$

where the periodicity, $f_{j+N} = f_j$ and $p_{j+N} = p_j$, is taking into account. The system of differential equations (11) and (12) is well-known as the periodic Toda lattice. Note that it is necessary $\mathrm{tr}T_3' = \frac{1}{2}\sum_{j=0}^{N-1} p_j' = 0$ due to the Nahm equations, which will be confirmed later. Eliminating the $p_j(s)$'s in (11) and (12), the equations for f_j's are found to be

$$\frac{d^2}{ds^2} \log f_j^2 = -f_{j+1}^2 + 2f_j^2 - f_{j-1}^2, \tag{13}$$

which are well-known form of Toda lattice.

Let us now find special solutions to (13) appropriate for the caloron Nahm data. By introducing τ-functions $\tau_j := \tau(s, j)$, $(j = 0, 1, \ldots, N-1)$ as

$$f_j^2 = -C^2 \frac{\tau_{j-1}\tau_{j+1}}{\tau_j^2}, \tag{14}$$

the differential equations for τ_j's from (13) turn out to be

$$\frac{d^2}{ds^2} \log \tau_j = C^2 \frac{\tau_{j-1}\tau_{j+1}}{\tau_j^2}, \tag{15}$$

where C is a constant defined below. Simultaneously, we find the expression for p_j's by the τ-functions

$$p_j = \frac{d}{ds}\left(\log \frac{\tau_j}{\tau_{j-1}}\right), \tag{16}$$

from (11). Now let us suppose the following form to the τ-functions in terms of elliptic, or Jacobi, theta functions $\vartheta_\nu(u,q)$, where $u \in \mathbb{C}$ and $\nu = 0,1,2,3$ and q is so-called the modulus parameter [1]. The definition of the elliptic theta functions are given in the literatures, see *e.g.*, [17]. The assumption we suppose is

$$\tau_j(s) = \exp\left(\frac{1}{2}\tilde{A}s^2 + bs + \tilde{b}j\right)\vartheta_\nu(\pm s + \kappa j + a, q), \tag{17}$$

where $\tilde{A}, b, \tilde{b}, \kappa$ and a are constants. This form of the τ-functions for the periodic Toda lattice was originally introduced by M.Toda in 1967 [16]. Substituting (17) into (15), we find a differential equation for the theta functions,

$$A + C^{-2}\left(\log \vartheta_\nu(s_j)\right)'' = \frac{\vartheta_\nu(s_j - \kappa)\vartheta_\nu(s_j + \kappa)}{\vartheta_\nu^2(s_j)}, \tag{18}$$

where $s_j := \pm s + \kappa j + a$, $A := \tilde{A}C^{-2}$, and we have omitted the modulus dependence. This differential equation (18) is solved if the constants are given by the special values of theta functions [15, 16],

$$C^{-2} = \left(\frac{\vartheta_1(\kappa)}{\vartheta_1'(0)}\right)^2, \tag{19}$$

$$A = \tilde{A}C^{-2} = \left(\frac{\vartheta_0(\kappa)}{\vartheta_0(0)}\right)^2 - \frac{\vartheta_0''(0)}{\vartheta_0(0)}\left(\frac{\vartheta_1(\kappa)}{\vartheta_1'(0)}\right)^2. \tag{20}$$

We therefore have found the elliptic theta function solution to the Nahm equations (11) and (12),

$$f_j^2 = -C^2\frac{\vartheta_\nu(s_{j-1})\vartheta_\nu(s_{j+1})}{\vartheta_\nu(s_j)^2}, \tag{21}$$

$$p_j = \frac{d}{ds}\left(\log\frac{\vartheta_\nu(s_j)}{\vartheta_\nu(s_{j-1})}\right). \tag{22}$$

Note that the exponential factors are not appeared in the expression. By taking the square root of (21), f_j's are rewritten as

$$f_j = \pm iC\frac{\sqrt{\vartheta_\nu(s_{j-1})\vartheta_\nu(s_{j+1})}}{\vartheta_\nu(s_j)}. \tag{23}$$

Without loss of generality, we can now take the plus sign in (23).

We now recall the periodicity of the theta functions $\vartheta_\nu(u+1) = \pm\vartheta_\nu(u)$, where the sign depends on ν [2]. Hence, we easily find that the periodicity $f_{j+N} = f_j$ and $p_{j+N} = p_j$ follow if we take $\kappa = 1/N$.

Let us now fix the hermiticity (2) of the theta function solutions given above. For this, it is necessary that $f_j, p_j \in \mathbb{R}$ as mentioned earlier. From (22) and (23), it is sufficient to take $\vartheta_\nu(s_j)$ is real and positive valued on the region $s \in [-\mu,\mu]$, and $C \in i\mathbb{R}$, *i.e.*, pure imaginary valued. For these to be enjoyed, we choose the modulus parameter q takes real values $0 < q < 1$, then we find $\vartheta_\nu(s_j)$ is positive real valued on $s \in [-\mu,\mu]$ for $\nu = 0$ and 3. On the other hand, for $\nu = 1$ or 2, $\vartheta_\nu(s_j)$ has a zero and then changes sign on the real axis, so that the numerical value

[1] ϑ_0 is also expressed as ϑ_4 in the literatures.

[2] Note that the period in [17] is π instead of 1, which is only due to the scaling of the variable u.

in the square root of (23) will be able to negative. One can find there does not exist the case in which all of the f_j's take real valued simultaneously for $\nu = 1$ or 2. Hence we eliminate $\nu = 1$ and 2 and concentrate hereafter on the solutions of $\nu = 0$ or 3, both of which are even functions. Next we fix the constant C to be pure imaginary. From (19), we find

$$C = \pm \frac{\vartheta_1'(0)}{\vartheta_1(\kappa)}, \tag{24}$$

where both the numerator and the denominator have a factor $q^{1/4}$,

$$\vartheta_1'(0) = 2\pi q^{1/4} \prod_{m=1}^{\infty} (1 - q^{2m})^3, \tag{25}$$

$$\vartheta_1(\kappa) = 2q^{1/4} \sin \kappa\pi \prod_{m=1}^{\infty} (1 - 2q^{2m} \cos 2\kappa\pi + q^{4m})^3. \tag{26}$$

Hence, if we take the relative branch of $q^{1/4}$ in (24) such that $C \in i\mathbb{R}$, then the hermiticity (2) holds.

3. The reality conditions

Let us now consider the reality conditions (3), *i.e.*, $T_j^t(s) = T_j(-s)$. It can easily be found that the bulk Nahm data obtained above do not satisfy these conditions. For example, the elements of the diagonal matrix $T_3(s)$ have to be even functions for the reality condition, however, the elements $p_j(s)$, (22), are not so in general. Therefore, we have to show that there is a basis of the bulk Nahm data in which the reality conditions are apparent. The situation is similar to the monopoles with Platonic symmetries [14]. The new basis are given by a unitary transformation U_N,

$$T_j \mapsto \tilde{T}_j = U_N T_j U_N^{-1}, \tag{27}$$

where

$$\tilde{T}_j^t(s) = \tilde{T}_j(-s). \tag{28}$$

In the new basis of the bulk data $\tilde{T}_j$, however, the C_N-symmetries are not apparent. We hereafter refer the primary T_j as the bulk data in the "C_N-symmetric basis", and $\tilde{T}_j$ as that in the "reality basis". In addition, the transformation of the boundary data from the C_N-symmatric basis to the reality basis is read from (4) as

$$\tilde{W}^\dagger = U_N W^\dagger, \ \tilde{W} = W U_N^{-1}, \tag{29}$$

where we define $\tilde{W}$ as the boundary data in the reality basis.

The next task is to find the unitary transformation from the C_N-symmetric basis to the reality basis. We can guess the transformation matrices heuristically as follows. First, we consider the odd $N = 2k + 1$ ($k = 1, 2, \dots$) cases. By defining $(2k + 1) \times (2k + 1)$ matrices

$$J_{2k+1} := \begin{bmatrix} & & 1 \\ & \diagup & \\ 1 & & \end{bmatrix}, \tag{30}$$

we find unitary matrices

$$U_{2k+1} = \frac{1}{\sqrt{2}}(1_{2k+1} + iJ_{2k+1}) = \frac{1}{\sqrt{2}} \begin{bmatrix} 1 & & & & & & i \\ & \diagdown & & & & \diagup & \\ & & 1 & & i & & \\ & & & 1+i & & & \\ & & i & & 1 & & \\ & \diagup & & & & \diagdown & \\ i & & & & & & 1 \end{bmatrix}. \tag{31}$$

For even $N = 2k+2$, defining $(2k+2) \times (2k+2)$ matrices $J_{2k+2} := 1 \oplus J_{2k-1}$, then we find unitary matrices

$$U_{2k+2} = \frac{1}{\sqrt{2}}(1_{2k+2} + iJ_{2k+2}) = \frac{1}{\sqrt{2}} \begin{bmatrix} 1+i & & & & & & & \\ & 1 & & & & & & i \\ & & \diagdown & & & & \diagup & \\ & & & 1 & & i & & \\ & & & & 1+i & & & \\ & & & i & & 1 & & \\ & & \diagup & & & & \diagdown & \\ & i & & & & & & 1 \end{bmatrix}. \tag{32}$$

By using these unitary matrices, we give a conjecture that the transformed bulk Nahm data $\tilde{T}_j$, (27), enjoys the reality conditions (28) manifestly. We have confirmed the conjecture is correct at least $N \leq 6$. The detail will be published elsewhere.

To illustrate, we consider the cases $N = 3$ and 4. The $N = 3$ bulk Nahm data in the C_3-symmetric basis from the previous section are

$$T_1 = \frac{1}{2} \begin{bmatrix} 0 & f_1 & f_0 \\ f_1 & 0 & f_2 \\ f_0 & f_2 & 0 \end{bmatrix}, \tag{33}$$

$$T_2 = \frac{i}{2} \begin{bmatrix} 0 & -f_1 & f_0 \\ f_1 & 0 & -f_2 \\ -f_0 & f_2 & 0 \end{bmatrix} \tag{34}$$

$$T_3 = \frac{1}{2} \text{ diag. } [p_1, p_2, p_0]. \tag{35}$$

where f_j and p_j are given by (23) and (22). Here we choose $\kappa = 1/3$ and $a = 0$ for $s_j = \pm s + \kappa j + a$ so that $\vartheta_\nu(s_{j+3}) = \vartheta_\nu(s_j)$. The ambiguity of the sign in s_j is fixed to be consistent with the matching conditions (4). Performing the unitary transformation into the reality basis

$$\tilde{T}_j = U_3 \, T_j \, U_3^{-1}, \tag{36}$$

by using a unitary matrix

$$U_3 = \frac{1}{\sqrt{2}}(1_3 + iJ_3) = \frac{1}{\sqrt{2}} \begin{bmatrix} 1 & 0 & i \\ 0 & 1+i & 0 \\ i & 0 & 1 \end{bmatrix}, \tag{37}$$

we find the $N = 3$ bulk data in the reality basis are

$$\tilde{T}_1 = \frac{1}{2}\begin{bmatrix} 0 & f_+ - if_- & f_0 \\ f_+ + if_- & 0 & f_+ - if_- \\ f_0 & f_+ + if_- & 0 \end{bmatrix}, \tag{38}$$

$$\tilde{T}_2 = \frac{1}{2}\begin{bmatrix} f_0 & -f_+ - if_- & 0 \\ -f_+ + if_- & 0 & f_+ + if_- \\ 0 & f_+ - if_- & -f_0 \end{bmatrix}, \tag{39}$$

$$\tilde{T}_3 = \frac{1}{4}\begin{bmatrix} -p_2 & 0 & i(p_0 - p_1) \\ 0 & 2p_2 & 0 \\ -i(p_0 - p_1) & 0 & -p_2 \end{bmatrix}, \tag{40}$$

where $f_\pm(s) = (f_1(s) \pm f_2(s))/2$. To confirm the reality conditions (3), we notice the behavior of the theta functions under the inversion $s \to -s$,

$$\vartheta_\nu(s_j) = \vartheta_\nu(\pm s + j/3) \longrightarrow \vartheta_\nu(\mp s + j/3) = \vartheta_\nu(\pm s - j/3) = \vartheta_\nu(s_{-j}), \tag{41}$$

then we observe $f_0(s), f_+(s)$ and $p_2(s)$ are even functions, $f_-(s)$ and $p_0(s) - p_1(s)$ are odd functions in s, respectively. We therefore find that the reality conditions hold for the bulk data (38)–(40). We show the profile to the components of the bulk data in the reality basis in Figure 1, where the case of the ϑ_0 solution with $q = 0.5$ is displayed.

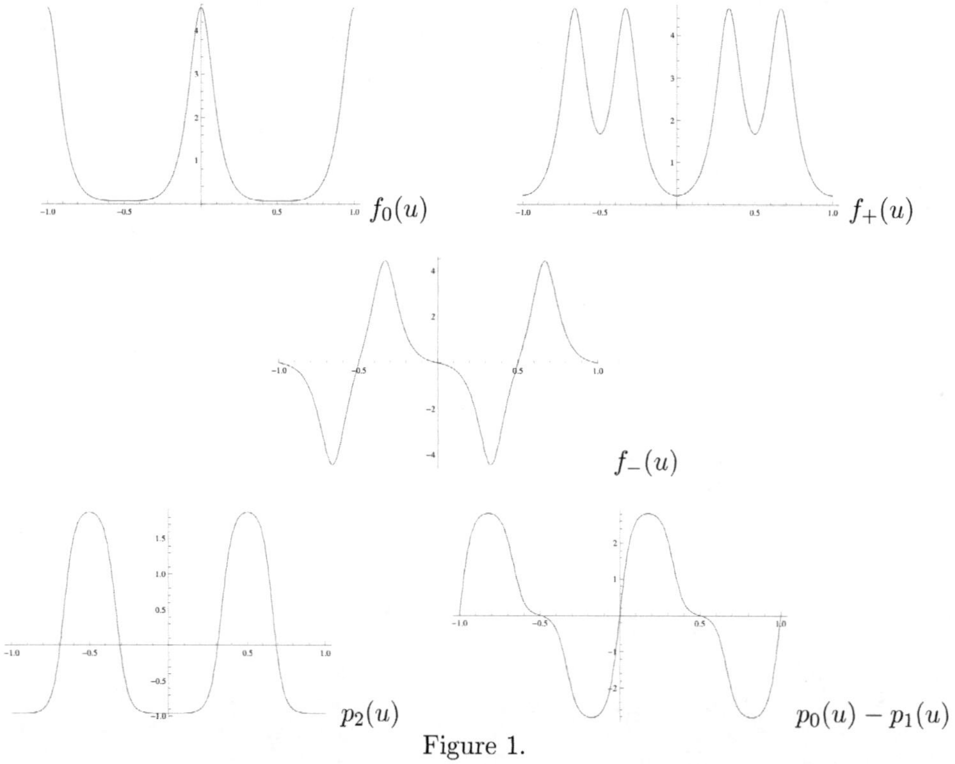

Figure 1.

For the case $N = 4$, the bulk Nahm data in the C_4-symmetric basis are

$$T_1 = \frac{1}{2}\begin{bmatrix} 0 & f_1 & 0 & f_0 \\ f_1 & 0 & f_2 & 0 \\ 0 & f_2 & 0 & f_3 \\ f_0 & 0 & f_3 & 0 \end{bmatrix}, \tag{42}$$

$$T_2 = \frac{i}{2}\begin{bmatrix} 0 & -f_1 & 0 & f_0 \\ f_1 & 0 & -f_2 & 0 \\ 0 & f_2 & 0 & -f_3 \\ -f_0 & 0 & f_3 & 0 \end{bmatrix}, \tag{43}$$

$$T_3 = \frac{1}{2}\,\mathrm{diag.}\,[p_1, p_2, p_3, p_0], \tag{44}$$

where f_j and p_j are given by (23) and (22), as in the $N = 3$ case. Here we choose $\kappa = 1/4$ and $a = -1/8$ in $s_j = \pm s + \kappa j + a$. By using a unitary matrix

$$U_4 = \frac{1}{\sqrt{2}}\begin{bmatrix} 1+i & 0 & 0 & 0 \\ 0 & 1 & 0 & i \\ 0 & 0 & 1+i & 0 \\ 0 & i & 0 & 1 \end{bmatrix}, \tag{45}$$

we perform the unitary transformation $\tilde{T}_j = U_4 T_j U_4^{-1}$, and find the transformed bulk data

$$\tilde{T}_1 = \begin{bmatrix} 0 & f_{+1} - if_{-1} & 0 & f_{+1} + if_{-1} \\ f_{+1} + if_{-1} & 0 & f_{+3} - if_{-3} & 0 \\ 0 & f_{+3} + if_{-3} & 0 & f_{+3} - if_{-3} \\ f_{+1} - if_{-1} & 0 & f_{+3} + if_{-3} & 0 \end{bmatrix}, \tag{46}$$

$$\tilde{T}_2 = \begin{bmatrix} 0 & f_{+1} + if_{-1} & 0 & -f_{+1} + if_{-1} \\ f_{+1} - if_{-1} & 0 & -f_{+3} - if_{-3} & 0 \\ 0 & -f_{+3} + if_{-3} & 0 & f_{+3} + if_{-3} \\ -f_{+1} - if_{-1} & 0 & f_{+3} - if_{-3} & 0 \end{bmatrix}, \tag{47}$$

$$\tilde{T}_3 = \frac{1}{4}\begin{bmatrix} 2p_1 & 0 & 0 & 0 \\ 0 & p_0 + p_2 & 0 & i(p_0 - p_2) \\ 0 & 0 & 2p_3 & 0 \\ 0 & -i(p_0 - p_2) & 0 & p_0 + p_2 \end{bmatrix}, \tag{48}$$

where we have defined

$$f_{\pm 1} := \frac{1}{4}(f_0 \pm f_1), \tag{49}$$

$$f_{\pm 3} := \frac{1}{4}(f_2 \pm f_3). \tag{50}$$

As in the case of $N = 3$, one can find f_{+1}, f_{+3}, p_1, p_3 and $p_0 + p_2$ are even fucntions, and f_{-1}, f_{-3} and $p_0 - p_2$ are odd functions, respectively, so that the bulk data (46)–(48) enjoy the reality conditions (3).

For larger N, we claim a conjecture that $\tilde{T}_j = U_N T_j U_N^{-1}$ satisfies the reality conditions (3) with an appropriate choice of κ and a in $s_j = \pm s + \kappa j + a$. Here the unitary matrix U_N is defined by (31) and (32) for odd and even N, respectively.

4. The boundary data

Let us now consider the boundary data. The boundary data are defined by the matching conditions (4), in which the contribution to the left hand side is only from the odd function parts of the bulk data. Thus, it is convenient to fix the boundary data in the reality basis considered in the previous section. The boundary data of the calorons with charge N in the reality basis are of the following form

$$\tilde{W} = (\lambda_1, \lambda_2, \dots, \lambda_N), \tag{51}$$

where λ_m $(m = 1, \dots, N)$ are quaternions with components $\lambda_m = (\lambda_{m0}, \lambda_{m1}, \lambda_{m2}, \lambda_{m3})$. Thus the matching conditions are

$$\begin{aligned}\tilde{T}_j(-\mu) - \tilde{T}_j(\mu) = \frac{1}{2}\mathrm{tr}_2\,\sigma_j \tilde{W}^\dagger \tilde{W} &= \frac{1}{2}\mathrm{tr}_2\,\sigma_j \begin{bmatrix} \lambda_1^\dagger\lambda_1 & \lambda_1^\dagger\lambda_2 & \dots & \lambda_1^\dagger\lambda_N \\ \lambda_2^\dagger\lambda_1 & \lambda_2^\dagger\lambda_2 & \dots & \lambda_2^\dagger\lambda_N \\ \vdots & \vdots & \ddots & \vdots \\ \lambda_N^\dagger\lambda_1 & \lambda_N^\dagger\lambda_2 & \dots & \lambda_N^\dagger\lambda_N \end{bmatrix} \\ &= \frac{1}{2}\mathrm{tr}_2\,\sigma_j \begin{bmatrix} 0 & \lambda_1^\dagger\lambda_2 & \dots & \lambda_1^\dagger\lambda_N \\ \lambda_2^\dagger\lambda_1 & 0 & \dots & \lambda_2^\dagger\lambda_N \\ \vdots & \vdots & \ddots & \vdots \\ \lambda_N^\dagger\lambda_1 & \lambda_N^\dagger\lambda_2 & \dots & 0 \end{bmatrix},\end{aligned} \tag{52}$$

where we have used the fact $\lambda_m^\dagger\lambda_m = \sum_{k=0}^{3}\lambda_{mk}^2 \in \mathbb{R}$ so that $\mathrm{tr}_2\sigma_j\lambda_m^\dagger\lambda_m = 0$. If we solve $\tilde{W}$ to (52) for a given bulk data $\tilde{T}_j(s)$, then the Nahm data $\{\tilde{T}_j(s), \tilde{W}\}$ gives a caloron of instanton charge N, however it has no specific symmetry in general. For the calorons to be C_N-symmetric, it is necessary that the boundary data enjoys (6) in addition to (52). To confirm the symmetry of the boundary data, we should make an investigation in the C_N-symmetric basis. From (29), the boundary data in the C_N-symmetric basis are given by

$$W^\dagger = U_N^{-1}\tilde{W}, \; W = \tilde{W}U_N. \tag{53}$$

For example, we consider the case $N = 3$ with C_3-symmetry. Let us first rename the quaternion entries of the boundary data for simplicity

$$\tilde{W} = (\lambda, \rho, \chi), \tag{54}$$

where λ, ρ and χ are quaternions with components $\lambda = (\lambda_0, \lambda_1, \lambda_2, \lambda_3)$, *etc.*. The matching conditions (52) turn out to be

$$\frac{1}{2}\mathrm{tr}_2\,\sigma_j \tilde{W}^\dagger \tilde{W} = \frac{1}{2}\mathrm{tr}_2\,\sigma_j \begin{bmatrix} 0 & \lambda^\dagger\rho & \lambda^\dagger\chi \\ \rho^\dagger\lambda & 0 & \rho^\dagger\chi \\ \chi^\dagger\lambda & \chi^\dagger\rho & 0 \end{bmatrix}. \tag{55}$$

From the odd function parts of the bulk data, we define

$$g(\mu) := \frac{1}{2}(f_-(-\mu) - f_-(\mu)) = -f_-(\mu), \tag{56}$$

$$\begin{aligned} h(\mu) &:= \frac{1}{4}\left\{(p_0(-\mu) - p_1(-\mu)) - (p_0(\mu) - p_1(\mu))\right\} \\ &= -\frac{1}{2}\left(p_0(\mu) - p_1(\mu)\right), \end{aligned} \tag{57}$$

then we find that the left hand sides of (52) are

$$\tilde{T}_1(-\mu) - \tilde{T}_1(\mu) = \begin{bmatrix} 0 & -ig(\mu) & 0 \\ ig(\mu) & 0 & -ig(\mu) \\ 0 & ig(\mu) & 0 \end{bmatrix}, \tag{58}$$

$$\tilde{T}_2(-\mu) - \tilde{T}_2(\mu) = \begin{bmatrix} 0 & -ig(\mu) & 0 \\ ig(\mu) & 0 & ig(\mu) \\ 0 & -ig(\mu) & 0 \end{bmatrix}, \tag{59}$$

$$\tilde{T}_3(-\mu) - \tilde{T}_3(\mu) = \begin{bmatrix} 0 & 0 & ih(\mu) \\ 0 & 0 & 0 \\ -ih(\mu) & 0 & 0 \end{bmatrix}, \tag{60}$$

respectively. We notice that the matching condition (52) is invariant under the multiplication of a unit quaternion h, *i.e.*, $h^\dagger h = 1$, to $\tilde{W}$ from the left. By using this degree of freedom, we can fix one of the quaternion component being zero. We choose, say, the real component of λ to be zero. Thus, the remaining components of the boundary data are

$$\begin{aligned} \lambda &= -i(\lambda_1\sigma_1 + \lambda_2\sigma_2 + \lambda_3\sigma_3), \\ \rho &= \rho_0 - i(\rho_1\sigma_1 + \rho_2\sigma_2 + \rho_3\sigma_3), \\ \chi &= \chi_0 - i(\chi_1\sigma_1 + \chi_2\sigma_2 + \chi_3\sigma_3). \end{aligned} \tag{61}$$

Then one can observe that a solution to (55) is given in terms of λ_1, λ_2 and λ_3, with $\lambda_1 \neq \lambda_2$ subject to a constraint

$$h(\mu) = \lambda_1^2 + \lambda_2^2 + \lambda_3^2 =: \Lambda^2, \tag{62}$$

as

$$\begin{aligned} &\rho_0 = -\frac{\lambda_1 + \lambda_2}{\Lambda^2} g(\mu), \quad \rho_1 = \rho_2 = \frac{\lambda_3}{\Lambda^2} g(\mu), \quad \rho_3 = \frac{\lambda_1 - \lambda_2}{\Lambda^2} g(\mu) \\ &\chi_0 = \lambda_3, \quad \chi_1 = -\lambda_2, \quad \chi_2 = \lambda_1, \quad \chi_3 = 0. \end{aligned} \tag{63}$$

This is the general boundary data without specific symmetry corresponding to the bulk data (38)–(40). There are two independent parameters in this boundary data due to the constraint (62). Note that the constraint (62) gives a restriction $h(\mu) > 0$ on the bulk data, which reads

$$h(\mu) = -\frac{1}{2}\left(p_0(\mu) - p_1(\mu)\right) > 0. \tag{64}$$

For this condition to be enjoyed, we have to take an appropriate sign in the argument of theta function, such as $s_j = s + j/3$ for the ϑ_3 solution, and $s_j = -s + j/3$ for the ϑ_0 solution, respectively.

Let us now consider the C_3-symmetric case, for which the boundary data in the C_3-symmetric basis satisfies

$$R_3 \otimes R_2\, W^\dagger = W^\dagger \hat{q}, \tag{65}$$

where $W = \tilde{W} U_3$, and R_3 and R_2 are the images of C_3-rotation. One can observe the boundary data (63) enjoys (65), if we impose additional constraints $\lambda_1 = \lambda_2$ and $\lambda_3 = 0$. In this case, the boundary data (63) reduces to

$$\begin{aligned} \lambda &= i\lambda_1(\sigma_1 + \sigma_2), \\ \rho &= \rho_0 = -\tfrac{2}{\lambda_1} g(\mu), \\ \chi &= -i\lambda_1(\sigma_1 - \sigma_2), \end{aligned} \tag{66}$$

with a constraint $h(\mu) = 2\lambda_1^2$. There is no independent parameter in the boundary data here. The exact form of R_2 and $\hat{q}$ are

$$R_2 = \hat{q} = \cos\left(\frac{\pi}{3}\right) + i\sin\left(\frac{\pi}{3}\right)\sigma_3. \tag{67}$$

We therefore have confirmed the existence of the C_3-symmetric caloron Nahm data.

5. Conclusion

To summarise, we have considered the caloron Nahm data with instanton charge N focusing on their spatial C_N-symmetries. Having applied the ansatz for the C_N-symmetric monopoles by Sutcliffe, we have reduced the bulk Nahm equations to the periodic Toda equations, which are solved by the elliptic theta functions. The hermiticity of the bulk data are also confirmed by a specific choice of the branch of the theta functions. We have also given a conjecture on the unitary transformation into a new basis of the Nahm data in which the reality conditions are manifest. In the new basis, as an illustration, the conditions on the boundary data have been written down explicitly for the $N = 3$ case, and the general solution is found. The boundary data of the $N = 3$ calorons with C_3-symmetry is also found, as a special case.

Acknowledgments Atsushi Nakamula is grateful to the conference organizers of ISQS-XX for their kind accommodation and hospitality.

References

[1] Atiyah M F, Drinfeld V G, Hitchin N J and Manin Yu I 1978 *Phys. Lett.* A **65** 185
[2] Nahm W 1984 *Springer Lecture Notes in Phys.* **201** 189
[3] Manton N and Sutcliffe P 2004 *Topological Solitons* (Cambridge: Cambridge University Press)
[4] Harrington B J and Shepard H K 1978 Phys. Rev. D **17** 2122; 1978 Phys. Rev. D **18** 2990
[5] K. -M. Lee and C. Lu, Phys.Rev.D **57** (1998) 5260;
[6] T. C. Kraan and P. van Baal, Nucl. Phys. B **533** (1998) 627;
[7] Ward R S 2004 *Phys.Lett.* B **582** 203
[8] D. Harland, J. Math. Phys. **48** (2007) 082905;
[9] A. Nakamula and J. Sakaguchi, J. Math. Phys. **51** (2010) 043503.
[10] Sutcliffe P M 1996 *Phys.Lett.* B **381** 129
[11] Sutcliffe P M 1997 *Nucl.Phys.* B **505** 517
[12] Hitchin N J 1983 *Comm.Math.Phys.* **89** 145
[13] Braden H W 2011 *Comm.Math.Phys.* **308** 303
[14] Hitchin N J, Manton N S and Murray M K 1995 *Nonlinearity* **8** 661
[15] Toda M 1989 *Nonlinear Waves and Solitons* (Tokyo: KTK Scientific Publishers/ Dordrecht, Boston, London: Kluwer Academic Publishers)
[16] Toda M 1967 *Journ. Phys. Soc. Japan* **22** 431
[17] Whittaker E T and Watson G N 1902 *A course of mordern analysis* (Cambridge: Cambridge University Press)

Towers with skeletons for the $(2+1)$-dimensional continuous isotropic Heisenberg spin model

Marcella Palese
Department of Mathematics, University of Torino, via C. Alberto 10, I-10123 Torino, Italy
E-mail: marcella.palese@unito.it

Abstract. We associate a tower with an infinitesimal algebraic skeleton to the (2 + 1)-dimensional (compact and noncompact) Heisenberg spin model. In particular, we construct the absolute parallelism defining the tower and the corresponding extension of the adjoint Lie algebra representation defining its skeleton.

2000 MSC: 58J70, 37K30.
Key words: nonlinear spin model, infinitesimal skeleton, tower, Cartan connection.

1. Introduction

The algebraic-geometric approach to integrability of nonlinear systems is based on the *request of the existence of conservation laws* which leads to the existence of symmetries expressed in terms of *algebraic structures*. The concept itself of integrability is a nontrivial matter and different definitions have been proposed in the literature on the subject, see *e.g.* [1]. We shall consider a concept of integrability as of having 'enough' conservation laws to exaustively describe the dynamics.

It is a well known fact that the study of equations at least locally variational (*i.e.* satisfying Helmholtz conditions) and thus arising as Euler–Lagrange equations of a (local) Lagrangian, enables one to precisely characterize conservation laws associated with symmetries of equations. From a physical point of view, in fact, field equations appear to be a fundamental object, since they describe the changing of the field in base space; symmetries of equations are transformations of the space leaving invariant the description of such a change. On the other hand the possibility of formulating a variational principle (*i.e.* a principle of *stationary* action) - from which both changing of fields and associated conservation laws (*i.e.* quantities not changing in the base space) could be obtained - enables to keep account of both what (*and how*) changes and what (*and how*) is conserved. In the variational calculus perspective Euler-Lagrange field equations are 'adjoint' to stationary principles up to conservation laws [5].

The question is now what can we say when we do not know, in principle, if a variational formulation, even local, of a given nonlinear model, is possible. It is clear that we would like to formulate a corresponding (maybe weaker) version of Helmholtz conditions of local variationality, being the Helmholtz morphism nothing but a quotient morphism of the exterior differential. In previous papers, we proposed an algebraic-geometric formulation in terms of integrable towers with infinitesimal algebraic skeletons suitably associated with nonlinear models, see *e.g.* [11, 13, 14]. The tower is constructed in such a way that both symmetries and conservation laws 'along equations' can be deduced by an integrability condition. In this paper we will consider algebraic structures also called 'open' Lie algebra structures, in the sense that not all the commutators (*i.e.* not all the Lie algebra structure constants) are determined, found by Wahlquist and Estabrook for the study of integrability properties of nonlinear dispersive systems, and related with the existence of an infinite set of associated conservation laws generated by pseudopotentials [16, 4]. Such algebras are different from freely generated infinite-dimensional Lie algebras and the geometric interpretation of them was the object of various studies, se *e.g.* [2, 3, 6, 9, 10, 11, 12, 13, 14, 15] and references therein.

In this note, we show that a tower with an infinitesimal algebraic skeleton can be associated with the $(2+1)$-dimensional continuous isotropic (compact and noncompact) Heisenberg spin model. In particular, we construct the absolute parallelism defining a tower, *i.e.* a connection 1-form with values in an infinite dimensional algebra associated with the model, and the corresponding extension of the adjoint Lie algebra representation defining its skeleton.

2. The $(2+1)$-dimensional Heisenberg spin model

In [9] algebraic properties of the $(1+1)$ dimensional (compact and noncompact) Heisenberg spin model, already well known to admit a Lax pair, were studied in both the direct and inverse prolongation structure procedure. In particular, the inverse prolongation (starting from the prolongation algebra) provided a whole family of spin models in $(1+1)$ dimension.

In this note, we shall investigate algebraic properties of a direct extension in more than one spatial dimensions: the continuous isotropic (compact and noncompact) Heisenberg model in $(2+1)$ dimension given by

$$(\Gamma \boldsymbol{S})_t = \boldsymbol{S} \times (\boldsymbol{S}_{xx} + \boldsymbol{S}_{yy}) \,, \qquad (1)$$

$$(\Gamma \boldsymbol{S}) \cdot \boldsymbol{S} = \gamma^2 \,,$$

$$\Gamma = diag(1,1,\gamma^2) \,, \qquad \gamma^2 = \pm 1 \,. \qquad (2)$$

Many variants of the model have been studied within different approaches; however the integrability properties of the present model obtained as the direct extension in $(2+1)$ dimension of the $(1+1)$-dimensional Heisenberg spin field model are not yet completely understood [7].

The construction of towers with skeletons associated with nonlinear field equations in $(2+1)$ dimension is highly non trivial [12, 13, 14]. We introduce some important concepts generalizing the concept of a homogeneous space and of a Cartan connection. To this aim let us represent this model by the following closed exterior differential system

$$\boldsymbol{\theta}_1 = d\boldsymbol{S} \wedge dy \wedge dt - \boldsymbol{S}_x dx \wedge dy \wedge dt \,,$$

$$\boldsymbol{\theta}_2 = d\boldsymbol{S} \wedge dx \wedge dt + \boldsymbol{S}_y dx \wedge dy \wedge dt \,,$$

$$\boldsymbol{\theta}_3 = d(\Gamma \boldsymbol{S}) \wedge dy \wedge dy + \boldsymbol{S} \times (d\boldsymbol{S}_x \wedge dy \wedge dt - d\boldsymbol{S}_y dx \wedge dt) \,,$$

$$\beta_1 = d(\Gamma \boldsymbol{S}) \cdot \boldsymbol{S}_x \wedge dy \wedge dt + (\Gamma \boldsymbol{S}) \cdot d\boldsymbol{S}_x \wedge dy \wedge dt \,,$$

$$\beta_2 = d(\Gamma \boldsymbol{S}) \cdot \boldsymbol{S}_y \wedge dx \wedge dt + (\Gamma \boldsymbol{S}) \cdot d\boldsymbol{S}_y \wedge dx \wedge dt \,.$$

According to [8], we generalize the notion of homogeneous spaces by defining an *algebraic skeleton* on a finite-dimensional vector space $\boldsymbol{V}$ as a triple $(\boldsymbol{E}, \boldsymbol{G}, \rho)$, with $\boldsymbol{G}$ a (possibly infinite-dimensional) Lie group, $\boldsymbol{E} = \boldsymbol{V} \oplus \mathfrak{g}$ is a (possibly infinite-dimensional) vector space *not necessarily equipped with a Lie algebra structure*, $\mathfrak{g}$ is the Lie algebra of $\boldsymbol{G}$, and ρ is a representation of $\boldsymbol{G}$ on $\boldsymbol{E}$ (infinitesimally of $\mathfrak{g}$ on $\boldsymbol{E}$) such that it reduces to the adjoint representation of $\mathfrak{g}$ on itself.

Within our perspective, it is important to stress that the Lie algebra $\mathfrak{g}$ and the representation ρ are the unknowns and that we want to formulate an approach which could enable us to determine them starting by the nonlinear model and an integrability condition for it. Once we know them, we can also construct $\boldsymbol{V}$.

Let us then introduce a manifold $\boldsymbol{P}$ on which a Lie group $\boldsymbol{G}$, with Lie algebra $\mathfrak{g}$, acts on the right; $\boldsymbol{P}$ is a principal bundle $\boldsymbol{P} \to \boldsymbol{Z} \simeq \boldsymbol{P}/\boldsymbol{G}$. By construction, we have that $\boldsymbol{Z}$ is a manifold of type $\boldsymbol{V}$, *i.e.* $\forall \boldsymbol{z} \in \boldsymbol{Z}$, $T_{\boldsymbol{z}}\boldsymbol{Z} \simeq \boldsymbol{V}$. A tower $\boldsymbol{P}(\boldsymbol{Z}, \boldsymbol{G})$ on $\boldsymbol{Z}$ with skeleton $(\boldsymbol{E}, \boldsymbol{G}, \rho)$ is an *absolute parallelism* Ω on $\boldsymbol{P}$ valued in $\boldsymbol{E}$, invariant with respect to ρ and reproducing elements of $\mathfrak{g}$ from the fundamental vector fields induced on $\boldsymbol{P}$. In general, the absolute parallelism *does not* define a Lie algebra homomorphism.

The representation ρ defines a left action of $\boldsymbol{G}$ on $\boldsymbol{E}$ so that we can construct a bundle $\boldsymbol{P}' = \boldsymbol{P} \times_\rho \boldsymbol{E}$, such that $V_{\boldsymbol{p}'}\boldsymbol{P}' \simeq \boldsymbol{E}$. This situation generalizes the standard construction of a principal bundle $\boldsymbol{P}' = \boldsymbol{P} \times_{\boldsymbol{G}} \boldsymbol{K}$, with $\boldsymbol{G}$ a closed subgroup of $\boldsymbol{K}$, or more precisely, if we consider the action of $\boldsymbol{G}$ on the Lie algebra $\mathfrak{k}$ of $\boldsymbol{K}$, $\boldsymbol{P}' = \boldsymbol{P} \times_{\boldsymbol{G}} \mathfrak{k}$. The generalization here is that $\boldsymbol{E}$ is a skeleton rather than a Lie algebra. We can think of a right action on $\boldsymbol{P}'$ induced by ρ in analogy with the homogeneous case; consequently, we consider a connection on $\boldsymbol{P}'$ as a section of $J_1\boldsymbol{P}' \to \boldsymbol{P}'$ invariant with respect to the induced right action. Accordingly, $J_1(\boldsymbol{P} \times_\rho \boldsymbol{E}) \to \boldsymbol{Z}$ can be thought as the bundle of invariant connections with respect to ρ.

Let $\mathfrak{k}$ be a Lie algebra and $\mathfrak{g}$ a Lie subalgebra of $\mathfrak{k}$. Let $\boldsymbol{G}$ be a Lie group with Lie algebra $\mathfrak{g}$ and $\boldsymbol{P}(\boldsymbol{Z}, \boldsymbol{G})$ be a principal fiber bundle with structure group $\boldsymbol{G}$ over a manifold $\boldsymbol{Z}$ as above. A *Cartan connection* in $\boldsymbol{P}$ of type $(\mathfrak{k}, \boldsymbol{G})$ is a 1–form Ω on $\boldsymbol{P}$ with values in $\mathfrak{k}$ such that $\omega|_{T_{\boldsymbol{p}}\boldsymbol{P}} : T_{\boldsymbol{p}}\boldsymbol{P} \to \mathfrak{k}$ is an isomorphism $\forall \boldsymbol{p} \in \boldsymbol{P}$, $R^*_g\omega = Ad(g)^{-1}\omega$ for $g \in \boldsymbol{G}$ and reproducing elements of $\mathfrak{g}$ from the fundamental vector fields induced on $\boldsymbol{P}$. Then it is clear that a Cartan connection $(\boldsymbol{P}, \boldsymbol{Z}, \boldsymbol{G}, \omega)$ of type $(\mathfrak{k}, \boldsymbol{G})$ is a special case of a tower on $\boldsymbol{Z}$.

It is well known that in the homogeneous case, there is a one-to-one correspondence between Cartan connections on $\boldsymbol{P}$ and principal connections on $\boldsymbol{P}'$ with certain properties. A Cartan connection on $\boldsymbol{P}$, that is a form ω on $\boldsymbol{P}$ valued in $\mathfrak{g}$, can be spread as a form over the whole $\boldsymbol{P}'$ $(\boldsymbol{P} \subset \boldsymbol{P}')$ as $\omega_{(p',g)} = Ad(g^{-1})\pi^*_{\boldsymbol{P}}\omega + \pi^*_{\boldsymbol{G}}\omega_{\boldsymbol{G}}$ where $\pi_p : \boldsymbol{P} \times \boldsymbol{G} \to \boldsymbol{P}$ and $\pi^*_{\boldsymbol{G}} : \boldsymbol{P} \times \boldsymbol{G} \to \boldsymbol{G}$ are canonical projections, and $\omega_{\boldsymbol{G}}$ is the left Maurer-Cartan form on $\boldsymbol{G}$; ω can be extended to a form on $\boldsymbol{P} \times \boldsymbol{G}$ as a pull-up of a connection form of a principal connection on $\boldsymbol{P}' = \boldsymbol{P} \times_{\boldsymbol{K}} \boldsymbol{G}$. Such a construction gives rise to principal connections on $\boldsymbol{P}'$, the horizontal bundle of which does not intersect the tangent bundle of $\boldsymbol{P}$ viewed as the subbundle of $\boldsymbol{P}'$; conversely the pull-back under the canonical inclusion $i : \boldsymbol{P} \to \boldsymbol{P} \times_{\boldsymbol{K}} \boldsymbol{G}$ of the connection form of a principal connection, which satisfies such a condition, is a Cartan connection on $\boldsymbol{P}$ with values in $\mathfrak{g}$.

Let us come back into the situation of a skeleton instead of a homogeneous space: we generalize this construction by requiring that $\boldsymbol{P}$ is a manifold of type $\boldsymbol{E}$, *i.e.* $T_{\boldsymbol{p}}\boldsymbol{P} \simeq \boldsymbol{E}$, and $V_{\boldsymbol{p}'}\boldsymbol{P}' \simeq T_{\boldsymbol{p}}\boldsymbol{P}$. A tower can be seen then as an invariant connection with respect to ρ on $\boldsymbol{P}'$ satisfying the latter condition. Under this perspective, $\boldsymbol{P}' = \boldsymbol{P} \times_\rho \boldsymbol{E}$ can be considered a sort of a gauge-natural bundle. and a tower $\boldsymbol{P}$ with skeleton $\boldsymbol{E}$ could be considered as a section of a subbundle of $\boldsymbol{P}'$. Let us now express locally the forms of the induced absolute parallelism

$$\Omega^k = H^k(\boldsymbol{S}, \boldsymbol{S}_x, \boldsymbol{S}_y; \xi)dx \wedge dy + F^k(\boldsymbol{S}, \boldsymbol{S}_x, \boldsymbol{S}_y; \xi)dy \wedge dt + G^k(\boldsymbol{S}, \boldsymbol{S}_x, \boldsymbol{S}_y; \xi)dx \wedge dt + \\ +(A^k_m dx + B^k_m dy + C^k_m dt) \wedge d\xi^m ,$$

where F^k, G^k, H^k are functions to be determined modulo $\boldsymbol{\theta}_1, \boldsymbol{\theta}_2, \boldsymbol{\theta}_3, \beta_1, \beta_2$ and A, B, C are invertible matrices [1].

Let $\mathcal{I}$ be the ideal generated by the forms $\boldsymbol{\theta}_i, \beta_l$, $i = 1, 2, 3$, $l = 1, 2$ and Ω^k, $k = 1, \ldots, N$. We say that $\mathcal{I}$ is closed if $d\Omega^k \in \mathcal{I}$ $(\boldsymbol{\theta}_i, \beta_l, \Omega^k)$.

[1] Notice that the absolute parallelism is actually given by $\omega^m = \Gamma^m_1 dx + \Gamma^m_2 dy + \Gamma^m_3 dt + d\xi^m$, so that $H^k = \Gamma^m_1 B^k_m - \Gamma^m_2 A^k_m$, $F^k = \Gamma^m_2 \delta^k_m - \Gamma^m_3 B^k_m$, $G^k = \Gamma^m_1 \delta^k_m - \Gamma^m_3 A^k_m$.

The integrability condition $d\Omega^k = 0\ (mod\,\mathcal{I})$ yields

$$\begin{aligned} H^k_{\boldsymbol{S}_x} &= 0\,, \quad H^k_{\boldsymbol{S}_y} = 0\,, \quad \Rightarrow H^k = H^k(\boldsymbol{S},\xi)\,, \\ F^k_{\boldsymbol{S}_x} &= -(\Gamma H^k_{\boldsymbol{S}}) \times \boldsymbol{S}\,, \quad F^k_{\boldsymbol{S}_y} = 0 \Rightarrow F^k = F^k(\boldsymbol{S},\boldsymbol{S}_x,\xi)\,, \\ G^k_{\boldsymbol{S}_y} &= (\Gamma H^k_{\boldsymbol{S}}) \times \boldsymbol{S}\,, \quad G^k_{\boldsymbol{S}_x} = 0 \Rightarrow G^k = G^k(\boldsymbol{S},\boldsymbol{S}_y,\xi)\,, \end{aligned}$$

and a further important constraint:

$$(F^k_{\boldsymbol{S}} \cdot \boldsymbol{S}_x - G^k_{\boldsymbol{S}} \cdot \boldsymbol{S}_y) dx \wedge dy \wedge dt + H^k_{\xi^m} d\xi^m \wedge dx \wedge dy + F^k_{\xi^m} d\xi^m \wedge dy \wedge dt + G^k_{\xi^m} d\xi^m \wedge dx \wedge dt = 0\,.$$

Without loosing of generality, we can assume $C^m_l = \delta^m_l$ so that, substituting

$$dt \wedge d\xi^m = \Omega^m - H^m dx \wedge dy - F^m dy \wedge dt - G^m dx \wedge dt - A^m_l dx \wedge d\xi^l - B^m_l dy \wedge d\xi^l$$

gives us the fundamental constraints

$$F^k_{\boldsymbol{S}} \cdot \boldsymbol{S}_x - G^k_{\boldsymbol{S}} \cdot \boldsymbol{S}_y + [G,F]^k = 0\,, \tag{3}$$

$$H^k_{\xi^l} + F^k_{\xi^m} A^m_l - G^k_{\xi^m} B^m_l = 0\,. \tag{4}$$

From the second equation we can easily infer that $[A,B] = 0$; furthermore, the functions H, G and F are related as follows (to simplify the notation, we omit the superscripts)

$$[G,F] = [\bar{B}H, \bar{B}F]\,. \tag{5}$$

Constraints above can be further recasted as follows

$$\begin{aligned} F &= -(\Gamma H_{\boldsymbol{S}}) \times \boldsymbol{S} \cdot \boldsymbol{S}_x + K(\boldsymbol{S};\xi)\,, \\ G &= (\Gamma H_{\boldsymbol{S}}) \times \boldsymbol{S} \cdot \boldsymbol{S}_y + \bar{K}(\boldsymbol{S};\xi)\,, \end{aligned}$$

thus, in particular, $F_{\boldsymbol{S}} \cdot \boldsymbol{S}_x = K_{\boldsymbol{S}} \cdot \boldsymbol{S}_x$ and $G_{\boldsymbol{S}} \cdot \boldsymbol{S}_y = \bar{K}_{\boldsymbol{S}} \cdot \boldsymbol{S}_y$. By substitution in (3) we thus obtain $H = \boldsymbol{X}(\xi) \cdot \boldsymbol{S} + Y(\xi)$, where $\boldsymbol{X} = (X_1, X_2, X_3)$ (therefore $H_{\boldsymbol{S}} = \boldsymbol{X}$), so that $F = -(\Gamma \boldsymbol{X}) \times \boldsymbol{S} \cdot \boldsymbol{S}_x + K(\boldsymbol{S};\xi)$, $G = (\Gamma \boldsymbol{X}) \times \boldsymbol{S} \cdot \boldsymbol{S}_y + \bar{K}(\boldsymbol{S};\xi)$. We can also verify that the main constraint also implies $K_{\boldsymbol{S}} = 0$ thus $K = K(\xi)$. On the other hand it also implies $\bar{K} = -S_1[X_2,X_3] + S_2[X_1,X_3] - \gamma^2 S_3[X_1,X_2] + Z(\xi)$. We can summarize that by

$$\begin{aligned} H &= \boldsymbol{X} \cdot \boldsymbol{S} + Y\,, \\ G &= (\Gamma \boldsymbol{X} \times \boldsymbol{S}) \cdot \boldsymbol{S}_y - S_1[X_2,X_3] + S_2[X_1,X_3] - \gamma^2 S_3[X_1,X_2] + Z\,, \\ F &= -(\Gamma \boldsymbol{X} \times \boldsymbol{S}) \cdot \boldsymbol{S}_x + K\,. \end{aligned}$$

To uniform the notation, let us put $Y = X_4$ and $Z = X_5$ (notice that K is is related with them according with the relationship (5)); by substitution and comparing terms of the same monomials, after long but simple algebraic manipulations, we obtain an algebra structure $\boldsymbol{E}$ generated by X_1, X_2, X_3, X_4, X_5:

$$[X_1,X_2] = 0\,, [X_1,X_3] = 0\,, [X_1,X_4] = X_6\,, [X_1,X_5] = X_7\,, \tag{6}$$

$$[X_2,X_3] = 0\,, [X_2,X_4] = X_8\,, [X_2,X_5] = X_9\,,$$

$$[X_3,X_4] = X_{10}\,, [X_3,X_5] = X_{11}\,, [X_4,X_5] = X_{12}\,,$$

$$\dots \quad \dots \quad \dots \quad \dots$$

By requiring the Jacobi identity to hold true, it is easy to verify that the structure does not close as a Lie algebra, since always new generators have to be introduced to name the commutators which are unknown. This open structure is infinite-dimensional and identifies an infinite dimensional vector space $\boldsymbol{E}$. It is different from a freely generated Lie algebra insomuch as there are some relations among some of the commutators. It can be provided of the structure of an infinitesimal algebraic skeleton on a finite dimensional space $\boldsymbol{V}$. We define a Lie algebra $\mathfrak{g}$ acting on $\boldsymbol{E}$ by the representation ρ obtained by means of the request of integrability for the *absolute parallelism* of a tower on $\boldsymbol{Z}$, with skeleton $(\boldsymbol{E}, \boldsymbol{V}, \mathfrak{g})$.

2.1. Skeletons homomorphic with a finite dimensional quotient Lie algebra

We can identify some special subalgebras and the corresponding extension of the adjoint Lie algebra representation defining skeletons as follows.

(i) Put $X_1 = 0\,, X_2 = 0$, *i.e.* $\boldsymbol{X} = (0, 0, X_3)$. We then also have $X_6 = X_7 = X_8 = X_9 = 0$. The resulting structure

$$[X_3, X_4] = X_{10}\,, [X_3, X_5] = X_{11}\,, [X_4, X_5] = X_{12}\,,$$
$$[X_3, X_{12}] = [X_4, X_{11}] - [X_5, X_{10}]\,,$$
$$\dots \quad \dots \quad \dots \quad \dots$$

identifies a tower with skeleton defined by

$$H = S_3 X_3 + X_4\,,$$
$$G = \gamma^2 (S_1 S_{2y} - S_2 S_{1y}) X_3 + X_5\,,$$
$$F = \gamma^2 (S_2 S_{1x} - S_1 S_{2x}) X_3 + K\,;$$

notice that $[X_5, K] = [\bar{B} X_4, \bar{B} K]$, so that, without loosing generality, we can take $K = X_{12}$.

(ii) Put $X_1 = 0\,, X_3 = 0$, *i.e.* $\boldsymbol{X} = (0, X_2, 0)$. We then also have $X_6 = X_7 = X_{10} = X_{11} = 0$. The resulting structure

$$[X_2, X_4] = X_8\,, [X_2, X_5] = X_9\,, [X_4, X_5] = X_{12}\,,$$
$$[X_2, X_{12}] = [X_4, X_9] - [X_5, X_8]\,,$$
$$\dots \quad \dots \quad \dots \quad \dots$$

identifies a tower with skeleton defined by

$$H = S_2 X_2 + X_4\,,$$
$$G = \gamma^2 (S_1 S_{3y} - S_3 S_{1y}) X_2 + X_5\,,$$
$$F = \gamma^2 (S_3 S_{1x} - S_1 S_{3x}) X_2 + X_{12}\,.$$

(iii) Put $X_2 = 0\,, X_3 = 0$, *i.e.* $\boldsymbol{X} = (X_1, 0, 0)$. We then also have $X_8 = X_9 = X_{10} = X_{11} = 0$. The resulting structure

$$[X_1, X_4] = X_6\,, [X_1, X_5] = X_7\,, [X_4, X_5] = X_{12}\,,$$
$$[X_1, X_{12}] = [X_4, X_7] - [X_5, X_6]\,,$$
$$\dots \quad \dots \quad \dots \quad \dots$$

identifies a tower with skeleton defined by

$$H = S_1 X_1 + X_4\,,$$
$$G = \gamma^2 (S_2 S_{3y} - S_3 S_{2y}) X_1 + X_5\,,$$
$$F = \gamma^2 (S_3 S_{2x} - S_2 S_{3x}) X_1 + X_{12}\,.$$

Such skeletons just differ for a renaming of some of the elements of the algebra (6) and then identify the same algebraic structure. The vector space $\boldsymbol{V}$ can be defined as the kernel of an homomorphism between the infinite dimensional algebra (6) and a finite-dimensional Lie algebra $\mathfrak{g} = \mathfrak{sl}(2, \mathbb{C})$.

Proposition 1 *There exists an homomorphism $\mathcal{H}$ between the algebraic structure defined above (and thus also between $\boldsymbol{E}$) and the $\mathfrak{sl}(2, \mathbb{C})$ Lie algebra.*

PROOF. The homomorphism $\mathcal{H}$ is defined by the closing conditions

$$X_{10} = 2i\lambda X_5 \,, \qquad X_{11} = -2i\lambda X_4 \,, \qquad X_{12} = 2i\lambda X_3 \,,$$

(analogously $X_8 = 2i\lambda X_5 \,, X_9 = -2i\lambda X_4 \,, X_{12} = 2i\lambda X_2$, or $X_6 = 2i\lambda X_5 \,, X_7 = -2i\lambda X_4 \,, X_{12} = 2i\lambda X_1$, for the other cases); here λ is a parameter. The $\mathfrak{sl}(2, \mathbb{C})$ Lie algebra is then given by $[X_i, X_j] = 2i\lambda\epsilon^{ijk}X_k$, with $i, j, k = 3, 4, 5$ (resp. $i, j, k = 2, 4, 5$ and $i, j, k = 1, 4, 5$).

Representations of such a quotient Lie algebra $\mathfrak{sl}(2, \mathbb{C})$ provide conservation laws in the form of linear spectral problems associated with the continuous isotropic (compact and noncompact) Heisenberg model in $(2 + 1)$ dimensions starting from the towers constructed in $(i) - (iii)$; in fact, each tower provides one of the components in which the model can be decomposed with respect to a base. Notice also that the same algebra can be obtained directly from $\boldsymbol{E}$ by setting $X_1 = X_2 = X_3 = -\frac{i}{2\lambda}X_{12}$.

Acknowledgements
Research supported by Department of Mathematics, University of Torino, project *Metodi Geometrici in Fisica Matematica e Applicazioni* (2011). The author thanks also the partial support of the Czech grant GA201/09/0981 *Globalni analyza a geometrie fibrovanych prostoru*, as well as the University of Brno and the University of Ostrava, where part of this work has been written. Thanks are due to E. Winterroth for useful remarks.

Dedicated to Hartwig.

References

[1] Ablowitz, M.J., Segur, H.: *Solitons and the inverse scattering transform*, SIAM Studies in Applied Math. 1981.

[2] Crampin, M. , Pirani, F.A.E., Robinson, D. C. : The soliton connection, *Lett. Math. Phys.* **2** (1) (1977) 15–19.

[3] Estabrook, F.B.: Moving frames and prolongation algebras. *J. Math. Phys.* **23** (1982) 2071–2076.

[4] Estabrook, F.B., Wahlquist, H.D.: Prolongation structures of nonlinear evolution equations. II, *J. Math. Phys.* **17** (1976) 1293–1297.

[5] M. Francaviglia, M. Palese, E. Winterroth: Variationally equivalent problems and variations of Noether currents, *Int. J. Geom. Methods Mod. Phys.* **10** (1) (2013) 1220024 (10 pages); – Locally variational invariant field equations and global currents: Chern–Simons theories, to appear in *Comm. Math.* (2012).

[6] Hermann, R.: Pseudopotentials of Estabrook and Wahlquist, the Geometry of Solitons, and the Theory of Connections, *Phys. Rev. Lett.*, **36** (15) (1976) 835–836.

[7] Lakshmanan, M.: The fascinating world of the Landau–Lifshitz–Gilbert equation: an overview, *Philos. Trans. R. Soc. Lond. Ser. A Math. Phys. Eng. Sci.***369**(no. 1939) (2011) 1280–1300; in particular, see also: Senthil Kumar, C., Lakshmanan, M., Grammaticos, B., Ramani, A.: Nonintegrability of $(2 + 1)$-dimensional continuum isotropic Heisenberg spin system: Painlevé analysis, *Phys. Lett. A* **356** (4-5) (2006) 339–345; as well as Ramani, A., Grammaticos, B., Tremblay, S.: Integrable systems without the Painlevé property, *J. Phys. A* **33** (15) (2000) 3045–3052.

[8] Morimoto, T.: Geometric structures on filtered manifolds, *Hokkaido Math. Jour.* **22** (1993) 263–347.

[9] Palese, M.: Prolongation structures of nonlinear field equations (Italian), *Master Thesis*, (University of Lecce, 1993); see also Alfinito, E., Leo, M., Leo, R. A., Palese, M., Soliani, G.: Algebraic properties of the $(1 + 1)$-dimensional Heisenberg spin field model, *Lett. Math. Phys.* **32** (3) (1994) 241–248.

[10] Palese, M., Winterroth, E.: Nonlinear $(2 + 1)$-dimensional field equations from incomplete Lie algebra structures, *Phys. Lett. B***532** (1-2) (2002) 129–134.

[11] Palese, M., Winterroth, E.: On the geometry of Bäcklund transformations, in *Nonlinear physics: theory and experiment II* World Sci. Publ. (2003) 254–257.

[12] Palese, M.: Bäcklund loop algebras for compact and noncompact nonlinear spin models in $2+1$ dimensions, *Theoret. and Math. Phys.* **144** (1) (2005) 1014–1021.

[13] Palese, M., Winterroth, E.: Infinitesimal Algebraic Skeletons for a $(2+1)$-dimensional Toda Type System, *Acta Polytechnica* **51** (1) (2011) 54–58

[14] Palese, M., Winterroth, E.: Constructing towers with skeletons from open Lie algebras and integrability, in 7th Int. Conf. Quantum Theory and Symmetries (QTS7) IOP Publishing, *Journal of Physics: Conference Series* **343** (2012) 012091.

[15] Pirani, F.A.E., Robinson, D.C., Shadwick, W.F.: *Local Jet Bundle Formulation of Bäcklund Transformations, Math. Phys. Stud.* D. Reidel Publishing Company, Dordrecht, Holland (1979).

[16] Wahlquist, H.D., Estabrook, F.B.: Prolongation structures of nonlinear evolution equations, *J. Math. Phys.* **16** (1975) 1–7.

Journal of Physics: Conference Series **411** (2013) 012025 doi:10.1088/1742-6596/411/1/012025

Higgs fields on spinor gauge-natural bundles

Marcella Palese and Ekkehart Winterroth

Department of Mathematics, University of Torino, via C. A.lberto 10, I-10124 Torino, Italy

E-mail: marcella.palese@unito.it, ekkehart.winterroth@unito.it

Abstract. We show that the Lie derivative of spinor fields is parametrized by Higgs fields defined by the kernel of a gauge-natural Jacobi morphism associated with the Einstein–Cartan–Dirac Lagrangian. In particular, the generalized Kosmann lift to the total bundle of the theory is constrained by *variational* Higgs fields on gauge-natural bundles.

2000 MSC: 58A20,58A32,58E30,58E40.
keywords: jet, gauge-natural bundle, reduced principal bundle, Cartan connection, spinor connection, Higgs field.

1. Introduction

Natural and gauge (classical) Lagrangian field theories have been framed within the geometric construction of a gauge-natural bundle, according to which classical physical fields are sections of bundles functorially associated with gauge-natural prolongations (also known as Ehresmann prolongations [2]) of principal bundles, by means of left actions of Lie groups on manifolds [1]. It is well known that, while the jet prolongation of a principal bundle is not a principal bundle, the gauge-natural prolongation of a principal bundle is provided with the structure of a principal bundle [2, 9]. We consider Lagrangian field theories which are assumed to be invariant with respect to the action of a gauge-natural group $W_n^{(r,k)}\mathbf{G}$ defined as the semidirect product of a k-th order differential group of the base manifold with the group of r-th order n-th velocities on the structure group $\mathbf{G}$ ($n = \dim \mathbf{X}$ is the dimension of the basis manifold).

Within such theories there is *a priori* no natural way of relating infinitesimal gauge transfomations with infinitesimal base transformations; we found that a *canonical* determination of Noether conserved quantities, without fixing any connection *a priori*, can be performed on a reduced bundle of $W^{(r,k)}\mathbf{P}$ determined by the original $W_n^{(r,k)}\mathbf{G}$-invariant variational problem. Connections can be characterized by means of such a canonical reduction [4, 5, 13, 17, 18]. Such conserved quantities can be characterized in terms of Higgs fields on gauge principal bundles having moreover the richer structure of a gauge-natural prolongation [19].

We consider the particular case of the Einstein–Cartan–Dirac Lagrangian and we show that, being the Lie derivative of fields constrained by Jacobi equations, the Kosmann lift to the total bundle of spin-tetrads, spin-connections and spinors is associated with a *variational* Higgs field on the underlying gauge-natural principal bundle.

2. Gauge-natural Jacobi fields and canonically conserved quantities

Let $J_s\mathbf{Y}$ of s–jet prolongations of (local) sections of a fibered manifold $\pi : \mathbf{Y} \to \mathbf{X}$, with $\dim \mathbf{X} = n$ and $\dim \mathbf{Y} = n + m$. The natural fiberings π^s_{s-1} are *affine* fiberings inducing a natural splitting $J_s\mathbf{Y} \times_{J_{s-1}\mathbf{Y}} T^*J_{s-1}\mathbf{Y} = J_s\mathbf{Y} \times_{J_{s-1}\mathbf{Y}} (T^*\mathbf{X} \oplus V^*J_{s-1}\mathbf{Y})$, which yields rising order decompositions: given a vector field $\Xi : J_s\mathbf{Y} \to TJ_s\mathbf{Y}$, we have a naturally induced decomposition in the sum of its *horizontal part* Ξ_H and its *vertical part* Ξ_V and analogously for the exterior differential on $\mathbf{Y}$ we have $(\pi_r^{r+1})^* \circ d = d_H + d_V$, where d_H and d_V are the *horizontal* and *vertical differential*, respectively.

Naturally induced is also a *sheaf splitting* $\mathcal{H}^p_{(s+1,s)} = \bigoplus_{t=0}^p \mathcal{C}^{p-t}_{(s+1,s)} \wedge \mathcal{H}^t_{s+1}$, where $\mathcal{H}^p_{(s,q)}$ and $\mathcal{H}^p_s$ are the sheaves of *horizontal forms* with respect to the projections π^s_q and π^s_0, respectively, while $\mathcal{C}^p_{(s,q)} \subset \mathcal{H}^p_{(s,q)}$ and $\mathcal{C}^p{}_s \subset \mathcal{C}^p_{(s+1,s)}$ are *contact forms* (see *e.g.* [7, 11]); the projection on the summand of lesser contact degree h is *the horizontalization.* We set $\Theta^*_s \doteq \ker h + d\ker h$, where $d\ker h$ is the sheaf generated by the corresponding presheaf. By quotienting the de Rham sequence with the contact structure so defined, we have *the Krupka variational sequence* $0 \to I\!R_Y \to \mathcal{V}^*_s$, where $\mathcal{V}^*_s = \Lambda^*_s/\Theta^*_s$. Let $\mathcal{E}_*$ denote its differential morphisms; a section $\lambda \in \mathcal{V}^n_s$ is a generalized Lagrangian and correspondingly a section $E_{d\lambda} \doteq \mathcal{E}_n(\lambda) \in \mathcal{V}^{n+1}_s$ is the *generalized higher order Euler–Lagrange type morphism* associated with λ [11].

Let $\mathbf{P} \to \mathbf{X}$ be a principal bundle with structure group $\mathbf{G}$. For $r \leq k$ integers consider the *gauge-natural prolongation of* $\mathbf{P}$ given by $\mathbf{W}^{(r,k)}\mathbf{P} \doteq J_r\mathbf{P} \times_{\mathbf{X}} L_k(\mathbf{X})$, where $L_k(\mathbf{X})$ is the bundle of k–frames in $\mathbf{X}$ [1, 9]; $\mathbf{W}^{(r,k)}\mathbf{P}$ is a principal bundle over $\mathbf{X}$ with structure group $\mathbf{W}^{(r,k)}_n\mathbf{G}$ which is the *semidirect* product with respect to the action of $GL_k(n)$ on $\mathbf{G}^r_n$ given by jet composition and $GL_k(n)$ is the group of k–frames in $I\!R^n$. Here we denote by $\mathbf{G}^r_n$ the space of (r,n)-velocities on $\mathbf{G}$. Let $\mathbf{F}$ be a manifold and $\zeta : \mathbf{W}^{(r,k)}_n\mathbf{G} \times \mathbf{F} \to \mathbf{F}$ be a left action of $\mathbf{W}^{(r,k)}_n\mathbf{G}$ on $\mathbf{F}$. There is a naturally defined right action of $\mathbf{W}^{(r,k)}_n\mathbf{G}$ on $\mathbf{W}^{(r,k)}\mathbf{P} \times \mathbf{F}$ so that we get in a standard way the associated *gauge-natural bundle* of order (r,k): $\mathbf{Y}_\zeta \doteq \mathbf{W}^{(r,k)}\mathbf{P} \times_\zeta \mathbf{F}$. All our considerations shall refer to a fibered manifold $\mathbf{Y}$ which has also the structure of a gauge-natural bundle.

Functorial linearity properties of a gauge-natural lift $\hat{\Xi}$ (for details, see *e.g.* [3, 9]) enabled us to define the *gauge-natural generalized Jacobi morphism* associated with a Lagrangian λ and the variation vector field $\hat{\Xi}_V$, *i.e.* the linear morphism $\mathcal{J}(\lambda, \hat{\Xi}_V) \doteq E_{\cdot\rfloor E_{j_s\hat{\Xi}\rfloor d\lambda}}$ [14]. The space $\mathfrak{K} \doteq \ker \mathcal{J}(\lambda, \hat{\Xi}_V)$ defines generalized gauge-natural Jacobi equations, the solutions of which we call *generalized Jacobi vector fields* and characterize *canonical* covariant conserved quantities [13].

Induced linearity properties of the Lie derivative of sections of gauge-natural bundles characterize the form $\omega(\lambda, \hat{\Xi}_V) \doteq -\pounds_{\hat{\Xi}}\rfloor\mathcal{E}_n(\lambda)$ as a new Lagrangian defined on an extended space. It is remarkable that when $\omega(\lambda, \hat{\Xi}_V)$ is an horizontal differential (*i.e.* a null Lagrangian) we get a conservation law which holds true along any section of the gauge natural bundle (not only along solutions of the Euler–Lagrange equations). It is also remarkable that the new Lagrangian ω, in principle, is not gauge-natural invariant; nevertheless, its restriction $\omega(\lambda, \mathfrak{K})$ is invariant and corresponding Noether conservation laws and Noether identities [20] can be obtained, so that a canonical determination of conserved quantities is given on a reduced bundle of $W^{(r,k)}\mathbf{P}$ determined by the original $W^{(r,k)}_n\mathbf{G}$-invariant variational problem [15, 16]; in particular, necessary conditions for the existence of global solutions of Jacobi equations associated with the existence of canonically defined global conserved quantities can be interpreted as topological conditions for the existence of a Cartan connection on the principal bundle $W^{(r,k)}\mathbf{P}$ [18].

3. Spinor gauge-natural Higgs fields

In the following we shortly recall the Einstein–Cartan–Dirac theory; details can be found *e.g.* in [3, 25]. In particular, we point out the gauge-natural structure of such a theory.

On a 4-dimensional manifold admitting Lorentzian structures ($SO(1,3)^e$-reductions) $\mathbf{X}$ consider a $SPIN(1,3)^e$-principal bundle $\mathbf{\Sigma} \to \mathbf{X}$ and a bundle map inducing a spin-frame on $\mathbf{\Sigma}$ given by $\tilde{\Lambda} : \mathbf{\Sigma} \to L(\mathbf{X})$ defining a metric g *via* the reduced subbundle $SO(\mathbf{X}, g) = \tilde{\Lambda}(\mathbf{\Sigma})$

of $L(\mathbf{X})$. A left action ρ of the group $W^{(0,1)}SPIN(1,3)^e$ on the manifold $GL(4, I\!R)$ is given so that the associated bundle $\mathbf{\Sigma}_\rho \doteq W^{(0,1)}\mathbf{\Sigma} \times_\rho GL(4, I\!R)$ is is a gauge-natural bundle of order $(0,1)$, the *bundle of spin-tetrads* θ. Let $\mathfrak{so}(1,3) \simeq \mathfrak{spin}(1,3)$ be the Lie algebra of $SO(1,3)$. One can consider the left action of $W_4^{(1,1)}SPIN(1,3)^e$ on the vector space $(I\!R^4)^* \otimes \mathfrak{so}(1,3)$. The associated bundle $\mathbf{\Sigma}_l \doteq W^{(1,1)}\mathbf{\Sigma} \times_l ((I\!R^4)^* \otimes \mathfrak{so}(1,3))$ is a gauge-natural bundle of order $(1,1)$, the *bundle of spin-connections* ϕ. If $\hat{\gamma}$ is the linear representation of $SPIN(1,3)^e$ on the vector space $\mathbb{C}^4$ induced by the choice of matrices γ we get a $(0,0)$-gauge-natural bundle $\mathbf{\Sigma}_{\hat{\gamma}} \doteq \mathbf{\Sigma} \times_{\hat{\gamma}} \mathbb{C}^4$, the *bundle of spinors*. A spinor connection $\tilde{\phi}$ is defined in a standard way in terms of the spin connection. We notice that a spin (as well as spinor) connection is induced by a principal connection on $\mathbf{\Sigma}$.

Within this picture, we assume that the total Lagrangian of a gravitational field interacting with spinor matter is $\lambda = \lambda_{EC} + \lambda_D$, where the Einstein–Cartan Lagrangian and the Dirac Lagrangian can be represented by the morphisms

$$\lambda_{EC} : \mathbf{\Sigma}_\rho \times_{\mathbf{X}} J_1\mathbf{\Sigma}_l \to \wedge^4 T^*\mathbf{X}, \qquad \lambda_D : \mathbf{\Sigma}_\rho \times_{\mathbf{X}} \mathbf{\Sigma}_l \times_{\mathbf{X}} J_1\mathbf{\Sigma}_{\hat{\gamma}} \to \wedge^4 T^*\mathbf{X},$$

respectively (local expressions can be found *e.g.* in [3]).

Let $\mathfrak{k}$ be the vector bundle defined by the Jacobi equations $\mathcal{J}(\lambda_{EC} + \lambda_D, \hat{\Xi}_V) = 0$, where $\hat{\Xi}$ is the gauge-natural lift to the associated total bundle of an infinitesimal principal automorphism of the principal bundle underlying the theory, *i.e.* of the $SPIN(1,3)^e$-principal bundle $\mathbf{\Sigma} \to \mathbf{X}$. Since for each gauge-natural lift we have the well known equality $\hat{\Xi}_V(\psi) = -\pounds_{\Xi}\psi$, in local fibered coordinates on the total bundle given by $(x^\mu, \theta^a_\mu, \phi^{ab}, \phi^{ab}_\mu, \psi)$ the gauge-natural Jacobi equations read

$$(-1)^{|\sigma|} d_\sigma \, (d_\mu(-\pounds_{\Xi}\psi)^{ab}(\partial_{cd}(\partial^\mu_{ab})\lambda - \sum_{|\alpha|=0}^{s-|\mu|}(-1)^{|\mu+\alpha|}\frac{(\mu+\alpha)!}{\mu!\alpha!} d_\alpha \partial^\alpha_{cd}(\partial^\mu_{ab}\lambda))) = 0 \,,$$

with $0 \leq |\sigma|, |\mu| \leq 1$, d_μ is the total derivative and we write for the total Lagrangian

$$\lambda \doteq \lambda_{EC} + \lambda_D = -\frac{1}{2k}\Phi_{ab} \wedge \epsilon^{ab} + (\frac{i\alpha}{2}(\bar{\psi}\gamma^a\nabla_a\psi - \nabla_a\bar{\psi}\gamma^a\psi) - m\bar{\psi}\psi)\epsilon \,,$$

where ϵ is a volume density on $\mathbf{X}$ and Φ the curvature form of the spin-connection ϕ, α and m are constants. Along $\mathfrak{k}$, we have $\bar{\Xi}^{ab}_v = -\tilde{\nabla}^{[a}\xi^{b]}$ (the so-called Kosmann lift [10]), where $\tilde{\nabla}$ is the covariant derivative with respect to the standard transposed connection on the bundle of spin-tetrads $\mathbf{\Sigma}_\rho$. We remark that, since the Lie derivative of spinor fields $\pounds_{\Xi}\psi$ can be written in terms of $\hat{\Xi}_{\tilde{h}}$ (the horizontal part of $\hat{\Xi}$ *with respect to the spinor-connection*) the spinor-connection $\tilde{\omega}$ is constrained [4, 26]. In the following we shall characterize this fact more precisely.

By an abuse of notation, we denote by $\mathfrak{k}$ the Lie algebra of generalized Jacobi vector fields. Let now $\mathfrak{h}$ be the Lie algebra of right-invariant vertical vector fields on $W^{(1,1)}\mathbf{\Sigma}$; the Lie algebra $\mathfrak{k}$ is characterized as a Lie subalgebra of $\mathfrak{h}$; the Jacobi morphism is self-adjoint and $\mathfrak{k}$ is of constant rank; the split structure $\mathfrak{h} = \mathfrak{k} \oplus Im\,\mathcal{J}$ is well defined and it is also reductive, being $[\mathfrak{k}, Im\,\mathcal{J}] = Im\,\mathcal{J}$ [17].

In particular, for each $\mathbf{p} \in W^{(1,1)}\Sigma$ by denoting $\mathcal{S} \doteq \mathfrak{h}_{\mathbf{p}}$, $\mathcal{R} \doteq \mathfrak{k}_{\mathbf{p}}$ and $\mathcal{V} \doteq Im\,\mathcal{J}_{\mathbf{p}}$ we have the reductive Lie algebra decomposition $\mathcal{S} = \mathcal{R} \oplus \mathcal{V}$, with $[\mathcal{R}, \mathcal{V}] = \mathcal{V}$. Notice that $\mathcal{S}$ is the Lie algebra of the Lie group $W\mathbf{G} \doteq W_4^{(1,1)}SPIN(1,3)^e$. For the purposes of this note, it is sufficient to know that the Lie algebra $\mathcal{R}$ exists and it is well defined; we shall not write down explicitly such a Lie algebra, although this question is of great interest and will be investigated extensively elsewhere.

As a consequence of the fact that $\mathcal{R}$ is a reductive Lie algebra of $\mathcal{S}$, there exists an isomorphism between $\mathcal{V} \doteq Im\mathcal{J}_{\mathbf{p}}$ and $T\mathbf{X}$ so that $\mathcal{V}$ turns out to be the image of an horizontal subspace. Thus we caracterize a principal bundle $\mathbf{S} \to \mathbf{X}$, with $dim\mathbf{S} = dim\mathcal{S}$ and such that $\mathbf{X} = \mathbf{S}/\mathbf{R}$, where $\mathbf{R}$ is a Lie group of the Lie algebra $\mathcal{R}$ and $\mathcal{R} = T_{\mathbf{q}}\mathbf{S}/\mathbf{R}$; the principal subbundle $\mathbf{S} \subset \mathbf{W}^{(1,1)}\mathbf{\Sigma}$ is then a reduced principal bundle.

In the following, to simplify the notation, we shall omit the orders of a gauge-natural prolongation; in particular, denote by $W\mathbf{G}$ a gauge-natural prolongation of a given appropriate order of the stucture group of the Einstein–Cartan–Dirac theory. The Lie group $\mathbf{R}$ of the Lie algebra $\mathcal{R}$ is in particular a closed subgroup of $W\mathbf{G}$ [18, 19]. We have the composite fiber bundle $W\mathbf{\Sigma} \to W\mathbf{\Sigma}/\mathbf{R} \to \mathbf{X}$, such that $W\mathbf{\Sigma}/\mathbf{R} = W\mathbf{\Sigma} \times_{W\mathbf{G}} W\mathbf{G}/\mathbf{R} \to \mathbf{X}$ is a gauge-natural bundle functorially associated with $W\mathbf{\Sigma} \times W\mathbf{G}/\mathbf{R} \to \mathbf{X}$ by the right action of $W\mathbf{G}$. The left action of $W\mathbf{G}$ on $W\mathbf{G}/\mathbf{R}$ is in accordance with the reductive Lie algebra decomposition.

Definition 1 We call a global section $h : \mathbf{X} \to W\mathbf{\Sigma}/\mathbf{R}$ a *spinor gauge-natural Higgs field.*

3.1. Higgs fields and the Lie derivative of spinors

Let ω be a principal connection on $\mathbf{W\Sigma}$ and $\bar{\omega}$ a principal connection on the principal bundle $\mathbf{S}$ *i.e.* a $\mathcal{R}$-invariant horizontal distribution defining the vertical parallelism $\bar{\omega} : V\mathbf{S} \to \mathcal{R}$ in the usual and standard way. It defines the splitting $T_{\mathbf{p}}\mathbf{S} \simeq_{\bar{\omega}} \mathcal{R} \oplus \hat{H}_{\mathbf{p}}$, $\mathbf{p} \in \mathbf{S}$. Since $\mathcal{R}$ is a subalgebra of the Lie algebra $\mathcal{S}$ and $dim\mathbf{S} = dim\mathcal{S}$, it is defined a principal Cartan connection of type $\mathcal{S}/\mathcal{R}$, such that $\hat{\omega}|_{V\mathbf{S}} = \bar{\omega}$. It is a connection on $\mathbf{W\Sigma} = \mathbf{S} \times_{\mathbf{R}} W\mathbf{G} \to \mathbf{X}$, thus a Cartan connection on $\mathbf{S} \to \mathbf{X}$ with values in $\mathcal{S}$ [5] and it splits into the $\mathcal{R}$-component which is a principal connection form on the $\mathcal{R}$-manifold $\mathbf{S}$, and the $\mathcal{V}$-component which is a displacement form [18].

A gauge-natural Higgs field, being a global section of $\hat{H}_p$, with $\mathbf{p} \in \mathbf{S}$, is related with the displacement form defined by the $\mathcal{V}$-component of the Cartan connection $\hat{\omega}$ above. The pull-back by h of the $\mathcal{R}$ valued component of a $\mathcal{S}$ valued pricipal connection ω on $W\mathbf{\Sigma}$ onto the reduced subbundle $\mathbf{S}$ is the connection form of a principal connection on $\mathbf{S}$. Given the composite fiber bundle $W\mathbf{\Sigma} \to W\mathbf{\Sigma}/\mathbf{R} \to \mathbf{X}$, we have the exact sequence

$$0 \to V_{W\mathbf{\Sigma}/\mathbf{R}}W\mathbf{\Sigma} \to VW\mathbf{\Sigma} \to W\mathbf{\Sigma} \times_{W\mathbf{\Sigma}/\mathbf{R}} VW\mathbf{\Sigma}/\mathbf{R} \to 0\,,$$

where $V_{W\mathbf{\Sigma}/\mathbf{R}}W\mathbf{\Sigma}$ denote the vertical tangent bundle of $W\mathbf{\Sigma} \to W\mathbf{\Sigma}/\mathbf{R}$. Every connection $\tilde{\omega}$ on the latter bundle determines a splitting $VW\mathbf{\Sigma} = V_{W\mathbf{\Sigma}/\mathbf{R}}W\mathbf{\Sigma} \oplus_{W\mathbf{\Sigma}/\mathbf{R}} \tilde{\omega}(W\mathbf{\Sigma} \times_{W\mathbf{\Sigma}/\mathbf{R}} VW\mathbf{\Sigma}/\mathbf{R})$, by means of which we can define a vertical covariant differential as a mapping $J^1W\mathbf{\Sigma} \to T^*\mathbf{X} \otimes_{W\mathbf{\Sigma}} V_{W\mathbf{\Sigma}/\mathbf{R}}W\mathbf{\Sigma}$. The covariant differential on $W\mathbf{\Sigma}_h$ relative to the pull-back connection $h^*(\tilde{\omega})$ can be expressed by means of this mapping in a known way; for coordinate expressions and further details see [12].

Remark 1 A geometric interpretation of the Kosmann lift as a reductive lift has been proposed for the definition of a $SO(1,3)^e$-reductive Lie derivative of spinor fields [8]. From a variational point of view the Kosmann lift is charaterized as the only gauge-natural lift satisfying the naturality condition $\mathcal{L}_{j_{s+1}\hat{\Xi}_H}[\mathcal{L}_{j_{s+1}\hat{\Xi}_V}\lambda] \doteq 0$ equivalent with Jacobi equations. Gauge-natural Jacobi equations state that Lie derivatives of spinors coincide with the vertical parts of gauge-natural lift of principal automorphisms *lying in* $\mathfrak{K}$, which can be expressed through the vertical covariant differential, defined for each global section h of $W\mathbf{\Sigma}/\mathbf{R} \to \mathbf{X}$; then we can say that the Lie derivative of gauge-natural spinors is constrained and it is parametrized by a Higgs field h defined by $\mathfrak{K}$. This condition implies a reduction of the structure group $W\mathbf{G}$ to $\mathbf{R}$. Each global section h of $W\mathbf{\Sigma}/\mathbf{R} \to \mathbf{X}$ affects spin and spinor connections induced functorially on the associated bundle. In particular, the Kosmann lift to the total bundle of spin-tetrads, spin-connections and spinors is constrained by *variational* Higgs fields on the spinor gauge-natural bundle.

Acknowledgements
Research supported by the local research project *Metodi Geometrici in Fisica Matematica e Applicazioni (2011)* of Department of Mathematics, University of Torino and by GNFM-INdAM. The authors would also like to thank the support of the University of Brno, where part of this work has been written within a scientific collaboration with J. Musilova.

References

[1] D.J. Eck: Gauge-natural bundles and generalized gauge theories, *Mem. Amer. Math. Soc.* **247** (1981) 1–48.
[2] C. Ehresmann: Les prolongements d'un espace fibré différentiable, *C. R. Acad. Sci. Paris* **240** (1955) 1755–1757.
[3] L. Fatibene, M. Francaviglia: *Natural and gauge natural formalism for classical field theories. A geometric perspective including spinors and gauge theories*; Kluwer Academic Publishers, Dordrecht, 2003.
[4] M. Ferraris, M. Francaviglia, M. Palese, E. Winterroth: Canonical connections in gauge-natural field theories, *Int. J. Geom. Methods Mod. Phys.* **5** (6) (2008) 973–988.
[5] M. Ferraris, M. Francaviglia, M. Palese, E. Winterroth: Gauge-natural Noether currents and connection fields, *Int. J. Geom. Methods Mod. Phys.* (2011).
[6] M. Ferraris, M. Francaviglia, M. Raiteri: Conserved Quantities from the Equations of Motion (with applications to natural and gauge natural theories of gravitation) *Class. Quant. Grav.* **20** (2003) 4043–4066.
[7] M. Francaviglia, M. Palese, R. Vitolo: Symmetries in finite order variational sequences, *Czech. Math. J.* **52(127)** (1) (2002) 197–213.
[8] M. Godina, P. Matteucci: Reductive G-structures and Lie derivatives, *J. Geom. Phys.* **47** (1) (2003) 66–86.
[9] I. Kolář, P.W. Michor, J. Slovák: *Natural Operations in Differential Geometry*, (Springer–Verlag, N.Y., 1993).
[10] Y. Kosmann: Dérivée de Lie de spineurs, *C. R. Acad. Sci. Paris Sér. A-B* **262** (1966) A289–A292; – Dérivée de Lie de spineurs. Applications, *C. R. Acad. Sci. Paris Sér. A-B* **262** (1966) A394–A397.
[11] D. Krupka: Variational Sequences on Finite Order Jet Spaces, *Proc. Diff. Geom. and its Appl.* (Brno, 1989); J. Janyška, D. Krupka eds.; World Scientific (Singapore, 1990) 236–254.
[12] L. Mangiarotti, G. Sardanashvily: *Connections in Classical and Quantum Field Theory*, (World Scientific, Singapore, 2000).
[13] M. Palese, E. Winterroth: Covariant gauge-natural conservation laws, *Rep. Math. Phys.* **54** (3) (2004) 349–364.
[14] M. Palese, E. Winterroth: Global Generalized Bianchi Identities for Invariant Variational Problems on Gauge-natural Bundles, *Arch. Math. (Brno)* **41** (3) (2005) 289–310.
[15] M. Palese, E. Winterroth: The relation between the Jacobi morphism and the Hessian in gauge-natural field theories, *Theoret. Math. Phys.* **152**(2) (2007), 1191–1200.
[16] M. Palese, E. Winterroth: Noether identities in Einstein–Dirac theory and the Lie derivative of spinor fields, *Proc. Diff Geom. Appl., Olomouc (2007)*, World Scientific Publ., Hackensack, NJ, (2008) 643–653.
[17] M. Palese, E. Winterroth: Lagrangian reductive structures on gauge-natural bundles, *Rep. Math. Phys.* **62** (2) (2008) 229–239.
[18] M. Palese, E. Winterroth: Invariant variational problems and Cartan connections on gauge-natural bundles, *AIP Conf. Proc.* **1191** (2009) 160–165.
[19] M. Palese, E. Winterroth: A variational perspective on classical Higgs fields in gauge-natural theories *Theoret. Math. Phys.*, **168**(1) (2011) 1002–1008.
[20] E. Noether: Invariante Variationsprobleme, *Nachr. Ges. Wiss. Gött., Math. Phys. Kl.* **II** (1918) 235–257.
[21] G. Sardanashvily: On the geometry of spontaneous symmetry breaking. J. Math. Phys. 33 (1992), no. 4, 15461549.
[22] G. Sardanashvily: Geometry of classical Higgs fields. Int. J. Geom. Methods Mod. Phys. 3 (2006), no. 1, 139148.
[23] D.J. Saunders: *The Geometry of Jet Bundles*, Cambridge Univ. Press (Cambridge, 1989).
[24] A. Trautman: Noether equations and conservation laws, *Comm. Math. Phys.* **6** (1967) 248–261.
[25] S. Weinberg: *Gravitation and cosmology: principles and applications of the general theory of relativity*, Wiley, New York, 1972.
[26] E.H.K. Winterroth: *Variational derivatives of gauge-natural invariant Lagrangians and conservation laws*, PhD thesis University of Torino, 2007.

Journal of Physics: Conference Series **411** (2013) 012026 doi:10.1088/1742-6596/411/1/012026

DSR-deformed relativistic symmetries in an expanding spacetime

G. ROSATI[1,2]

[1] Dipartimento di Fisica, Università di Roma "La Sapienza", P.le A. Moro 2, 00185 Roma, EU
[2] INFN, Sez. Roma1, P.le A. Moro 2, 00185 Roma, EU

Abstract. We present here a perspective on results obtained in collaboration with Amelino-Camelia, Marcianó and Matassa, reported in Ref. [1]. We characterize a relativistic theory of worldlines of particles with 3 nontrivial relativistic invariants: a large speed scale ("speed-of-light scale"), a large distance scale (inverse of the "expansion-rate scale'), and a large momentum scale ("Planck scale"). This is particularly relevant in relation to the opportunities for testing Planck-scale-deformed Lorentz symmetry scenarios with analyses, from a signal-propagation perspective, of observations of bursts of particles from cosmological distances. We address some of the challenges that had obstructed success for previous attempts by exploiting the recent understanding of the connection between deformed Lorentz symmetry and relativity of spacetime locality.

1. Introduction

Besides the conceptual motivations[1] , the interest attracted over the last decade by the possibility of Planck-scale-deformed relativistic kinematics, as conceived within the proposal "DSR" (doubly-special, or, for some authors, deformed-special relativity) introduced in Refs. [2, 3], comes from the opportunities in phenomenology. Some of the novel effects that can be accommodated within a DSR-relativistic kinematics can be tested through observations of bursts of particles from cosmological distances [10, 11]. The key for these analyses are the implications of DSR deformations for the propagation of signals, and the sought Planck-scale sensitivity is reached thanks to the huge amplification afforded by the cosmological distances.

DSR-deformed relativistic frameworks have been investigated so far only for flat (Minkowskian) spacetimes, and this is not the case relevant for the analysis of signals received from sources at cosmological distances, for which the curvature/expansion of spacetime is very tangible. With our work [1], we provide a significant step toward filling this gap by exhibiting an explicit example of Planck-scale-deformed relativistic symmetries of a spacetime with constant rate of expansion (deSitterian). This led us to introduce the first ever example[2] of a relativistic theory of worldlines of particles with 3 nontrivial relativistic invariants: a large speed scale

[1] Properties for the Planck length introduced as observer-independent laws [2, 3]; it reflects the content of results rigorously established for 3D quantum gravity properties (see, *e.g.*, Refs. [4, 5, 6]) and for some 4D noncommutative spacetimes (see, *e.g.*, Refs. [7, 8]); it fits the indications emerging from some compelling semiheuristic arguments based on 4D Loop Quantum Gravity [4, 9].

[2] Studies such as those in Refs. [12, 13, 14] did contemplate the possibility of 3 invariants, but did not go as far as giving a consistent relativistic picture of worldlines of particles.

("speed-of-light scale"), a large distance scale (inverse of the "expansion-rate scale'), and a large momentum scale ("Planck scale").

There had been previous attempts of investigating the interplay between DSR-type deformation scales and spacetime expansion (see, *e.g.*, Refs. [13, 14]), but without ever producing a fully satisfactory picture of how the worldlines of particles should be formalized and interpreted. In retrospect we can now see that these previous difficulties were due to the fact that the notion of relative locality had not yet been understood, and without that notion the interplay between DSR-deformation scale and expansion-rate scale remains unintelligible.

Relative locality is the spacetime counterpart of the DSR-deformation scale ℓ just in the same sense that relative simultaneity is the spacetime counterpart of the special-relativistic scale c (scale of deformation of Galilean Relativity into Special Relativity). This was understood only very recently, in studies such as the ones in Refs. [15, 16, 17]. Awareness of the possibility of relative locality is already very important in making sense of the implications of DSR-deformations in a flat/non-expanding spacetime. And, as we showed in [1], it plays an even more crucial role in the consistency of the spacetime picture emerging from the interplay between DSR-deformation scale and expansion-rate scale.

We work at leading order in the DSR-deformation scale ℓ, and in a 2D spacetime (one time and one spatial dimension).

2. DSR-deformed de-Sitter-relativistic symmetries and physical velocity

The de Sitter relativistic symmetries can themselves be viewed as a deformation of the special-relativistic symmetries of Minkowski spacetime such that the expansion-rate parameter H is an invariant. For our analysis of particles worldlines within ℓ, H-deformed relativistic symmetries, we take as starting point the analysis in Refs. [15, 22]. We start by specifying that our ℓ, H-deformed relativistic symmetries shall leave invariant the following combination of the energy E, momentum p and boost $\mathcal{N}$ charges of particles:

$$\mathcal{C}_{H,\alpha,\beta} = E^2 - p^2 + 2H\mathcal{N}p + \ell\left(\alpha E^3 + \beta E p^2\right) \ . \tag{1}$$

Evidently for $\ell \to 0$ this reproduces the standard invariant of de Sitter symmetries.

The following ℓ-deformed (2D) de Sitter algebra of charges is compatible with the invariance of $\mathcal{C}_{H,\alpha,\beta}$

$$\begin{gathered}\{E,p\} = Hp - \ell\alpha H E p \ , \\ \{\mathcal{N},E\} = p + H\mathcal{N} - \ell\alpha E(p + H\mathcal{N}) - \ell\beta E p \ , \\ \{\mathcal{N},p\} = E + \frac{1}{2}\ell\alpha E^2 + \frac{1}{2}\ell\beta p^2 \ .\end{gathered} \tag{2}$$

One easily sees that for $\ell \to 0$ this reproduces the standard properties [25] of the classical de Sitter algebra of charges while for $H \to 0$ it reduces to a ℓ-deformation of the Poincaré algebra (see Ref. [1]).

We introduce spacetime coordinates η, x with Poisson bracket[3]

$$\{\eta, x\} = -\ell x \ , \tag{3}$$

[3] This choice of coordinates is motivated by studies of quantum-spacetime pictures that can be analyzed in relation to such DSR scenarios. The most notable case is "κ-Minkowski" (see, *e.g.*, Ref. [18, 19]) non-commutative spacetime, with $[\hat{x},\hat{t}] = i\ell\hat{x}$. Some authors (see, *e.g.*, Ref. [21, 20]) have suggested that in cases in which the deformations of Lorentz symmetry is inspired by non-commutative spacetime scenarios such as the one of "κ-Minkowski", one should adopt spacetime coordinates with non-vanishing Poisson brackets $\{t, x\} = -\ell x$. It is worth noticing however, as shown in Refs. [22, 23, 1], that the results of analyses of travel times give the same results both assuming $\{t, x\} = -\ell x$ and assuming $\{\tilde{t}, \tilde{x}\} = 0$. This is a simple consequence of the fact that one can obtain coordinates $\{\tilde{t}, \tilde{x}\} = 0$, from our coordinates ($\{t, x\} = -\ell x$) by posing $\tilde{t} = t$, $\tilde{x} = x - \ell E x$. (The point is [22, 23] that x and $x - \ell E x$ coincide in the origin, so their difference can never affect the determination of when a particle reaches (or is emitted) from the origin of the spacetime coordinates of an observer.

and we define the ℓ, H-deformed phase space

$$\begin{aligned}\{E,\eta\} &= 1-H\eta-\ell\alpha(1-H\eta)E\ ,\\ \{E,x\} &= -Hx-\ell\,(1-\beta+H\eta)\,p+\ell\alpha HxE\\ \{p,\eta\}=0\ , \qquad \{p,x\} &= -1-\ell\frac{E-Hxp}{1-H\eta}\ .\end{aligned} \tag{4}$$

In these coordinates the charge associated to boost transformations (i.e. the boost transformations generator) can be represented as

$$\begin{aligned}\mathcal{N} &= -\eta p + xE + \frac{1}{2}H\left(\eta^2 - x^2\right)p\\ &+\ell\left(\left(\frac{1}{2}\alpha-\frac{1}{1-H\eta}\right)xE^2+\left(\eta+\frac{Hx^2}{1-H\eta}\right)Ep+\left(\frac{1}{2}\beta-H\eta\right)xp^2\right)\end{aligned} \tag{5}$$

The set of Poisson brackets defined by (2-5) between $E, p, \mathcal{N}, \eta, x$ satisfy the Jacobi identities.

As done in Ref. [15, 22, 1], we derive the worldlines adopting a covariant formulation of classical relativistic mechanics. Evolution is coded in a pure-constraint Hamiltonian; specifically, the evolution in the auxiliary parameter on the worldline is governed by a Hamiltonian constraint: $\mathcal{C}_{H,\alpha,\beta} - m^2 = 0$. Of course the fact that $\mathcal{C}_{H,\alpha,\beta}$ is an invariant of the (deformed-)relativistic symmetries implies that the charges that generate the symmetry transformations are conserved over this evolution:

$$\dot{E}=\{\mathcal{C}_{H,\alpha,\beta},E\}=0\ ,\ \ \dot{p}=\{\mathcal{C}_{H,\alpha,\beta},p\}=0\ ,\ \ \dot{\mathcal{N}}=\{\mathcal{C}_{H,\alpha,\beta},\mathcal{N}\}=0\,. \tag{6}$$

One finds [15, 22, 1], for the worldlines of massless particles (specifying $p > 0$)

$$x_{m=0,p>0}\,(\eta) = x_0 + \eta - \eta_0\ . \tag{7}$$

By construction these worldlines are covariant under the (deformed-)relativistic transformations generated by the charges $E, p, \mathcal{N}$, as one can also easily verify explicitly.

As already established in previous studies of the case without spacetime expansion [15, 22] the DSR-deformed relativistic symmetries introduce (as most significant among many other novel features) a dependence on energy of the travel time of a massless particle from a given source to a given detector. At this stage of the analysis the physical content of these worldlines is still hidden behind the relativity of locality. A warning that this might be the case is seen in the fact that we are analyzing a case where the on-shell relation, in light of (1), is ℓ, and α/β, dependent, whereas our worldlines are independent of ℓ (and α/β). Most importantly the coordinate velocity one infers from those wordlines is independent of ℓ (and α/β). But one of the main known manifestations of relative locality is a mismatch [15, 22, 24] between coordinate velocity[4] and physical velocity of particles.

For what concerns specifically the analysis so far reported, the main challenge resides in the fact that we are used to read velocities off the formulas for worldlines, but this implicitly

[4] Examples of velocity artifacts were of course known well before the understanding of relative locality for theories with deformed Lorentz symmetry: for example in de Sitter spacetime (and any expanding spacetime) the coordinate velocity of a particle distant from the observer can be described by the observer as a velocity greater than the speed of light, even though in classical de Sitter spacetime the physical velocity measured by an observer close to the particle is of course always no greater than the speed of light. Since these previously known velocity artifacts are connected with spacetime curvature, the recent realization that some of our currently investigated theories formulating (one form or another of) "deformed Minkowski spacetime" are subject to relative locality, with some associated velocity artifacts [15, 22, 24], was largely unexpected.

assumes that translation transformations are trivial. Essentially we take the worldline written by a certain observer Alice to describe both the emission of a particle "at Alice" (in Alice's origin) and the detection of the particle far away from Alice. The observer/detector Bob that actually detects the particle, since he is distant from Alice, should be properly described by acting with a corresponding translation on Alice's worldline. And the determination of the "arrival time at Bob" (crucial for determining the physical velocity [22]) should be based on Bob's description of the worldline, just as much as the "emission time at Alice" should be based on Alice's description of the worldline. When translations are trivial (translation generators conjugate to the spacetime coordinates) we can go by without worrying about this more careful level of discussion, since the naive argument based solely on Alice's worldline gives the same result as the more careful analysis using Alice's worldline for the emission and Bob's description of that same worldline for the detection. But when translations are nontrivial, and one has associated features of relativity of locality, this luxury is lost. One way to have "relative locality" is indeed the case here of interest, with the translation generators E, p acting on spacetime coordinates by the Poisson brackets (4).

Following Ref. [22], let us probe the difference between coordinate velocity and physical velocity considering the simultaneous emission "at Alice" of two massless particles, one "soft" (with momentum p_s small enough that ℓ-deformed terms in formulas fall below the experimental sensitivity available) and one "hard" (with momentum p_h big enough that at least the leading ℓ-deformed terms in formulas fall within the experimental sensitivity available). Alice describes the two particles according to

$$x^A_{p_s}(\eta^A) = \eta^A \ , \tag{8}$$

$$x^A_{p_h}(\eta^A) = \eta^A \tag{9}$$

where we specified $x^A_0 = \eta^A_0 = 0$, so that the emission is at $(0,0)^A$. Since translations are a relativistic symmetry of our novel framework, we already know that the same two worldlines will be described by the distant observer Bob in the following way

$$x^B_{p_s}(\eta^B) = x^B_{0;s} + \eta^B - \eta^B_{0;s} \ , \tag{10}$$

$$x^B_{p_h}(\eta^B) = x^B_{0;h} + \eta^B - \eta^B_{0;h} \ , \tag{11}$$

i.e. the same type of worldlines but with a difference of parameters here codified in $x^B_{0;s}, \eta^B_{0;s}, x^B_{0;h}, \eta^B_{0;h}$. Indeed, Alice's worldlines (8)-(9) and Bob's worldlines (10)-(11) have exactly the same form, but for the ones of Alice we had by construction (by having specified simultaneous emission at Alice) that $x^A_{0;h} = \eta^A_{0;h} = x^A_{0;s} = \eta^A_{0;s} = 0$ whereas Bob's values of the parameters, $x^B_{0;s}, \eta^B_{0;s}, x^B_{0;h}, \eta^B_{0;h}$, should be determined by establishing which (ℓ-deformed) translation transformation connects Alice to Bob.

This is the same task performed in some of the previous studies (such as Ref. [22]) involving relative locality in DSR-deformed relativistic theories without spacetime expansion. But here we have to deal with spacetime expansion. Let us then proceed determining the translation transformation that connects Alice to Bob. As usual we shall give the action on phase space coordinates $k \equiv (E, p, \eta, x)$ of a worldline by a finite transformation $T_{G;a}$ generated by the generic element G of the relativistic-symmetry algebra in terms of its exponential representation as

$$T_{G;a} \triangleright k = e^{-aG} \triangleright k \equiv \sum_{n=0}^{\infty} \frac{(-a)^n}{n!} \{G, k\}_n \ , \tag{12}$$

where a is the transformation parameter and $\{G, k\}_n$ is the n-nested Poisson bracket defined by the relation

$$\{G, k\}_n = \{G, \{G, k\}_{n-1}\} \ , \qquad \{G, k\}_0 = k \ . \tag{13}$$

Before proceeding, we must also implement something else that can be specified about observer Bob. The worldline parameters $x^A_{0;h}$, $\eta^A_{0;h}$, $x^A_{0;s}$, $\eta^A_{0;s}$ of observer Alice are fully specified ($x^A_{0;h} = \eta^A_{0;h} = x^A_{0;s} = \eta^A_{0;s} = 0$) by its being at the point of simultaneous emission of the two particles. We have introduced observer Bob as one that detects the particles (the particles worldlines should cross Bob spatial origin) but we have so far left completely unspecified its worldline parameters $x^B_{0;s}, \eta^B_{0;s}, x^B_{0;h}, \eta^B_{0;h}$. We assume, without loss of generality, that the soft particle reaches Bob in his spacetime origin so that $x^B_{0;s} = \eta^B_{0;s} = 0$. This is particularly convenient because it involves the "soft particle", *i.e.* the one whose momentum p_s has been chosen to be small enough to render the ℓ-deformed effects inappreciable within the experimental sensitivities available to Alice and Bob.

The fact that both $x^A_{0;s} = \eta^A_{0;s} = 0$ and $x^B_{0;s} = \eta^B_{0;s} = 0$ for a soft particle leads us to focus on the case of the observer Bob connected to Alice by the following transformation

$$X^B = \left(e^{a_x P} \triangleright e^{-a_\eta E} \triangleright X\right)^A , \tag{14}$$

with

$$a_x = \frac{1 - e^{-Ha_\eta}}{H} . \tag{15}$$

Through this we are essentially exploiting the fact that the deformation is ineffective on the soft particle as a way for us to focus on a distant observer Bob whose relationship to Alice (translation parameters connecting Alice to Bob) can be specified using only known results on the undeformed/standard relativistic properties[5]. Indeed in classical (relativistically undeformed) de Sitter spacetime one easily finds that a massless particle emitted in the origin of some observer Alice will cross the origin of all observers connected to Alice by a spatial translation of parameter a_x followed by a conformal-time translation of parameter a_η with the request that $a_x = H^{-1}[1 - e^{-Ha_\eta}]$.

Using our translation generators (4) one finds that for points on the worldline of the hard particle the map from Alice to Bob is such that

$$\begin{aligned}
\eta^B_h &= \frac{1-e^{Ha_\eta}}{H} + e^{Ha_\eta}\eta^A_h + \ell\,\alpha\,a_\eta e^{Ha_\eta}(1 - H\eta^A_h)\left(E^A_h - Ha_x p^A_h\right),\\
x^B_h &= e^{Ha_\eta}(x^A_h - a_x) - \ell\beta\frac{\sinh(Ha_\eta)}{H}p^A_h - \ell\alpha a_\eta e^{Ha_\eta}H(x^A_h - a_x)\left(E^A_h - Ha_x p^A_h\right)\\
&\quad+\ell\frac{(1 - e^{-Ha_\eta})\left(1 + e^{Ha_\eta}H\eta^A_h\right)}{H}p^A_h - \ell\left(a_x - \left(1 - e^{Ha\eta}\right)x^A_h\right)\frac{E^A_h - Hx^A_h p^A_h}{1 - H\eta^A_h},\\
p^B_h &= e^{-Ha_\eta}\left(p^A_h + \ell\alpha Ha_\eta p^A_h\left(E^A_h - Ha_x p^A_h\right)\right) .
\end{aligned}$$

Crucial for us is the fact that, in light of this result for the laws of transformation from Alice's η^A_h, x^A_h, p^A_h to Bob's η^B_h, x^B_h, p^B_h, we can deduce that the worldline of the hard particle emitted at Alice is described by Bob as follows

$$x^B_{p_h}(\eta^B) = \eta^B - \ell|p^B_h|\left(\alpha a_\eta + \beta\frac{e^{2Ha_\eta} - 1}{2H}\right), \tag{16}$$

where we made use of all the specifications discussed above, including $a_x = \frac{1-e^{-Ha_\eta}}{H}$.

In turn this allows us to obtain the sought result for the dependence on energy(/momentum) of the travel times of massless particles: by construction of the worldlines and of the Alice→Bob

[5] This also implicitly requires [22] that the clocks at Alice and Bob are synchronized by exchanging soft massless particles.

Journal of Physics: Conference Series **411** (2013) 012026 doi:10.1088/1742-6596/411/1/012026

transformation the soft massless particle emitted in Alice's spacetime origin reaches Bob's spacetime origin, whereas from (16) we see that the hard massless particle also emitted in Alice's spacetime origin reaches Bob at a nonzero conformal time. Specifically the difference in conformal travel times derivable from (16) is

$$\Delta\eta^B = \eta^B\Big|_{x^B_{p_h}=0} = \ell|p^B|\left(\alpha a_\eta + \beta\frac{e^{2Ha_\eta}-1}{2H}\right) . \tag{17}$$

We summarize the relativistic properties of this travel-time analysis, in conformal coordinates, in Fig. 1.

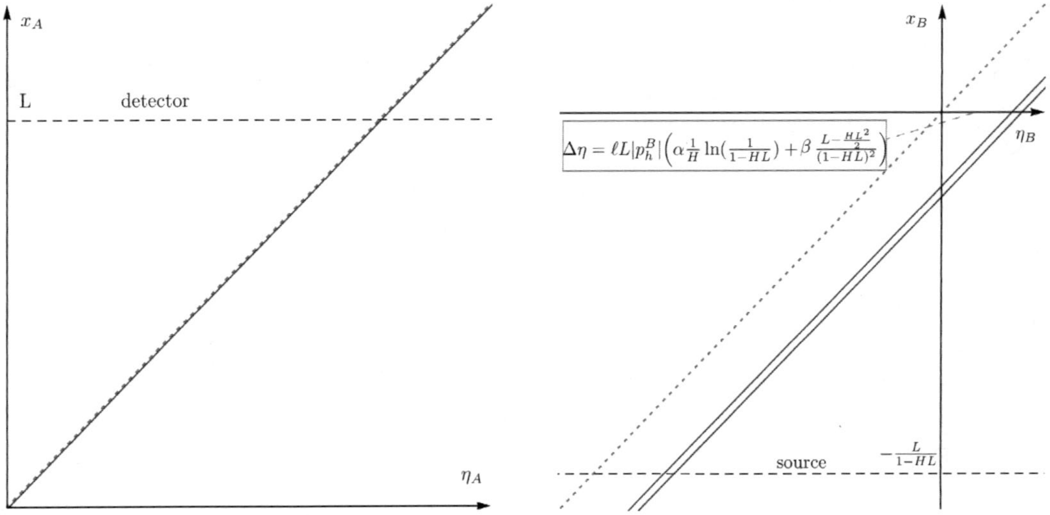

Figure 1. We illustrate the results for travel times of massless particles derived in this section, adopting conformal coordinates. We consider the case of two distant observers, Alice and Bob, connected by a pure translation, and two massless particles, one soft (dashed red) and one hard (solid blue and solid violet), emitted simultaneously at Alice. And we consider two combinations of values of α and β producing the same $\alpha+\beta$: the case $\alpha = 2/3, \beta = 2/3$ (Bob's hard worldline in blue) and the case $\alpha = 1/3, \beta = 1$ (Bob's hard worldline in violet). We have here that in Bob's coordinatization (right panel) the emission of the particles at Alice appears not to be simultaneous. Similarly for the difference in times of arrival at the distant detector Bob Alice finds in her coordinatization (left panel) that the photons arrive at the detector simultaneously. Comparison of the blue and violet worldlines shows that in the case with spacetime expansion the travel time does depend individually on α and β (not just on $\alpha+\beta$ as in the case without spacetime expansion). For visibility we assumed here unrealistic values for the scales involved.

Differently to the non expanding case [1], with spacetime expansion, there are tangible differences between Ep^2 deformations and E^3 deformations.

3. Implications for phenomenology

We close this manuscript mentioning some results we stumbled upon during our work [1], which are of rather sizable significance for phenomenology. A more detailed discussion can be found in [1]. The results we derived in the previous section are easily reformulated as the following prediction for the differences in detection times of photons of different energies(/momenta) emitted simultaneously by a source at redshift z

$$\Delta t = \ell|p| \left(\alpha \frac{\ln(1+z)}{H} + \beta \frac{z + \frac{z^2}{2}}{H} \right) . \tag{18}$$

Here again Δt is the difference in detection times[6] between a hard gamma-ray photon of momentum p and a reference ultrasoft photon emitted simultaneously to the hard photon at the distant source.

A first aspect of phenomenological relevance which must be noticed in our result (18) is the dependence on the parameters α and β, *i.e.* the difference between Ep^2 deformations and E^3 deformations. Related to this observation is the comparison with analogous studies of scenarios where relativistic symmetries are actually "broken" (allowing for a preferred/"aether" frame), rather than DSR-deformed. Of course, there is a crucial difference between the properties of ℓ and the properties of λ_{LIV}, the characteristic scale of Lorentz symmetry breaking: ℓ, as characteristic scale of a deformed-symmetry picture, takes the same value for all observers, while λ_{LIV}, as characteristic scale of a broken-symmetry picture, takes a certain value in the preferred frame and different values in frames boosted with respect to the preferred frame. Setting momentarily these differences aside we can compare our results for DSR-deformed symmetries with constant rate of expansion, with the special case of the broken-symmetry formula, on which a rather universal consensus has been achieved throughout several years of investigations [27, 28, 29, 29, 30, 33], obtained for constant rate of expansion (more details are given in [1]).

We find that fixing $\alpha = 0$ in (18), one gets a formula (valid in all reference frames) which is the same as the formula of the broken-symmetry case in the preferred frame. So if $\alpha = 0$ the differences between deformed-symmetry and broken-symmetry cases would be tangible only by comparing studies of travel times of massless particles between two telescopes with a relative boost: the difference there would be indeed that ℓ takes the same value for studies conducted by the two telescopes whereas for λ_{LIV} the two telescopes should give different values.

We must stress however that within our deformed-symmetry analysis we found no reason to focus specifically on the choice $\alpha = 0$. And if $\alpha \neq 0$ in the deformed symmetry case even studies conducted by a single telescope could distinguish between the case of symmetry deformation and the case of symmetry breakdown.

In Fig. 2 we compare the dependence on redshift of our deformed-symmetry effect among two limiting cases of balance between α and β, and we also compare these results to bounds on travel-time anomalies [31, 32, 33] obtained in studies of sources at redshift smaller than 1 (where the assumption of a constant rate of expansion is not completely misleading).

In Fig. 3 we illustrate the "constraining power" of our results and of foreseeable generalizations of our analysis gaining access to cases with non-constant rate of expansion. Analyzing data from sources at redshift of $\simeq 1$ assuming a picture with constant rate of expansion cannot produce conservative experimental bounds, but the content of the right panel of Fig. 3 serves the purpose of providing evidence of the fact that full Planck-scale sensitivity will be within reach of improved versions of our analysis, extending our results to the case of expansion at non-constant rate.

[6] Notice that with Δt we denote the comoving time but in the relevant timing sequences of detections at telescopes the difference between comoving and conformal time is intangible (see [1]).

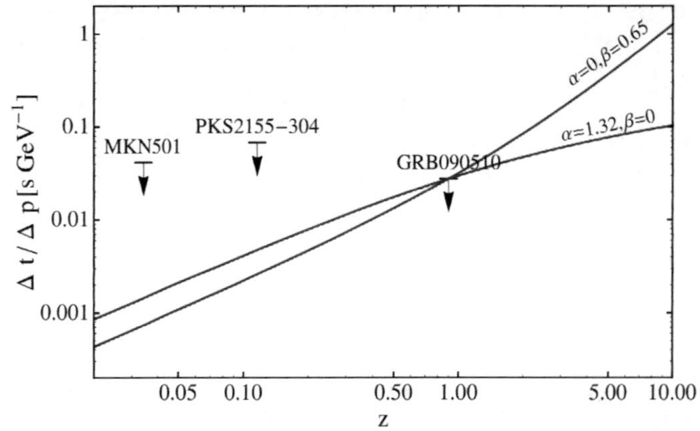

Figure 2. Here we show the dependence on redshift of the expected time-of-arrival difference divided by the difference of energy of the two massless particles. Two such functions are shown, one for the case $\alpha = 0, \beta = 0.65$ (violet) and one for the case $\alpha = 1.3, \beta = 0$ (blue). We also show the upper limits that can be derived from data reported in Refs. [31, 32, 33], setting momentarily aside the fact that our analysis adopted the simplifying assumption of a constant rate of expansion (whereas a rigorous analysis of the data reported in Refs. [31, 32, 33] should take into account the non-constancy of the expansion rate). The values $\alpha = 0, \beta = 0.65$ and $\alpha = 1.3, \beta = 0$ have been chosen so that we have consistency with the tightest upper bound, the one established in Ref. [33]. The main message is coded in the fact that at small values of redshift the blue and the violet lines are rather close, but at large values of redshift they are significantly different (this is a log-log plot). In turn this implies that at high redshift the difference between adding correction terms of form Ep^2 and adding correction terms of form E^3 can be very tangible.

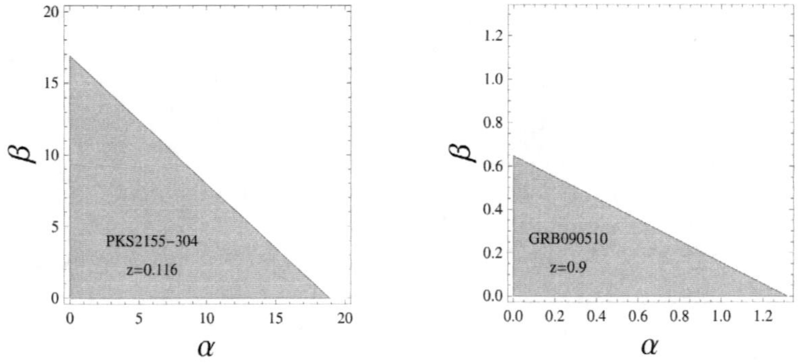

Figure 3. Here in the left panel we show the constraint on the α, β parameter space that can be obtained from Ref. [32], concerning a source at the relatively small redshift of $z = 0.116$, where we can confidently apply our results for constant rate of expansion as a reliable first approximation. Through this we show that even within the confines of our analysis Planck-scale sensitivity (values of $|\alpha|$ and $|\beta|$ smaller or comparable to 1) is not far. In the right panel we show the much tighter (indeed "Planckian") constraint on the α, β parameter space which would be within our reach if we could assume our analysis to apply also to redshifts close to 1, as for GRB090510 observed by the Fermi telescope [33]. By comparing the left and the right panel one also finds additional evidence of how the difference between adding correction terms of form Ep^2 and adding correction terms of form E^3 becomes more significant at higher redshifts: in the left panel (data on source at small redshift of $z = 0.116$) the bound on the α parameter is nearly as strong as on the β parameter, whereas in the right panel (data on source at redshift of $z \simeq 0.9$) the constraint on the alpha parameter is significantly weaker than the constraint on the β parameter.

[1] G. Amelino-Camelia, A. Marcianó, M.Matassa and G. Rosati, arXiv:1206.5315 [hep-th].
[2] G. Amelino-Camelia, gr-qc/0012051, Int. J. Mod. Phys. D 11 (2002) 35.
[3] G. Amelino-Camelia, hep-th/0012238, Phys. Lett. B 510 (2001) 255.
[4] G. Amelino-Camelia, L. Smolin and A. Starodubtsev, hep-th/0306134, Class. Quant. Grav. 21 (2004) 3095.
[5] L. Freidel, J. Kowalski-Glikman and L. Smolin, hep-th/0307085, Phys. Rev. D 69 (2004) 044001.
[6] L. Freidel and E. R. Livine, hep-th/0502106, Class. Quant. Grav. 23 (2006) 2021.
[7] G. Amelino-Camelia and S. Majid, hep-th/9907110, Int. J. Mod. Phys. A 15 (2000) 4301.
[8] J. Kowalski-Glikman and S. Nowak, hep-th/0204245, Int. J. Mod. Phys. D12 (2003) 299.
[9] L. Smolin, hep-th/0501091, Nucl. Phys. B742 (2006) 142.
[10] G. Amelino-Camelia, J. Kowalski-Glikman, G. Mandanici and A. Procaccini, gr-qc/0312124, Int. J. Mod. Phys. A 20 (2005) 6007.
[11] G. Amelino-Camelia and L. Smolin, arXiv:0906.3731, Phys. Rev. D80 (2009) 084017.
[12] J. Kowalski-Glikman and L. Smolin hep-th/0406276, Phys. Rev. D70 (2004) 065020.
[13] S. Mignemi, arXiv:0807.2186 [gr-qc].
[14] A. Marciano, G. Amelino-Camelia, N. R. Bruno, G. Gubitosi, G. Mandanici and A. Melchiorri, arXiv:1004.1110, cosmology," JCAP 1006 (2010) 030.
[15] G. Amelino-Camelia, M. Matassa, F. Mercati and G. Rosati, arXiv:1006.2126, Phys. Rev. Lett. 106 (2011) 071301.
[16] L. Smolin, relativistic spacetimes," arXiv:1007.0718 [gr-qc].
[17] G. Amelino-Camelia, L. Freidel, J. Kowalski-Glikman and L. Smolin, arXiv:1101.0931, Phys. Rev. D 84 (2011) 084010.
[18] S. Majid and H. Ruegg, Phys. Lett. B334 (1994) 348.
[19] J. Lukierski, H. Ruegg and W.J. Zakrzewski, Ann. Phys. 243 (1995) 90.
[20] J. Lukierski and A. Nowicki, q-alg/9702003.
[21] M. Daszkiewicz, K. Imilkowska, and J. Kowalski-Glikman, Phys. Lett. A 323 (2004) 345.
[22] G. Amelino-Camelia, N. Loret and G. Rosati, arXiv:1102.4637, Phys. Lett. B 700 (2011) 150.
[23] G. Amelino-Camelia, M. Arzano, J. Kowalski-Glikman, G. Rosati and G. Trevisan, arXiv:1107.1724, Class. Quant. Grav. 29 (2012) 075007
[24] L. Freidel and L. Smolin, arXiv:1103.5626.
[25] S. Cacciatori, V. Gorini and A. Kamenshchik, arXiv:0807.3009, Annalen Phys. 17 (2008) 728.
[26] G. Rosati, "Special relativity in classical de Sitter spacetime" (in preparation).
[27] J. R. Ellis, N. E. Mavromatos, D. V. Nanopoulos, A. S. Sakharov and E. K. G. Sarkisyan, astro-ph/0510172, Astropart. Phys. 25 (2006) 402.
[28] M. Rodriguez Martinez and T. Piran, astro-ph/0601219, JCAP 0604 (2006) 006.
[29] U. Jacob and T. Piran, arXiv:0712.2170, JCAP 0801 (2008) 031.
[30] J. R. Ellis, N. E. Mavromatos, D. V. Nanopoulos, A. S. Sakharov and E. K. G. Sarkisyan, Astropart. Phys. 29 (2008) 158.
[31] J. Albert *et al.* [MAGIC Collaboration] and J.R. Ellis, N.E. Mavromatos, D.V. Nanopoulos, A.S. Sakharov and E.K.G. Sarkisyan, arXiv: 0708.2889, nucleus Markarian 501 observed by the MAGIC telescope Phys. Lett. B 668 (2008) 253.
[32] F. Aharonian *et al.* [H.E.S.S. collaboration], Galaxy PKS 2155-304, Phys. Rev. Lett. 101 (2008) 170402.
[33] A. A. Abdo *et al.* [Fermi LAT/GBM Collaborations], Nature 462 (2009) 331.
[34] A. Albrecht and J. Magueijo, astro-ph/9811018, Phys. Rev. D59 (1999) 043516.
[35] S. Alexander and J. Magueijo, cosmology," hep-th/0104093; S. Alexander, R. Brandenberger and J. Magueijo, hep-th/0108190, Phys. Rev. D 67 (2003) 081301.
[36] U. D. Machado and R. Opher, arXiv:1102.4828, Class. Quant. Grav. 29 (2012) 065003.

Journal of Physics: Conference Series **411** (2013) 012027 doi:10.1088/1742-6596/411/1/012027

On representations of Yangian of Lie Superalgebra $A(n,n)$ type

Vladimir Stukopin

Department of Applied Math., DSTU, Gagrin sq., 1, Rostov-on-Don, Russia; South Mathematical Institute, Vladikavkaz, Russia

E-mail: stukopin@mail.ru

Abstract. The finite-dimensional irreducible representations of Yangian of Lie Superalgebra of $A(n,n)$ type is described in terms of Drinfel'd polynomials. The necessarily and sufficient conditions of finite-dimensionality of irreducible representation are formulated and proved. The Poincare-Birkoff-Witt theorem for Yangian of $A(n,n)$ Lie Superalgebra is proved also.

1. Introduction

The description of irreducible representations of Yangians of Lie Superalgebras be an important problem for the theory of exactly solvable models of Statistical Mechanics and Quantum Field Theory. For example, construction of the transfer-matrix based on the determination of image of universal R-matrix of quantum double of Yangian by the action of tensor product of irreducible and identical representations. Computation of spectrum of Hamiltonian and correlation functions is based also on the representation theory of Yangians by using universal R-matrix formula. Therefore representations theory of Yangians of simple and reductive Lie algebras is a developed theory, which began to develop until to appearance of term "Yangian"(see [1], [10], [3], [4], [15], [25], [26]). In contrast with these theory, representation theory of Yangians of Lie Superalgebras is a young discipline, which appear at the beginning of 90-th years of 20 century. In the last years the number of applications of Representation Theory of Yangians of Lie Superalgebras rise, to began research problems of Yang-Mills fields and Quantum Superstring Theory using Yangian Theory technique (see [2], [6], [19]).

In this paper we research the finite-dimensional irreducible representations of Yangian of Lie Superalgebra of $A(n,n)$ type (see [14], [11]). The main result of this work is a theorem on classification of such representations. Let's note that classification of the finite-dimensional irreducible representations of Yangian of Lie Superalgebra of $\mathfrak{gl}(m+1,n+1)$ was received in the middle of 90-th by R.B. Zhang (see [27], [28]) and classification of the the finite-dimensional irreducible representations of Yangian of Lie Superalgebra of $A(m,n), m \neq n$ was received in [24] (see also, [23]). The case $m = n$ is a more difficult, because in these case Cartan generators aren't linearly independent.

[0] This work is partially supported by Federal Program "Scientific and scientific-educational personnel of innovative Russia event 1.2.2 (contract number P116) and by Federal Program "Scientific and scientific-educational personnel of innovative Russia"(agreement № 14.A18.21.0356 on application 2012-1.1-12-000-1003-029).

Let's note that Yangians together with quantized enveloping algebras and elliptic quantum algebras are one of the three most important examples of quantum algebras. The notion Yangian was introduced by V. Drinfel'd in honour to C.N. Yang. But algebras isomorphic to Yangian were used for investigation exactly solvable models by means of Quantum Inverse Scattering Method in the 80-th years by V. Tarasov (see, for example, [25], [26]).

It should be mentioned that representation theory of Lie Superalgebra $\mathfrak{sl}(m+1, n+1)$ differs from representation theory of simple Lie algebra $\mathfrak{sl}(n+1)$ that Lie Superalgebra has together typical representations also so called atypical representations. This pathology reveals itself in the representation theory of Yangian of Lie superalgebra $A(m, n)$ also.

Let's note that Yangian with quantized enveloping algebras and quantum elliptic algebras are most important examples of quantum algebras. The notion of quantum algebra was introduced by V. Drinfel'd in the middle of 80-th years of 20 century (see [7], [8]). But earlier algebras isomorphic to Yangian are used for investigation exactly solvable models by Quantum Inverse Scattering Method (QISM) (see [25], [26]). Yangians of simple and reductive Lie algebras are investigated in detail in present time (see books [3], [15]). Yangians of Lie Superalgebras began investigate at the 90-th years of last century (see [16], [20]). In the papers [21], [22] was proved formula for universal R-matrix of Yangin Double of Yangin for special linear Lie Superalgebra. Let us note that using of results papers [21], [22], [24] and this work it can be systematically describe all possible quantum R-matrices and transfer matrices connecting with Yangian $Y(A(m, n))$ including examples important for exactly solvable models of Quantum Field Theory.

Some words about structure od this works. In the second section we recall definition of the Yangian of Lie Superalgebra $A(n, n)$ in terms of generators and defining relations. We define it in terms of current system of generators following the author's work [20]. In the third section we recall construction of PBW bases and give a sketch of the proof of this theorem. In the forth section we recall the main definitions from representation theory of Yangians of Lie Superalgebras and formulate and prove the main result of work the theorem on classification of finite dimensional irreducible representations of Yangian of Lie Superalgebra $A(n, n)$.

We'll use the following notations. We denote by C the field of complex numbers, by $M_n(K)$ the ring of $N \times N$ - matrices with elements from ring K; by $K[u]$, $K[[u]]$ the ring of polynomials and ring of formal power series, correspondingly, with coefficients from ring K; by $Z_+ = \{0, 1, 2, \ldots\}$ the set of non negative integers, which is union of zero and set of natural numbers N. The end of proof we denoted by symbol $\square$.

2. Definition of Yangian of Lie Superalfebra $A(n, n)$

2.1. Lie Superalfebra $A(n, n)$

Let me recall basic definitions from Lie Superalgebra theory, related to the Lie Superalgebra $A(n, n)$. This Lie superalgebra has rank equals $2n-2$ and dimension equals $4n^2-2$, number of simple roots equals $2n-1$ and differs from rank. Lie Superalgebra $A(n, n)$ is generated the generators $h_i, x_i^{\pm}, i \in \{1, 2, \ldots, n, n+1, \ldots, 2n+1\}$. The Cartan matrix (distinguished Cartan matrix) determining the system of defining relations has following form:

$$A = (a_{i,j})_{i,j=1}^{2n+1} = \begin{pmatrix} 2 & -1 & \ldots & 0 & 0 & \ldots & 0 & 0 \\ -1 & 2 & \ldots & 0 & 0 & \ldots & 0 & 0 \\ \ldots & \ldots & \ldots & \ldots & \ldots & \ldots & \ldots & \ldots \\ 0 & 0 & \ldots & 0 & 1 & \ldots & 0 & 0 \\ 0 & 0 & \ldots & -1 & 2 & \ldots & 0 & 0 \\ \ldots & \ldots & \ldots & \ldots & \ldots & \ldots & \ldots & \ldots \\ 0 & 0 & \ldots & 0 & 0 & \ldots & -1 & 2 \end{pmatrix}.$$

The symmetric Cartan matrix is follows:

$a_{ii} = 2, i < n+1$, $a_{n+1,n+1} = 0$, $a_{i,i} = -2, n+1 < i \leq 2n+1$, $a_{i,i+1} = a_{i+1,i} = -1$ for $i < n+1$, и $a_{i,i+1} = a_{i+1,i} = 1$ for $i \geq n+1$. The all other matrix elements are equal zero. As a rule we'll use the symmetric Cartan matrix.

The system of defining relations of $A(n,n)$ has the following form:

$$[h_i, h_j] = 0,$$
$$i \in I = \{1, 2, \dots, n, n+1, \dots, 2n+1\}; \quad (1)$$
$$[h_i, x_j^{\pm}] = \pm a_{ij} x_j^{\pm}, i, j \in I; \quad (2)$$
$$[x_i^{+}, x_j^{-}] = \delta_{ij} h_i, i, j \in I; \quad (3)$$
$$[x_{n+1}^{\pm}, x_{n+1}^{\pm}] = 0; \quad (4)$$
$$[x_i^{\pm}, x_j^{\pm}] = 0; |i-j| > 1; \quad (5)$$
$$[x_i^{\pm}, [x_i^{\pm}, x_j^{\pm}]] = 0, |i-j| = 1, i, j \in I; \quad (6)$$
$$[[x_n^{\pm}, x_{n+1}^{\pm}], [x_{n+1}^{\pm}, x_{n+2}^{\pm}]] = 0. \quad (7)$$

Let's note that the generators $x_{n+1}^{\pm}$ are odd and other generators are even, or $p(x_{n+1}^{\pm}) = 1, p(x_i^{\pm}) = p(h_j) = 0$, where p be a parity function, $j \in I, i \in I \backslash \{n+1\}$ (see [14]). Let be below $\mathfrak{g} = A(n,n) = \mathfrak{sl}(n+1, n+1)$.

2.2. Definition of Yangian

First, we recall the definition of Yangian $Y(\mathfrak{g}) = Y(A(n,n))$ (see also [20], [12]).

Yangian, as other quantum algebras, is defined as a result of deformation (in the case of Yangian) of universal enveloping (super)algebra of current Lie superalgebra (wih values in Lie superalgebra) in the class of Hopf superalgebras. In the paper [20] was shown that this definition is equivalent other definition using so called current system of generators. This system of generators is convenient for representation theory of Yangians. Therefore we'll consider it further.

Definition 1 *Let $Y(\mathfrak{g})_\hbar$ be a superalgebra (over ring of formal power series $C[[\hbar]]$), generated by generators $h_{i,k}, x_{i,k}^{\pm}, i \in I = \{1, 2, \dots, 2n+1\}, k \in Z_+$ $(p(x_{n+1,k}^{\pm}) = 1, p(x_{i,k}^{\pm}) = p(h_{j,k}) = 0, i \in I \backslash \{n+1\}, j \in I, k \in Z_+)$, which satisfied the following defining relations (compare with relations from paper [18], definition 2 of paper [20], and also relations from proposition 9.1 of work [12]):*

$$[h_{i,k}, h_{j,l}] = 0, i, j \in I, k \in Z_+; \quad (8)$$
$$[x_{i,k}^{+}, x_{j,l}^{-}] = \delta_{i,j} h_{i,k+l}, i, j \in I, k, l \in Z_+; \quad (9)$$
$$[h_{i,k+1}, x_{j,l}^{\pm}] = [h_{i,k}, x_{j,l+1}^{\pm}] \pm (a_{ij}/2)\hbar(h_{i,k} x_{j,l}^{\pm} + x_{j,l}^{\pm} h_{i,k}),$$
$$i, j \in I, k, l \in Z_+; i \neq n \quad \textit{или} \quad j \neq n \quad (10)$$
$$[h_{n+1,k+1}, x_{n+1,l}^{\pm}] = 0; \quad (11)$$
$$[h_{i,0}, x_{j,l}^{\pm}] = \pm a_{ij} x_{j,l}^{\pm}, i, j \in I, l \in Z_+; \quad (12)$$
$$[x_{i,k+1}^{\pm}, x_{j,l}^{\pm}] = [x_{i,k}^{\pm}, x_{j,l+1}^{\pm}] \pm (a_{ij}/2)\hbar(x_{i,k}^{\pm} x_{j,l}^{\pm} + x_{j,l}^{\pm} x_{i,k}^{\pm}),$$
$$i, j \in I, k, l \in Z_+; i \neq n \quad \textit{or} \quad j \neq n; \quad (13)$$
$$[x_{n+1,k}^{\pm}, x_{n+1,l}^{\pm}] = 0; \quad (14)$$
$$[x_{i,k_1}^{\pm}, [x_{i,k_2}^{\pm}, x_{j,k_3}^{\pm}]] + [x_{i,k_2}^{\pm}, [x_{i,k_1}^{\pm}, x_{j,k_3}^{\pm}]] = 0,$$
$$|i-j| = 1, i, j \in I, k_1, k_2, k_3 \in Z_+; \quad (15)$$

$$[x^{\pm}_{i,k}, x^{\pm}_{j,l}] = 0, |i-j| > 1; \tag{16}$$

$$[[x^{\pm}_{n,r}, x^{\pm}_{n+1,0}], [x^{\pm}_{n+1,0}, x^{\pm}_{n+2,s}]] = 0. \tag{17}$$

Easy to check that for different $\hbar$, (aren't equal zero), Hopf Superalgebras $Y(\mathfrak{g})_\hbar$ are isomorphic.

Yangian $Y(\mathfrak{g})$ is called Hopf Superalgebra $Y(\mathfrak{g})_1$ over field of cjmplex numbers C.

Let us introduce the following generating functions of Yangian generators:

$$h_i(u) = 1 + \sum_{k=0}^{\infty} h_{i,k} \cdot u^{-k-1} = \sum_{k=-1}^{\infty} h_{i,k} \cdot u^{-k-1}, i = 1, 2; \tag{18}$$

$$x_i^{\pm} = \sum_{k=0}^{\infty} x^{\pm}_{i,k} \cdot u^{-k-1}, i = 1, 2. \tag{19}$$

Easy to check that relation (10) for Yangian generators is equivalent the following relation for generating functions:

$$[h_i(u), x_j^{\pm}(t)] = \mp \frac{a_{ij}}{2} \frac{(h_i(u)(x_j^{\pm}(t) - x_j^{\pm}(u)) + (x_j^{\pm}(t) - x_j^{\pm}(u))h_i(u))}{(u-t)}. \tag{20}$$

3. Root generators. Poincare-Birkgoff-Witt theorem.

3.1.

In this subsection we formulate Poincare-Birkgoff-Witt theorem. This result is a quantum analogue of classical Poincare-Birkgoff-Witt theorem for universal enveloping algebras (see [13], [3]).

Let below $\mathfrak{g} = A(n, n)$.

Let degree of generators $a_{i,k}, a_{i,k} \in \{x^{+}_{i,k}, x^{-}_{i,k}, h_{i,k}\}$ be its second index.

A degree of monom from generators is called the sum of degrees of multipliers. A degree of polynomial is called maximum of degrees of monoms of this polynomial. Let also degree of tensor products of monoms be the sum of degrees of tensor multipliers. A degree of tensor polynomial is called maximum of degrees of tensor monoms forming this polynomial.

Let's denote the space of elements of $Y(\mathfrak{g})$ of degreee less or equal k by $Y_k = Y_k(\mathfrak{g})$). We get on $Y(\mathfrak{g})$ the following filtration:

$$Y_0 \subset Y_1 \subset Y_2 \subset ... \subset Y_n \subset ....$$

Let's note that on modulo of terms of lesser degree (that is on modulo $Y(\mathfrak{g})_{k+l-1}$) commutation relations in $Y(\mathfrak{g})$ has the following form:

$$[h_{\alpha,\bar{k}}, x_{\beta,\bar{l}}] = (\alpha, \beta) x_{\beta,\bar{k}+\bar{l}}; \quad [h_{\alpha,\bar{k}}, h_{\beta,\bar{l}}] = 0; \tag{21}$$

$$[x_{\alpha,\bar{k}}, x_{\beta,\bar{l}}] = \begin{cases} 0, & \alpha+\beta \notin \Delta \cup \{0\}, \\ h_{\alpha,\bar{k}+\bar{l}}, & \beta = -\alpha, \\ N(\alpha,\beta) x_{\alpha+\beta,\bar{k}+\bar{l}}, & \alpha+\beta \in \Delta \end{cases} \tag{22}$$

where $N(\alpha, \beta)$ is defined from the following relation in $U(\mathfrak{g})$:
$[x_\alpha, x_\beta] = N(\alpha, \beta) x_{\alpha+\beta}$, для $\alpha + \beta \in \Delta$.

Let's construct root vectors for $Y(\mathfrak{g})$. Let α_{n+1} be an odd root and Δ_+ be a set of positive roots of Lie superalgebra $A(n,n)$. Let's define on Δ_+ so called convex order by induction. At first we fix ordering on the set of positive roots, corresponding its numeration: $\alpha_1 \prec \alpha_2 \prec ... \prec \alpha_{2n+1}$. If $\alpha, \beta \in \Delta_+$, $\alpha \prec \beta$ and its sum is defined $\alpha+\beta$, then we define ordering by condition $\alpha \prec \alpha+\beta \prec \beta$. Let's fix monotonous mapping $\alpha : [1, 2, ..., N] \to \Delta_+$ with respect to order: $\alpha(k) \prec \alpha(n)$ if $k < n$. (Let's note that $N = \frac{(2n+1)(2n+2)}{2}$.)

Let $\overline{(i,k)} = (i_1, k_1, i_2, k_2, \dots, i_{2n+1}, k_{2n+1})$ be a vector, $|\overline{(i,m)}| = \sum_{j=1}^{m+n+1}(i_j + m_j)$ be sum of components of its vector.

Let Y_- be an associative subalgebra of Yangian $Y(A(m,n))$, generated by generators $\{x^-_{i,k} = x^-_{\alpha(i),k} | 1 \leq i \leq N, k \in Z_+\}$. Let's describe ordering base in Yangian $Y(\mathfrak{g})$, as in vector space, i.e Poincare-Birkgoff-Witt base (PBW – base). Let $\overline{k} = (k_1, ..., k_r), k_i \geq 0$, be a vector, $k = |\overline{k}| = \sum_{i=1}^{r} k_i$. Let also

$$x(\overline{k}, \overline{w}; \overline{s}, \overline{p}; \overline{r}, \overline{t}) = (x^+_{\alpha(1),k_1})^{w_1} \cdot \ldots \cdot (x^+_{\alpha(N),k_N})^{w_N} \cdot$$
$$(h_{1,s_1})^{p_1} \cdot \ldots \cdot (h_{2n+1,s_{2n+1}})^{p_{2n+1}} \cdot (x^-_{\alpha(1),r_1})^{t_1} \cdot \ldots\ldots \cdot (x^+_{\alpha(N),r_N})^{t_N}.$$

Let's define lexicographic ordering on the set of vectors $\{x(\overline{k}, \overline{w}; \overline{s}, \overline{p}; \overline{r}, \overline{t})\}$. Let $\alpha = \alpha_s + \alpha_{s+1} + \ldots + \alpha_t; \alpha, \alpha_s, \alpha_{s+1}, \ldots, \alpha_t \in \Delta_+$, and $\alpha_s, \alpha_{s+1}, \ldots, \alpha_t$ be simple roots, $s < t, l = t - s + 1$. Let $x^\pm_{s,k_1}, \ldots, x^\pm_{t,k_l} \in Y(G)$, $k = k_1 + k_2 + ... + k_l$. Let's define root vectors by formulas:

$$x_{\pm\alpha,k} = [x^\pm_{s,k_1}, [x^\pm_{s+1,k_2}, [\ldots, x^\pm_{t,k_l}] \cdots]].$$

Easy to check that if $(k'_1, \ldots, k'_l)$ be another decomposition of number k, then $x_{\alpha,k} = [x^\pm_{n,k'_1}, [x^\pm_{n+1,k'_2}, \ldots, x^\pm_{t,k'_l}] \ldots]]$.

Let's note that onto modulo of terms of lesser degree (i.e. onto modulo $Y(\mathfrak{g})_{k-1}$) commutation relations in $\oplus_{k=1}^{\infty} Y(\mathfrak{g})_k / Y(\mathfrak{g})_{k-1} \oplus Y(\mathfrak{g})_0$ are coincide with commutation relations in universal enveloping superalgebra of current Lie superalgebra $U(\mathfrak{g}[t])$, i.e have the following form:

$$[h_{i,k}, x^\pm_{j,l}] = \pm(\alpha_i, \alpha_j) x^\pm_{j,k+l},$$
$$[x^+_{i,k}, x^-_{j,l}] = \delta_{ij} h_{i,k+l},$$
$$[h_{i,k}, h_{j,l}] = 0,$$
$$[x^\pm_{i,l}, [x^\pm_{i,r}, x^\pm_{j,s}]] = 0,$$
$$[[x^\pm_{n,r}, x^\pm_{n+1,0}], [x^\pm_{n+1,0}, x^\pm_{n+2,s}]] = 0.$$

It should be note that these relations enough for determination linear order on elements of base of polynomial current superalgebra $\mathfrak{g}[t] = A(n,n)[t]$, when odd root generators take part in relations. This can be checked by direct calculations. Foe every number k let's fix vector $(k_1, k_2, ..., k_l), k = k_1 + ... + k_l$, defines partition of its number.

Above we introduce the linear order onto set of roots Δ and inducing by its order an order on root vectors. Let's note by Ω the set of ordering monoms from $x_\alpha, h_i, \alpha \in \Delta, i \in I$. We use notation $x_\alpha = x^+_\alpha, x_{-\alpha} = x^-_\alpha, \alpha \in \Delta_+$.

Theorem 1 $(\Omega, \prec)$ *is Poincare-Birkgoff-Witt base in Yangian* $Y(\mathfrak{g})$.

3.2.

Let's prove Poincare-Birkgoff-Witt theorem for Yangian $Y(A(n,n))$. We give a sketch of proof.

At first let's prove completeness of $\Omega(<)$.

For monom M belong to $\Omega(<)$ let's define its length $l(M)$ as number of multipliers or generators are contained in M. After reordering multipliers from (21), (22) it follows that we'll get additional summands either lesser degree or equal degree and lesser length. Using induction we get that $Y(\mathfrak{g})$ coincide with linear span of $\Omega(<)$. Note when we reording monoms contain odd root generators we use relation (17).

Let us prove that monoms from $\Omega(<)$ are linear independent. Our proving is based on existence of such representation ρ of Yangian $Y(\mathfrak{g})$ that $\rho(x^{+}_{i,0})$, $\rho(x^{-}_{i,0})$, $\rho(h_{i,0})$ are are linear independent.

After them we reduce to contradiction assumption that elements of $\Omega(<)$ are linear dependent. Suppose that monoms from $\Omega(<)$ aren't linear independent. Then exist such numbers $c_1, \ldots, c_s \in C \setminus \{0\}$ and monoms $M_1, \ldots, M_s \in \Omega(<)$, that

$$\sum_{1\leq j\leq s} c_j M_j = 0 \tag{23}$$

It can be show that this assumption reduced to contradiction (see [24]).

□

4. Representations of Yangian $Y(A(n,n))$

4.1.

Let V be a module over Yangian $Y(A(n,n))$.

Definition 2 *Let V be a module over Yangian $Y(\mathfrak{g})$ of Lie Superalgebra $\mathfrak{g} = A(m,n)$, $\bar{d} = \{d_{i,r} | i \in I, r \in Z_+\}$ be a collection of complex numbers. Let's denote by $V_{\bar{d}}$ and call weight subspace of module V the space*

$$V_{\bar{d}} = \{v \in V : h_{i,r}v = d_{i,r}v\} \tag{24}$$

Collection of numbers $\bar{d} = \{d_{i,r}\}$ is called weight of Yangian module.

We are going to describe a structure of finite dimensional modules over Yangian $Y(\mathfrak{g})$ and to formulate necessary and sufficient conditions finite dimensionality of irreducible module.

The vector $v \in V$ is called *primitive*, if $v \in V_{\bar{d}}$ and $x^{+}_{i,r} \cdot v = 0$ for all $i \in I, r \in Z_+$.

The module V is called *highest weight module* if it is generated by primitive vector, that is $V = Y(A(n,n)) \cdot v$ for some primitive vector $v \in V_{\bar{d}}$.

At first we show that every finite dimensional representation of Yangian $Y(\mathfrak{g})$ has highest (primitive) vector.

Let's note that every highest weight module should be constructed as a factor of *Verma module*. Verma module should be constructed in usual manner as f factor module of Yangian $Y(\mathfrak{g})$ on ideal generated by vectors $x^{+}_{i,r}$ and $h_{i,k} - d_{i,k}$. In this case highest vector is a unite of Yangian 1. Let's consider weight $\sum_i \lambda_i \cdot h_{i,0}$, where λ is a fundamental weight of Lie Superalgebra $\mathfrak{g}$. Then weight subspace of Yangian Verma module with such weight is 1-dimensional. Easy to show that it follows Verma module has unique irreducible factor module is denoted as $V(\bar{d})$, i.e. $V(\bar{d}) = M(\bar{d})/N(\bar{d})$, where $N(\bar{d})$ be a maximal submodule of Verma module $M(\bar{d})$. Let $\pi : M(\bar{d}) \to V(\bar{d})$ be a canonical projection.

Theorem 2 *Every finite dimensional representation of Yangian $Y(\mathfrak{g})$ (simple $Y(\mathfrak{g})$-module) V contains unique (up to scalar multiplier) highest vector v.*

The proof of this theorem is based on the following lemma. Let $V_0 = \{v \in V | x^{+}_{i,k}v = 0, \forall k \in Z_+\}$.

Lemma 1 *1) $h_{i,k}V_0 \subset V_0$.*
2) $V_0 \neq 0$.

it Proof 1) Let $v \in V_0$. Then $x^+_{i,k} h_{j,l} v = h_{j,l} x^+_{i,k} v + [h_{i,l+1}, x^+_{i,k-1}] v + \frac{1}{2}(h_{i,k} x^+_{i,l} + x^+_{i,l} h_{i,k}) v$. Let rearrange generators $x^+_{i,s}$ и $h_{j,r}$ using commutation relations (defining relations of Yangian) we get $x^+_{i,k} h_{j,l} v = 0$ or $h_{j,l} v \in V_0$ для $v \in V_0$, that is $h_{i,l} V_0 \in V_0$.

2) Let $v' \in V_0$. Let act on vector v' by elements of PBW-base. At first let consider the case $m = n = 1$, i.e. case of $Y(\mathfrak{sl}(1,1))$ and after we consider a general case. In the case of $Y(\mathfrak{sl}(1,1))$ exist such number $m \in N$, that $(x^+_0)^m v' \neq 0$, $(x^+_0)^{m+1} v' = 0$. Let $v_2 = (x^+_0)^m v'$. We'll act by elements of PBW base in correspondence with order on the PBW base. Easy to check that exist such integer r that $v_r = (x^+_r)^{p_r} \cdot \ldots \cdot (x^+_0)^{p_0} v \neq 0$, but $x^+_t \cdot v_r = 0$ for $\forall t \in Z_+$. Then $v_r \in V_0$. WE get that $V_0 \neq \{0\}$.

Let consider the general case of $Y(A(n,n))$. Consider linear order on vectors of PBW base from Y_+. Let's denote these vectors by $\{x^+(j)\}_{j=0}^{\infty}$, and let $x^+(j) \prec x^+(l)$ for $j < l$. Then exist such integer $m \in Z_+$, that $x^+(m) v' \neq 0$, buto $x^+(m+1) v' = 0$ for $v' \in V_0$. □

Lemma 2 *V_0 be 1-dimensional space (that is every two vectors from V_0 are proportional).*

Proof Let $v', v" \in V_0, v' \neq v"$. Then acting on these vectors by elements of Yangian $Y(A(n,n))$, we get two submodules $Y(A(m,n))v'$ and $Y(A(m,n))v"$ of module V. These fact contradicts to irreducibility of module V in the case when $v', v"$ aren't proportional. □

The theorem follows from these two lemmas.

Let us introduce class of highest weight modules are analogues of Verma modules. Let $V_0 = Cv^\Lambda_+$ be 1-dimensional vector space, Y^+_0 be a subsuperalgebra of Yangian $Y(A(n,n))$, generated by generators $x^+_{i,k}, i \in \{1, 2, ..., m+n+1\}, k \in Z_+$; $Y^0_0 = \langle h_{i,k} | i \in I, k \in Z_+ \rangle$ is a linear span of generators $\{h_{i,k} | i \in I, k \in Z_+\}$; $Y^+ = Y^+_0 \cdot Y^)_0$. Let also $h_{i,k} v^\Lambda_+ = d_{i,k} v^\Lambda_+, Y^+_0 \cdot v^\Lambda_+ = 0$. (We'll use also notation $\Lambda = \bar{d}$ in order to stress analogy with highest weights of modules of Lie Superalgebras. We'll use also notation $v_+ = v^\Lambda_+$ for highest vector). Then V_0 became 1-dimensional $Y(A(n,n))$-module. Let's define free module M_Λ with highest weight Λ by formula:

$$M_\Lambda = Y(\mathfrak{g}) \otimes_{Y^+} v^\Lambda_+. \tag{25}$$

Evidently, that M_Λ isomorphic to $Y^-_0 \otimes v^\Lambda_+ : M_\Lambda \cong Y^-_0 \otimes v^\Lambda_+$ as a vector space. We'll write $x^-_{i,k} v^\Lambda_+$ instead of $x^-_{i,k} \otimes v^\Lambda_+$. Clearly, that module M_Λ is infinite dimensional. Standard arguments show that module M_Λ contains maximal submodule N_Λ. Then module $V_\Lambda = M_\Lambda / N_\Lambda$ is irreducible module with highest weight Λ. Using the standard methods of representation theory easy to check that every two modules with equal highest weights are isomorphic (see also [?]). Namely, let $V_1(\Lambda), V_2(\Lambda)$ be two irreducible modules with equal highest weights $\Lambda = \{\lambda_k\}_{k=0}^{\infty}$ and u^Λ_+, w^Λ_+ be its highest vectors, correspondingly. Let $W = V_1(\Lambda) \oplus V_2(\Lambda)$, $v^\Lambda_+ = (u^\Lambda_+, w^\Lambda_+)$ be a highest vector of some submodule $V(\Lambda)$ with highest weight Λ generated in W by vector v^Λ_+ (under action of Yangian $Y(A(n,n))$). Let's define a projections $P_i : V(\Lambda) \to V_i(\Lambda), i = 1,2 : P_1(v_1, v_2) = (v_1, 0), P_2(v_1, v_2) = (0, v_2), v_i \in V_i(\Lambda)$. Easy to check that $P_i, i = 1,2$ are module homomorphisms and $P_1(v^\Lambda_+) = u^\Lambda_+, P_2(v^\Lambda_+) = w^\Lambda_+$. It follows that $Im P_i = V_i(\Lambda)$ from irreducibility of modules $V_i(\Lambda), i = 1,2$. Then $Ker P_1 = \{0\}, Ker P_2 = \{0\}$ (or $Ker P_1 = V_2(\Lambda), Ker P_2 = V_1(\Lambda)$). But last equality is impossible because vectors $(0, w^\Lambda_+), (u^\Lambda_+, 0) \notin W$. Therefore $P_i, i = 1,2$ are module isomorphisms and $V(\Lambda) \cong V_1(\Lambda) \cong V_2(\Lambda)$.

So then we prove the following theorem

Theorem 3 *For every highest weight $\Lambda = \{\lambda_i\}_{i=0}^{\infty}$ exist unique irreducible $Y(\mathfrak{g})$-module $V(\Lambda)$ with highest weight Λ.*

4.2.
Let's define on module $M(\bar{d})$ the following two filtrations. As the module $M(\bar{d})$ as a vector space naturally isomorphic to $Y(\mathfrak{g})_-$, we can define at first filtrations on $Y(\mathfrak{g})_-$. After them we can define this filtrations on $M(\bar{d})$ using isomorphism between $Y(\mathfrak{g})_-$ and $M(\bar{d})$. Let $(Y(\mathfrak{g})_-)_k$ $((Y(\mathfrak{g})_-)^k)$ be a linear span of monoms od degree lesser or equal k. The degree of monom from generators , as usual, is sum of degrees of these generators form this monom. A degree of generators is a value of second index of this generators in first case and is number equals value of second index plus 1 in the second case. In he first case we'll denote degree of monom x by $d_1(x)$ and by $d_1(x)$ in the second case. Also as above we extend our definition on polynomials defining a degree of polynomial as a maximum of degrees of monoms form this polynomial. Let $Y(\mathfrak{g})_k = \{x \in Y(\mathfrak{g}) : d_1(x) \leq k\}$, a $Y(\mathfrak{g})^k = \{x \in Y(\mathfrak{g}) : d_2(x) \leq k\}$. Let's consider constriction of these filtrations on $Y(\mathfrak{g})_-$. Thus we get the following two filtrations on $Y(\mathfrak{g})_-$:

$$C \subset (Y(\mathfrak{g})_-)_0 \subset (Y(\mathfrak{g})_-)_1 \subset \cdots (Y(\mathfrak{g})_-)_k \subset \ldots \tag{26}$$
$$\{0\} \subset C \subset (Y(\mathfrak{g})_-)^0 \subset (Y(\mathfrak{g})_-)^1 \subset \cdots (Y(\mathfrak{g})_-)^k \subset \ldots \tag{27}$$

Let $M(\bar{d})_k = (Y(\mathfrak{g})_-)_k \cdot v^{\bar{d}}$, $M(\bar{d})^k = (Y(\mathfrak{g})_-)^k \cdot v^{\bar{d}}$. As irreducible module $V(\bar{d})$ is a factor-module of Verma module $M(\bar{d})_k$, the the above defined filtrations determine filtration on irreducible module $V(\bar{d})$. It should be mentioned that properties of these two filtrations are various. Let's note also that the every space with index k of the first filtration contains all spaces of second filtration with indices lesser or equals k.

Let's describe conditions of finite dimensionality of irreducible highest weight Yangian module. Let note by $\bar{B}_{n+1,t}, \bar{B}_{i,k}, i \in I\backslash\{n+1\}$ the linear spans of following vectors

$$\bar{B}_{n+1,s} := \langle (x^-_{n+1,k_1})^t \cdot \ldots \cdot x^-_{n+1,k_r} \cdot v_+ | (k_1+1) + \ldots + (k_r+1) \leq s \rangle$$
$$\bar{B}_{i,s} := \langle (x^-_{i,k_1})^{t_1} \cdot \ldots \cdot (x^-_{i,k_r})^{y_r} \cdot v_+ | t_1(k_1+1) + \ldots t_r(k_r+1) \leq s \rangle, \quad i \in I\backslash\{n+1\}$$

Lemma 3 *If $\bar{B}_{n+1,p} = \bar{B}_{n+1,p+1}$, то и $\bar{B}_{n+1,p} = \bar{B}_{n+1,p+k}$ for every integer $k \in N$.*

Proof. Let $a \in \bar{B}_{n+1,p}$. At first let show that every vector from $\bar{B}_{n+1,p+k}$ can be represent as image of $\bar{B}_{n+1,p}$ under action Cartan subalgebra $\mathfrak{h} = \langle h_{n+1,k}, h_{n+2,k} | k \in Z_+ \rangle$, namely, $\bar{B}_{n+1,p+k} \subset \mathfrak{h} \cdot \bar{B}_{n+1,p}$. For $k = 1$ this fact it follows from condition of lemma. Let now $k = 2$. We need the following relation
$h_{i,1} \cdot x^-_{j,l} \cdot v_+ = x^-_{j,l} \cdot h_{i,1} \cdot v_+ + [h_{i,1}, x^-_{j,l}] \cdot v_+ = x^-_{j,l} \cdot h_{i,1} \cdot v_+ + [h_{i,0}, x^-_{j,l+1}] \cdot v_+$
$+ a_{i,j}/2(h_{i,0}x^-_{j,l} + x^-_{j,l}h_{i,0}) \cdot v_+ = d_{i,1}x^-_{j,l} \cdot v_+ + a_{i,l}x^-_{j,l+1} \cdot v_+ +$
$a_{i,j}/2(h_{i,0}x^-_{j,l} + x^-_{j,l}h_{i,0}) \cdot v_+ = a_{i,j}x^-_{j,l+1} \cdot v_+ + (d_{i,1} + (a_{i,j}/2)(a_{i,j} + 2d_{i,0}))x^-_{j,l} \cdot v_+.$

Using this relation let prove the lemma. Let $a \in \bar{B}_{m+1,p+2}$, then $a = \sum_{s=1}^r x^-_{i_s,k_s} b_s v_+$. Let represent an element a in the form

$$a = \sum_{s=1}^{r} a^{-1}_{i_s-1,i_s}[h_{i_s-1,1}, x^-_{i_s,k_s-1}]b_s v_+ \tag{28}$$

Repeating these arguments we can represent the element a as a sum of elements from $\bar{B}_{n+1,p+1}$ and product $x^-_{n+1,1}$ and elements from $\bar{B}_{n+1,p+1}$. As every element from $\bar{B}_{n+1,p+1}$ is contained in $\bar{B}_{n+1,p}$ we get that $a \in \bar{B}_{n+1,p+1}$ and after them it follows from condition of lemma that $a \in \bar{B}_{n+1,p}$. Lemma is proved. □

Lemma 4 *1) There are the strict inclusions*
$B_{i,k} \subset B_{i,k+1}, \forall k < n_i, k \geq 0, i \in I$;
2) $B_{i,n_i} = B_{i,n_i+1} = \cdots$.

Proof The case $i \in I\backslash\{n+1\}$ it follows from results of representation theory of Yangian $Y(\mathfrak{sl}(2))$ (see [8], [25]). The 1)it follows from definitions. The point 2) it follows from previous lemma in the case $i = n+1$. □

4.3.
Let's formulate and prove the main result of our work – the following theorem. We'll suppose, that $n \geq 1$.

Theorem 4 *1) Every irreducible finite dimensional $Y(A(n,n))$-module V be a highest weight module with highest weight d :$V = V(d)$.*
2) The module $V(d)$ be finite dimensional iff when exist the polynomials P_i^d, $i \in \{1,2,\ldots,n,n+2,\ldots 2n\} = I\backslash\{n+1\}$, and polynomials P_{n+1}^d, Q_{n+1}^d, such that:
a) all these polynomials with highest coefficients equal 1;
b)

$$\frac{P_i^d(u+1)}{P_i^d(u)} = 1 + \sum_{k=0}^{\infty} d_{i,k} \cdot u^{-k-1}, \quad i \in I\backslash\{n+1\}, \tag{29}$$

$$\frac{P_{n+1}^d(u)}{Q_{n+1}^d(u)} = 1 + \sum_{k=0}^{\infty} d_{n+1,k} \cdot u^{-k-1}. \tag{30}$$

Proof. Let's note that our proof is a slight modification analogous result from work [24].

The point 1) of theorem is proved already. Let deduce the point 2) from above proved lemmas. At first let's prove the necessary in theorem, that is let's prove that if the irreducible module $V(d)$ be finite dimensional then exist mentioned in point 2) of theorem the polynomials satisfying the conditions a), b) of theorem.

At first, we consider particular case when $\mathfrak{g} = A(0,1)$. More precisely, we consider inclusion $A(0,1) \to A(n,n)$, inducing by the mapping of roots $\alpha_1 \to \alpha_{n+1}, \alpha_2 \to \alpha_{n+2}$. These inclusion unduces embedding $Y(A(0,1))$ into $Y(A(n,n))$: $Y(A(0,1)) \to Y(A(n,n))$ (in category of topological superalgebras) defined by mapping of root generators (which is induced by above mentioned mapping of roots).

In the first part of the proof we'll deal with particular case of irreducible representation of Yangian $Y(A(0,1))$ with weight $\bar{d}(u) = (d_{n+1}(u), d_{n+2}(u))$. After them we'll reduce the general case to considering particular case.

Let as above $x_i^-(u) = \sum_{k=0}^{\infty} x_{i,k}^- \cdot u^{-k-1}, i \in I$. From proved above lemmas it follows that $x_i^-(u) \cdot v_+ = \sum_{s=0}^{N} \beta_s^i(u) \cdot v_{i,s}$, где $\{v_{i,s}\}$ form a base in $\bar{B}_{i,N}, i \in I$. Further we'll get explicit form for $\beta_s^i(u)$. Her it should be note that we can get all results for even root generators considering inclusions of $\mathfrak{sl}(2)-$triples into $Y(A(n,n))$. These results are known as they are can be reduced to well known results on representations of Yangian $\mathfrak{sl}(2)$ (see [25] and also [26], [8], [9], [4]). Therefore we prove in detail relation (30) relating to odd part of Yangian $Y(A(n,n))$.

We need the relations (20) in the following particular case:

$$[h_{n+1}(u), x_{n+2}^-(t)] =$$

$$\frac{1}{2(u-t)}(h_{n+1}(u)(x_{n+2}^-(t) - x_{n+2}^-(u)) + (x_{n+2}^-(t) - x_{n+2}^-(u))h_{n+1}(u)); \tag{31}$$

$$[h_{n+1}(u), x_{n+1}^-(t)] = 0. \tag{32}$$

Let $v_{i,k} = x_{i,k}^- v_+$. Then from lemma 4 it follows following equality

$$x_k^-(u)v_+ = \sum_{i=0}^{N} \beta_i^k(u) v_{k,i}. \tag{33}$$

For brevity we'll use the notation:

$$\beta_i(u) = \beta_i^{n+2}(u). \tag{34}$$

Note that
$\sum_{k=0}^{\infty} x_{n+2,k}^- u^{-k-1} v_+ = \sum_{k=0}^{N} x_{n+2,k}^- u^{-k-1} v_+ + \sum_{k=N+1}^{\infty} x_{n+2,k}^- u^{-k-1} v_+ =$
$\sum_{k=0}^{N} x_{n+2,k}^- u^{-k-1} v_+ + \sum_{k=0}^{N} u^{-N-1} \sum_{i=0}^{\infty} \varphi_k^i u^{-i-1} v_{n+2,k}.$

Let's denote

$$\varphi_k(u) = \textstyle\sum_{i=0}^{\infty} \varphi_k^i u^{-i-1}, \tag{35}$$
$$\tilde{\varphi}_k(u) = u^{-k-1} + \varphi_k(u). \tag{36}$$

Let's rewrite relation (33) in the following form:

$$x_{n+2}^-(u) v_+ = \sum_{i=0}^{N} \beta_i(u) v_{n+2,i} = \sum_{i=0}^{N} (u^{-i-1} + u^{-N-1} \varphi_i(u) v_{n+2,i}). \tag{37}$$

Then from equality (37) it follows that

$$x_{n+2,N+k+1}^- v_+ = \sum_{i=0}^{k} \varphi_i^k x_{n+2,i}^- v_+. \tag{38}$$

Let's act on the left and right sides of equality (38) by element $h_{n+2,1}$.
We get the following relations equating coefficients of equal linear independent vectors:

$$\varphi_k^{i+1} = \varphi_{k-1}^i + \varphi_N^i \cdot \varphi_k^0 \tag{39}$$
$$\varphi_0^{k+1} = 0. \tag{40}$$

Let's introduce the following notations:

$$\delta_i(u) = 1 + u^{i-N} \varphi_i(u). \tag{41}$$
$$\bar{d}_i(u) = \textstyle\sum_{k=0}^{\infty} d_{i,k} u^{-k-1}, i \in \{1, \cdots, 2n+1\} \tag{42}$$

Easy to see that from relations (39), (40), (41) it follows the equality :

$$\delta_i(u) = \delta_N(u)(1 - \sum_{j=i}^{N} \varphi_j^0 u^{j-N-1}). \tag{43}$$

Let

$$\beta_i(u) = u^{-i-1} \delta_i(u), i = 1, 2, \ldots, N. \tag{44}$$

Then for some matrices $A^k(u) = (A_{i,j}^k)_{i,j=0}^N(u) \in M_{N+1}(C[[u^{-1}]]), k, r = n+1, n+2$ is fulfilled equality:

$$h_r(u) x_k^-(t) v_+ = \sum_{0 \le i,j \le N} A_{i,j}^r(u) \beta_j^r(t) v_{k,i}. \tag{45}$$

Journal of Physics: Conference Series **411** (2013) 012027 doi:10.1088/1742-6596/411/1/012027

Considering relations (45) и (31-32) jointly we get the following relations

$$\sum_{0\le i,j\le N} A^{n+1}_{i,j}(u)\beta_j(t)v_{n+2,i} = x^-_{n+2}(t)\bar{d}_{n+1}(u)v_+ +$$
$$\frac{1}{2(u-t)}(h_{n+1}(u)(x^-_{n+2}(t)-x^-_{n+2}(u))+(x^-_{n+2}(t)-x^-_{n+2}(u))h_{n+1}(u))v_+; \tag{46}$$
$$\sum_{0\le i,j\le N} A^{n+1}_{i,j}(u)\beta_j(t)v_{n+1,i} = x^-_{n+1}(t)\bar{d}_{n+1}(u)v_+. \tag{47}$$

These relations taking into account (33) can be rewrite in the following form

$$\sum_{0\le i,j\le N} A^{n+1}_{i,j}(u)\beta_j(t)v_{n+2,i} = \bar{d}_{n+1}(u)(\sum_{i=0}^N\beta_i(u)v_{n+2,i})+$$
$$\frac{1}{2(u-t)}(\bar{d}_{n+1}(u)(\sum_{i=0}^N(\beta_i(t)-\beta_i(u))v_{n+2,i})+$$
$$\sum_{0\le i,j\le N} A^{n+1}_{i,j}(u)(\beta_j(t)-\beta_j(u))v_{n+2,i})); \tag{48}$$
$$\sum_{0\le i,j\le N} A^{n+1}_{i,j}(u)\beta_j(t)v_{n+1,i} = \bar{d}_{n+1}(u)(\sum_{i=0}^N\beta_i(t)v_{n+1,i}). \tag{49}$$

Let's equate coefficients by linear independent vectors $v_{n+2,i}$ and we get that from relation (49) it follows the following equality:

$$\sum_{j=0}^N A^{n+1}_{i,j}(u)(\beta_j(t)-\frac{1}{2(u-t)}(\beta_j(t)-\beta_j(u)))v_{n+2,i} =$$
$$\bar{d}_{n+1}(u)((\beta_i(u)v_{n+2,i})+\frac{1}{2(u-t)}(\beta_i(t)-\beta_i(u)))v_{n+2,i} \tag{50}$$

Let's note the following easy checking fact. We can to select such constants $u_0,\mu_0,\ldots,\mu_N$, such that the sum $\sum_{i=0}^N\mu_i\beta_i(u_0)$ isn't equals zero. Let's put in equality (50) $u=u_0$ and multiply equality on μ_i. After addition getting equalities (for $0\le i\le N$ we get equality. From this last equality we can explicitly express $d_{n+1}(u)$. Namely, we get that fulfilled equality

$$\bar{d}_{n+1}(u) = \frac{\sum_{i=0}^N(a_iu+b_i)\beta_i(u)}{\sum_{j=0}^N(c_ju+d_j)\beta_j(u)} \tag{51}$$

for some constants $a_i,b_i,c_i,d_i, i=0,\cdots,N$.

From relation (51), taking into account relations (43), (44), it follows proving equality (30).

As above we can deduce that relation (29) it follows from relations (46) for other indexes k, corresponding to other simple even roots of Lie Superalgebra $A(n,n)$. But, as above mentioned, the relation 29) follows from results of papers [25], [9]. Now, we finish the proof of theorem by standard arguments founding on the fact that inclusions of $Y(\mathfrak{sl}(1,2))$ $Y(\mathfrak{sl}(2))$ in to $Y(A(n,n))$ are morphisms of Hopf Superalgebras. Thus, necessary condition is proved.

Let's prove sufficient condition. The second part of proof can be get conversing above given proving. Really, let $\bar{d}_{n+1}(u)$ satisfies to relations (29 – 30). Let's prove that the simple module $V(d)$ be finite dimensional. We, using theorem 1, get estimate of dimension of this module. We'll act on highest vector v_+ by monoms in the following form $x^-(t)=x^-_{1,k_1}\ldots x^-_{1,k_r}, k_1+\ldots+k_r = t, k_r<\cdots<k_1$. Let' note that from (30), $d_{n+1}(u)$ be a ratio of two polynomials. Let us show that from commutation relations it follows that only finite number of vectors $x^-(n)v_+$ are linearly independent. We get more strong result, namely, we prove that dimension of module $V(\bar{d})$ is limited (from above) by number $2^N, N=n_1+\ldots+n_k$, where $n_i=deg(P_i(u))$ are the degrees of polynomials determine the highest weight. Let $\bar{d}_{n+1}(u)=1+\sum_{k=0}^\infty d_{n+1,k}\cdot u^{-k-1}$, and $\bar{d}_{n+1}(u)=\frac{P^d_{n+1}(u)}{Q^d_{n+1}(u)}$ be polynomials in a numerator and denominator, correspondingly, and these polynomials haven'n common multipliers. Let us construct simple highest weight $Y(\mathfrak{sl}(1,1))$–module $V(\bar{d})$ with highest weight $\bar{d}(u)$. Let us show that dimension of this module is limited

(from above) by number 2^M, where $M = m_{d+1} = deg(P_{n+1}(u)) = deg(Q_{n+1}(u))$. Let us consider the linear span

$$M_1 = \langle \{v_+; x^-(r_1) \dots x^-(r_s) \cdot v_+ | 1 \leq r_s \leq \dots \leq r_1 \leq N\} \rangle \tag{52}$$

Let us show that

$$V(\bar{d}) \subseteq M_1 \tag{53}$$

Sufficiently to show that

$$x_k^+(u) \cdot x^-(n) = 0, \forall n > N. \tag{54}$$

Let's note that (54) it follows from following commutation relation in Yangian, which it follows from relations (9), (30):

$$\begin{aligned} [x_k^+(u), x_{m+1}^-(v)]v_+ = \delta_{k,m+1} \tfrac{h_{m+1}(u) - h_{m+1}(v)}{u-v} v_+ = \\ \delta_{k,n+1} \tfrac{P_{n+1}(u)Q_{n+1}(v) - P_{n+1}(v)Q_{n+1}(u)}{Q_{n+1}(u)Q_{n+1}(v)(u-v)} v_+. \end{aligned} \tag{55}$$

Like, we can prove that dimensions of corresponding $Y(\mathfrak{sl}(2))$– modules $V(d_k)$ with highest weights $d_k(u)$ are limited by numbers 2^{m_k}, where $m_k = deg(P_k(u)$ is a degree of polynomial $P_k(u)$.

As embeddings of Yangians $Y(\mathfrak{sl}(2))$, $Y(\mathfrak{sl}(1,1))$ in to Yangian $Y(A(n,n))$ are morphisms of Hopf superalgebras they induce structure of $Y(A(n,n))$–module on every modules $V(d_k)$ and also on they tensor products: $V(d_1) \otimes \dots \otimes V(d_{m+n})$. Easy to check that $V(\bar{d}) \subseteq V(d_1) \otimes \dots \otimes V(d_{m+n})$ and it is submodule of tensor product of modules. From last assertion it follows that module $V(\bar{d})$ be a finite dimensional and we get estimate of its dimension. Sufficient condition is proved. Theorem is proved also.
□

[1] T. Arakawa Drinfeld functor and finite-dimensional representations of Yangian *Comm. Math. Phys.* 1999 **205 no 1** 1 – 18 (*Preprint: math/9807144*
[2] G. Arutyunov, S. Frolov 2009 Foundations of the $AdS^5 \times S^5$ Superstring. Part 1 (*Preprint: hep-th/0901.4937*)
[3] Chari, V., Pressley, A. A guide to quantum groups *Cambridge: Camb.Univ.Press*
[4] Chari, V., Pressley, A. 1990 *L'Enseignment Mathematique* **36** 267 – 302
[5] Chari, V., Pressley, A. 1991 *J. Reine Angew. Math.* **417** 87 – 128
[6] L. Dolan, Ch. Nappi, E. Witten 2004 Yangian Symmetry in D=4 Superconformal Yang-Mills theory *Preprint: hep-th/0401243*
[7] Drinfeld V. 1988 Quantum groups *Berkley: Proc. Int. Cong. Math.* **1** 789–820.
[8] Drinfeld V 1985 *Soviet Math. Dokl.* **32** 1060 – 1064 (in Russian).
[9] Drinfeld V 1988 *Soviet Math. Dokl.* **36** 212 – 216 (in Russian).
[10] Drinfeld V 1986 *Funct. Anal. and its Applic.* **20 no 1** 69 – 70 (in Russian).
[11] Frappat L., Sorba P. 2000 Dictionary on Lie Superalgebras *London: Academic Press*
[12] L. Gow 2007 Gauss decomposition of the Yangian $Y(gl(m|n))$ *Comm. Math. Phys.* **276** no 3 799 – 825 *Preprint: math/0605219*
[13] Hamphreys J.E. 1978 Introduction to Lie Algebras and Representations Theory *New-York: Springer-Verlag*
[14] Kac V. 1977 A Sketch of Lie Superalgebra Theory *Commun.Math.Phys.* **53** 31-64
[15] Molev A. Yangians and their applications *Preprint: math. QA/0211288*
[16] Nazarov M. 1991 Quantum Berezinian and the classical Capelly identity *Lett. Math. Phys.* **21** 123-131
[17] Nazarov, M., Tarasov, V. 1998 On irreducibility of tensor products of Yangian modules *Internat. Math. Research Notices* 125-150 *Preprint: q-alg/9712004.*
[18] F. Spill Weakly coupled N = 4 Super Yang-Mills and N = 6 Chern-Simons theories from u(2|2) Yangian symmetry *Preprint: hep-th/0810.3897*
[19] F. Spill, A. Torrielli 2008 On Drinfeld's second realization of the *AdS/CFT* $\mathfrak{su}(2|2)$ Yangian *Preprint: hep-th/0803.3194*

[20] Stukopin V 1994 *Funct. Anal. and its Applic.* **28**, **no 3** 217 – 219 (in Russian).
[21] Stukopin V 2006 *Funct. Anal. and its Applic.* **42**, **no 2** 81 – 84 (in Russian).
[22] Stukopin V 2005 *Fundamental and Applyed Mathematics* **11**, **no 2** 185 – 208 (in Russian).
[23] Stukopin V 2002 *Asymptotic Combinatorics with Appic. to Math. Phys. Kluwer Academic Publishers* 255 – 265
[24] Stukopin V 2012 *Izvestija RAN. Serija matem.* (to appear).
[25] Tarasov V. 1984 *Theor. and math. phys.* **61 no 2** 163–173 (in Russian).
[26] Tarasov V. 1985 *Theor. and math. phys.* **63 no 2** 175–196 (in Russian).
[27] Zhang R.B. 1995 Representations of super Yangian *J. Math. Phys.* **36** 3854-3865 *Preprint: hep-th/9411243*
[28] Zhang R.B. 1996 The $\mathfrak{gl}(M,N)$ super Yangian and its finite-dimensional representations *Lett. Math. Phys.* **37** 419 – 434 *Preprint: hep-th/9507029*

Exotic Chekanov tori in toric symplectic varieties

Nikolay A. Tyurin

Bogolyubov Lab. of Theoretical Physics, JINR (Dubna), Joint Institute for Nuclear research, Joliot - Curie, 6, Moscow region, Dubna 141980, Russia and Bogomolov Lab. of Algebraic Geometry, HSE (Moscow), National Research University Higher School of Economics, 20 Myasnitskaya Ulitsa, Moscow 101000, Russia

Abstract. We present a generalization of toric structures on compact symplectic manifolds called pseudotoric structure. In the present talk we show that every toric manifold admits pseudotoric structures and then we show that the construction of exotic Chekanov tori can be peformed in terms of pseudotoric structures.

Let (X, ω) be a symplectic manifold of real dimension $2n$ so it can be understood as the phase space of a classical mechanical system. Lagrangian geometry of X is focused on the questions about lagrangian submanifolds of X namely: which homology classes from $H_n(X, \mathbb{Z})$ can be realized by smooth lagrangian submanifolds; what are the topological types of these lagrangian submanifolds; classification up to lagrangian deformations of lagrangian submanifolds of the same topological type and homology class; classification up to Hamiltonian isotopy of lagrangian submanifolds of the same deformation type; unification of all lagrangian submanifolds in an appropriate category.

Recall that $S \subset X$ is lagrangian if the restriction $\omega|_S$ vanishes identically and real dimension of S is maximal, equal to n. Thus at least the homology class of S must be perpendicular to the cohomology class $[\omega]$. Two lagrangian submanifolds $S_0, S_1 \subset X$ are of the same deformation type if there is a family of lagrangina submanifolds $S_t, t \in [0, 1]$ which ends at S_0 and S_1. Hamiltonian isotopy of lagrangian submanifold $S_0 \subset X$ is given by a time dependent Hamiltonian function $H(x, t) : X \times \mathbb{R} \to \mathbb{R}$ which generates the flow ϕ_H^t, and $S_t = \phi_H^t(S_0)$ is the corresponding isotopy.

Toy example: dim $= 2$. Let Σ be a Riemann surface equipped with a symplectic form. Then since every loop is lagrangian (dimensional reason): every primitive homology class from $H_1(\Sigma, \mathbb{Z})$ is realizable by a smooth lagrangian submanifold; every smooth lagrangian submanifold is isomorphic to S^1; two loops from the same homology class are deformation equivalent; two loops are Hamiltonian isotopic if the symplectic area of the oriented film bounded by the loops is zero; the Fukaya category for a curve of any genus exists.

Thus for this case the problem is completely solved!

But making one new step we face already highly nontrivial situation. Consider "the simplest and basic" 4- dimensional compact symplectic manifold — the projective plane $\mathbb{CP}^2$. Fot the projective plane we have that: since the cohomology group $H^2(\mathbb{CP}^2, \mathbb{Z}) = \mathbb{Z}$, any lagrangian submanifold must present trivial homology class; there are no lagrangian 2 - spheres (M. Gromov), Riemannian surfaces of genus $g > 1$ (M. Audin), Klein bottles (S. Nemirovsky and V. Shevchishin) — all these types are not realized by smooth lagrangian submanifolds of the projective plane; it was believed that well known Clifford tori are unique examples of lagrangian tori in $\mathbb{CP}^2$ since in 1996 Yu. Chekanov proposed a construction of lagrangian torus which is not

Hamiltonian isotopic to a Clifford torus — and nobody knows are there other types of lagrangian tori; nevertheless certain constructions of appropriate categories exist (Fukaya - Seidel).

Thus even for this basic case in dimension 4 the problem is not solved yet.

Why we are interested in lagrangian geometry? Lagrangian geometry is very important in Mathematical physics; f.e. several approaches to Geometric Quantization are based on Lagrangian geometry. In these approaches lagrangian submanifolds represent quantum states so an old idea of P.M. Dirac, stated that the phase space of classical mechanical system should contain the ingredients of a natural quantization procedure, is realized. Thus it is natural to study all possible states so it is reasonable to find all types of lagrangian submanifolds (see, f.e. [1]).

F.e. in ALAG - programme (abelian algebraic lagrangian geometry, see [2]) the Chekanov result ensures that the moduli space of half weighted Bohr - Sommerfeld lagrangian cycles of level 3, $\mathcal{B}^{hw,r}_{S,3}$, has at least two disjoint components, and may be in a future one will find certain connecting space with a tunneling effect between these components.

As well in a popular modern subject of Mathematical physics — Homological Mirror symmetry — one should try to describe all objects in the Fukaya category, so all types of nonisotopic lagrangian tori.

Well known Clifford tori in $\mathbb{CP}^2$ comes from the toric geometry: the projective plane carries two real Morse functions in involution with respect to the Poisson brackets induced by the Kahler form of the standard Fubini - Study metric. These functions can be explicitly expressed as:

$$f_1 = \frac{|z_1|^2 - |z_2|^2}{\sum_{i=0}^2 |z_i|^2}, f_2 = \frac{|z_0|^2 - |z_1|^2}{\sum_{i=0}^2 |z_i|^2}, \{f_1, f_2\}_\omega = 0$$

in homogeneous coordinates $[z_0 : z_1 : z_2]$; the degeneration set

$$\Delta(f_1, f_2) = \{df_1 \wedge df_2 = 0\} \subset \mathbb{CP}^2$$

is formed by three lines $l_i, l_i = \{z_i = 0\}$; the action map $F = (f_1, f_2) : \mathbb{CP}^2 \to P_{\mathbb{CP}^2} \subset \mathbb{R}^2$ sends $\Delta(f_1, f_2)$ to the boundary component $\partial P_{\mathbb{CP}^2}$ of the convex polytop $P_{\mathbb{CP}^2}$, and the preimage of any inner point $p \in P_{\mathbb{CP}^2}$ is a smooth lagrangian torus, labeled by values of f_1, f_2. Thus the Clifford tori are just Liouville tori for this completely integrable system. And it is the standard picture for any toric manifold.

In 1996 Yu. Chekanov in [3] proposed the construction of exotic lagrangian tori by the first version to $\mathbb{R}^4$. The construction looks rather simple: fix a complex structure, so we have $\mathbb{C}^2$ with a coordinate system (z_1, z_2); choose a smooth contractible loop $\gamma \subset \mathbb{C}^*$ which lies in a half plane so Re$\gamma > 0$; consider two - dimensional subset given in the coordinates by the explicit formula $(z_1, z_2) = (e^{i\phi}\gamma, e^{-i\phi}\gamma)$ — and it is a lagrangian torus! Note however that if the loop γ is not contractible, we get a torus which is equivalent to the standard one. Furthemore, since the projective plane without projective line $\mathbb{CP}^2 \backslash l$ is symplectomorphic to an open ball in $\mathbb{R}^4$ one implements the construction to the projective plane. Using certain special Hofer's capacity technique, Chekanov proved this torus is not equivalent to the standard one.

This exotic torus was called **the Chekanov torus**; the forthcoming paper by Yu. Chekanov and F. Schlenk contains the details how to construct these nonstandard tori in the projective space $\mathbb{CP}^n$ for certain n, the products $S^1 \times ... \times S^1$, and some other cases, see [4].

An alternative description of the Chekanov tori based on the notion of pseudotoric structure. We can produce the torus taking the pencil $\{Q_w\}$ such that $Q_w = \{z_1 z_2 = w\} \subset \mathbb{C}^2$ — one dimensional complex family of quadratic surfaces given in the coordinate system (z_1, z_2) by the quadratic equation which depends on complex parameter $w \in \mathbb{C}$. Then one takes real Morse function $F = |z_1|^2 - |z_2|^2$ and observes that the Hamiltonian vector field X_F of this function F preserves each quadric Q_w from the family. Then one fixes a smooth contractible loop $\gamma' \subset \mathbb{C}^*_w$

Journal of Physics: Conference Series **411** (2013) 012028 doi:10.1088/1742-6596/411/1/012028

where $\mathbb{C}_w$ parameterizes our family $\{Q_w\}$. The choice of the value for our function F marks the level set on each quadratic surface which is a loop, so taking smooth loops $S_w = \{F = 0\} \cap Q_w$ on every quadratic surface $Q_w, w \in \gamma'$ and collecting all these loops along γ' one gets a torus:

$$T(\gamma') = \bigcup_{w \in \gamma'} S_w;$$

it is not hard to see, that we again get the Chekanov torus from the previous construction, if we put $\gamma = \sqrt{\gamma'}$.

Let's repeat the construction for the projective plane. To do this consider pencil of quadrics $\{Q_p\}$, $p \mapsto [\alpha : \beta] \subset \mathbb{CP}^1_{\alpha,\beta}$ where $Q_p = \{\alpha z_1 z_2 = \beta z_0^2\} \subset \mathbb{CP}^2$. Take real Morse function F explicitly given by the formula

$$F = \frac{|z_1|^2 - |z_2|^2}{\sum_{i=0}^{2} |z_i|^2}$$

in homogenoues coordinates $[z_0 : z_1 : z_2]$. It can be checked directly that its Hamiltonian vector field X_F preserves each element of the pencil, so we can proceed as in the previous noncompact case. Let's choose a smooth contractible loop $\gamma \subset \mathbb{CP}^1_{\alpha,\beta} \backslash \{[1:0],[0:1]\}$ since the last points are covered by singular quadrics; then on each quadric $Q_p, p \in \gamma$ we can take the level set $S_p = \{F = 0\} \cap Q_p$, and this level set is a smooth loop. Then we collect the level sets S_p along the loop γ getting again a lagrangian torus $T(\gamma) = \bigcup_{p \in \gamma} S_p$. The point is that the resulting torus is exactly the Chekanov torus, given by the identification of symplectic ball in $\mathbb{R}^4$ and $\mathbb{CP}^2 \backslash$line. On the other hand if $\gamma \subset \mathbb{CP}^1_{\alpha,\beta}$ was taken non contractible then the resulting torus would be equivalent to a Clifford torus.

Therefore we get certain correspondence between the equivalence classes of lagrangian tori and the fundamental group of the punctured projective line $\pi_1(\mathbb{CP}^1_{\alpha,\beta} \backslash \{[0:1],[1:0]\})$ without the north and the south poles.

What is the difference between toric and pseudo toric considerations?

We illustrate it on the ideal level by the following diagramme:

$\mathbb{R}$	$\mathbb{C}$
real Morse function f	Lefschetz pencil $\{Q_p\}$
$\Vert$	$\Vert$
$f : X \to \mathbb{R}$	$\psi : X \backslash B \to \mathbb{CP}^1, Q_p = \overline{\psi^{-1}(p)}$
$\downarrow$	$\downarrow$
toric case	pseudtoric case
(f_1, f_2) on $\mathbb{CP}^2$	$(f, \{Q_p\})$ on $\mathbb{CP}^2$
such that $\{f_1, f_2\}_\omega = 0$	such that $X_f \parallel Q_p$
$\uparrow$	$\uparrow$
standard commutation rel.	new commutation rel.

A complex analog of a real Morse function is a Lefschetz pencil, roughly speaking it is just a complex (or symplectic) map to the compactified complex space. The question is how to relate the real data and new complex data, so what does it mean that a real function and a Lefschetz pencil commute? We propose the following new commutation relation: pencil $\{Q_p\}$ commutes with real function f if the Hamiltonian vector field X_f is parallel to each element Q_p of the pencil at each point. Geometrically (or dynamically) this means that the Hamiltonian flow generated by f preserves the "level sets" of the Lefschetz pencil — but it is exactly the same as for the real functions!

Leaving aside other speculative arguments, we summurize with the following

Definition ([5]): Pseudotoric structure on a compact symplectic manifold (X, ω_X) consists of

· *(real data)* $\{f_1, ..., f_k\}$ — algebraically independent almost everywhere real Morse functions in involution, $\{f_i, f_j\}_\omega = 0$;
· *(complex data)* family of compact symplectic $2k$ -dimensional submanifolds $\{Q_p\}, Q_p \subset X$, parameterized by a compact toric symplectic manifold $(Y, \omega_Y) \ni p$ (or, equivalently, a map with symplectic fibers

$$\psi : X \backslash B \to Y, \quad Q_p = \overline{\psi^{-1}(p)}, \quad B = \text{base set}$$

such that the following *commutation relations* hold:
· the Hamiltonian vector field X_{f_i} of each Morse function f_i from (*real data*) is parallel to each Q_p at each point (or, equivalently, each f_i preserves the fibers of ψ by the Hamiltonian action);
· for each smooth function $h \in C^\infty(Y, \mathbb{R})$ bi- vector field
$X_{\psi^* h} \wedge \nabla_\psi X_h \equiv 0$ — identically vanishes on $X \backslash B$.

In the last expression $X_{\psi^* h} \wedge \nabla_\psi X_h$ we take two vector fields for a function $h \in C^\infty(Y, \mathbb{R})$ taken on the base manifold Y, namely for the lifted function $\psi^* h \in C^\infty(X \backslash B, \mathbb{R})$ on $X \backslash B$ one takes the Hamiltonian vector field with respect to the symplectic form ω_X; on the other hand one takes the lift $\nabla_\psi X_h$ of the Hamiltonian vector field X_h defined by the symplectic form ω_Y on Y, and ∇_ψ is the symplectic connection defined by ψ since this map has symplectic fibers:

$$\nabla_\psi : \Gamma(TY) \to \Gamma(T(X \backslash B)).$$

The last condition looks too horrible but in practice one avoids all the difficulties, taking in mind the following remark: if X and Y are complex, $k = n - 1$, and ψ is complex then the last commutation relation is automatically satisfied.

It's easy to see that the base set B of the family $\{Q_p\}$ must be contained by the degeneration locus $\Delta(f_1, ..., f_k) = \{df_1 \wedge ... \wedge df_k = 0\}$; the singular points of any fiber Q_p must be contained by the degeneration locus $\Delta(f_1, ..., f_k)$ as well; any fiber $Q_p = \psi^{-1}(p) \cup B$ endowed with restrictions $(f_1|_{Q_p}, ... f_k|_{Q_p})$ — is a completely integrable system (= toric (perhaps non smooth) symplectic manifold). Therefore pseudotoric structures supply us with the solutions of the following problem: for non completely integrable Hamiltonian system with the integrals $(f_1, ..., f_k)$ find toric leaves of the Hamiltonian action.

The simplest (and the trivial) example of pseudotoric structure arises if one takes the direct product of two toric manifolds $Y_1 \times Y_2$. This structure is topologically trivial, as the product vector bundle. In analogy with the theory of vector bundles we introduce the following

Definition ([5]): *the number* $n - k = \frac{1}{2} dim Y$ *is called the rank of pseudotoric structure* $(f_1, ..., f_k, \psi, Y)$.

Clearly it is parallel to the notion of the rank of vector bundle.

If singular points of fiber Q_p lies in the base set B we say that the fiber Q_p is regular; if generic fiber Q_p is regular then we say that pseudotoric structure is regular; it's not hard to see that in the regular case the image $\psi(\Delta(f_1, ..., f_k) \backslash B) = D_{\text{sing}} \subset Y$ is a proper compact symplectic submanifold. This submanifold measures topological non triviality of the pseudotoric structure; this subset is empty if and only if the peudotoric structure is topologically trivial so it is the product of toric manifolds.

The main reason for the introduction of this new structure is the possibility to construct lagrangian fibrations on whole X starting with lagrangian fibrations on the base toric manifolds and using the toric nature of the fibers. If we choose a system $(h_1, ..., h_{n-k})$ of commuting moment maps on Y (since Y by the definition is toric) we get a lagrangian fibration on the base Y but at the same time we have the following

Theorem ([6]): *Choice of moment maps* $(h_1, ..., h_{n-k})$ *on the base* Y *of a regular pseudotoric structure* $(f_1, ..., f_k, \psi, Y)$ *on a given* X *defines a lagrangian foliation on* X *whose generic fiber is a smooth lagrangian torus.*

The dimensional reduction which happens on $\Delta(h_1, ..., h_{n-k}) \subset Y$ is reflected by the fact that the collection of fibers over $\Delta(h_1, ..., h_{n-k})$ must be cutted from X, and then the resting part

$X \backslash (\bigcup_{p \in \Delta(h_1,...,h_{n-k})} Q_p)$ carries lagrangian fibration. This lagrangian fibration is only generically smooth (so generic fiber is a smooth lagrangian torus), but the singular fibers have singularities which are not of generic type. The type of the singularities is controlled and can be described as follows. A Liouville torus in a completely integrable system carries periodic orbits and unbounded real lines (if we consider irrational motion along the torus). Our singular tori admit additionally trajectories of the separatrix type: take a periodic loop on a torus and contract it to a point — then the torus turns to be singular and instead of periodic loop one gets a stable point. This is the type of singularities which appear in our lagrangian fibration.

General scheme can be summarized by the following diagramme

$$\begin{array}{ccccl} (f_1, ..., f_k): & B & \longrightarrow & \partial P_{Q_p} & -\text{boundary component} \\ & \cap & & \cap & \\ (f_1, ..., f_k): & X & \longrightarrow & P_{Q_p} & -\text{moment pol.for gen.fib.} \\ & \psi \downarrow & & \times & \\ (h_1, ..., h_{n-k}): & Y & \longrightarrow & P_Y & -\text{moment pol.for base} \\ & \cup & & \cup & \\ (h_1, ..., h_{n-k}): & D_{\text{sing}} & \longrightarrow & N & -\text{hypersurface in} P_Y \end{array}$$

— here singular lagrangian tori in the fibration are parameterized by certain "incidence cycle" appears from irregular singular fibers of the pseudotoric structure.

Now we slightly generalize the discussion concerning not lagrangian fibrations but lagrangian tori. For this case we can say about possible lifting of lagrangian tori from the base manifold the of pseudotoric structure, namely

Theorem ([7]) *Let $(f_1, ..., f_k, \psi, Y)$ be a regular pseudotoric structure on a compact symplectic manifold X. Let $S \subset Y$ be a smooth lagrangian torus which doesn't intersect $D_{sing} \subset Y$. Then the choice of non critical values $(c_1, ...c_k)$ of $f_1, ..., f_k$ defines a smooth lagrangian torus $T(S, c_1, ..., c_k) \subset X$.*

Shortly, the proof of the theorem based on the same procedure we've applied above to construct the Chekanov tori. Taking a lagrangian torus on the base, we collect lagrangian tori from the toric fiber — it can be done simulteneously thanks to the global functions $f_1, ..., f_k$ — and the commutation relations from the very definition of pseudotoric structure ensure that the resulting figure is a lagrangian torus in X.

The last theorem shows that lagrangian tori from the base manifold after lifting could give different types of lagrangian tori in whole X. We can take the homology group $H_{n-k}(Y \backslash D_{\text{sing}}, \mathbb{Z})$ of the "punctured" base manifold and then attach to smooth lagrangian tori in the punctured base manifold different classes from the group. Conjecturelly the different classes from $H_{n-k}(Y \backslash D_{\text{sing}}, \mathbb{Z})$ can give different types of lagrangian tori in X. For example, it is true for the projective plane — as we've seen for Clifford and Chekanov tori in $\mathbb{CP}^2$ the following alternative appears:

$$H_1(\mathbb{CP}^1 \backslash ([1:0], [0:1]), \mathbb{Z}) \quad \begin{array}{l} \nearrow \text{Clifford type = primitive elem.} \\ \searrow \text{Chekanov type = trivial elem.} \end{array}$$

But is it possible to construct such exotic lagrangian tori for any compact toric variety? The answer is an affirmative in view of the following

Theorem ([8]): *Any smooth compact toric symplectic manifold admits regular pseudotoric structure $(f_1, ..., f_{n-1}, \psi, \mathbb{CP}^1)$ of rank one. For this structure the singular divisor $D_{sing} \subset \mathbb{CP}^1$*

consists of exactly two distinct points, $p_N, p_S \subset \mathbb{CP}^1$. The primitive and the trivial elements of $H_1(\mathbb{CP}^1 \backslash (p_N \cup p_S), \mathbb{Z})$ generates lagrangian tori of the standard type and of the Chekanov type respectively.

Suppose additionaly that our given toric (X, ω_X) is monotone, this means that the cohomology class of the symplectic form is proportional to the canonical class of the associated almost complex structure on X: $K_X = k[\omega_X] \subset H^2(X, \mathbb{Z})$; so f.e. Fano varieties in algebraic geometry are monotone. Then the main conjecture we would like to propose in this circumstances is based on the following remark: if there is a standard monotone lagrangian torus then there exists a monotone lagrangian torus of the Chekanov type. And for the future work we have as the major aim the following

Main conjecture. *These monotone tori are not Hamiltonian isotopic.*

The Theorem above which states the existence of pseudotoric structures on toric symplectic manifolds can be proved as follows ([8]). Let's take for a given toric X the set of commuting Morse moment maps $(f_1, ... f_n)$, which give the action map by "action coordinates" $F = (f_1, ..., f_n) : X \to P_X$ to convex moment polytop $P_X \subset \mathbb{R}^n$; then for the components D_i of the boundary divisor $D = F^{-1}(\partial P_X)$ one can find an integer combination $\sum \lambda_i D_i$ equals to zero. This sum can be rearranged to the form

$$\sum_{\lambda_i > 0} \lambda_i D_i = \sum_{\lambda_j < 0} |\lambda_j| D_j, \quad D_i \neq D_j;$$

therefore we have two divisors from the same linear system

$$D_+ = \sum_{\lambda_i > 0} \lambda_i D_i, \quad D_- = \sum_{\lambda_j < 0} |\lambda_j| D_j \in |\sum_{\lambda_i > 0} \lambda_i D_i|.$$

Then one takes the pencil $< D_+, D_- >$ spanned by two divisors $D_\pm$ with the base set $B = D_+ \cap D_-$, and it would be our pencil ψ from the definition of pseudotoric structure, and for generic point $p \in \mathbb{CP}^1$, $p \neq [1:0](\mapsto D_+)$, $[0:1](\mapsto D_-)$, the divisor $\overline{\psi^{-1}(p)} \subset X$ is smooth outside the base set B. The same linear combination $\sum \lambda_i D_i$ after substitution of linear forms l_i which correspond to D_i in $\mathbb{R}^n$ gives a linear relation on x_i — and this relation derive our real data $f'_1, ... f'_{n-1}$ from $f_1, ..., f_n$ just implying the corresponding linear condition.

Example: $\mathbb{CP}^2_3$ **— del Pezzo surface of degree 6** is usually given by the blow up procedure applied to the projective plane $\mathbb{CP}^2$ at three points. Explicitly it can be realized in the direct product

$$\mathbb{CP}^2_x \times \mathbb{CP}^2_y \supset \mathcal{U} = \{x_0 y_0 = x_1 y_1 = x_2 y_2\}$$

by the last equation, then the projection to the projective plane is given by $p_x : \mathcal{U} \to \mathbb{CP}^2_x$, $p_x(x_i, y_j) = [x_0 : x_1 : x_2]$. Thus $p^0_x : \mathcal{U} \backslash \text{three lines} \simeq \mathbb{CP}^2_x \backslash \text{three points}$, but $(p^0_x)^{-1}(T_{\text{Ch}}) \subset \mathcal{U}$ is **not lagrangian** — so we can not lift the Chekanov torus using this standard algebro geometric construction. But fortunately we can lift the corresponding pseudotoric structure!

Indeed, let's take the pencil

$$\{Q_{\alpha,\beta}\} = \{\alpha x_0 x_1 y_2^2 = \beta x_2^2 y_0 y_1\} \subset \mathbb{CP}^2_x \times \mathbb{CP}^2_y,$$

on the whole direct product, and then the intersections $Q_{\alpha,\beta} \cap \mathcal{U}$ would give us the Lefschetz pencil ψ on $\mathcal{U}$. Further, the real Morse function

$$F = \frac{|x_0|^2 - |x_1|^2}{\sum_{i=0}^2 |x_i|^2} + \frac{|y_1|^2 - |y_0|^2}{\sum_{i=0}^2 |y_i|^2}$$

preserves by the Hamiltonian action our surface $\mathcal{U}$ and at the same time does do it for each element $Q_{\alpha,\beta}$ of the pencil, therefore $f = F|_{\mathcal{U}}$ must preserve the intersections and thus would give the real data of the desired psedudotoric structure. The choice of a smooth loop $\gamma \subset \mathbb{CP}^1 \backslash ([1:0],[0:1])$ gives a lagrangian torus $T(0,\gamma) = \bigcup_{p\in\gamma}\{f|_{\overline{\psi^{-1}(p)}} = 0\}$, and if γ is contractible, we get a Chekanov torus in $\mathbb{CP}^2_3$

At the end I would like to mention several applications of this generalized notion, pseudotoric structure.

Lots of methods in Mathematical Physics are invented and realized with great success in the case of toric varities. In Geometric Quantization (see, f.e., [9]) we know what one should do in the case when the phase space of classical mechanical system carries a real polarization, namely one takes the Bohr - Sommerfeld fibers and spann the Hilbert space. In Homological Mirror Symmetry (see, f.e., [10]) one takes the canonical fibration on lagrangian tori, counts the fibers with non trivial Floer cohomologies — and then builds on these fibers the corresponding Fukaya category. But all these methods can not be applied in non toric case since if one takes a non toric variety — nobody knows in general how to slice it on lagrangian fibers.

Pseudotoric structure on a symplectic manifold X gives way to apply these methods in more general setup. Indeed, the theorems above ensure that we can construct aslmost canonically certain lagrangian fibrations in the presence of pseudotoric structures. What is the difference with the "regular" toric case? It is in the appearence of singular lagrangian fibers. But as we've seen the types of singularities which appear in the fibers are very special: for example the notion of Bohr - Sommerfeld lagrangian cycle is still valid for these singular lagrangian tori without any modification. One hopes that the definition of the Floer cohomology can be modified as well. Then we can use the powerfull methods not only in toric geometry but in much more wider context — since there are compact symplectic manifolds which are not toric but nevertheless which admit pseudotoric structures. It is natural to call such a manifold **pseudotoric**: the examples are complex quadrics and certain complete intersections in $\mathbb{CP}^n$, the flag variety F^3 and conjecturelly certain complex Grasmmanians. And coming back to the main subject of the present talk we shoud say that in all these cases it is possible to construct lagrangian tori of different type using the pseudtoric structure.

Thus the natural problem arises: **which symplectic manifolds are pseudotoric?** Toric geometry itself concerns this question since in the framework of toric geometry one meets the problem with induced objects. F.e. if one takes the projectivization of the (co)tangent bundle of a toric variety – it is not longer toric (F^3 — the flag variety — is the projectivization of the tangent bundle of the "most toric" one — the projective plane) but it admits the Hamitonian action of an incomplete set of integrals lifted from the toric base. As well it could happen for certain moduli spaces over toric varieties. The construction of a pseudotoric structure is a solution of the toric leaves problem in general; and if this solution exists we can adopt the strong methods from toric geometry to this case.

As a byproduct we've touched a classical problem from mechanics — the study of non completely integrable systems. Again as we've seen the problem could be solved in the case when the phase space admits a psuodotoric structure. Then the solutions can be described in terms of the "action - angle" variables of the base manifold and the toric fibers. The difference with the completely integrable case is in the appearance of singular lagrangian tori of the Liouville type — and it is not problematic since it just means that some additional types of trajectories — separatrices — are presented in the story.

Acknowledgments

I would like to thank Prof. C. Burdik for the warm invitiation to participate in the conference and all his team in Praha who organized this inetersting meeting. The author was partially supported by AG Laboratory NRU-HSE, RF government grant, ag. 11.G34.31.0023. This work

Journal of Physics: Conference Series **411** (2013) 012028 doi:10.1088/1742-6596/411/1/012028

was supported in parts by the RFBR (grant no.11-01-00980-?) and by the NRU HSE Academic Fund Program (grant no.12-09-0064).

References

[1] N. A. Tyurin, *"Geometric quantization and algebraic lagrangian geometry"*, Surveys in Geometry and Number theory, LMS Lect. Note Ser. 338 (2007), pp. 279 - 318;

[2] A. L. Gorodentsev, A. N. Tyurin, *"Abelian lagrangian algebraic geometry"*, Izvestiya: Math., 65:3 (2001), pp. 437 -467;

[3] Yu. Chekanov, *"Lagrangian tori in symplectic vector space and global symplectomorphisms"*, Math. Zeitschrift, 223: 1 (1996), pp. 547 - 599;

[4] Yu. Chekanov, F. Schlenk, *"Notes on monotone lagrangian twist tori"*, arXiv:1003.5960;

[5] N. A. Tyurin, *"pseudotoric lagrangian fibrations of toric and non toric Fano varieties"*, Theor. Math. Physics, 162:3 (2010), pp. 307 -333;

[6] S. A. Belyov, N. A. Tyurin, *"On nontoric lagrangian fibrations of toric Fano varieties"*, Math. Notes, 87: 1 (2010), pp. 48 -59;

[7] S. A. Belyov, N. A. Tyurin, *"Liftings of lagrangian tori"*, Math. Notes, 91:5 (2012), pp. 784 - 786;

[8] S. A. Belyov, *"Pseudotoric structures on toric varieties"*, in preparation;

[9] N. Woodhouse, *"Geometric Quantization"*, The Clarendon press Oxford University press, New York, 1980;

[10] K. Fukaya, Y.-G. Oh, H. Ohta, K. Ono, *"Lagrangian Floer theory on compact toric manifolds I"*, Duke. Math. J. 151 (2010), pp. 23-174.

XXth International Conference on Integrable Systems and Quantum Symmetries (ISQS-20) IOP Publishing
Journal of Physics: Conference Series **411** (2013) 012029 doi:10.1088/1742-6596/411/1/012029

Statistical approach of modulational instability beyond Gaussian approximation

Anca Visinescu, Dan Grecu

Department of Theoretical Physics,
Horia Hulubei National Institute of Physics and Nuclear Engineering,
30 Reactorului, P.O.B. MG-6, 077125 Magurele, Romania, EU

E-mail: avisin@theory.nipne.ro

Abstract. A statistical description of the modulation instability is discussed for the NLS equation. The kinetic equation for the two-point correlation function is obtained by using a decoupling procedure beyond the usual Gaussian approximation for the averaged values of products of four and six field variables. The induced evolution generated by a plane wave perturbation are calculated. They comprise both an evolution of the background and of the second harmonics.

1. Introduction

The nonlinear Schrödinger type equations are generic equations describing the evolution of (quasi) monochromatic waves in weak nonlinear media. A general feature of (quasi) monochromatic plane waves with an amplitude dependent dispersion relation (a Stokes wave) is their instability with respect to slow modulations of their amplitudes.

The modulation instability (MI) is one of the most frequent instability in nature. It was first discussed by Benjamin and Feir [1] in the context of hydrodynamics and by Bespalov and Talanov [2] in plasma physics. Is studied in electrodynamics and nonlinear optics also. In a very simple way the MI is the interaction of a strong carrier wave of frequency ω and small sidebands of frequencies $\omega_{1,2} = \omega \pm \Omega$ with the fulfillment of resonance condition on the corresponding wave vectors $k_1 + k_2 = 2k$ also. All these approaches can be termed as deterministic, as they generally study the evolution of small perturbations of the amplitude of the carrier wave [3], [4], [5], [6], [7]. The early history of the phenomenon was recently reviewed by Zakharov and Ostrovsky [3].

An alternative and complementary approach was based on the study of wave-wave energy transfer within a broad spectrum in a nearly homogeneous medium, especially in hydrodynamics [8], [9]. A bridge between the deterministic and the random school was elaborated by Alber [10]. In his approach the field variable is satisfying a deterministic nonlinear evolution equation (Davey-Stewardson equation in this case), but the attention is focused on a statistical quantity, the two-point correlation function, and no on the solutions of the evolution equation. Writing an evolution equation for the two-point correlation function, a linear stability analysis was investigated, based on the Wigner-Moyal transform [11], and the instability regions were determined for various backgrounds of the unperturbed initial state. A similar approach was successfully applied on the case of incoherent light propagation in nonlinear media [12],

[13], theory of surface gravity waves in deep oceans [14], wave propagation in non-stationary inhomogeneous plasma [15], dynamics of charged beams in accelerators [16], dynamics of Bose-Einstein condensate [17]. Recently in a series of papers the SAMI was studied for several NLS type equations (discrete nonlinear lattices, derivative NLS equation, Manakov's system, cylindrical/spherical NLS equation) [18]-[20].

In the present paper starting from the NLS equation the kinetic equation for the two-point correlation functions was obtained using a decoupling procedure beyond the usual Gaussian decoupling for averaged values of products of four and six field variable. The induced evolutions generated by a plane wave perturbation are calculated. The induced second harmonic component of Wigner's function and of the second harmonic component of the density were written down.

2. SAMI for NLS. Basic equation.

We consider the NLS equation

$$i\partial_t\Psi + \frac{1}{2}\partial_x^2\Psi + \alpha|\Psi|^2\Psi = 0 \tag{1}$$

and the following 2-point correlation function

$$W(1,2) =< \Psi(x_1,t)\Psi^*(x_2,t) >=< \Psi_1\Psi_2^* > .$$

It has symmetry properties $W(2,1) = (W1,2)^*$. A kinetic equation for the time evolution of $W(1,2)$ can be derived in the usual way [10]

$$i\frac{\partial W(1,2)}{\partial t} + \frac{1}{2}\left(\frac{\partial^2}{\partial x_1^2} - \frac{\partial^2}{\partial x_2^2}\right)W(1,2) + 2\alpha(n(1)-n(2))W(1,2) + \alpha P(1,2) = 0, \tag{2}$$

where $P(1,2)$ is a higher order 2-point correlation function, namely

$$P(1,2) =< (|\Psi_1|^2 - |\Psi_2|^2)\Psi_1\Psi_2^* > -2(n(1)-n(2))W(1,2).$$

In the case of a Gaussian decoupling of higher order products of four Ψ variables one has

$$< |\Psi(1)|^2\Psi(1)\Psi^*(2) >\simeq 2n(1)W(1,2)$$

and consequently $P(1,2) \equiv 0$.

Following the Wigner-Moyal transform [11] we introduce the center of mass coordinate $X = \frac{1}{2}(x_1 + x_2)$ and the relative coordinate $x = x_1 - x_2$ and perform a Fourier transform of $W(1,2)$ with respect to the relative coordinate

$$\rho(X,k,t) = \frac{1}{2\pi}\int e^{-ikx} < \Psi(X+\frac{x}{2})\ \Psi(X-\frac{x}{2}) >dx,$$

$\rho(X,k,t)$ is usually called the Wigner function. It is easily seen to be a real function.

Denoting $n(X,t) =< |\Psi(X,t)|^2 >$, ($n(X,t)$ is the pulse intensity in optics and the fluid density in hydrodynamics), it is easily seen that

$$n(X,t) = \int \rho(X,k,t)dk.$$

Then we can perform the Fourier transform of the exact kinetic equation (2). Taking into account that the Fourier transform of $P(1,2)$ is a purely imaginary quantity $i\Pi(X,k,t)$, with $\Pi(X,k,t)$ a real one, we get

$$\frac{\partial\rho(X,k,t)}{\partial t} + k\frac{\partial\rho}{\partial X} + 4\alpha n(X,t)\sin\left(\frac{1}{2}\frac{\overleftarrow{\partial}}{\partial X}\frac{\overrightarrow{\partial}}{\partial k}\right)\rho(X,k,t) + \alpha\Pi(X,k,t) = 0, \tag{3}$$

$$\sin\left(\frac{1}{2}\frac{\overleftarrow{\partial}}{\partial X}\frac{\overrightarrow{\partial}}{\partial k}\right) = \sum_{j=0}^{\infty}\frac{(-1)^j}{(2j+1)!}\left(\frac{1}{2}\frac{\overleftarrow{\partial}}{\partial X}\frac{\overrightarrow{\partial}}{\partial k}\right)^{2j+1}.$$

Here the sin(...) operator is defined by its Taylor expansion and the arrows are indicating the direction in which the derivatives are acting.

The next step is to write a kinetic (approximate) equation for P. We have

$$i\frac{\partial P(1,2)}{\partial t} =< \left(i\frac{\partial|\Psi_1|^2}{\partial t} - i\frac{\partial|\Psi_2|^2}{\partial t}\right)\Psi_1\Psi_2^* > + < (|\Psi_1^2 - |\Psi_2|^2)i\frac{\partial(\Psi_1\Psi_2^*)}{\partial t} > -$$

$$-2\left(i\frac{\partial n(1)}{\partial t} - i\frac{\partial n(2)}{\partial t}\right)W(1,2) - 2(n(1)-n(2))i\frac{\partial W(1,2)}{\partial t}.$$

We transform this using

- the conservation of number density

$$i\frac{\partial|\Psi|^2}{\partial t} = \frac{1}{2}\left(\Psi\frac{\partial^2\Psi^*}{\partial x^2} - \Psi^*\frac{\partial^2\Psi}{\partial x^2}\right) = \frac{1}{2}\frac{\partial}{\partial x}\left(\Psi\frac{\partial\Psi^*}{\partial x} - \Psi^*\frac{\partial\Psi}{\partial x}\right),$$

- the equation

$$i\frac{\partial(\Psi_1\Psi_2^*)}{\partial t} = -\frac{1}{2}\left(\frac{\partial^2}{\partial x_1^2} - \frac{\partial^2}{\partial x_2^2}\right)\Psi_1\Psi_2^* - \alpha(|\Psi_1|^2 - |\Psi_2|^2)\Psi_1\Psi_2^*,$$

- and the equations for $\frac{\partial n}{\partial t}$, $\frac{\partial W}{\partial t}$.

At this stage it is necessary to adopt a certain decoupling procedure for the higher correlation functions appearing in the resulting equation. We consider

$$< \Psi_1\frac{\partial^2\Psi_1^*}{\partial x_1^2}\Psi_1\Psi_2^* >\simeq 2 < \Psi_1\frac{\partial^2\Psi_1^*}{\partial x_1^2} > W(1,2),$$

$$< \Psi_1^*\frac{\partial^2\Psi_1}{\partial x_1^2}\Psi_1\Psi_2^* >\simeq< \Psi_1^*\frac{\partial^2\Psi_1}{\partial x_1^2} > W(1,2) + n(1)\frac{\partial^2}{\partial x_1^2}W(1,2),$$

and for the products of six Ψ functions the simple decoupling

$$< (|\Psi_1|^2 - |\Psi_2|^2)^2\Psi_1\Psi_2^* >\simeq$$

$$2(n(1)-n(2))\left[P(1,2) + 2(n(1)-n(2))W(1,2)\right] + (n(1)-n(2))^2 W(1,2).$$

We remark that this decoupling of the product of six Ψ-variables is a very simple one, and better decouplings are possible; the following results are mainly dependent on this simple decoupling procedure.

Then the kinetic equation for $P(1,2)$ takes a very simple form

$$i\frac{\partial P(1,2)}{\partial t} + \alpha(n(1)-n(2))^2 W(1,2) = 0. \tag{4}$$

and its Fourier transform is

$$\frac{\partial\Pi}{\partial t} + 4\alpha n(X,t)\sin\left(\frac{1}{2}\frac{\overleftarrow{\partial}}{\partial X}\frac{\overrightarrow{\partial}}{\partial k}\right)\rho(X,k,t)\sin\left(\frac{1}{2}\frac{\overleftarrow{\partial}}{\partial k}\frac{\overrightarrow{\partial}}{\partial X}\right)n(X,t) = 0.$$

Journal of Physics: Conference Series **411** (2013) 012029 doi:10.1088/1742-6596/411/1/012029

The quantity $\Pi(X,k,t)$ can now be eliminated from the equation (3) satisfied by $\rho(X,k,t)$. We finally remain with

$$\frac{\partial}{\partial t}\left[\frac{\partial \rho}{\partial t}+k\frac{\partial \rho}{\partial X}+4\alpha n(X,t)\,\sin\left(\frac{1}{2}\frac{\overleftarrow{\partial}}{\partial X}\frac{\overrightarrow{\partial}}{\partial k}\right)\rho\right]=$$

$$=4\alpha n(X,t)\,\sin\left(\frac{1}{2}\frac{\overleftarrow{\partial}}{\partial X}\frac{\overrightarrow{\partial}}{\partial k}\right)\rho\,\sin\left(\frac{1}{2}\frac{\overleftarrow{\partial}}{\partial k}\frac{\overrightarrow{\partial}}{\partial X}\right)n(X,t). \tag{5}$$

3. Linear stability analysis

The initial (unperturbed) state is assumed to be stationary, homogeneous and isotropic. Consequently the two-point correlation function in the initial state will be a function of $|x|$, $W_0(|x|)$, and its Fourier transform $\rho_0(k)$ is an even function of k.

A linear stability analysis can be done assuming that the perturbed state $\rho(X,k,t)$ can be written as

$$\rho(X,k,t)=\rho_0(k)+\epsilon\rho_1(X,k,t)+\epsilon^2\rho_2(X,k,t)$$
$$n(X,t)=n_0+\epsilon n_1(X,t)+\epsilon^2 n_2(X,t) \tag{6}$$

where

$$n_0=\int_{-\infty}^{+\infty}\rho_0(k)dk,\qquad n_1(X,t)=\int_{-\infty}^{+\infty}\rho_1(X,k,t)dk,\ldots$$

Here $\epsilon\ll 1$ and moreover the present discussion is valid for small time values where $\epsilon e^{\Omega_i t}\ll 1$, Ω_i being the instability increment. In the first order in ϵ we have

$$\frac{\partial \rho_1}{\partial t}+k\frac{\partial \rho_1}{\partial X}+4\alpha n_1(X,t)\sin\left(\frac{1}{2}\frac{\overleftarrow{\partial}}{\partial X}\frac{\overrightarrow{\partial}}{\partial k}\right)\rho_0(k)=0,$$

and looking for plane wave solutions

$$\rho_1(X,k,t)=g_1(k)e^{i(QX-\Omega t)}+g_1^*(k)e^{-i(QX-\Omega^* t)}$$

$$n_1(X,t)=N_1\left(e^{i(QX-\Omega t)}+e^{-i(QX-\Omega^* t)}\right),$$

$$N_1=\int g_1(k)dk,$$

one obtains the known dispersion relation of the Gaussian decoupling case [19]

$$1+\frac{2\alpha}{Q}\int\frac{\rho_0(k+\frac{Q}{2})-\rho_0(k-\frac{Q}{2})}{k-\frac{\Omega}{Q}}dk=0. \tag{7}$$

It can be solved for different initial distributions $\rho_0(k)$. We consider both the case of a δ-distribution $\rho_0(k)=n_0\delta(k)$, and a Lorentzian one, $\rho_0(k)=\frac{n_0}{\pi}\frac{p}{k^2+p^2}$, where n_0 is the mean density of the unperturbed state. One obtains $\Omega=i\Omega_i$ (purely imaginary) $\Omega_i=Q\sqrt{2\alpha n_0-\frac{Q^2}{4}}$ for the δ-distribution, and $\frac{\Omega_i}{Q}=\sqrt{2\alpha n_0-\frac{Q^2}{4}}-p$ for the Lorentzian distribution. The instability is associated with the positive imaginary part of Ω, and for a δ-distribution the instability domain is restricted to the long wave length region

$$0\le Q\le 2\sqrt{2\alpha n_0}$$

$\Omega_i = 0$ for $Q = 0$ and $Q = 2\sqrt{2\alpha n_0}$ and has a maximum $2\alpha n_0$ at $Q_0 = 2\sqrt{\alpha n_0}$.

For the Lorentzian distribution the instability domain is more restricted

$$0 \leq Q \leq 2\sqrt{2\alpha n_0 - p^2}$$

and disappears for $p^2 \geq 2\alpha n_0$.

In the second order in ϵ we have

$$\frac{\partial}{\partial t}\left[\frac{\partial \rho_2}{\partial t} + k\frac{\partial \rho_2}{\partial X} + 4\alpha n_2(X,t)\sin\left(\frac{1}{2}\frac{\overleftarrow{\partial}}{\partial X}\frac{\overrightarrow{\partial}}{\partial k}\right)\rho_0(k)\right] =$$

$$= -\frac{\partial}{\partial t}\left[4\alpha n_1(X,t)\sin\left(\frac{1}{2}\frac{\overleftarrow{\partial}}{\partial X}\frac{\overrightarrow{\partial}}{\partial k}\right)\rho_1(X,k,t)\right] + \tag{8}$$

$$+4\alpha^2 n_1(X,t)\sin\left(\frac{1}{2}\frac{\overleftarrow{\partial}}{\partial X}\frac{\overrightarrow{\partial}}{\partial k}\right)\rho_0(k)\sin\left(\frac{1}{2}\frac{\overleftarrow{\partial}}{\partial k}\frac{\overrightarrow{\partial}}{\partial X}\right)n_1(X,t),$$

which is a driven non-homogeneous linear partial differential equation for $\rho_2(X,k,t)$. with the r.h.s. depending only on $\rho_1(X,k,t)$ and $n_1(X,t)$.

We evaluate the r.h.s. assuming $\Omega = i\Omega_i$, N_1 real, and

$$\rho_1(X,k,t) = e^{\Omega_i t}\left(g_1(k)e^{iQX} + g_1^*(k)e^{-iQX}\right)$$

$$n_1(X,t) = e^{\Omega_i t}N_1\left(e^{iQX} + e^{-iQX}\right)$$

$$g_1 = 2\alpha N_1\frac{\rho_0(k+\frac{Q}{2}) - \rho_0(k-\frac{Q}{2})}{\Omega - kQ}.$$

Then the r.h.s. of the non-homogeneous linear partial differential equation (8) contains
- a time dependent, spatial uniform term

$$A_0(k)e^{2\Omega_i t} = 2\left(G_0(k) - \Omega_i Im G_1(k)\right)e^{2\Omega_i t}$$

- and a time and space dependent term

$$\left(A_2(k)e^{2iQX} + A_2^* e^{-2iQX}\right)e^{2\Omega_i t}$$

$$A_2(k) = -\left(G_0(k) + i\Omega_i G_1(k)\right).$$

Here we denoted

$$G_0(k) = \alpha^2 N_1^2\left[\rho_0(k+Q) - 2\rho_0(k) + \rho_0(k-Q)\right]$$

$$G_1(k) = 4\alpha N_1\left[g_1(k+\frac{Q}{2}) - g_1(k-\frac{Q}{2})\right].$$

This allow us to seek $\rho_2(X,k,t)$ of the form

$$\rho_2(X,k,t) = \rho_2^{(0)}(k,t) + \left(\rho_2^{(2)}(k,t)e^{2iQX} + c.c.\right) \tag{9}$$

where $\rho_2^{(0)}(k,t)$ - real, $\rho_2^{(2)}(k,t)$-complex and

$$n_2(X,t) = n_2^{(0)}(t) + \left(n_2^{(0)}(t)e^{2iQX} + c.c.\right)$$

$n_2^{(0)}(t)$ - modulation of the uniform background, $n_2^{(2)}(t)$ - amplitude of the generated second harmonic, $n_2^{(0)}(t)$ - real, $n_2^{(2)}(t)$ - eventually complex.

$$n_2^{(0)}(t) = \int \rho_2^{(0)}(k,t)dk, \qquad n_2^{(2)}(t) = \int \rho_2^{(2)}(k,t)dk.$$

$\rho_2^{(0)}(k,t)$ and $\rho_2^{(2)}(k,t)$ will satisfy the equations

$$\frac{\partial^2 \rho_2^{(0)}(k,t)}{\partial t^2} = A_0(k)e^{2\Omega_i t}$$

$$\frac{\partial}{\partial t}\left\{\frac{\partial \rho_2^{(2)}(k,t)}{\partial t} + 2iQk\rho_2^{(2)} + 2i\alpha n_2^{(2)}\left(\rho_0(k+\frac{Q}{2}) - \rho_0(k-\frac{Q}{2})\right)\right\} = A_2(k)e^{2\Omega_i t}.$$

These have to be solved with the initial conditions

$$\rho_2^{(0)}(t=0) = \rho_2^{(2)}(t=0) = 0$$

and assuming that in the limit $\Omega_i \to 0$ the solutions are non-singular.

We consider

$$\rho_2^{(0)}(k,t) = a_0(k) + b_0(k)t + \frac{A_0(k)}{4\Omega_i^2}e^{2\Omega_i t},$$

$$a_0(k) = -\frac{A_0(k)}{4\Omega_i^2}, \qquad b_0(k) = -\frac{1}{\Omega_i}G_0(k).$$

Then

$$\rho_2^{(0)}(k,t) = G_0(k)\frac{e^{2\Omega_i t} - (1+2\Omega_i t)}{2\Omega_i^2} - ImG_1(k)\frac{e^{2\Omega_i t} - 1}{2\Omega_i}, \tag{10}$$

$$n_2^{(0)}(t) = \frac{e^{2\Omega_i t} - (1+2\Omega_i t)}{2\Omega_i^2}\int G_0(k)dk - \frac{e^{2\Omega_i t} - 1}{2\Omega_i}\int G_1(k)dk$$

and both conditions mentioned above are satisfied.

For a δ-distribution

$$Im\ g_1(k) = -2\alpha N_1 \frac{\Omega_i}{Q^2}\frac{\delta(k+\frac{Q}{2}) - \delta(k-\frac{Q}{2})}{k^2 + \frac{\Omega_i^2}{Q^2}}$$

$$\int G_0(k)dk = 0, \qquad \int Im\ G_1(k)dk = 0.$$

and consequently

$$n_2^{(0)} \equiv 0.$$

For reasons to be seen later we consider

$$\rho_2^{(2)}(k,t) = f(t)G_0(k) + g_2^{(2)}(e^{2\Omega_i t} - 1) \tag{11}$$

$$f(0) = 0.$$

Then because $\int G_0(k)dk = 0$ for any distribution of $\rho_0(k)$ we have

$$n_2^{(2)}(t) = N_2(e^{2\Omega_i t} - 1)$$

$$N_2 = \int g_2^{(2)}(k)dk.$$

Introducing into the equation satisfied by $\rho_2^{(2)}(k,t)$ after several manipulations we get

$$\left(\frac{d^2f}{dt^2} - 2\Omega_i\frac{df}{dt} + e^{2\Omega_i t}\right)\frac{1}{2\Omega_i}\int\frac{G_0(k)}{\Omega_i + iQk}dk + \left(N_2 + \frac{i}{2}\int\frac{G_1(k)}{\Omega_i + iQk}dk\right)e^{2\Omega_i t} = 0.$$

For a δ-distribution

$$i\alpha\int\frac{\delta(k+\frac{Q}{2}) - \delta(k-\frac{Q}{2})}{\Omega_i + iQk}dk = -\frac{1}{2},$$

$$-\frac{i}{2}\int\frac{G_1(k)}{\Omega_i + iQk}dk = \frac{6\alpha^2 N_1^2 Q^2}{\Omega_i^2 + Q^4},$$

but

$$-\frac{1}{2\Omega_i}\int\frac{G_0(k)}{\Omega_i + iQk}dk = \alpha^2 N_1^2 n_0\,\frac{Q^4}{\Omega_i^2(\Omega_i^2 + Q^4)}$$

is singular when $\Omega_i \to 0$. To eliminate this singularity we determine $f(t)$ from

$$\frac{d^2f}{dt^2} - 2\Omega_i\frac{df}{dt} + e^{2\Omega_i t} = 0. \tag{12}$$

One obtains ($\Omega_i < 1$)

$$f(t) = 2e^{2\Omega_i t}\sin\sqrt{1-\Omega_i^2}\,t.$$

Finally we get

$$\rho_2^{(2)}(k,t) = 2e^{2\Omega_i t}\sin\sqrt{1-\Omega_i^2}\,t + g_2^{(2)}(k)\left(e^{2\Omega_i t} - 1\right), \tag{13}$$

for the induced second harmonic component of Wigner's function and

$$N_2 = \frac{6\alpha^2 N_1^2 Q^2}{\Omega_i^2 + Q^4}, \tag{14}$$

for the induced second harmonic component of the density.

4. Conclusions

The main results of the present paper can be summarized as follows:

1) A statistical approach of the modulation instability was realized, going beyond the usual Gaussian approximation. The exact kinetic equation (2) for the two-point correlation function $W(1,2)$ was supplemented with an approximate kinetic equation (4) for the higher order correlation function $P(1,2)$. This was obtained using the usual Gaussian approximation for the decoupling of products of four Ψ-functions, but a special decoupling procedure for products of six Ψ-functions. We expect that an improved decoupling procedure will lead to better results. Finally an improved kinetic equation (5) for the Fourier transform $\rho(X,k,t)$ is obtained.

2) A linear stability analysis of eq. (5) around a stationary initial state is performed. In first order the dispersion relation of the usual Gaussian approximation is obtained and this was not changed in higher orders. The second order represents a driven non-homogeneous partial differential equation for $\rho_2(X,k,t)$ (eq.(8)), and the solution (9) is written as a superposition of a modulated $\rho_2^{(0)}(k,t)$ of the initial Wigner function $\rho_0(k)$ and a new generated second harmonic $(\rho_2^{(2)}(k,t)e^{2iQX} + cc)$. These were determined for a zero initial condition and assuming that in the limit $\Omega_i \to 0$ the solutions are non-singular.

3) We emphasize again that these results were obtained using a simple decoupling of the products of six Ψ-functions, and probably will be essentially changed for a better decoupling. This fact is for the moment under investigation.

4.1. Acknowledgments
This work is supported by a joint Romanian-LIT, JINR, Dubna Research Project, theme no. 05-6-1060-2005/2013.

References
[1] Benjamin T B and Feir J E 1967 *J. Fluid Mech.*, **27** 4175; Benjamin T B 1967 *Proc. Roy. Soc. London* A **299** 59
[2] Bespalov V I and Talanov V I 1966 *Sov. Phys. JETP Lett.* **3** 307
[3] Zakharov V E and Ostrovsky L E 2009 *Physica D* **238**, 540-548
[4] Dodd R K, Eilbeck J C, Gibbon J D and Morris H C 1982 *Solitons and Nonlinear Wave Equations*, Academic Press, London
[5] Remoissenet M 1999 *Waves Called Solitons, Concepts and Experiments*, Springer, Berlin
[6] Drazen P G and Reid W H 1981 *Hydrodynamics Stability* Cambridge Univ. Press
[7] Scott A 2003 *Nonlinear Science, Emergence and Dynamics of Coherent Structures* Oxford Univ. Press
[8] Hasselman K 1962 *J. Fluid Mech.* **12** 481; Hasselman K 1963 *J. Fluid Mech.* **15** 273; Willebrand J 1975 *J. Fluid Mech.* **70** 113
[9] Longuet-Higgins M S 1976 *Proc. Roy. Soc. London* A **347** 311; Jansen P.A.E.M. 2003 *J.Phys.Ocean.* **33** 863
[10] Alber I E 1978 *Proc. Roy. Soc. London A* **263**, 525
[11] Wigner E 1932 *Phys. Rev.* **40** 749; Moyal JE 1949 *Proc. Cambridge Phil. Soc.* **45** 99
[12] Hall B, Lisak M, Anderson D, Fedele R and Semenov VE 2002 *Phys. Rev. E* **65**, 035602 (R)
[13] Kivshar Y S and Agarwal G P - 2003 *Optical Solitons. From Fibers to Photonic Crystals*, Elsevier
[14] Onorato M, Osborne A, Fedele R and Serio M 2003 *Phys. Rev.* E **67**, 046305; Shukla PK, Kourakis I, Elliasson B, Marklund M and Stenflo L 2006 *Phys. Rev. Lett.* **97**, 044501
[15] Tsintsadze N L and Mendosa J T 1998 *Plasma Phys.*, **5**, 3609 ; 2000 *Phys. Rev.* E, **62**, 427 (2000).
[16] Fedele R, Anderson D and Lisak M 2000 *Physica Scripta* T **84**, 27; Fedele R and Anderson D 2000 *J. Opt. B: Quantum Semiclassical Opt.*, **2**, 207
[17] Gardiner SA, Jaksch D, Dum R, Cirac JI and Zeller P 2000 *Phys. Rev.* A, **62**, 023612
[18] Shukla P K, Marklund M and Stenflo L2006 *Pis'ma ZhETF*, **84**, 764; nlin.CD/0611039
[19] Grecu D and VisinescuA 2003*Eur. Phys. J.* B **34**, 225 ; 2004 in *Nonlinear Waves: Classical and Quantum Aspects* p. 151, editors . Abdullaev F Kh and Konotop V V, Kluwer; Visinescu A and Grecu D 2004 in *Global Analysis and Applied Mathematics*, AIP **729**, editors Tas K et al., p.386-395; Grecu D and Visinescu A 2005 *Theor. Math. Phys.*, **144**, 92 (2005); Grecu A T, Grecu D and Visinescu A 2007 *Int. J. Theor. Phys.*, **46**, 1190
[20] Fedele R, De Nicola S, Grecu D, Shukla P K and Visinescu A 2008 *Frontier in Modern Plasma Physics*, AIP Conf. Proc. **1061**, editors Shukla P K, Eliasson B and Stenflo L 273, Grecu A T, Grecu D and Visinescu A 2009 *Romanian Reports in Physics* **61** 3, 467

Journal of Physics: Conference Series **411** (2013) 012030 doi:10.1088/1742-6596/411/1/012030

Hidden symmetries on Kerr-NUT-(A)dS metrics of Einstein-Sasaki type

Mihai Visinescu

Department Theoretical Physics, Horia Hulubei National Institute for Physics and Nuclear Engineering, Magurele MG-6, Bucharest, Romania

E-mail: mvisin@theory.nipne.ro

Abstract. The hidden symmetries of higher dimensional Euclideanised Kerr-NUT-(A)dS metrics are investigated. In certain scaling limits these metrics are related to the Einstein-Sasaki ones. The complete set of Killing-Yano tensors of the Einstein-Sasaki spaces are presented. For this purpose the Killing forms of the Calabi-Yau cone over the Einstein-Sasaki manifold are constructed. Two new Killing forms on Einstein-Sasaki manifolds are identified associated with the complex volume form of the cone manifolds. As a concrete example we present the complete set of Killing-Yano tensors on the five-dimensional Einstein-Sasaki $Y(p,q)$ spaces. The corresponding hidden symmetries are not anomalous and the geodesic equations are superintegrable.

1. Introduction

The usual spacetimes symmetries are represented by isometries connected with the Killing vector fields. Slightly more generally, the conformal Killing vector field preserve a given conformal class of metrics. For each of the (conformal) Killing vector fields there exists a conserved quantity for the (null) geodesic motions.

Besides them a spacetime may also possess hidden symmetries generated by higher order symmetric or antisymmetric tensor fields. The symmetric Stäckel-Killing tensors give rise to conserved quantities of higher order in particle momenta. A natural generalization of (conformal) Killing vector fields is given by the antisymmetric (conformal) Killing-Yano tensors. Killing-Yano tensors are also called Yano tensors or Killing forms, and conformal Killing-Yano tensors are sometimes referred as conformal Yano tensors, conformal Killing forms or twistor forms.

In physics, Yano tensors play a fundamental role being related to the separability of field equations with spin, pseudoclassical spinning models, the existence of quantum symmetry operators, supersymmetries, etc.

In this paper we want to take a closer look at the Killing forms of Euclideanised Kerr-NUT-(A)dS metrics which are related to Einstein-Sasaki metrics. Recently Einstein-Sasaki geometries have been the object of much attention in connection with the supersymmetric backgrounds relevant to the AdS/CFT correspondence. On the other hand a lot of interest focuses on higher dimensional black hole spacetimes. The search of hidden symmetries generated by the Killing forms in rotating black hole geometries has an important role for describing the properties of black holes in various dimensions.

The Kerr-NUT-AdS metrics in all dimensions have been constructed in [1]. The general Kerr-NUT-AdS metrics have $(2n-1)$ non-trivial parameters where the spacetime dimension is $(2n+1)$

in the odd-dimensional case and $(2n)$ in the even dimensional case. It was also considered the BPS, or supersymmetric, limits of these metrics. After Euclideanisation, these limits yield in odd dimensions new families of Einstein-Sasaki metrics, whereas the even-dimensional metrics result in the Ricci-flat Kähler manifolds. An alternative procedure was proposed in [2] generalizing the scaling limit of Martelli and Sparks [3]. More precisely, in a certain limit one gets an Einstein-Kähler metric from an even-dimensional Kerr-NUT-(A)dS spacetime and the Einstein-Sasaki space is constructed as a $U(1)$ bundle over this metric. On the other hand, performing the scaling limit of the odd-dimensional Kerr-NUT-(A)dS spacetimes one gets directly the same Einstein-Sasaki space obtained as a $U(1)$ bundle over the Einstein-Kähler metric [2].

The Kerr-NUT-(A)dS metrics possess explicit and hidden symmetries encoded in a series of Stäckel-Killing tensors and Killing vectors [1]. These symmetries allow one constructs a set of quantities conserved along geodesics. Moreover they are functionally independent, in involution and guarantee complete integrability of the geodesic motions [4, 5, 6].

The structure of the hidden symmetries for a Sasaki space is derived from the characteristic Sasakian 1-form. Thus a tower of Killing-Yano and conformal Killing-Yano tensors can be constructed [2]. The corresponding hidden symmetries are purely geometrical, irrespective of the fact whether the Einstein equations are satisfied or not.

The main purposes of this paper is to point out the special case of the higher dimensional Kerr-NUT-(A)dS metrics which are related to the Einstein-Sasaki ones. In this case there are two additional Killing-Yano tensors taking into account that the metric cone is Calabi-Yau [7, 8]. These two exceptional Killing forms can be also described using the Killing spinors of an Einstein-Sasaki manifold [9].

In Section 2 we review some basic facts about the Einstein-Sasaki spaces and their cone manifolds. In the next Section we discuss the Killing forms on Einstein-Sasaki spaces which proceed from Euclideanised Kerr-NUT-(A)dS metrics in certain scaling limits. We identity two new Killing forms associated with the complex volume form of the cone manifold. In Section 4 we restrict to the five-dimensional $Y(p,q)$ manifolds and present the complete set of Killing forms. Finally we give our conclusions in Section 5.

2. Mathematical preliminaries

For convenience we shall only give the briefest account of the mathematical concepts and results needed to study the hidden symmetries on Einstein-Sasaki spaces.

2.1. Killing vector fields and their generalizations

A vector field X on a Riemannian manifold (M,g) is said to be a *Killing vector field* if the Lie derivative of the metric g with respect to X vanishes or, equivalently, if the Levi-Civita connection ∇ of g satisfies

$$g(\nabla_Y X, Z) + g(Y, \nabla_Z X) = 0\,,$$

for all vector fields Y, Z on M. A natural generalization of Killing vector fields is given by the *conformal Killing vector fields*, i.e. vector fields with a flow preserving a given conformal class of metrics [10]. On the other hand, a *conformal Killing-Yano tensor* of rank p on a Riemannian manifold (M,g) is a p-form ω which satisfies:

$$\nabla_X \omega = \frac{1}{p+1} X \lrcorner\, d\omega - \frac{1}{n-p+1} X^* \wedge d^*\omega\,, \tag{1}$$

for any vector field X on M, where ∇ is the Levi-Civita connection of g, n is the dimension of M, X^* is the 1-form dual to the vector field X with respect to the metric g, $\lrcorner$ is the operator dual to the wedge product and d^* is the adjoint of the exterior derivative d. If ω is co-closed in (1), then we obtain the definition of a *Killing-Yano tensor* [10]. It is easy to see that for

$p = 1$, they are dual to Killing vector fields. Moreover, a Killing form ω is said to be a *special Killing form* if it satisfies for some constant c the additional equation

$$\nabla_X(d\omega) = cX^* \wedge \omega\,, \tag{2}$$

for any vector field X on M.

Besides the antisymmetric generalization of the Killing vectors one might also consider higher order symmetric tensors. A symmetric tensor $K_{(i_1 \dots i_k)}$ obeying the equation

$$K_{(i_1 \dots i_k;j)} = 0\,,$$

is called a *Stäckel-Killing tensor*. For any geodesic with a tangent vector u^i the following object

$$P_K = K_{i_1 \dots i_k} u^{i_1} \cdots u^{i_k}\,,$$

is conserved.

These two generalizations of the Killing vectors could be related. Given two Killing-Yano tensors $\omega^{i_1,\dots,i_k}$ and $\sigma^{i_1,\dots,i_k}$ it is possible to associate with them a Stäckel-Killing tensor of rank 2:

$$K_{ij}^{(\omega,\sigma)} = \omega_{ii_2\dots i_k}\sigma_j{}^{i_2\dots i_k} + \sigma_{ii_2\dots i_k}\omega_j{}^{i_2\dots i_k}\,. \tag{3}$$

Therefore a method to generate higher order integrals of motion is to identify the complete set of Killing-Yano tensors. The existence of enough integrals of motion leads to complete integrability or even superintegrability of the mechanical system when the number of functionally independent constants of motion is larger than its number of degrees of freedom. Let us mention that when a Stäckel-Killing tensor is of the form (3), there are no quantum anomalies thanks to an integrability condition satisfied by the Killing-Yano tensors [11, 12, 13].

2.2. Kähler and Sasakian manifolds

An almost Hermitian structure on a smooth manifold M is a pair (g, J), where g is a Riemannian metric on M and J is an almost complex structure on M, which is compatible with g, i.e.

$$g(JX, JY) = g(X, Y)\,,$$

for all vector fields X, Y on M. In this case, the triple (M, J, g) is called an *almost Hermitian manifold.* Moreover, if J is parallel with respect to the Levi-Civita connection ∇ of g, then (M, J, g) is said to be a *Kähler manifold.* We remark that on a Kähler manifold, the associated Kähler form, i.e the alternating 2-form Ω defined by

$$\Omega(X, Y) = g(JX, Y)\,,$$

is closed. In local holomorphic coordinates $(z^1, \dots, z^m)$, the associated Kähler form Ω can be written as

$$\Omega = ig_{j\bar{k}}dz^j \wedge d\bar{z}^k = \sum X_j^* \wedge Y_j^* = \frac{i}{2}\sum Z_j^* \wedge \bar{Z}_j^*\,,$$

where $(X_1, Y_1, \dots, X_m, Y_m)$ is an adapted local orthonormal field (i.e. such that $Y_j = JX_j$), and $(Z_j, \bar{Z}_j)$ is the associated complex frame given by

$$Z_j = \frac{1}{2}(X_j - iY_j),\ \bar{Z}_j = \frac{1}{2}(X_j + iY_j)\,.$$

There is an intimate connection between its Kähler form and the volume form (which is just the Riemannian volume form determined by the metric) as follows

$$d\mathcal{V} = \frac{1}{m!}\Omega^m\,,$$

where $d\mathcal{V}$ denotes the volume form of M, Ω^m is the wedge product of Ω with itself m times, m being the complex dimension of M [14]. Hence the volume form is a real (m, m)-form on M.

On the other hand, if the volume of a Kähler manifold is written as

$$d\mathcal{V} = dV \wedge d\bar{V} \,,$$

then dV is the complex volume holomorphic $(m, 0)$ form of M. It is now clear that the complex volume form of a Kähler manifold can be written in a simple way with respect to any orthonormal basis, using complex vierbeins $e_i + Je_i$. In fact, the complex volume form of a Kähler manifold M is, up to a power factor of the imaginary unit i, the exterior product of these complex vierbeins.

2.3. The Kähler cone of an Einstein-Sasaki manifold

Let M be a smooth manifold equipped with a triple (φ, ξ, η), where φ is a field of endomorphisms of the tangent spaces, ξ is a vector field and η is a 1-form on M. If we have:

$$\varphi^2 = -I + \eta \otimes \xi \,, \quad \eta(\xi) = 1 \,,$$

then we say that (φ, ξ, η) is an *almost contact structure* on M [15].

A Riemannian metric g on M is said to be *compatible* with the almost contact structure (φ, ξ, η) if and only if the relation

$$g(\varphi X, \varphi Y) = g(X, Y) - \eta(X)\eta(Y) \,,$$

holds for all pair of vector fields X, Y on M.

An almost contact metric structure (φ, ξ, η, g) is a *Sasakian structure* if and only if the Levi-Civita connection ∇ of the metric g satisfies

$$(\nabla_X \varphi) Y = g(X, Y)\xi - \eta(Y) X \,, \tag{4}$$

for all vector fields X, Y on M [16].

A Sasakian structure may also be reinterpreted and characterized in terms of the metric cone as follows. The metric cone of a Riemannian manifold (M, g) is the Riemannian manifold $C(M) = (0, \infty) \times M$ with the metric given by

$$\bar{g} = dr^2 + r^2 g \,,$$

where r is a coordinate on $(0, \infty)$. Then M is a Sasaki manifold if and only if its metric cone $C(M)$ is Kähler [17]. In particular, the cone $C(M)$ is equipped with an integrable complex structure J and a Kähler 2-form Ω, both of which are parallel with respect to the Levi-Civita connection $\bar{\nabla}$ of $\bar{g}$. Moreover, M has odd dimension $2n+1$, where $n+1$ is the complex dimension of the Kähler cone. We note that the Sasakian manifold (M, g) is naturally isometrically embedded into the cone via the inclusion

$$M = \{r = 1\} = \{1\} \times M \subset C(M) \,,$$

and the Kähler structure of the cone $(C(M), \bar{g})$ induces an almost contact metric structure (ϕ, ξ, η, g) on M satisfying (4).

A *Einstein-Sasaki manifold* is a Riemannian manifold (M, g) that is both Sasaki and Einstein, i.e. a Sasakian manifold satisfying the Einstein condition

$$Ric_g = \lambda g \,,$$

for some real constant λ, where Ric_g denotes the Ricci tensor of g. Einstein manifolds with $\lambda = 0$ are called *Ricci-flat manifolds*. Similarly, an *Einstein-Kähler manifold* is a Riemannian manifold (M, g) that is both Kähler and Einstein. The most important subclass of Einstein-Kähler manifolds are the *Calabi-Yau manifolds*, which are Kähler and Ricci-flat.

It is also very important to note that a Sasaki manifold M is Einstein if and only if the cone metric $C(M)$ is Kähler Ricci-flat. In particular the Kähler cone of an Einstein-Sasaki manifold has trivial canonical bundle and the restricted holonomy group of the cone is contained in $SU(m)$, where m denotes the complex dimension of the Kähler cone [7, 18].

2.4. Progression from Einstein-Kähler to Einstein-Sasaki to Calabi-Yau manifolds

Suppose we have an Einstein-Sasaki metric g_{ES} on a manifold M_{2n+1} of odd dimension $2n+1$. An Einstein-Sasaki manifold can always be written as a fibration over an Einstein-Kähler manifold M_{2n} with the metric g_{EK} twisted by the overall $U(1)$ part of the connection [19]

$$ds^2_{ES} = (d\psi_n + 2A)^2 + ds^2_{EK}\,, \tag{5}$$

where dA is given as the Kähler form of the Einstein-Kähler base. This can be easily seen when we write the metric of the cone manifold $M_{2n+2} = C(M_{2n+1})$ as

$$ds^2_{cone} = dr^2 + r^2 ds^2_{ES} = dr^2 + r^2\left((d\psi_n + 2A)^2 + ds^2_{EK}\right)\,.$$

The cone manifold is Calabi-Yau and its Kähler form can be written as

$$\Omega_{cone} = r dr \wedge (d\psi_n + 2A) + r^2 \Omega_{EK}\,, \tag{6}$$

and the Kähler condition $d\Omega_{cone} = 0$ implies

$$dA = \Omega_{EK}\,, \tag{7}$$

where Ω_{EK} is Kähler form of the Einstein-Kähler base manifold M_{2n}.

The Sasakian 1-form of the Einstein-Sasaki metric is

$$\eta = 2A + d\psi_n\,, \tag{8}$$

which is a special unit-norm Killing 1-form obeying for all vector fields X [8]

$$\nabla_X \eta = \frac{1}{2} X \lrcorner\, d\eta\,, \quad \nabla_X(d\eta) = -2X^* \wedge \eta\,.$$

3. Killing forms on Kerr-NUT-(A)dS space in a certain scaling limit

In recent time new Einstein-Sasaki spaces have been constructed by taking certain BPS [20] or scaling limits [2, 3] of the Euclideanised Kerr-de Sitter metrics.

In even dimensions, performing the scaling limit on the Euclideanised Kerr-NUT-(A)dS spaces, the Einstein-Kähler metric g_{EK} and the Kähler potential A are [2]:

$$g_{EK} = \frac{\Delta_\mu dx_\mu^2}{X_\mu} + \frac{X_\mu}{\Delta_\mu}\left(\sum_{j=0}^{n-1} \sigma_\mu^{(j)} d\psi_j\right)^2\,,$$

with

$$X_\mu = -4\prod_{i=1}^{n+1}(\alpha_i - x_\mu) - 2b_\mu\,, \quad A = \sum_{k=0}^{n-1} \sigma^{(k+1)} d\psi_k\,,$$

and

$$\Delta_\mu = \prod_{\nu\neq\mu}(x_\nu - x_\mu)\,, \quad \sigma_\mu^{(k)} = \sum_{\substack{\nu_1<\dots<\nu_k \\ \nu_i\neq\mu}} x_{\nu_1}\dots x_{\nu_k}\,, \quad \sigma^{(k)} = \sum_{\nu_1<\dots<\nu_k} x_{\nu_1}\dots x_{\nu_k}\,.$$

Here, coordinates $x_\mu\,(\mu = 1,\dots,n)$ stands for the Wick rotated radial coordinate and longitudinal angles and the Killing coordinates $\psi_k\,(k=0,\dots,n-1)$ denote time and azimuthal angles with Killing vectors $\xi^{(k)} = \partial_{\psi_k}$. Also $\alpha_i\,(i=1,\dots,n+1)$ and b_μ are constants related to the cosmological constant, angular momenta, mass and NUT parameters [1].

We mention that in the case of odd-dimensional Kerr-NUT-(A)dS spaces the appropriate scaling limit leads to the same Einstein-Sasaki metric (5).

The hidden symmetries of the Sasaki manifold M_{2n+1} are described by the special Killing $(2k+1)-$forms [8]:

$$\Psi_k = \eta \wedge (d\eta)^k \quad , \quad k = 0,1,\cdots,n-1\,. \tag{9}$$

Semmelmann obtained in [8] that special Killing forms on a Riemannian manifold M are exactly those forms which translate into parallel forms on the metric cone $C(M)$. Therefore, the metric cone being either flat or irreducible, the problem of finding all special Killing forms is reduced to a holonomy problem [21]. In the case of holonomy $U(n+1)$, i.e. the cone $M_{2n+2} = C(M_{2n+1})$ is Kähler, or equivalently M_{2n+1} is Sasaki, it follows that all special Killing forms are spanned by the forms Ψ_k (9). Besides these Killing forms, there are n closed conformal Killing forms (also called $*$-Killing forms)

$$\Phi_k = (d\eta)^k \quad , \quad k = 1,\cdots,n\,. \tag{10}$$

In the case of holonomy $SU(n+1)$, i.e. the cone $M_{2n+2} = C(M_{2n+1})$ is Kähler and Ricci-flat, or equivalently M_{2n+1} is Einstein-Sasaki, there are *two additional* Killing forms of degree n on the manifold M_{2n+1}. These additional Killing forms are connected with the additional parallel forms of the Calabi-Yau cone manifold M_{2n+2} given by the complex volume form and its conjugate [8].

In order to write explicitly these additional Killing forms, we introduce the complex vierbeins on the Einstein-Kähler manifold M_{2n}. First of all we shall write the metric g_{EK} in the form

$$g_{EK} = o^{\hat\mu} o^{\hat\mu} + \tilde{o}^{\hat\mu}\tilde{o}^{\hat\mu}\,,$$

and the Kähler 2-form

$$\Omega = dA = o^{\hat\mu} \wedge \tilde{o}^{\hat\mu}\,,$$

where

$$o^{\hat\mu} = \sqrt{\frac{\Delta_\mu}{X_\mu(x_\mu)}}\,dx_\mu\,, \quad \tilde{o}^{\hat\mu} = \sqrt{\frac{X_\mu(x_\mu)}{\Delta_\mu}}\sum_{j=0}^{n-1}\sigma_\mu^{(j)} d\psi_j\,.$$

We introduce the following complex vierbeins on Einstein-Kähler manifold M_{2n} [22]:

$$\zeta_\mu = o^{\hat\mu} + i\tilde{o}^{\hat\mu} \quad , \quad \mu = 1,\cdots,n\,.$$

On the Calabi-Yau cone manifold M_{2n+2} we take $\Lambda_\mu = r\zeta_\mu$ for $\mu = 1,\cdots,n$ and

$$\Lambda_{n+1} = dr + i\,r\,\eta\,.$$

The standard complex volume form of the Calabi-Yau cone manifold M_{2n+2} is [22]

$$dV = \Lambda_1 \wedge \Lambda_2 \wedge \cdots \wedge \Lambda_{n+1}\,.$$

Journal of Physics: Conference Series **411** (2013) 012030 doi:10.1088/1742-6596/411/1/012030

As real forms we obtain the real respectively the imaginary part of the complex volume form. For example, writing

$$\Lambda_j = \lambda_{2j-1} + i\lambda_{2j},\ j = 1, ..., n+1\,,$$

we obtain that the real part of the complex volume is given by

$$\mathcal{R}e\ dV = \sum_{p=0}^{[\frac{n+1}{2}]} \sum_{\substack{1\le i_1<i_2<...<i_{n+1}\le 2n+2 \\ (C)}} (-1)^p \lambda_{i_1}\wedge\lambda_{i_2}\wedge ... \wedge\lambda_{i_{n+1}}\,, \tag{11}$$

where the condition (C) in (11) means that in the second sum are taken only the indices $i_1, ..., i_{n+1}$ such that $i_1+...+i_{n+1} = (n+1)^2+2p$ and $(i_k, i_{k+1}) \neq (2j-1, 2j)$, for all $k \in \{1, ..., n\}$ and $j \in \{1, ..., n+1\}$.

On the other hand, we obtain that the imaginary part of the complex volume is given by

$$\mathcal{I}m\ dV = \sum_{p=0}^{[\frac{n}{2}]} \sum_{\substack{1\le i_1<i_2<...<i_{n+1}\le 2n+2 \\ (C')}} (-1)^p \lambda_{i_1}\wedge\lambda_{i_2}\wedge ... \wedge\lambda_{i_{n+1}}\,, \tag{12}$$

where the condition (C') in (12) means that in the second sum are considered only the indices $i_1, i_2, ..., i_{n+1}$ such that $i_1 + ... + i_{n+1} = (n+1)^2 + 2p + 1$ and $(i_k, i_{k+1}) \neq (2j-1, 2j)$, for all $k \in \{1, ..., n\}$ and $j \in \{1, ..., n+1\}$.

To find the additional Killing forms on the Einstein-Sasaki spaces we connect them with the additional parallel forms on the metric cone. As real forms they are given by the real and respectively imaginary part of the complex volume form. For this purpose we make use of the fact that for any p-form ω^M on the space M_{2n+1} we can define an associated $(p+1)$-form ω^C on the cone $C(M_{2n+1})$

$$\omega^C := r^p dr\wedge\omega^M + \frac{r^{p+1}}{p+1} d\omega^M\,. \tag{13}$$

Moreover ω^C is parallel if and only if ω^M is a special Killing form (2) with constant $c = -(p+1)$ [8]. Therefore the 1-1-correspondence between special Killing p-forms on M_{2n+1} and parallel $(p+1)$-forms on the metric cone $C(M_{2n+1})$ allows us to describe the additional Killing forms on Einstein-Sasaki spaces.

In order to find the additional Killing forms on the Einstein-Káhler manifold M_{2n+1} we must identify the ω^M form in the complex volume form of the Calabi-Yau cone. An explicit example is presented in the next Section.

4. $Y(p,q)$ manifolds

The infinite family $Y(p,q)$ of Einstein-Sasaki metrics on $S^2\times S^3$ [3, 20, 23, 24]. Such manifolds provide supersymmetric backgrounds relevant to the AdS/CFT correspondence. The total space $Y(p,q)$ of an S^1-fibration over $S^2\times S^2$ with relative prime winding numbers p and q is topologically $S^2\times S^3$.

The starting point is the explicit local metric of the 5-dimensional $Y(p,q)$ manifold given by the line element [23, 24, 25]

$$\begin{aligned} ds^2_{ES} = {} & \frac{1-c\,y}{6}(d\theta^2+\sin^2\theta\, d\phi^2) + \frac{1}{w(y)q(y)}dy^2 + \frac{q(y)}{9}(d\psi-\cos\theta\, d\phi)^2 \\ & + w(y)\left[d\alpha + \frac{ac-2y+c\,y^2}{6(a-y^2)}[d\psi-\cos\theta\, d\phi]\right]^2, \end{aligned} \tag{14}$$

Journal of Physics: Conference Series **411** (2013) 012030 doi:10.1088/1742-6596/411/1/012030

where

$$w(y) = \frac{2(a - y^2)}{1 - cy}, \quad q(y) = \frac{a - 3y^2 + 2cy^3}{a - y^2},$$

and a, c are constants. A detailed discussion of the range of these parameters is given in [24] in connection with the regularity properties of the $Y(p, q)$ metrics. For $c = 0$ the metric takes the local form of the standard homogeneous metric on $T^{1,1}$ [26]. Otherwise the constant c can be rescaled by a diffeomorphism and in what follows we assume $c = 1$.

The coordinate change $\alpha = -\frac{1}{6}\beta - \frac{1}{6}c\psi'$, $\psi = \psi'$ takes the line element (14) to the following form

$$ds^2_{ES} = \frac{1-y}{6}(d\theta^2 + \sin^2\theta\, d\phi^2) + \frac{1}{p(y)}dy^2 + \frac{1}{36}p(y)(d\beta + \cos\theta\, d\phi)^2 \\ + \frac{1}{9}[d\psi' - \cos\theta\, d\phi + y(d\beta + \cos\theta\, d\phi)]^2,$$

with $p(y) = w(y)\, q(y)$.

From (8) in the case of the $Y(p, q)$ space the Sasakian 1-form is

$$\eta = \frac{1}{3}d\psi' + 2A, \tag{15}$$

with

$$A = \frac{1}{6}[-\cos\theta\, d\phi + y(d\beta + \cos\theta\, d\phi)],$$

connected with local Kähler form Ω_{EK} as in (7).

The form of the metric (14) with the 1-form (15) is the standard one for a locally Einstein-Sasaki metric with $\frac{\partial}{\partial\psi'}$ the Reeb vector field. Note also that the holomorphic $(2,0)$-form of the Einstein-Kähler base manifold is

$$dV_{EK} = \sqrt{\frac{1-y}{6p(y)}}(d\theta + i\sin\theta\, d\phi) \wedge \left[dy + i\frac{p(y)}{6}(d\beta + \cos\theta\, d\phi)\right].$$

From the isometries $SU(2) \times U(1) \times U(1)$ the momenta P_ϕ, P_ψ, P_α and the Hamiltonian describing the geodesic motions are conserved [25, 27]. P_ϕ is the third component of the $SU(2)$ angular momentum, while P_ψ and P_α are associated with the $U(1)$ factors. Additionally, the total $SU(2)$ angular momentum given by

$$J^2 = P_\theta^2 + \frac{1}{\sin^2\theta}(P_\phi + \cos\theta P_\psi)^2 + P_\Psi^2, \tag{16}$$

is also conserved.

In what follows we are looking for further conserved quantities specific for motions in Einstein-Sasaki spaces. First of all, according to (9), the Killing 1-form η (15) together with the third rank form

$$\Psi = \eta \wedge d\eta \\ = \frac{1}{9}[(1-y)\sin\theta\, d\theta \wedge d\phi \wedge d\psi' + dy \wedge d\beta \wedge d\psi' \\ + \cos\theta\, dy \wedge d\phi \wedge d\psi' - \cos\theta\, dy \wedge d\beta \wedge d\phi + (1-y)y\sin\theta\, d\beta \wedge d\theta \wedge d\phi],$$

are special Killing forms (2) with constants $c = -2$ and $c = -4$ respectively. Let us note also that Φ_k (10) with $k = 1, 2$ are closed conformal Killing forms.

On the Calabi-Yau manifold $C(M_{2n+2})$ the Kähler form (6) with the Sasakian 1-form (15) is

$$\begin{aligned}\Omega_{cone} &= r^2\frac{1-cy}{6}\sin\theta\, d\theta\wedge d\phi + \frac{r^2}{6}dy\wedge(d\beta+\cos\theta\, d\phi)\\ &+\frac{1}{3}rdr\wedge[y\, d\beta + d\psi' - (1-cy)\cos\theta\, d\phi],.\end{aligned}$$

The complex volume holomorphic (3,0) form on the metric cone is [26, 28]

$$\begin{aligned}dV_{cone} &= e^{i\psi'}r^2 dV_{EK}\wedge(dr + ir\wedge\eta)\\ &= e^{i\psi'}r^2\sqrt{\frac{1-cy}{6p(y)}}(d\theta + i\sin\theta\, d\phi)\\ &\wedge\left[dy + i\frac{p(y)}{6}(d\beta+\cos\theta\, d\phi)\right]\\ &\wedge\left[dr + i\frac{r}{3}[y\, d\beta + d\psi' - (1-cy)\cos\theta\, d\phi]\right].\end{aligned} \tag{17}$$

Extracting from the complex volume form (17) the form ω^M on the Einstein-Sasaki space according to (13) for $p=2$ we get the following additional Killing 2-forms of the $Y(p,q)$ spaces written as real forms:

$$\begin{aligned}\Xi = \mathrm{Re}\,\omega^M &= \sqrt{\frac{1-y}{6\,p(y)}}\\ &\times\left(\cos\psi'\left[-dy\wedge d\theta + \frac{p(y)}{6}\sin\theta\, d\beta\wedge d\phi\right]-\right.\\ &\left.-\sin\psi'\left[-\sin\theta\, dy\wedge d\phi - \frac{p(y)}{6}d\beta\wedge d\theta + \frac{p(y)}{6}\cos\theta\, d\theta\wedge d\phi\right]\right),\end{aligned}$$

$$\begin{aligned}\Upsilon = \mathrm{Im}\,\omega^M &= \sqrt{\frac{1-y}{6\,p(y)}}\\ &\times\left(\cos\psi'\left[-\sin\theta\, dy\wedge d\phi - \frac{p(y)}{6}d\beta\wedge d\theta + \frac{p(y)}{6}\cos\theta\, d\theta\wedge d\phi\right]-\right.\\ &\left.+\sin\psi'\left[-dy\wedge d\theta + \frac{p(y)}{6}\sin\theta\, d\beta\wedge d\phi\right]\right),\end{aligned}$$

The Stäckel-Killing tensors associated with the Killing forms Ψ, Ξ, Υ are constructed as in (3). The list of the non vanishing components of these Stäckel-Killing tensors is quite long and will be given elsewhere. Together with the Killing vectors P_ϕ, P_ψ, P_α and the total angular momentum J^2 (16) these Stäckel-Killing tensors provide the superintegrability of the $Y(p,q)$ geometries.

5. Concluding remarks

In this paper we presented the complete set of Killing forms on Einstein-Sasaki spaces associated with Euclideanised Kerr-NUT-(A)dS spaces in a certain scaling limit. The multitude of Killing-Yano and Stäckel-Killing tensors makes possible a complete integrability of geodesic equations. In the case of geodesic and Klein-Gordon equations, the existence of separable coordinates is connected with Stäckel-Killing tensors. On the other hand from (conformal) Killing-Yano tensors one can construct first order differential operators which commute with Dirac operators

[29]. In [30, 31] it was shown that the solutions of Dirac equation in general higher dimensional Kerr-NUT-(A)dS spacetimes can be found by separating variables and the resulting ordinary differential equations can be completely decoupled. It is interesting to study separability and eigenvalues of Dirac operators on Einstein-Sasaki manifolds. Let us note also that in the higher dimensional Kerr-NUT-(A)dS spacetimes the stationary string equations are completely integrable [32]. An important open question is a separability problem for the gravitational perturbations in higher dimensional rotating black holes spacetimes, some preliminary results being achieved recently [33].

Another important direction of research is whether the Killing forms are also intrinsically linked to other higher spin perturbations. It is still an open question whether massless field equations, e. g. the Maxwell field, allow separation of variables in Kerr-NUT-(A)dS spaces.

These remarkable properties of higher dimensional black hole solutions offer new perspectives in investigation of hidden symmetries of other spacetimes structures. A possible extension of these techniques can be performed on the spaces with mixed 3-structures which appear in many modern studies [17, 22]. Finally we mention some recent extensions of the Killing-Yano symmetry in the presence of skew-symmetric torsion. Preliminary results [34, 35] indicate that Killing forms in the presence of torsion preserve most of the properties of the standard Killing forms.

As an exemplification of the general framework we have presented the complete set of Killing forms on five-dimensional Einstein-Sasaki $Y(p,q)$ spaces. The multitude of Stäckel-Killing tensors associated with these Killing forms implies the superintegrability of the geodesic motions.

In connection with the third rank Killing-Yano tensors on the $Y(p,q)$ spaces let us note an interesting geometrical interpretation of the Lax representation [36, 37, 38].

These remarkable properties of the Killing forms offer new perspectives in the investigation of the supersymmetries, separability of Hamilton-Jacobi, Klein-Gordon, Dirac equations on Einstein-Sasaki spaces.

Acknowledgments

This work is supported in part by a joint Romanian-LIT, JINR, Dubna Research Project, theme no. 05-6-1060-2005/2013. M.V. is partially supported by program PN-II-ID-PCE-2011-3-0137, Romania.

References

[1] Chen W, Lü H and Pope C N 2006 General Kerr-NUT-AdS metrics in all dimensions *Class. Quantum Grav.* **23** 5323 (*Preprint* hep-th/0604125)
[2] Kubizňák D 2009 On the supersymmetric limit of Kerr-NUT-AdS metrics *Phys. Lett. B* **675** 110 (*Preprint* 0902.1999)
[3] Martelli D and Sparks J 2005 Toric Sasaki-Einstein metrics on $S^2 \times S^3$ *Phys. Lett. B* **621** 208 (*Preprint* hep-th/0505027)
[4] Page D N, Kubizňák D, Varsudevan M and Krtouš P 2007 Complete integrability of geodesic motion in general Kerr-NUT-AdS spacetimes *Phys. Rev. Lett.* **98** 061102 (*Preprint* hep-th/0611083)
[5] Krtouš P, Kubizňák D, Page D N and Varsudevan M 2007 Constants of geodesic motion in higher-dimensional black hole spacetimes *Phys. Rev. D* **76** 084034 (*Preprint* 0707.0001)
[6] Sergyeyev A and Krtouš P 2008 Complete set of commuting symmetry operators for the Klein-Gordon equation in generalized higher-dimensional Kerr-NUT-(A)dS spacetimes *Phys. Rev. D* **77** 044033 (*Preprint* 0711.4623)
[7] Boyer C and Galicki K 2010 Sasakian geometry, holonomy, and supersymmetry *Handbook of pseudo-Riemannian geometry and supersymmetry* (IRMA Lect. Math. Theor. Phys.) vol 16 (Zürich: Eur. Math. Soc.) p 39 (*Preprint* math/0703231)
[8] Semmelmann U 2003 Conformal Killing forms on Riemannian manifolds *Math. Z.* **245** 503 (*Preprint* math/0206117)
[9] Bär C 1993 Real Killing spinors and holonomy *Comm. Math. Phys.* **154** 509
[10] Yano K 1952 Some remarks on tensor fields and curvature *Ann. of Math.* **55** 328

[11] Carter B 1977 Killing tensor quantum numbers and conserved currents in curved spaces *Phys. Rev. D* **16** 3395

[12] Cariglia M 2004 Quantum mechanics of Yano tensors: Dirac equation in curved spacetime 2004 *Class. Quantum Grav.* **21** 1051 (*Preprint* hep-th/0305153)

[13] Visinescu M 2010 Hidden conformal symmetries and quantum-gravitational anomalies *EPL Europhys. Lett.* **90** 41002

[14] Ballmann W 2006 *Lectures on Kähler Manifolds* (ESI Lectures in Mathematics and Physics) (Zürich: Eur. Math. Soc.)

[15] Sasaki S 1960 On differentiable manifolds with certain structures which are closely related to almost contact structure I *Tohoku Math. J.* **12** 459

[16] Blair D E 1976 *Contact manifolds in Riemannian geometry* (Lecture Notes in Mathematics) vol 509 (Berlin: Springer)

[17] Boyer C and Galicki K 1999 3-Sasakian manifolds *Surveys in Differential Geometry: Essays on Einstein Manifolds* vol VI (Boston: Int. Press) p 123 (*Preprint* hep-th/9810250)

[18] Sparks J 2011 Sasaki-Einstein Manifolds *Surv. Diff. Geom.* **16** 265 (*Preprint* 1004.2461)

[19] Gibbons G W, Hartnoll S A and Pope C N 2003 Bohm and Einstein-Sasaki metrics, black holes and cosmological event horizons *Phys. Rev. D* **67** 084024 (*Preprint* hep-th/0208031)

[20] Cvetič M, Lü H, Page D N and Pope C N 2009 New Einstein-Sasaki and Einstein spaces from Kerr-de Sitter *J. High Energy Phys.* JHEP07(2009)082 (*Preprint* hep-th/0505223)

[21] Besse A 1987 *Einstein manifolds* (Berlin: Springer)

[22] Visinescu M and Vîlcu G E 2012 Hidden symmetries of Euclideanised Kerr-NUT-(A)dS metrics in certain scaling limit *SIGMA* **8** 058 (*Preprint* 1205.6036)

[23] Gauntlett J P, Martelli D, Sparks J and Waldram D 2004 Supersymmetric AdS_5 solutions of M-theory *Class. Quant. Grav.* **21** 4335 (*Preprint* hep-th/0402153)

[24] Gauntlett J P, Martelli D, Sparks J and Waldram D 2004 Sasaki-Einstein metrics on $S^2 \times S^3$ *Adv. Theor. Math. Phys.* **8** 711 (*Preprint* hep-th/0403002)

[25] Rubín de Celis E and Santillán O P 2012 Massless geodesics in $AdS_5 \times Y(p,q)$ as a superintegrable system *Preprint* 1205.3256

[26] Martelli D and Sparks J 2006 Toric geometry, Sasaki-Einstein manifolds and a new infinite class of AdS/CFT duals *Commun. Math. Phys.* **262** 51 (*Preprint* hep-th/0411238)

[27] Benvenuti S and Kruczenski M 2006 Semiclassical strings in Sasaki-Einstein manifolds and long operators in $\mathcal{N} = 1$ gauge theories *J. High Energy Phys.* JHEP10(2006)051 (*Preprint* hep-th/0505046)

[28] Visinescu M 2012 Killing forms on the five-dimensional Einstein-Sasaki $Y(p,q)$ spaces *Preprint* 1207.2581

[29] Carter B and McLenagham R G 1979 Generalized total angular momentum operator for Dirac equation in curved space-time *Phys. Rev. D* **19** 1093

[30] Oota T and Yasui Y 2008 Separability of Dirac equation in higher dimensional Kerr-NUT-deSitter spacetime *Phys. Lett. B* **659** 688 (*Preprint* 0711.0078)

[31] Cariglia M, Krtouš P and Kubizňák D 2011 Dirac equation in Kerr-NUT-(A)dS spacetimes: Intrinsic characterization of separability in all dimensions *Phys. Rev. D* **84** 024008 (*Preprint* 1104.4123)

[32] Kubizňák D and Frolov V P 2008 Stationary strings and branes in the higher-dimensional Kerr-NUT-(A)dS spacetimes *J. High Energy Phys.* JHEP02(2008)007 (*Preprint* 0711.2300)

[33] Oota T and Yasui Y 2008 Separability of Dirac gravitational perturbation in generalized Kerr-NUT-deSitter spacetime *Int. J. Mod. Phys. A* **25** 3055 (*Preprint* 0812.1623)

[34] Houri T, Kubizňák D, Warnick C and Yasui Y 2010 Symmetries of the Dirac operator with skew-symmetric torsion *Class. Quantum Grav.* **27** 185019 (*Preprint* 1002.3616)

[35] Houri T, Kubizňák D, Warnick C and Yasui Y 2010 Generalized hidden symmetries and the Kerr-Sen black hole *J. High Energy Phys.* JHEP07(2010)055 (*Preprint* 1004.1032)

[36] Rosquist K and Goliath M 1998 Lax pair tensors and integrable spacetimes *Gen. Rel. Grav.* **30** 1521 (*Preprint* gr-qc/9707003)

[37] Goliath M, Karlovini M and Rosquist K 1999 Lax pair tensors in arbitrary dimensions *J. Phys. A: Math. Gen.* **32** 3377 (*Preprint* solv-int/9810011)

[38] Visinescu A and Visinescu M 2012 Killing-Yano tensors of rank three and Lax pair tensors *Rom. J. Phys.* **57** 1002

Journal of Physics: Conference Series **411** (2013) 012031 doi:10.1088/1742-6596/411/1/012031

Hopf Algebra Structure of Continual Virasoro Algebra Deformations

Alexander Zuevsky
National University of Ireland, Galway, Ireland
E-mail: zuevsky@yahoo.com

Abstract. The Hopf algebra structure of a few parametric deformations of the continual Virasoro algebra is proven.

1. Introduction

For various purposes of q-deformed structure construction [6] it is important to study deformations of the Virasoro algebra [1], [4], [3]. In [4] four deformations of the continual version of the Witt algebra (the centerless Virasoro algebra) were introduced, and to make connections with Hopf algebras corresponding comultiplications were found, although antipode and counit were missing. In [3], the Hopf algebra structure of the discrete form of one of the q-deformed Virasoro algebra [1] was established. In this paper we unify these two approaches to deformations of the Virasoro algebra and determine the full Hopf algebra structure [2] for all four deformations. We formulate the definition of such deformations with central extensions in terms of a family of deformed algebras parametrized by four parameters. The actual parameter is the difference $\theta = \beta_1 - \beta_2$ of the first two parameters.

Apart from pure algebraic interest for studies of quantum groups and generalizations of vertex algebras, possible applications of the deformations under consideration can be found in the domain of quantum exactly solvable models [5]. Namely, to every such deformation one can associate a pair of q-deformed operators with the zero curvature condition applied giving non-linear equations which could be solvable under some further assumptions on underlying continual Lie algebra.

2. Deformations of Continual Version of Virasoro Algebra

2.1. Continual Virasoro Algebra

In this section we recall two known families of q-deformations [4] of the Virasoro algebra in a continual Lie algebra form [7]. Let us introduce notations. In addition to ordinary commutator we will use the standard (in the sense of [4], [3]) deformed commutator

$$[a, b]_z \equiv zab - z^{-1}ba,$$

as well as a commutator with inserted operation:

$$[a,^{z\cdot} b] \equiv a\ z.b - b\ z.a,$$

where z can be an operator valued in corresponding algebra. Next we define

$$[z] \equiv \frac{q^z - q^{-z}}{q - q^{-1}},$$

$$\langle z \rangle \equiv q^z + q^{-z},$$

for $z \in \mathbb{C}$.

Let E be a commutative associative algebra over a field $\mathcal{K}$. The continual form of the Virasoro algebra, a central extension (denoted by $Vir(E)$) of the Witt algebra [4], [7], is defined by the commutation relations for the generators $\{C(\phi, \psi), X(\phi) | \phi, \psi \in E\}$:

$$\begin{aligned} [X(\phi), X(\psi)] &= X(K(\phi, \psi)) + C(\phi, \psi), \\ [C(\phi, \psi), X(\chi)] &= 0, \end{aligned} \tag{1}$$

$$K(\phi, \psi) = \phi \partial \psi - \psi \partial \phi, \tag{2}$$

where the symbol ∂ stands for a formal differentiation with respect to a parameter. Note that (2) can be written as $X\left([\phi,^\partial \psi]\right)$. Jocobi identies for generators of a continual Lie algebra dictate cocycle property for the central element $C(\phi, \psi)$,

$$C(\phi, C(\psi, \chi)) + C(\psi, C(\chi, \phi)) + C(\chi, C(\phi, \psi)) = 0,$$

for any ϕ, ψ, $\chi \in E$. The continual Virasoro algebra (1) can be Fourier decomposed to give the commutation relations of the Virasoro algebra Vir with discrete generators $\{c, L_m | m \in \mathbb{Z}\}$, and ordinary relations

$$[L_m, L_n] = (m - n) L_{n+m} + c(m, n), \qquad [c, L_n] = 0 \tag{3}$$

where

$$c(m, n) = \delta_{m,-n} \frac{m^3 - m}{12} c. \tag{4}$$

2.2. Deformations

Next we consider the universal enveloping algebra of parametric deformations $Vir_q(E)$ of the continual Lie algebra $Vir(E)$ defined by commutation relations (1) with the mapping (2). Let $U(Vir_q(E))$ be an associative unital algebra over $\mathcal{K}$ with parameters $(\alpha, \beta_1, \beta_2, \gamma)$, $\alpha \in \mathbb{R}$, $\alpha \neq 0$, $\beta_i \in \{-1, 0, 1\}$, $\gamma \in \{0, 1\}$, and determined by generators $\left\{C_q(\phi, \psi), T^{\pm 1},\ X(\phi) | \phi, \psi, \chi \in E\right\}$ subject to the commutation relations:

$$\begin{aligned} T\, T^{-1} &= T^{-1} T = 1, && (5) \\ T\, C_q(\phi, \psi) &= C_q(\phi, \psi)\, T^{-1}, && (6) \\ C_q(\phi, \psi)\, X(\chi) &= X(\chi)\, C_q(\phi, \psi) && (7) \\ T X(\phi) &= X(q^{-\theta \partial} \phi) T, && (8) \end{aligned}$$

where θ is a parameter. The commutation relations for the generators $X(\phi)$ have the form:

$$\begin{aligned} X(q^{\beta_1 \partial} \phi) X(q^{\beta_2 \partial} \psi) &- X(q^{\beta_1 \partial} \psi) X(q^{\beta_2 \partial} \phi) \\ &= T^{\beta_2 \delta_{\beta_1, 0}} X(K(\phi, \psi)) T^{\beta_1 \delta_{\beta_2, 0}} + C_q(\phi, \psi), \end{aligned} \tag{9}$$

where the mapping is given by

$$\begin{aligned} K(\phi,\psi) &= q^{\gamma(\partial+\alpha)}\phi \cdot [\partial+\alpha+\beta_1\delta_{\beta_2,0}]_q\psi \\ &- q^{\gamma(\partial+\alpha)}\psi \cdot [\partial+\alpha+\beta_1\delta_{\beta_2,0}]_q\phi \\ &= m_E \circ (q^{\gamma(\partial+\alpha)} \otimes [\partial+\alpha+\beta_1\delta_{\beta_2,0}]_q) \cdot [\phi,^{\otimes}\psi], \end{aligned} \tag{10}$$

and m_E is the multiplication $E \otimes E \to E$. We take $\theta = \beta_1 - \beta_2$ in (8) as an effective parameter regulating commutation relations (8), (13) and definitions of the the Hopf algebra structure (see (18)–(32)). In [4] four specific deformations *1.–4.* of the continual form of the centerless Virasoro algebra (1) – the Witt algebra, were determined by the following choices of parameters,

$$(\beta_1,\beta_2,\gamma) = \{(1,-1,1),(1,-1,0),(0,1,1),(1,0,1)\},$$

so that $\theta = \{2,2,-1,1\}$ correspondingly. In (9) and (10) it is assumed that the formal series $q^{\beta\partial}$ of powers of differential operators acts on on the second and the first argument of $X(\phi)$ generators in the ordinary bracket. In the limit $q \to 1$, the generators $T^{\pm1}$ commute with $X(\phi)$ for every $\phi \in E$, and $C_q(\phi,\psi)$ reduces to the central element $C(\phi,\psi)$ of the undeformed continual Virasoro algebra (1). Thus these deformations $Vir_q(E)$ reduce to $Vir(E)$ in the limit $q \to 1$. Note that in the case of the deformations *1.* and *2.* the left hand side bracket can be written as $[X(\phi),X(\psi)]_{q^{\beta\partial}}$, while for the deformations *3.* and *4.* as a bracket $[X(\phi),^{q^{\beta\partial}} X(\psi)]$.

In discrete generator form the deformations $Vir_q(E)$ (9)–(10) of [4] correspond to q-deformed versions Vir_q of the Virasoro algebra Vir (3) [3], [1]. They are determined by the generators $\{c_q, T^{\pm1}, l_n | n \in \mathbb{Z}\}$ subject to the commutation relations (in addition to the relation (5))

$$[c_q,T]_q = 0, \tag{11}$$

$$[l_n,c_q]_{q^n} = 0, \tag{12}$$

$$[T,l_n]_{q^{\frac{\theta}{2}(n+1)}} = 0, \tag{13}$$

$$\left[l_m,^{q^{wt}} l_n\right]_{q^{\beta_1 m+\beta_2 n}} = k_i(m,n)l_{n+m}T^{\delta_{\beta_1,0}+\delta_{\beta_2,0}} + \mathcal{C}[m,n], \tag{14}$$

where the operator wt takes the value of the deformed Virasoro mode index, i.e., $wt(l_n) = n$, and

$$k_1(m,n) = [m-n]_q, \quad k_2(m,n) = [m+\alpha]_q - [n+\alpha]_q, \tag{15}$$

$$k_3(m,n) = [m-n]_q, \quad k_4(m,n) = q^{-1}[m-n]_q, \tag{16}$$

for deformations *1.–4.* correspondingly, and

$$\mathcal{C}[m,n] \equiv c[m,n]\, c = \delta_{m,-n}\frac{[m][m-1][m+1]}{[2][3]\langle m\rangle} c_q. \tag{17}$$

Note that $m-n$ and (17) are limits of k_i and $\mathcal{C}[m,n]$ when $q \to 1$.

3. Hopf Algebra Structure

In this section we define the comultiplication Δ, antipode S and counity ε for the deformations *1.–4.* (9)–(10) of the continual version of the Virasoro algebra $Vir(E)$ and in the discrete form Vir_q (11)–(17), and prove their Hopf algebra structure. For Vir_q these operations are given by (for the deformation *1.*, $\theta = 2$, they were introduced in [3])

$$\Delta(T) = T \otimes T, \tag{18}$$
$$\Delta(c_q) = c_q \otimes 1 + 1 \otimes c_q, \tag{19}$$
$$\Delta(l_n) = l_n \otimes T^n + T^n \otimes l_n, \tag{20}$$
$$\varepsilon(T) = q^{\pm(2-|\theta|)\partial}, \tag{21}$$
$$\varepsilon(c_q) = 0, \tag{22}$$
$$\varepsilon(l_n) = 0, \tag{23}$$
$$S(T) = T^{-1}, \tag{24}$$
$$S(c_q) = -c_q, \tag{25}$$
$$S(l_n) = -T^{(1-\theta)n} l_n T^{(1-\theta)n}. \tag{26}$$

The $+/-$ sign in (21) corresponds to the left/right action of $\varepsilon(T)$ in the tensor product. For the continual form of (5)–(9), in addition to (18), (21), (24) we introduce

$$\Delta(C_q(\phi,\psi)) = C_q(Q_1\phi, Q_1\psi) \otimes T^{2-|\theta|} + T^{2-|\theta|} \otimes C_q(Q_2\phi, Q_2\psi), \tag{27}$$
$$\Delta(X(\phi)) = X(Q_1\phi) \otimes T^{2-|\theta|} + T^{2-|\theta|} \otimes X(Q_2\phi), \tag{28}$$
$$\varepsilon(C_q(\phi,\psi)) = 0, \tag{29}$$
$$\varepsilon(X(\phi)) = 0, \tag{30}$$
$$S(C_q(\phi,\psi)) = -T^{-(2-|\theta|)} C_q(Q_0\phi, Q_0\psi) T^{-(2-|\theta|)}, \tag{31}$$
$$S(X(\phi)) = -T^{-(2-|\theta|)} X(Q_0\,\phi) T^{-(2-|\theta|)}, \tag{32}$$

for any ϕ, $\psi \in E$. Here the operators $Q_i = Q_i(q)$, $i = 0, 1, 2$ are distributions on E, commuting with the formal differentiation ∂ with respect to q. As in [4] we demand (see the proof of Proposition 3.1) the commutativity and projectivity of the operators Q_i, $i = 1, 2$, i.e.,

$$X(Q_i^2\phi) = X(Q_i\phi), \qquad X([Q_1, Q_2]\phi) = 0, \tag{33}$$

and for $\beta \in \mathbb{Z}$,

$$Y(Q_i q^{\beta\partial}) = \pm Y(Q_i q^{-\beta\partial}). \tag{34}$$

where $Y \in \{C_q(\phi,\psi), X(\phi) | \phi, \psi \in E\}$, and $\lim_{q\to 1} X(Q_i\phi) = X(\phi)$, for any $\phi \in E$. We also require that $Y(Q_0^2) = Y(1)$, for $i = 1, 2$, and

$$Y(Q_a) = Y(Q_0 Q_{\bar{a}}), \tag{35}$$

where $\bar{a}$ is the opposite to a in the set $\{1, 2\}$, and $Y(Q_1 q^{-(2-|\theta|)^2\partial}) = Y(1) = Y(q^{(2-|\theta|)^2\partial} Q_2)$, for $Y = C_q$. The actual form of the distributions $Q_i(q)$, $i = 0, 1, 2$ can be determined [4] for each deformation *1.–4.* by the coinvariance condition of the comultiplication Δ and antipode S with respect to the defining relations (11)–(9) on generators of $Vir_q(E)$. Then taking into account conditions (33)–(35), we formulate the following

Proposition 3.1 *The universal enveloping algebra*

$$\mathcal{U}(Vir_q(E)) = [T; T^{-1}]_{\mathcal{K}} \otimes_{\mathcal{K}} U(Vir_q(E)),$$

of $(\alpha, \beta_1, \beta_2, \gamma)$-parameter q-deformation $Vir_q(E)$ defined by the generators $\{C_q(\phi,\psi), T^{\pm 1}, X(\phi) \mid \phi, \psi \in E\}$, subject to the commutation relations (5)–(9) with the mapping (10) (and for deformations Vir_q with discrete generators $\{c_q, T^{\pm 1}, l_n | n \in \mathbb{Z}\}$, relations (11)–(14), and mappings (15)–(16)) possesses a non-commutative co-commutatvie Hopf algebra structure given by (18), (21), (24), (27)–(32) ((18)–(26) correspondingly).

Proof. We first prove the Hopf algebra structure in the discrete form (11)–(17) $U(Vir_q)$. Note that the case *1.* was proven in [3]. It is clear that ε is an algebraic homomorphism, $\varepsilon(ab) = \varepsilon(a)\varepsilon(b)$, and $\mu(1\otimes\varepsilon)\Delta = \mu(\varepsilon\otimes 1)\Delta = \mathrm{Id}$, where μ is the multiplication. One can easily verify the coassoativity of Δ: $(\Delta\otimes 1)\Delta = (1\otimes\Delta)\Delta$. We next check coinvariance of the relations (11)–(14) with respect to Δ. This is easy to see that for the relations (5), (11), (12) and (13). Then we apply the comultiplication Δ to the commutation relations (14) of generators l_n:

$$\begin{aligned}\Delta(l_n)\Delta(l_m) &= l_n l_m \otimes T^{n+m} + T^{n+m} \otimes l_n l_m \\ &+ l_n T^m \otimes T^n l_m + T^n l_m \otimes l_n T^m \\ &= l_n l_m \otimes T^{n+m} + T^{n+m} \otimes l_n l_m \\ &+ q^{-\theta n(m+1)} l_n T^m \otimes l_m T^n \\ &+ q^{-\theta n(m+1)} l_m T^n \otimes l_n T^m,\end{aligned}$$

where $\theta = 2$ for the deformations *1.–2.*, and $\theta = -1$, $\theta = 1$ for the deformations *3.–4.*. Therefore for all deformations we obtain

$$\begin{aligned}q^{\beta_1\, wt}\Delta(l_m)q^{\beta_2\, wt}\Delta(l_n) &- q^{\beta_1\, wt}\Delta(l_n)q^{\beta_2\, wt}\Delta(l_m) \\ &= \left(q^{\beta_1\, wt}l_m q^{\beta_2\, wt}l_n - q^{\beta_1\, wt}l_n q^{\beta_2\, wt}l_m\right) \otimes T^{n+m} \\ &\quad + T^{n+m} \otimes \left(q^{\beta_1\, wt}l_m q^{\beta_2\, wt}l_n - q^{\beta_1\, wt}l_n q^{\beta_2\, wt}l_m\right) \\ &= \Delta\left(q^{\beta_1\, wt}l_m q^{\beta_2\, wt}l_n - q^{\beta_1\, wt}l_n q^{\beta_2\, wt}l_m\right) \\ &= \Delta\left(\left[l_m, {}^{q^{wt}} l_n\right]_{q^{\beta_1 m+\beta_2 n}}\right) \\ &= k_i(m,n)\Delta(l_{n+m})\Delta(T^{\delta_{\beta_1,0}+\delta_{\beta_1,0}}) + c[m,n]\Delta(c_q).\end{aligned}$$

We conclude that the commutation relations (14) are preserves by the comultiplication. Thus Δ is an algebraic homomorphism.

It is also clear that $\mu(1\otimes S)\Delta = \mu(S\otimes 1)\Delta = \eta.\varepsilon$, for some η, and $S^2 = 1$, where μ is the multiplication in $\mathcal{U}(Vir_q)$. Next we check that S is an algebraic anti-homomorphism. The antipode S preserves (5) and (11)–(13),

$$[S(c_q), S(T^n)]_{q^n} = 0, \qquad [S(c_q), S(l_n)]_{q^n} = 0,$$

for any $n \in \mathbb{Z}$. We also have

$$\begin{aligned}q^{\beta_1 m+\beta_2 n}S(l_m l_n) &= q^{\beta_1 m+\beta_2 n}S(l_n)S(l_m) \\ &= q^{\beta_1 m+\beta_2 n}T^{(1-\theta)n}l_n T^{(1-\theta)(n+m)}l_m T^{(1-\theta)m} \\ &= q^{-\beta_2 n-\beta_1 m}T^{-(1-\theta)(n+m)}l_n l_m T^{-(1-\theta)(n+m)}.\end{aligned}$$

Thus

$$\begin{aligned}\left[S(l_n), {}^{q^{wt}} S(l_m)\right]_{q^{\beta_1 m+\beta_2 n}} &= S\left([l_n, {}^{q^{wt}} l_m]_{q^{\beta_1 m+\beta_2 n}}\right) \\ &= k_i(m,n)\, S\left(T^{\delta_{\beta_1,0}+\delta_{\beta_2,0}}\right) S(l_{n+m}) \\ &\quad + c[n,m]S(c_q).\end{aligned}$$

Therefore the commutation relations (14) are preserved by the antipode S. Thus we conclude that $\mathcal{U}(Vir_q)$ is a Hopf algebra.

Now we show that the universal enveloping algebra $\mathcal{U}(Vir_q(E))$ in the continual form of generators has a Hopf algebra structure. The algebraic homomorphisity of ϵ is obvious. The coassociativity of the comultiplication, $(1 \otimes \Delta)\Delta = (\Delta \otimes 1)\Delta$, and the condition $S^2 = 1$ (since $Y(Q_0^2) = Y(1)$), lead to the commutativity and projectivity of operators Q_i, $i = 1$, 2. It is easy to check $\mu(1 \otimes \varepsilon)\Delta = \mu(\varepsilon \otimes 1)\Delta = \mathrm{Id}$, using the projectivity of Q_i, $i = 1$, 2, i.e., $C_q(Q_i\phi, Q_i\psi) = Q_i C_q(\phi, \psi)$, (35), and the left or right action of (21). Next for $\mu(1 \otimes S)\Delta = \mu(S \otimes 1)\Delta = \eta.\varepsilon$, we use (6) and (35). Indeed,

$$\begin{aligned} \mu(1 \otimes S)\Delta(C_q(\phi, \psi)) &= C_q(Q_1\phi, Q_1\psi)\, T^{-(2-|\theta|)} \\ &\quad - C_q(Q_0Q_2\phi, Q_0Q_2\psi)\, T^{-(2-|\theta|)} \\ &= 0 = \varepsilon(C_q(\phi, \psi)), \end{aligned}$$

$$\begin{aligned} \mu(S \otimes 1)\Delta(C_q(\phi, \psi)) &= -T^{-(2-|\theta|)}\, C_q(Q_0Q_1\phi, Q_0Q_1\psi) \\ &\quad + T^{-(2-|\theta|)}\, C_q(Q_2\phi, Q_2\psi) \\ &= 0 = \varepsilon(C_q(\phi, \psi)), \end{aligned}$$

$$\begin{aligned} \mu(1 \otimes S)\Delta(X(\phi)) &= X(Q_1\phi)\, T^{-(2-|\theta|)} - X(Q_0Q_2\phi)\, T^{-(2-|\theta|)} \\ &= 0 = \varepsilon(X(\phi)), \end{aligned}$$

$$\begin{aligned} \mu(S \otimes 1)\Delta(X(\phi)) &= -T^{-(2-|\theta|)}\, X(Q_0Q_1\phi) + T^{-(2-|\theta|)}\, X(Q_2\phi) \\ &= 0 = \varepsilon(X(\phi)). \end{aligned}$$

Using (6) we also have $S(TC_q(\phi, \psi)) = S(C_q(\phi, \psi)T^{-1})$, and similarly for (7) and (8). Applying the comultiplication to the commutation relations (9) we obtain

$$\begin{aligned} & \Delta(X(q^{\beta_1\partial}\phi))\, \Delta(X(Q_1q^{\beta_2\partial}\psi)) - \Delta(X(q^{\beta_1\partial}\psi))\, \Delta(X(Q_1q^{\beta_2\partial}\phi)) \\ = \; & X(Q_1q^{\beta_1\partial}\phi)\, X(Q_1q^{\beta_2\partial}\psi) \otimes T^{2(2-|\theta|)} - X(Q_1q^{\beta_1\partial}\psi)\, X(Q_1q^{\beta_2\partial}\phi) \otimes T^{2(2-|\theta|)} \\ + \; & T^{2(2-|\theta|)} \otimes X(Q_2q^{\beta_1\partial}\phi)\, X(Q_2q^{\beta_2\partial}\psi) - T^{2(2-|\theta|)} \otimes X(Q_2q^{\beta_1\partial}\psi)\, X(Q_2q^{\beta_2\partial}\phi) \\ & + T^{2-|\theta|}\, X(Q_1q^{\beta_1+(2-|\theta|)\theta)\partial}\phi) \otimes X(Q_2q^{\beta_2-(2-|\theta|)\theta)\partial}\psi)\, T^{2-|\theta|} \\ & \quad + T^{2-|\theta|}\, X(Q_1q^{\beta_2\partial}\psi) \otimes X(Q_2q^{\beta_1\partial}\phi)\, T^{2-|\theta|} \\ & \quad - T^{2-|\theta|}\, X(Q_1q^{\beta_1+(2-|\theta|)\theta)\partial}\psi) \otimes X(Q_2q^{\beta_2-(2-|\theta|)\theta)\partial}\phi)\, T^{2-|\theta|} \\ & \qquad - T^{2-|\theta|}\, X(Q_1q^{\beta_2\partial}\phi) \otimes X(Q_2q^{\beta_1\partial}\psi)\, T^{2-|\theta|} \\ = \; & T^{\beta_2\delta_{\beta_1,0}}\, X(Q_1K(\phi, \psi))\, T^{\beta_1\delta_{\beta_2,0}} \otimes T^{2-|\theta|} \\ & + T^{2-|\theta|} \otimes T^{\beta_2\delta_{\beta_1,0}}\, X(Q_2K(\phi, \psi))\, T^{\beta_1\delta_{\beta_2,0}} \\ & \quad + T^{2-|\theta|} \otimes C_q(Q_2\phi, Q_2\psi) + C_q(Q_1\phi, Q_1\psi) \otimes T^{2-|\theta|} \\ = \; & \Delta(T^{\beta_2\delta_{\beta_1,0}}\, X(K(\phi, \psi))\, T^{\beta_1\delta_{\beta_2,0}} + C_q(\phi, \psi)), \end{aligned} \tag{36}$$

where it was assumed that Q_i, $i = 1$, 2 commutes with ∂. Cross terms in (36) cancel when the condition (34) is satisfied.

By similar computations one checks that S given by (31)–(32) is an algebraic anti-homomorphism. Indeed, when the distribution Q_0 satisfies $X(K(Q_0\phi, Q_0\psi)) = X(Q_0K(\phi, \psi))$, for any $\phi,\ \psi \in E$, then

$$\begin{aligned}
&S(X(q^{\beta_1\partial}\phi)\, X(q^{\beta_2\partial}\psi)) - S(X(q^{\beta_1\partial}\psi)\, X(q^{\beta_2\partial}\phi)) \\
&= T^{-2(2-|\theta|)}\, X(q^{(\beta_2-(2-|\theta|)\theta)\partial}Q_0\psi)\, X(q^{(\beta_1+(2-|\theta|)\theta)\partial}Q_0\phi)\, T^{-2(2-|\theta|)} \\
&\qquad -T^{-2(2-|\theta|)}\, X(q^{(\beta_2-(2-|\theta|)\theta)\partial}Q_0\phi)\, X(q^{(\beta_1+(2-|\theta|)\theta)\partial}Q_0\psi)\, T^{-2(2-|\theta|)} \\
&= T^{-(2-|\theta|)}\, X(q^{\beta_1\partial}Q_0\psi)\, X(q^{\beta_2\partial}Q_0\phi)\, T^{-2(2-|\theta|)} \\
&\qquad -T^{-2(2-|\theta|)}\, X(q^{\beta_1\partial}Q_0\phi)\, X(q^{\beta_2\partial}Q_0\psi)\, T^{-2(2-|\theta|)} \\
&= S\left(T^{\beta_1\delta_{\beta_2,0}}\, X(K(\phi,\psi))\, T^{\beta_2\delta_{\beta_1,0}} + C_q(\phi,\psi)\right).
\end{aligned}$$

We proved that the commutation relations (5)–(9) are preserved by the comultiplication and antipode actions when operators $Q_i(q)$ satisfy the property (34). Thus $\mathcal{U}(Vir_q(E))$ possesses a Hopf algebra structure.

□

[1] Aizawa N and Sato H 1991 *Phys. Lett. B* **256** 185
[2] Chari V and Pressley A 1994 *A guide to quantum groups* (Cambridge University Press)
[3] Hu N 1998 *Lett. Math. Phys.* **44** 99
[4] Devchand Ch and Saveliev M V 1991 *Phys. Lett. B* **258** 364
[5] Leznov A N and Saveliev M V 1992 *Group-Theoretical Methods for Integration of Non–Linear Dynamical Systems. Progress in Physics Series, v.15* (Basel: Birkhauser–Verlag)
[6] Li H 2006 *Internat. J. Math.* **17** 441
[7] Saveliev M V and Vershik A M 1989 *Commun. Math. Phys.* **126**, 367

AUTHOR INDEX